Electrochemical Techniques

for Corrosion Engineering

R. Baboian, Editor

Published by
National Association of Corrosion Engineers
1440 South Creek Drive
Houston, Texas 77084

Library of Congress Catalog Card Number: 86-63423
ISBN 0-915567-23-7

Foreword

The papers in this book were presented at the symposium on Electrochemical Techniques for Corrosion at the NACE CORROSION/85 meeting held in Boston, Massachusetts, March 25-29, 1985. The symposium and the book were sponsored by Unit Committee T-3L on Electrochemical and Electrical Techniques for Corrosion Measurement and Control. The symposium was chaired by Robert Baboian and consisted of 6 sessions and 34 papers, all of which are a part of this book. Two papers have been added to complete the coverage of the subject.

Preface

Electrochemical techniques for corrosion engineering have gained increased importance over the past three decades. During recent years, these techniques have become recognized as powerful tools in both basic and applied work. It is important, however, that these techniques be properly understood in order to avoid their misuse in corrosion investigations.

This symposium was organized to focus attention on the present technological status of electrochemical techniques for corrosion engineering. The interest of this book is to review the techniques used in both basic and applied corrosion work. It is organized into seven sections: Fundamental Studies, Basic Techniques, Alloy Development, Coatings and Inhibitors, Electrochemical Tests, Localized Corrosion, and Corrosion Monitoring.

The section on Fundamental Studies begins with Barnartt's discussion on the electrochemical nature of corrosion and continues with some novel and standard techniques in electrode studies by Pickering and co-authors. Isaacs presents the principles, advantages, and applications of the scanning vibrating electrode technique, while Kruger describes some newer approaches to the study of localized corrosion. Cyclic voltametric studies to determine the passive-transpassive behavior of steel are described by Hoare while Nisancioglu deals with the error in polarization resistance and capacitance measurements and provides correction factors.

The Basic Techniques section begins with an introduction to potentiostatic testing by Hack. Today's status on polarization resistance measurements is reviewed by Mansfeld. Silverman provides a primer on the AC impedance technique whereby theory and applications are described. A review of the rapid-scan potentiodynamic polarization technique along with examples of its use are presented by Siebert. Corrosion rate measurements through galvanostatic polarization transients are described by Giuliani.

In the Alloy Development section, Verink describes the application of electrochemical techniques in the development of alloys based on the development of potential versus pH diagrams. Bakama and Heidersbach continue in this section with electrochemical methods for determining potential-pH diagrams. Silverman describes the derivation and application of thermodynamic EMF-pH diagrams pointing out the problems and use of a consistent data base. Bond and co-authors discuss the development of corrosion resistant alloys based on their experience in this area. The specific study of aluminum-lithium alloys is covered by Tsao and Pizzo.

The section on Coatings and Inhibitors begins with a summary of ellipsometric and electrochemical studies of the delamination of organic coatings from iron and steel by Ritter. Kendig and co-authors present the application of impedance spectroscopy to the evaluation of corrosion protection by inhibitors and polymer coatings. The evaluation of pipeline coatings by computer-controlled AC impedance measurements is described by Kellner. Morrissey describes the effects of deposit porosity on galvanic corrosion in precious metal electrodeposits. The effect of ion implantation on the corrosion behavior of high density sintered tungsten alloy is discussed by Chang, Levy, and Lin. Cameron and Chiu cover the use of electrochemical techniques for corrosion inhibitor studies.

The Electrochemical Tests section begins with a thorough review of this subject by Dean. The interaction of microcomputers with electrochemical corrosion testing methods is covered by Tipton and Melton, who include several examples of computer data acquisition systems in corrosion testing applications. Majidi and Streicher present four nondestructive tests for detecting sensitization in AISI 304 and 304L stainless steels, while Lee and Dianatkhah discuss a modification of the oxalic acid test for field nondestructive testing of welded AISI 304 stainless steel pipes. Chance and Walker cover recent applications of electrochemical techniques for corrosion research in the automotive industry. Baboian reviews electrochemical methods of predicting galvanic corrosion at the end of this section.

Localized corrosion, an important subject in this area, is the focus of the next section. It starts with a review on the techniques to study crevice corrosion in natural seawater by Kain, Lee, and Oldfield. The study of pitting and crevice corrosion resistance of stainless steels is covered by Lau and Bernhardsson. Morris describes the measurement of pit initiation and propagation for iron-chromium-nickel alloys in acid chloride environments. Horowitz presents a history of the problem of polythionic acid stress corrosion cracking of sensitized stainless steels and discusses the mechanism.

The last section on Corrosion Monitoring begins with a review of on-line corrosion monitoring by Bovankovich bringing the readers up to date on this subject. Haagenrud, Henriksen, and Wyzisk describe an atmospheric corrosion monitor which provides late data by recording and integrating the current vs time in electrolytic cells.

Hirozawa describes corrosion monitoring by galvanostaircase polarization and at the end of this section, Cooper discusses the selection of the proper electrical-resistance probe in corrosion monitoring.

R. Baboian
Editor

Table of Contents

Section 1 – Fundamental Studies

Section 2 – Basic Techniques

Section 3 – Alloy Development

Section 4 – Coatings and Inhibitors

Section 5 – Electrochemical Tests

Section 6 – Localized Corrosion

Section 7 – Corrosion Monitoring

Section 1

Fundamental Studies

Electrochemical Nature of Corrosion

S. BARNARTT*

Introduction

The electrochemical nature of corrosion would be of only minor or academic interest to the corrosion engineer if it provided merely an additional method of measuring average corrosion rate. The standard gravimetric method of weighing the amount of metal reacted during a specified time is sufficiently accurate when good descaling procedures are used. The gravimetric method is generally satisfactory and is still widely used in investigating corroding systems of industrial importance. Corrosion rate measured by metal weight loss is, however, an average value over a relatively long reaction time.

The electrochemical nature of corrosion provides a means of determining an almost instantaneous corrosion rate. Over the time interval required to obtain a gravimetric corrosion rate, one can follow electrochemically how the rate varies with time. The mean corrosion rate, obtained by integration of the rate/time curve, should be equal to the rate obtained by descaled weight loss.

Consider the reaction of iron in sulfuric acid to form ferrous sulfate:

$$Fe + H_2SO_4 = FeSO_4 + H_2 \qquad (1)$$

This is electrochemical in nature; thus, it can be followed by the rate of the anodic process:

$$Fe = Fe^{2+} + 2e^- \qquad (2)$$

or by the rate of the cathodic process:

$$2H^+ + 2e^- = H_2 \qquad [2(a)]$$

Any such electrochemical process has an equivalent electric current I in amperes (A) given by:

$$I = F\dot{n} \qquad (3)$$

Here, F is the Faraday (96,487 coulombs equiv^{-1}) and n is the number of equivalent weights reacted per second. If $\dot{m}$ is the corrosion rate of the metal in grams per second, and M its equivalent weight, Equation (3) can be written as:

$$\dot{m} = IM/F \qquad [3(a)]$$

In terms of units more familiar to the corrosion engineer, with current density i in μA cm^{-2}, the reaction rate given by Equation (3) becomes:

$$\text{rate} = 0.0895\,iM \quad mg\,dm^{-2}\,d^{-1} \qquad [3(b)]$$

$$\text{rate} = 0.129\,iM/D \quad mil\,y^{-1} \qquad [3(c)]$$

where D is the metal density in g cm^{-3}, d = day, and y = year.

Unfortunately, an ammeter cannot simply be attached to any naturally corroding metal and measure the current equivalent to the instantaneous corrosion rate, because this current is internally shorted. There are, however, three fundamental concepts that can be considered in the development of electrochemical techniques for measuring corrosion rate.

Fundamental Concepts

Tafel Equation

In 1905, Tafel[1] found that the current I equivalent to the rate of a *single* electrode reaction on a metal surface is related to the potential of the metal by:

$$E = a + b\log I \qquad (4)$$

★Previously published in Electrochemical Techniques for Corrosion, R. Baboian, Ed., National Association of Corrosion Engineers, Houston, Texas, p. 1, 1977.

*Present address: Hillsdale Drive, Monroeville, Pennsylvania.

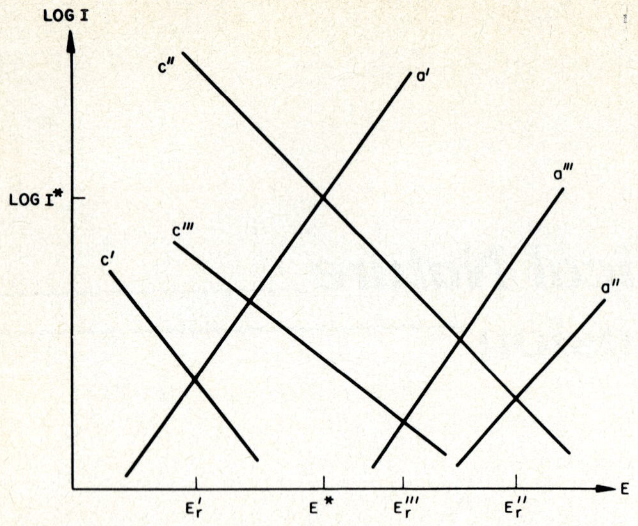

FIGURE 1 — Partial currents at a corroding electrode.

where E is the potential measured with respect to a selected reference electrode, with a and b being constants. The Tafel equation has been confirmed experimentally for a wide variety of individual anodic and cathodic reactions. Constants a and b can be predicted theoretically for many reaction mechanisms.[2] For the purpose of this report, the Tafel slope, $b \equiv dE/d \log I$, is considered an experimental quantity characteristic of a particular anodic process (b_a) or cathodic process (b_c), not a theoretical quantity calculated from the reaction model.

Additivity Principle

A naturally corroding metal always has more than one electrochemical process occurring on its surface at a measurable rate. These must include at least one anodic and one cathodic process. If external current is applied to the corroding metal to change its potential, the potential change affects the rates of all electrochemical processes at the metal surface. This is depicted graphically in Figure 1.

The system represented in Figure 1 is a generalized corroding metal on which there are three reactions designated by prime, double prime, and triple prime, the three reactions having reversible potentials E_r', E_r'', and E_r'''. The lines represent linear Tafel (E vs log I) plots for the anodic and cathodic processes of each reaction. The current for any one of these processes is referred to as a partial current, and it represents the rate of that process.

An example of a corroding system having partial currents distributed as in Figure 1 is zinc corroding in an acid solution in the presence of oxygen. The Tafel lines correspond to the following six processes.

$$Zn \underset{c'}{\overset{a'}{\rightleftharpoons}} Zn^{2+} + 2e^- \qquad [5(a)]$$

$$2H_2O \underset{c''}{\overset{a''}{\rightleftharpoons}} O_2 + 4H^+ + 4e^- \qquad [5(b)]$$

$$H_2 \underset{c'''}{\overset{a'''}{\rightleftharpoons}} 2H^+ + 2e^- \qquad [5(c)]$$

An important principle that applies to any such electrode is that the partial currents are algebraically additive. The generality of this principle was elucidated in 1938 by Wagner and Traud.[3] In terms of the partial currents in Figure 1 (all currents have positive sign), the additivity principle requires that, for any selected potential E,

$$I = \pm\{I_a' + I_a'' + I_a''' - (I_c' + I_c'' + I_c''')\} \qquad (6)$$

where I is the external meter current that establishes electrode potential E. The plus sign before the brackets applies when the meter current to the electrode is anodic, the negative sign when it is cathodic.

The naturally corroding metal state is indicated by superscript asterisk and refers to I = 0, so that

$$E^* : I_a' + I_a'' + I_a''' = I_c' + I_c'' + I_c''' \qquad (7)$$

If the metal corrodes via reaction prime only, then the corrosion rate is given by

$$E^* : I^* = I_a' - I_c' = (I_c'' - I_a'') + (I_c''' - I_a''') \qquad (8)$$

General Rate Equation for Corroding Metals

The corroding system shown in Figure 1 comprises three anodic partial currents dominated by I_a' at all potentials, and three cathodic currents dominated by I_c''. Many practical corroding systems conform to the situation where the other anodic currents are of negligible magnitude compared with I_a', and other cathodic currents are negligible with respect to I_c''. For these, Equations (6) and (8) reduce to:

$$E : I = \pm(I_a' - I_c'') \qquad [6(a)]$$

$$E^* : I^* = I_a' = I_c'' \qquad [8(a)]$$

Current I_a' at potential E is given by the Tafel equation for line a' in Figure 1. When potentials are referred to the corrosion potential, the Tafel equation takes the form

$$E - E^* = b_a' \log \frac{I_a'}{I^*}, \text{ or}$$

$$\frac{I_a'}{I^*} = \exp \frac{2.3\,(E - E^*)}{b_a'} \qquad (9)$$

Similarly, I_c'' is given by

$$\frac{I_c''}{I^*} = \exp \frac{2.3\,(E - E^*)}{b_c''} \qquad (10)$$

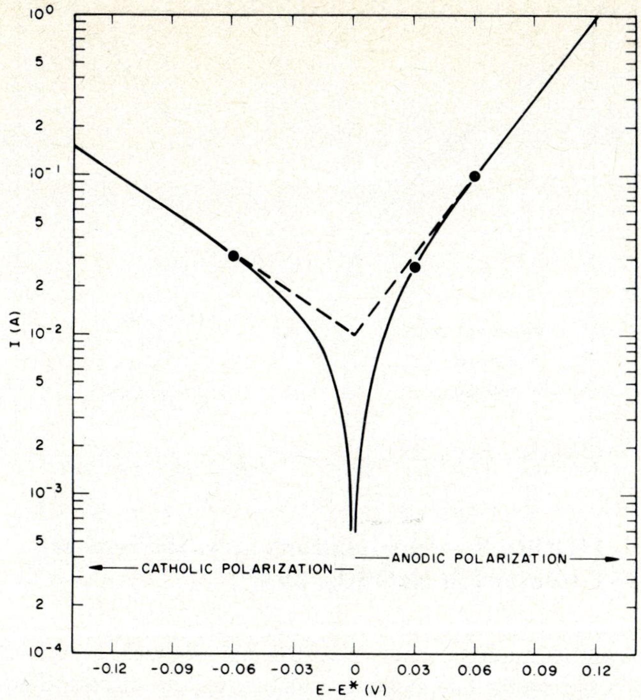

FIGURE 2 — Polarization curves from Equation (11), b_a = 0.0592 V, b_c = −0.1183 V, and I^* = 10^{-2} A.

Note that b_c'' is a negative number and b_a' positive.

Substitution for I_a' and I_c'' in Equation [6(a)] yields

$$\frac{I}{I^*} = \pm \left\{ \exp \frac{2.3\,(E - E^*)}{b_a'} - \exp \frac{2.3\,(E - E^*)}{b_c''} \right\} \quad (11)$$

which is a useful, dimensionless form of the general rate equation developed by Wagner and Traud.[3]

Equation (11) relates measured meter current I to electrode potential E. Figure 2 shows an example of log I vs E plots (solid polarization curves) obtained from Equation (11) for rather typical numerical values of the parameters (25 C):

$$I^* = 10^{-2}\,A, b_a' = 2.3\,RT/F = 0.0592\,V,$$

$$b_c'' = -2.3\,(2RT/F) = -0.1183\,V$$

Shown as broken lines are the Tafel lines for the partial anodic and cathodic currents. Note that the measured anodic polarization curve almost coincides with the anodic partial current Tafel line when $(E - E^*) > 2.3$ (RT/F). And similar coincidence occurs on the cathodic side when $(E - E^*) < -2.3$ (RT/F).

Many real corroding systems comply very closely with Equation (11); this means that only two of the possible electrode reactions have significant partial currents, which in turn requires that the reversible potentials[(1)] for these two reactions be far from E^*, namely $|E_r - E^*| >$ 2.3 RT/F.

There is a common type of corroding system in which one of the two dominant reactions occurs at constant rate. For example, this condition is found in the zinc/acid system [Equation (5)] when oxygen diffusion to the zinc electrode is restricted and the corrosion is diffusion controlled. Here the partial cathodic current I_c'' is a constant, independent of potential. This type of system can also be represented by the general rate Equation (11), by setting $b_c'' = -\infty$. The rate equation then simplifies to:

$$\frac{I}{I^*} = \pm \left\{ \exp \frac{2.3\,(E - E^*)}{b_a'} - 1 \right\} \quad [11(a)]$$

The general applicability of Equation (11) to practical corroding electrodes is well established experimentally; this relation is the basis for most electrochemical methods of determining corrosion rate. These methods are described below.

Electrochemical Determination of Corrosion Rate

Rigorous General Method

There are three parameters in the rate equation relating meter current to potential, namely, I^*, b_a', and b_c''. A rigorous method of deriving corrosion current from E/I measurements may also yield values of the Tafel slopes. One such procedure is the three-point method,[4] developed in 1970, which is general and applicable within any potential range. In this method, three related I/E data points are measured; the first at a selected value of $\triangle E$ $(= E - E^*)$, the second at $2\triangle E$, and the third at $-2\triangle E$. A typical set of selected measurements is illustrated by the 3 points shown on Figure 2, where $\triangle E$ is selected to be +30 mV. These three measurements are all that is needed to evaluate I^*, b_a', and b_c''.

In analyzing the experimental data, two ratios are formed:

$$r_1 = I_{2\triangle E}/I_{-2\triangle E} \quad [12(a)]$$

$$r_2 = I_{2\triangle E}/I_{\triangle E} \quad [12(b)]$$

It is readily shown[4] that these ratios are interrelated through the quadratic equation:

$$u^2 - r_2 u + \sqrt{r_1} = 0 \quad (13)$$

the two roots of which are

$$\exp \frac{2.3\,(E - E^*)}{b_a'} \text{ and } \exp \frac{2.3\,(E - E^*)}{b_c''}$$

[(1)] The rate equation becomes more complicated than Equation (11) in a system where one of the reversible potentials is close to E^*. This complication is not treated here, for brevity and clarity.

TABLE 1 — Determination of Corrosion Current and Tafel Slopes from the Three Points in Figure 2

△E (mV)	I (mA)	I* (mA)	$b_a{}'$ (mV)	$b_c{}''$ (mV)
Set 1 — I Rounded to Nearest 0.1 mA				
+30	26.6			
+60	100.0	10.07	+59.4	−119.
−60	31.2			
Set 2 — I Rounded to Nearest mA				
+30	27.			
+60	100.	10.6	+60.6	−124.
−60	31.			
Set 3 — I Rounded to Nearest 5 mA				
+30	25.			
+60	100.	8.5	+55.5	−107.
−60	30.			
True values:		10.00	+59.2	−118.3

The first root gives the numerical value of $b_a{}'$, the second of $b_c{}''$. With the Tafel slopes known, any one of the three I/E data points is substituted into Equation (11) to obtain the corrosion current.

The method was applied to the three points marked on Figure 2, and the results are listed in Table 1. The currents were rounded off as indicated, to simulate the effect of measurement precision on the calculated values of the three parameters. Since the three-point method involves no approximations, the calculated parameters must become increasingly more accurate as the accuracy of the meter current increases, and Table 1 demonstrates this. The table also shows that the measurement accuracy required for the three-point analysis is not stringent.

An application of the three-point method to experimental measurements is shown in Figure 3 for the system: $Fe/1M\ NaHSO_4/H_2$ atm.[5] The points are galvanostatic measurements, and were analyzed by first drawing preliminary smooth curves through them and picking off the required trio of points for a selected △E. This procedure was repeated at several △E values and mean values of the derived parameters were determined, namely $i^* = 8.7\ A\ m^{-2}$, $b_a = 0.076$ V, and $b_c = -0.104$ V. These were then used to construct the solid polarization curves shown in Figure 3. The very close fit of the calculated curves and experimental points attest to the good precision of the measured currents. Broken lines a and c are the calculated partial currents for anodic iron dissolution and cathodic hydrogen evolution, respectively.

The three-point method is general and is based on

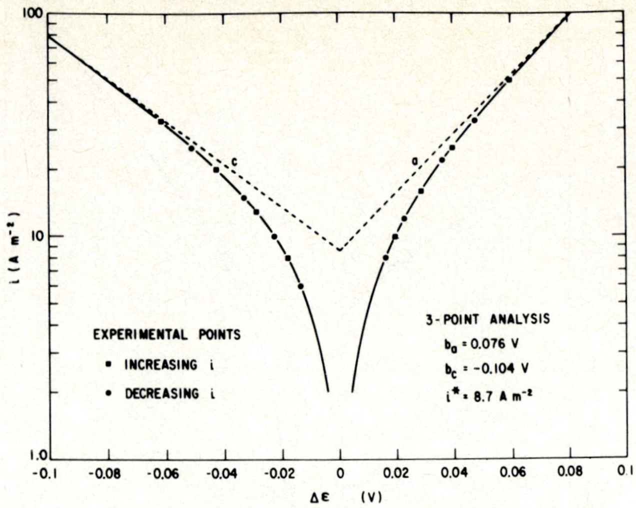

FIGURE 3 — Polarization curves for Ferrovac E iron in 1 M $NaHSO_4$, 25 C.[5]

rigorous analysis. However, it is still relatively new and should be further evaluated with a variety of corroding systems.

Rigorous Method, Known Tafel Slopes

The anodic and cathodic Tafel slopes are frequently known to remain constant over the set of conditions under investigation, as for example, in studying corrosion rate as a function of time and concentration for a given metal in a particular salt solution. In such cases, the Tafel slopes may be determined once, say by the three-point method. Thereafter, the determination of corrosion rate requires just a single current measurement at any value of E. Simple substitution of the measured values of I and E, as well as E*, into Equation (11) yields the corrosion current.

This procedure is simple, straightforward, and rigorous, but to date, surprisingly little use has been made of it in routine corrosion rate measurements. Procedures that have been applied extensively utilize either a large potential or a small potential approximation to Equation (11).

Large Potential Approximation

The rate Equation (11) contains two exponential terms on the righthand side, one of which has a positive exponent and the other a negative exponent. The latter term becomes negligible for large values of $|E - E^*|$, where the polarization curves are then given by:

$$\text{Anodic:}\quad \frac{I}{I^*} = \exp \frac{2.3\,(E - E^*)}{b_a{}'} \qquad [14(a)]$$

$$\text{Cathodic:}\quad \frac{I}{I^*} = \exp \frac{2.3\,(E - E^*)}{b_c{}''} \qquad [14(b)]$$

These are also the equations for the partial current Tafel lines. That the large potential approximation to the polarization curve coincides with the partial current curve has already been demonstrated graphically in Figure 2.

Thus, the anodic Tafel slope b_a' is determined as the slope of a linear semilogarithmic plot of large potential I/E measurements. Extrapolation of this line to $E = E^*$ yields the corrosion current. Similarly cathodic I/E measurements yield b_c'' and I^*.

The large potential method has been utilized widely for measuring Tafel slopes. It fails in some corroding systems because of an interfering electrochemical process. For example, anodic metal dissolution is often followed at high potentials by passivation, which can distort the log I/E plot before the linear Tafel region is established. The cathodic Tafel slope and corrosion current can still be obtained from the cathodic polarization curve, provided the latter exhibits a linear log I/E section. In this case, however, it is best to use the three-point method at small values of $E - E^*$, to obtain both Tafel slopes as well as I^*.

Small Potential Approximation

A simple procedure for determining corrosion rate was developed in 1938 by Wagner and Traud[3] based on a small potential approximation. This procedure has become well established experimentally (Stern[6,7] and later authors) and is generally known as the linear polarization method. It is based on the fact that Equation (11) yields direct proportionality between meter current and $(E - E^*)$ as the latter becomes very small.

Taking the derivative dI/dE from Equation (11), we find:

$$E = E^*: \quad \frac{dI}{dE} = \left(\frac{2.3}{b_a'} + \frac{2.3}{|b_c''|}\right) I^* \qquad (15)$$

for either anodic or cathodic polarization. Since $I = 0$ at $E = E^*$, Equation (15) can be restated in terms of meter current at potential E:

$$E \to E^*: \; I^* \cong \frac{1}{k}\frac{I}{E - E^*};$$

$$k \equiv \frac{2.3}{b_a'} + \frac{2.3}{|b_c''|} \qquad [15(a)]$$

Before making corrosion rate measurements on a given corroding system by the linear polarization method, it is necessary to determine the two Tafel slopes and obtain k from them. Thereafter, a single I/E measurement is sufficient to derive the instantaneous corrosion current. For good accuracy, the experimental values of I and $\triangle E$ should not be too small; hence, the question arises, how large a value of $\triangle E$ can be used without incurring an unacceptable error (say >5%) from the linear polarization assumption?

This question has been examined in some detail,[8] the answer being that the maximum $\triangle E$ varies considerably with the corroding system. This is illustrated in Figure 4, which shows the variation with $\triangle E$ of the fractional deviation δ of the linear polarization line from the true polarization curve, where $\delta \equiv [I(\text{linear}) - I(\text{true})]/I(\text{true})$. The curves in Figure 4 refer to corroding electrodes at 25 C, each with cathodic slope $b_c'' = -2.303$

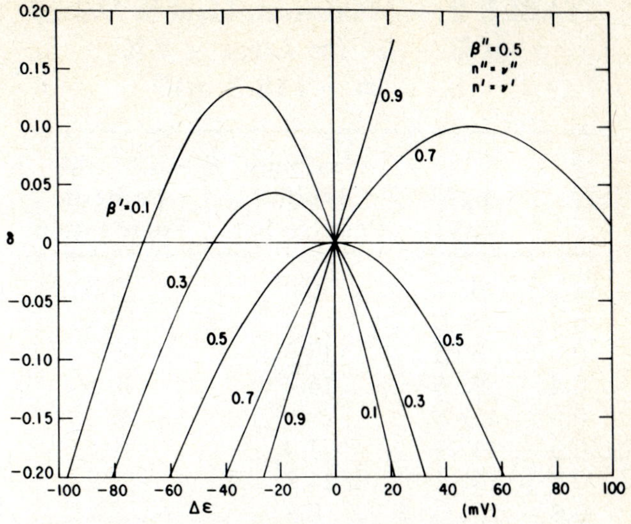

FIGURE 4 – Fractional deviation δ as a function of potential change.[8]

RT/0.5F = −118.3 mV. These curves represent various values of anodic slope given by $b_a' = 2.303\ RT/(1 - \beta')F$, with the values of β' as shown on the curves (β = transfer coefficient, n = number of electrons, and ν = stoichiometric number).

An important characteristic demonstrated in Figure 4 is that the error from the linear polarization assumption is asymmetric, that is, greater for anodic polarization than for cathodic or vice versa, except for the 0.5 curve ($b_a' = |b_c''|$). The asymmetry can be very pronounced, as shown numerically in Table 2. Listed here is the maximum value of $E - E^*$, below which the linear polarization error remains less than 5%. The maximum $\triangle E$ is seen to vary considerably, from 6 to 57 mV, as the anodic Tafel slope is changed.

Basic information of this type can help the corrosion engineer decide whether to select anodic or cathodic polarization for linear polarization measurements on a given corroding metal. There is a third technique, however, namely the dual electrode technique—an important variation of the small potential approximation that combines both anodic and cathodic polarization in a single measurement. A two-electrode cell is used, in which the electrodes are duplicate specimens of the test metal. When a potential difference $\triangle E$ is applied across the two electrodes, each experiences a potential change of 0.5 $\triangle E$ if the two are identical and $\triangle E$ is small. Therefore, the slope of the linear polarization curve $(I/\triangle E)$ for dual electrodes is half of the slope for anode or cathode alone. It follows that Equation [15(a)] must be changed to:

$$E \to E^*: \; I^* \cong \frac{2}{k}\frac{I}{\triangle E} \qquad [15(b)]$$

to calculate corrosion current from dual-electrode measurements.

The dual electrode technique was first developed in 1963 by Marsh,[10] who utilized polarity reversal of the ap-

TABLE 2 — Maximum Usable Value of E — E* for Error ≤ 5%, b_c'' = −118.3 mV[8]

β'	b_a' (mV)	Maximum E − E*, mV Anodic Polarization	Cathodic Polarization
0.1	65.7	6.0	7.0
0.2	74.0	7.7	10.3
0.3	84.5	10.6	56.0
0.4	98.6	16.2	42.5
0.5	118.3	28.6	28.6
0.6	147.9	57.0	18.0
0.7	197.2	14.8	12.2
0.8	296.	8.8	8.4
0.9	592.	6.4	6.5

plied potential so that both a forward current I_f and a reverse polarity current I_r were measured. The average of these meter readings is inserted into Equation [15(b)] to calculate I*. In application of this method to real corroding metals, allowance must be made for the fact that duplicate test electrodes seldom yield identical corrosion potentials or corrosion currents. As Marsh showed,[10] it is necessary to determine the starting potential difference between the two test electrodes 1 and 2, $(E_2^* - E_1^*)$, and to calculate the mean current as $I = 1/2\,(I_f + I_r)$ when $|E_2^* - E_1^*| < |\Delta E|$ and as $I = 1/2\,|I_r - I_f|$ when $|E_2^* - E_1^*| > |\Delta E|$.

Marsh's reversal method has been widely applied in engineering studies of corrosion resistant metals. In fact, in the past decade, more electrochemical corrosion rate data have been generated by dual-electrode polarization than by anodic or cathodic polarization singly.

A simpler dual-electrode procedure, which eliminates the polarity reversal feature of Marsh's method, was recently reported.[11] It is termed the "direct" method because I* is calculated directly from a single current I, measured after increasing the starting potential difference $(E_2^* - E_1^*)$ by an amount ΔE_{de}. The corrosion current is now:

$$E \rightarrow E^*:\ I^* \cong \frac{2}{k}\frac{I}{\Delta E_{de}} \qquad [15(c)]$$

The applied ΔE_{de} can be selected in the range 20 to 50 mV. This is a higher range than that which can be reliably employed with anodic or cathodic polarization alone. The higher range is justified by the extended linear polarization curves characteristic of dual electrodes.[11]

The direct method yields a mean corrosion current I_m^* which is not the arithmetic mean of the corrosion currents at the individual test electrodes (I_1^* and I_2^*), but is given by:[11]

$$2/I_m^* = 1/I_1^* + 1/I_2^* \qquad (16)$$

Note that mean I* obtained by the reversal method also obeys Equation (16).

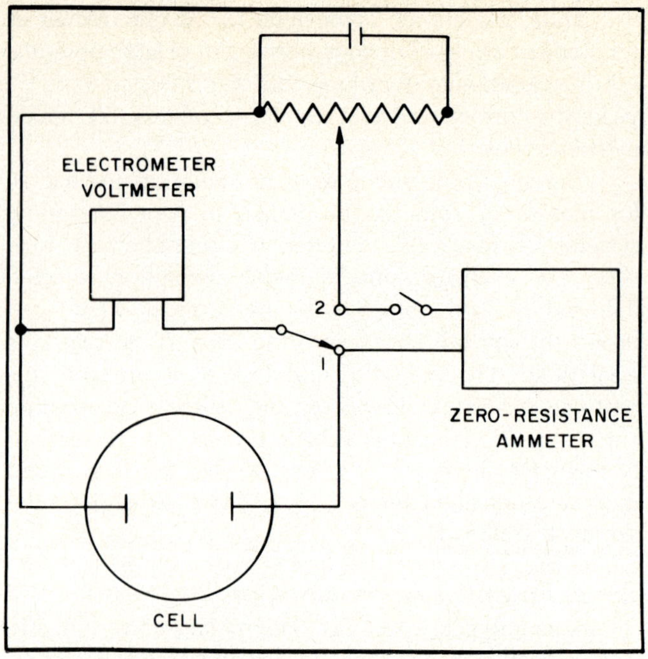

FIGURE 5 — Circuit for dual-electrode measurements by the direct method.

Figure 5 gives a convenient circuit and procedure for the direct method. While the current circuit is open, cell emf is measured with the voltmeter in position 1 and the desired applied voltage is set in position 2. Then the circuit is closed and the current measured with a zero resistance ammeter (see Reference 12).

The direct method represents a major improvement over the reversal method when $(E_2^* - E_1^*)$ is relatively large. An illustration of this is given in Table 3, based on experimental anodic and cathodic polarization curves for Ferrovac E iron in 1M $NaHSO_4$ (25 C).[11] Dual-electrode currents that would be obtained by the reversal method and by the direct method are listed for the case of electrodes differing in starting potential by 60 mV. Column 4 shows that, as a result of the large starting potential difference, the best value for mean corrosion current density achievable by the reversal method has over 30% error. The direct method is more accurate at any applied ΔE, and its error can be made negligible by applying a sufficiently small ΔE.

It may be concluded that linear polarization methods yield corrosion rate with adequate precision when the Tafel slopes are known. But why has the corrosion engineer preferred this method over the rigorous procedure previously discussed? There are several answers, the main one involving corroding systems for which the Tafel slopes were determined with poor or uncertain precision. Under these conditions, the linear polarization constant k for a particular corroding system may be in error by a large factor, but all values of I* calculated from it are erroneous by the same factor. This is useful for comparative purposes, even if the error factor is not established. On the other hand, constant error factor is not obtained with the rigorous method, based on substitution of an I/E

TABLE 3 — Comparison of Reversal and Direct Dual-Electrode Methods Data[11] for Ferrovac E Iron/1M $NaHSO_4$/25 C with $(E_2^* - E_1^*) = 0.060$ V

$\Delta E^{(1)}$ (V)	$i^{(2)}$ (A m^{-2})	$i_m^{*(3)}$ (A m^{-2})	% Error
	Reversal Method		
0.01	1.26	4.81	32
0.02	2.52	4.82	32
0.03	3.86	4.90	35
0.04	5.24	5.00	37
0.05	6.81	5.20	43
	Direct Method		
0.01	0.97	3.70	1.6
0.02	2.04	3.90	7.1
0.03	3.21	4.09	12.
0.04	4.50	4.30	18.
0.05	5.97	4.56	25.

(1) ΔE is the actual impressed dual-electrode voltage in the reversal method; it is the incremental voltage ΔE_{de} in the direct method.
(2) In the reversal method, $i \equiv 0.5\ |i_f - i_r|$, whereas in the direct method, i is the single measured current density.
(3) True value of i_m^*, from $i_1^* = 8.70$ and $i_2^* = 2.30$, is 3.64 A m^{-2} from Equation (16), where i_1^* and i_2^* are the corrosion current densities of isolated electrodes 1 and 2.

measurement pair into Equation (11). In fact, incorrect values of Tafel slopes generate errors that vary enormously with the potential selected for the measurement.

Localized Corrosion Rates

Fundamental Relations

A pure metal or a relatively homogeneous alloy tends to corrode over the entire exposed surface in a macroscopically uniform manner, although considerable microscopic roughening occurs. Microscopic roughening indicates that the anodic metal corrosion reaction, at any given instant, is accelerated at specific spots on the surface. Simultaneously, the cathodic reaction may be accelerated over other areas of the surface. To result in macroscopic uniformity, the location of the anodic areas must move relatively quickly while the anodic area fraction changes with time, both in unpredictable ways. Such corroding metals are treated, however, as if the average anodic current is distributed uniformly over the entire metal surface.

There are some types of corrosion where the anodic reaction remains stationary and highly localized. Stationary anodic areas can be induced by structural factors in the metal (at grain boundaries, for example) and a metal coated with a more noble metal can exhibit stationary anodic reaction at pores in the coating.

Corroding metals of this kind will now be examined, assuming total area A with stabilized anodic area fraction f_a and stabilized cathodic fraction f_c, where $f_a + f_c = 1$. We wish to know what happens to the localized corrosion penetration and to the corrosion potential when the anodic area fraction is changed. These two fundamental relations were developed in 1958 by Stern.[9] They are derived here in modified form to simplify both graphical representation and treatment of experimental data.

If the anodic Tafel line a′ in Figure 1 is described by the partial currents at potentials E_r' and E^*, its Tafel equation takes the form:

$$E^* - E_r' = b_a' \log (I^*/I_o') = b_a' \log (I^*/f_a A i_o') \quad (17)$$

where I_o' is the current at the reversible potential, and i_o' the corresponding current density over anodic area $f_a A$. Similarly, the Tafel equation for line c′′ in Figure 1 can be given as:

$$E^* - E_r'' = b_c'' \log (I^*/f_c A i_o'') \quad (18)$$

The desired relation between corrosion potential and area fraction is obtained from Equations (17) and (18) by equating the expressions they yield for I^*. The result is readily shown to be:

$$f_a \neq 0 \text{ or } 1:\ E^* = E^*_{0.5} - \left(\frac{I}{b_a'} + \frac{I}{|b_c''|}\right)^{-1} \log \frac{f_a}{f_c} \quad (19)$$

Potential $E^*_{0.5}$ is the corrosion potential when the anodic and cathodic areas are equal ($f_a = f_c = 0.5$). This potential depends on the reversible behavior of the two dominant reactions on the corroding surface, and is given by:

$$E^*_{0.5} = \left(\frac{I}{b_a'} + \frac{I}{|b_c''|}\right)^{-1} \times \left(\frac{E_r'}{b_a'} + \frac{E_r''}{|b_c''|} + \log \frac{i_o''}{i_o'}\right) \quad (20)$$

The fundamental relation between corrosion current and anodic area is obtained from Equations (17) and (18) by elimination of E^*. The result may be put in the form:

$$f_a \neq 0 \text{ or } 1:\ I^* = 2I^*_{0.5}\,(f_a)^{1-q}\,(f_c)^q$$

$$q \equiv \frac{|b_c''|}{b_a' + |b_c''|} \quad (21)$$

where $I^*_{0.5}$ is the corrosion current for equal anodic and cathodic areas.

The corrosion current given by Equation (21) goes through a maximum. In general, the current will not be at the maximum value when anodic and cathodic areas are equal. The area ratio that yields maximum current is found by setting $d\,I^*/d\,f_a = 0$, which leads to:

$$I^* = \max: \; f_a/f_c = b_a'/|b_c''| \qquad (22)$$

the general condition for current maximum reported by Mansfeld.[13,14] Thus, the corrosion current is a maximum when the areas are in the same ratio as the corresponding Tafel slopes.

More significant than the corrosion current is the metal penetration rate, which is proportional to the corrosion current density $i^* = I^*/f_a$ A. This quantity, as obtained from Equation (21), is given by:

$$f_a \neq 0 \text{ or } 1: \; i^* = i^*_{0.5}\,(f_c/f_a)^q \qquad (23)$$

Here, $i^*_{0.5}$ is the corrosion current density when the anodic and cathodic areas are equal. Its magnitude depends on the equilibrium partial current densities at the reversible potentials:

$$i^*_{0.5} = (i_o')^{1-q}\,(i_o'')^q \exp \frac{2.3\,(E_r'' - E_r')}{b_a' + |b_c''|} \qquad (24)$$

It should be noted that i^* is related to E^* by the anodic Tafel equation, which takes the form

$$\frac{i^*}{i^*_{0.5}} = \exp \frac{2.3\,(E^* - E^*_{0.5})}{b_a'} \qquad (25)$$

when the reference state is taken to be that of equal anodic and cathodic areas. This relation emerges directly from Equations (19) and (23) upon eliminating (f_a/f_c).

Practical applications of Equations (19) and (23), which give the area variations of potential and of corrosion penetration, respectively, are described next.

Corrosion Potential and Anodic Area

In accordance with Equation (19), experimental measurements of corrosion potential as a function of anodic area should yield a linear plot of E^* vs $\log(f_a/f_c)$. Figure 6 shows the lines that would be obtained with four corroding electrodes, which are described by the following Tafel slopes at 25 C ($\epsilon \equiv 2.3$ RT/F).

Electrode:	**A**	**B**	**C**	**D**
b_a':	2ϵ	2ϵ	ϵ	0.5ϵ
$\|b_c''\|$:	ϵ	2ϵ	2ϵ	2ϵ
$(1/b_a' + 1/\|b_c''\|)^{-1}$, mV:	39.4	59.2	39.4	23.7

The last row gives the absolute value of the slope of each line in Figure 6. This slope is the change in corrosion potential for a tenfold increase in anodic/cathodic area ratio. Its measured value can be used to determine one of the Tafel slopes of the corroding system if the other one is known.

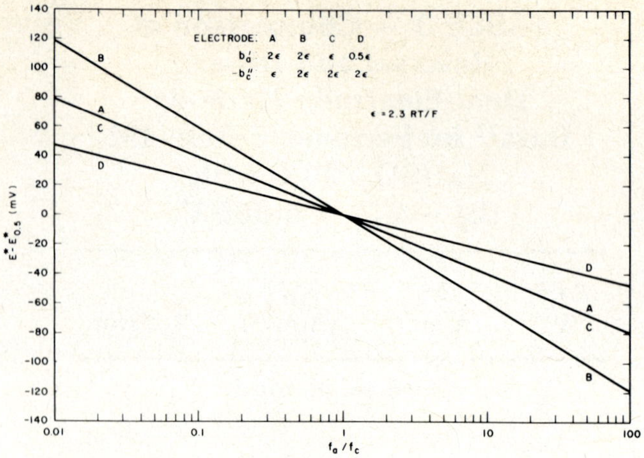

FIGURE 6 — Dependence of corrosion potential on anodic/cathodic area ratio, 25 C.

Figure 6 illustrates the important generalization that the corrosion potential goes more positive (or less negative) when the anodic reaction is confined to a smaller area. The potential changes are seen to be symmetrical about the condition of equal anodic and cathodic areas; that is, an increase in anodic area to n times the cathodic area would cause the same potential change as would an increase in cathodic area to n times the anodic, except that the two changes are in opposite directions.

If both Tafel constants b_a' and b_c'' are known, the corrosion potential E^* is a measure of the fractional anodic area. It is necessary first to determine E^* at one known area ratio in order to establish the value of $E^*_{0.5}$. This suffices as a calibration of the corrosion system. Thereafter, a single measurement of E^* yields f_a.

Morrissey[15] measured the variation of E^* with anodic area in a study of porosity in electroplated gold deposits on copper. His calibration curve was established by the use of thin copper wires shorted to large gold surfaces. He showed that, for 0.1M NH_4Cl solution, a plot of E^* vs $\log f_a$ yielded excellent linearity. The anodic area fractions used were very small, $f_a = 10^{-3}$ to 10^{-6}, so that $f_c \cong 1$. Therefore, these data represent partial experimental confirmation of Equation (19). This work demonstrates that corrosion potential measurements provide a simple way of determining porosity as a function of deposit thickness or of electroplating procedure.

Corrosion Rate and Anodic Area

As the anodic area becomes smaller, the corrosion per unit area goes up. This is seen from Equation (23), which may be rewritten as:

$$\log i^* = \log i^*_{0.5} - q \log (f_a/f_c) \qquad [23(a)]$$

Thus, a log-log plot of measured i^* vs (f_a/f_c) yields a

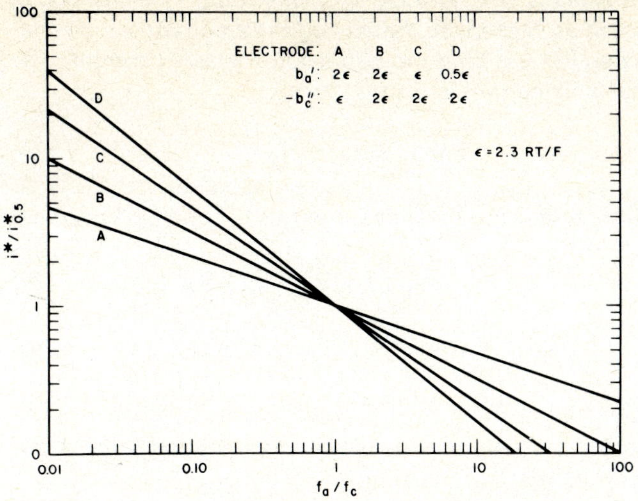

FIGURE 7 — Dependence of corrosion current density on anodic/cathodic area ratio, 25 C.

straight line of slope $-q = -|b_c''|/(b_a' + |b_c''|)$. Figure 7 presents plots of this kind for the same four electrodes as in Figure 6. The lines have slopes given by:

Electrode:	**A**	**B**	**C**	**D**
q:	1/3	1/2	2/3	4/5

Figure 7 demonstrates that the localized corrosion penetration rate, represented by i^*, increases considerably as the anodic area becomes smaller. As an example, a decrease in anodic area from $f_a = f_c$ to $f_a = f_c/100$ increases local penetration tenfold when $|b_c''| = b_a'$ (line B), and increases it fortyfold when $|b_c''| = 4\ b_a'$ (line D). The lines in Figure 7 are symmetrical about the condition of equal anodic and cathodic areas. This means that an increase in anodic area to n times cathodic changes i^* by the same factor as does an n-fold decrease in area ratio.

Experimental data of Morrissey[16] on gold-covered copper corroding in 0.1 M NH_4Cl yielded an accurately linear plot of log i^* vs log f_a. His measurements were all at very small values of f_a, namely 10^{-4} to 10^{-7}, so that $f_c \cong 1$. Therefore, these experiments contribute partial confirmation of Equation (23).

General Analysis of Localized Corrosion

The general problem of localized corrosion when the anodic and cathodic Tafel slopes are unknown is considered here, and a short procedure for calibrating and analyzing the corroding system is presented.

The measurements required are just two E^*/I^* determinations, one at each of two measured values of f_a (any two convenient values); each I^* may be measured by the three-point Tafel extrapolation or three-electrode linear polarization method. These suffice to calibrate the system, and thereafter, one need measure only the corrosion potential of an unknown electrode (unknown anodic area). From E^*, both the area fraction f_a and the corrosion penetration i^* are obtained. In addition, the two calibration data points are sufficient to evaluate the anodic and cathodic Tafel slopes. The following analysis is suggested.

First, the two E^*/f_a measurements are plotted as E^* vs log $f_a/(1 - f_a)$ to evaluate $E^*_{0.5}$ and the slope of the line (Figure 6). The two I^* measurements, converted to i^* with the known values of anodic area, are plotted as in Figure 7 to obtain $i^*_{0.5}$ and the slope of the line. Also, the two i^*/E^* data pairs yield the anodic Tafel slope b_a' from Equation (25). With this value of b_a', the experimental slope of the E^* vs log (f_a/f_c) plot yields b_c''. Alternatively, b_c'' may be evaluated from b_a' and the experimental slope of the log i^* vs log (f_a/f_c) plot.

Thus, the above measurements at two values of f_a are sufficient to determine constants $E^*_{0.5}$, $i^*_{0.5}$, b_a', and b_c''. It may be desirable, however, to make further measurements at other values of f_a, to refine the potential/area ratio and current density/area ratio plots and to obtain more precise values of the four constants.

With the corroding system thus standardized, any unknown state of the system requires simply a measurement of E^* to characterize it fully. From E^*, the corresponding value of anodic area fraction is read off the E^* vs log (f_a/f_c) graph, or is calculated from Equation (19). Then from f_a, the corresponding corrosion penetration rate is obtained either from Equation (23) or from the log i^* vs log (f_a/f_c) graph.

Galvanic Stimulation of Corrosion

A corroding metal is stimulated to corrode faster by galvanic coupling to a more noble metal. This effect is depicted schematically in Figure 8. The broken lines a′ and c′′ give the partial currents of the anodic and cathodic reactions on the corroding metal before coupling. The solid lines represent the anodic polarization curve for the corroding metal and the cathodic polarization curve for the noble metal. These curves cross at the

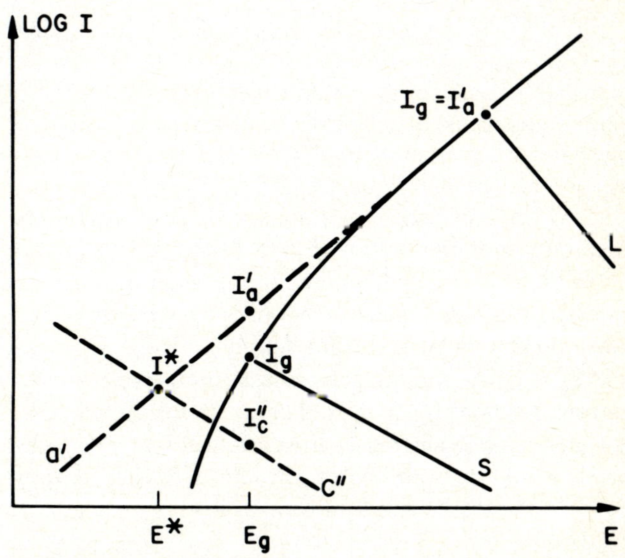

FIGURE 8 — Schematic polarization curves for galvanically stimulated corrosion.

galvanic current I_g, corresponding to the potential E_g of the two metals after they are electrically shorted, assuming negligible IR-drop between anode and cathode. The simplest procedure for measuring I_g is to short the two metals externally through a zero resistance ammeter.

Two cathodic polarization curves are shown in Figure 8. The upper one (L) refers to a large noble metal cathode which, upon coupling to the corroding metal, shifts the potential a relatively large amount, i.e., $(E_g - E^*) > 2.3\ RT/F$. The lower one (S) represents galvanic coupling to a smaller area of noble metal or to a metal on which the cathodic process is slow, so that $(E_g - E^*) < 2.3\ RT/F$. The latter is the general case; however, before treating this case, it is helpful to treat the special case of large coupled cathode (curve L) first, plus one other special case.

Special Case of Large Stimulation of Corrosion Rate

Considering curve L in Figure 8, it is seen that the measured galvanic current I_g yields the stimulated corrosion rate directly: $I_g = I_a{'}$. If the uncoupled corrosion current I^* is determined, then the galvanic current measurement also yields a value for the anodic Tafel slope from the relation:

$$I_g/I^* \doteq \exp\left[2.3\,(E_g - E^*)/b_a{'}\right] \tag{26}$$

Special Case of Corroding Metal with Diffusion Limited Cathodic Reaction

Here the lower cathodic polarization curve (S) is considered, but with the cathodic reaction at the uncoupled corroding metal occurring at a constant diffusion controlled rate. Constant cathodic rate means that line c'' in Figure 8 must be horizontal and $I_c{''} = I^*$. If the corrosion current I^* before coupling is determined, then the galvanic current measurement yields the stimulated corrosion current directly since

$$I_a{'} = I_g + I_c{''} = I_g + I^* \tag{27}$$

Mansfeld[17] showed that stimulated corrosion rates of aluminum alloys, galvanically coupled to a variety of more noble metals in aerated 3.5% NaCl solutions, were in good accord with Equation (27).

As an application of Equation (27), consider the case of a corroding metal of area A coupled to a more noble metal of area nA where the same diffusion limited cathodic reaction (oxygen dissolution, for example) occurs on each metal at current density i_x. Here, $I^* = i_x A$ and $I_g = i_x nA$, and the galvanically stimulated corrosion current becomes $I_a{'} = i_x A\,(1 + n)$. In this simple case, galvanic coupling increases the corrosion rate by a factor of $(1 + n)$, since the corrosion rate is directly proportional to the total cathodic area available.

General Determination of Stimulated Corrosion Rate

In the general case, Tafel line c'' in Figure 8 is not horizontal, but has slope $b_c{''}$. Also, $(E_g - E^*)$ can be as small as desired, so that the galvanic current can be considerably less than the stimulated corrosion current. The galvanic current is always given by:

$$E_g\colon\ I_g = I_a{'} - I_c{''} \tag{28}$$

At potential E_g, the partial currents $I_a{'}$ and $I_c{''}$ become:

$$I_a{'} = I^* \exp\left[2.3\,(E_g - E^*)/b_a{'}\right] \tag{29}$$

$$I_c{''} = I^* \exp\left[2.3\,(E_g - E^*)/b_c{''}\right] \tag{30}$$

Combining these with Equation (28), a relation is obtained that permits the calculation of the stimulated corrosion current $I_a{'}$, namely:

$$I_g/I_a{'} = 1 - \exp\left[-k(E_g - E^*)\right] \tag{31}$$

where k is the same constant as is used in determining corrosion current from a linear polarization measurement, defined in Equation [15(a)]. Equation (31) is the same as that given by Mansfeld.[17]

It should be emphasized that Equation (31) is general and exact. Although constant k appears in it, the linear polarization approximation is not invoked. In applying this relation to a corroding metal for which k is known, a single measurement of the galvanic current I_g at potential E_g yields the stimulated corrosion current $I_a{'}$. It is necessary to measure the uncoupled corrosion potential E^* but not I^*. It is also not necessary to know the kinetics of the reaction occurring on the noble metal of the couple, that is, its cathodic Tafel slope.

When k has not been determined, it is still possible to derive galvanically stimulated corrosion rates. The following method is recommended. One measurement of $I_a{'}$ is made by coupling to a sufficiently large specimen of noble metal so that $I_a{'} = I_g$. It is necessary that the uncoupled value of I^* be measured also. Then the anodic Tafel slope $b_a{'}$ is calculated by use of Equation (29). Thereafter, a single measurement of couple potential E_g (the measurement of galvanic current is unnecessary) yields the stimulated corrosion current by use of Equation (29).

Concluding Remarks

In the kinetic treatment of electrochemical corrosion presented here, the focus is on the most useful relations for determining corrosion rate. These relations all stem from Equation (6), which expresses meter current in terms of the partial currents at the electrode, combined in compliance with the additivity principle. For clarity, corroding systems having only two significant partial currents are treated above. But many complex systems involving three or more large partial currents can be treated also. The general procedure is to start with Equation (6), substitute the Tafel expression similar to Equation (9) for each of the partial currents, and then solve for the corrosion current. The simpler cases solved in this paper suffice to demonstrate how corroding systems, by virtue of their electrochemical nature, can be investigated by quick and easy current and potential measurements that yield essentially instantaneous corrosion rates.

References

1. J. Tafel, Z. Physik. Chem., Vol. 50, p. 641, 1905.
2. J. O'M. Bockris, A. K. N. Reddy, Modern Electrochemistry, Plenum Press, New York, New York, 1973.
3. C. Wagner, W. Traud, Z. Elektrochem., Vol. 44, p. 391, 1938.
4. S. Barnartt, Electrochimica Acta, Vol. 15, p. 1313, 1970.
5. S. Barnartt, Corrosion, Vol. 27, p. 467, No. 11, 1971.
6. M. Stern, A. L. Geary, J. Electrochem. Soc., Vol. 104, p. 56, 1957.
7. M. Stern, Corrosion, Vol. 14, No. 9, p. 440t, 1958.
8. S. Barnartt, Corros. Sci., Vol 9, p. 145, 1969.
9. M. Stern, Corrosion, Vol. 14, No. 7, p. 329t, 1958.
10. G. A. Marsh, "The Measurement of Instantaneous Corrosion Rates," Proc. Second Int. Congress on Metallic Corrosion, National Association of Corrosion Engineers, Houston, Texas, p. 936, 1963.
11. S. Barnartt, Corrosion, Vol. 31, No. 5, p. 164, 1975.
12. G. Lauer, F. Mansfeld, Corrosion, Vol. 26, No. 11, p. 504, 1970.
13. F. Mansfeld, Corrosion, Vol. 27, No. 10, p. 436, 1971.
14. K. B. Oldham, F. Mansfeld, J. Applied Electrochem., Vol. 2, p. 183, 1972.
15. R. J. Morrissey, J. Electrochem. Soc., Vol. 117, p. 742, 1970.
16. R. J. Morrissey, J. Electrochem. Soc., Vol. 119, p. 446, 1972.
17. F. Mansfeld, Corrosion, Vol. 29, No. 10, p. 403, 1973.

Some Novel and Standard Techniques in Electrode Studies

H. PICKERING, H. SHIH, H. HACK, and J. FRITZ*

Abstract

Recent results on three research topics will be presented to illustrate a range of electrochemical techniques useful in corrosion studies; potentiostatic and dynamic polarization, rotating disk electrodes, and impedance measurements will be included. The first topic considers the small amplitude linear potential scanning technique (SALPST). Because of recent theoretical analysis, it has become an attractive method for polarization resistance measurement of alloys in aqueous solutions. The technique has been used successfully for observing protective layer formation on copper alloys in 3.4 wt% NaCl solution. The second topic illustrates the effective use of electrochemical methods for clarifying the role of the adherent corrosion products in reducing the corrosion rate of the cupronickels in aqueous NaCl solutions. The third topic considers some electrochemical methods that are useful for a restricted number of alloys in identifying characteristic and sometime unique features of the individual component polarization curves of alloys.

Introduction

Modern studies of electrode reactions in which degradation of the electrode occurs involve a broad spectrum of electrochemical techniques and include post analyses of the electrode by a multitude of sophisticated microscopy, diffraction, and spectroscopy methods. However, it is often the *in situ* electrochemical experiment or series of experiments that provides the greatest amount of new understanding. A few examples of the value of electrochemical measurements are illustrated in three recent studies of the corrosion of Cu alloys, as follows.

*Metallurgy Program, Pennsylvania State University, University Park, Pennsylvania; Current address for H. Hack: David W. Taylor Naval Research and Development Center, Annapolis, Maryland; and for H. Shih: Dept. Material Sciences, University of Southern California, Los Angeles, California.

Measurement of R_p and C_p of Cu Alloys Using SALPST

The small amplitude linear potential scanning technique (SALPST) has been analyzed by Macdonald[1] on the basis of the transform method. Gabrimm and coworkers[2] presented a theoretical analysis using Heaviside calculus and an application of the measurements of polarization resistance and interfacial capacitance for steel in acidic solution. Shih and Pickering[3] furthered the theoretical analysis using the transform method and Heaviside calculus, set up the mathematical modeling needed for transient measurements of polarization resistance and interfacial capacitance, and applied the method to copper alloys in seawater.

The corrosion behavior of copper-zinc alloys and copper-tin-zinc alloys has been studied by Sugawara and Shimodaira.[4-6] For Cu-Al-Zn alloys, Castle, et al.,[7,8] proposed a corrosion mechanism in their annual report. Recently, Kato, et al.,[9-12] investigated the corrosion mechanism of Cu-9.4Ni-1.7Fe alloy in aerated 3.4 wt% NaCl solution with and without H_2S. They pointed out that the good corrosion resistance of cupronickels in sea water is due to the poor catalytic nature of the corrosion-product surfaces for the oxygen-reduction reaction, rather than to mechanisms based on protective Cu_2O films.[13] Ijsseling used the polarization resistance method to check surface film formation on cupronickels in seawater.[14] Many other authors have also investigated surface film formation on cupronickels and copper alloys.[15-23]

The goal here is to outline the SALPST measurement procedure for the polarization resistance and interfacial capacitance of corroding alloys. This technique appears attractive and with some advantages when compared with the charging curve, AC impedance, and weight loss methods. To this end this paper presents new mathematical modeling of SALPST and its application to the measurements of R_p and C_p of copper alloys in 3.4 wt% NaCl solutions, in particular, during the initial transient period of surface film formation.

The measurements can be done either automatically or manually and essentially consist of measuring the transient response as a function of voltage scan rate of an electrode/solution interface to a small amplitude triangular voltage excitation. In automatic experiments, the computer system includes a microcomputer with a GPIB, single drive floppy disc, video monitor, and a graphic printer. A potentiostat with an interface is connected to the computer system and a plotter is interfaced to the computer. The whole system is connected to a mainframe computer by an automodem. The function generator system includes an oscilloscope, a function generator, and superlow frequency signal generator in order to satisfy the needs of frequency domain work. The experimental work is computer controlled and the data are printed, stored on disc, plotted or recorded, and treated by regression analysis.

In manual mode, the experimental setup includes an electrolytic cell, signal generator, a function recorder, and a corrosion potential compensator. The hysteresis loops are plotted directly on the function recorder for various scanning rates or for different time periods at the same scanning rate.

SALPST with computer control can give R_p and C_p values for copper alloys in a relatively easy and quick fashion. By means of mathematical methods including Heaviside calculus and curve fitting and from a series of values of measurements at different voltage scan rates, the true R_p and C_p values can be obtained. The techniques have been found to be successful for copper alloys at different stages of protective film formation.

Mechanism of Cu-Ni Corrosion in 3.4 wt% NaCl Solution

The corrosion mechanism of copper in aqueous chloride-containing solutions has been studied extensively. The thermodynamically stable species under conditions applicable to seawater exposures may be determined from the Pourbaix diagram to be cuprous oxide, cupric hydroxy-chloride, and cuprous chloride ions.[24] The anodic process of the corrosion mechanism for fresh copper surfaces under these conditions is believed to occur in two stages. Bacarella and Greiss[25] and others suggest that the first stage consists of several fast reactions whose net result is to combine copper with two chloride ions to form cuprous chloride ion in the solution at the copper surface. The second and rate-controlling stage involves the diffusion of this ion through the diffusion boundary layer to the bulk solution. This mechanism has been found to be consistent with experimental evidence on the chloride reaction order, electrode rotation speed dependence, and the Tafel slope.[25] Thus, diffusion of the cuprous chloride ion away from the surface determines the anodic reaction rate of a corroding fresh copper surface. The rate of diffusion depends on the surface concentration of the ions, which depends on potential; this in turn, is affected by the cathodic reaction kinetics. Thus, the anodic reaction rate can be controlled by the cathodic reaction kinetics.

Until recently, the corrosion mechanism of copper-nickel alloys was not as well understood. The formation of a corrosion product film has been found to greatly reduce the corrosion rate of these alloys.[26] However, the role of this film in reducing corrosion was unclear. The corrosion product was known to consist of a loosely adherent porous-layered structure consisting primarily of cupric hydroxychloride over a thin tightly adherent layer of cuprous oxide.[9,10] The iron and nickel from the alloy are concentrated in the inner part of the porous layer.[10] Until recently, copper-nickel was believed to obtain its corrosion resistance from the cuprous oxide layer in the following manner.

Cuprous oxide has a cubic, copper-deficient lattice, making it a p-type semiconductor[27,28] with low ionic resistance that results from highly mobile vacancies and low electronic resistance that is due to holes with a low activation energy for movement.[29] Additions of nickel and iron were thought to improve corrosion performance by substitution for copper in the lattice, thus reducing the hole concentration, or by filling in vacancies, thus reducing the concentrations of both vacancies and holes.[27] Reduction of the number of mobile charge carriers would increase film resistance,[29] providing a more effective barrier to the corrosion process.

More recently, Kato, et al.,[9,11] have developed a more highly supported mechanism for the corrosion resistance of copper-nickel alloys based primarily on electrochemical measurements, including use of rotating disc electrodes and catalytic Pd deposits on the corrosion product layer. Anodic polarization near the corrosion potential is only slightly affected by the presence of the layer, while cathodic polarization (oxygen reduction) is greatly affected, indicative of a corrosion process that is cathodically controlled on alloys covered with a corrosion product layer. The low rate of oxygen reduction persisted even upon stripping off the outer porous layers. Thus, the inner cuprous oxide either provides a diffusion barrier to the cathodic reaction[27] or is (along with the outer layer) a poor catalyst for the oxygen reduction reaction.[9,11] Coating the inner cuprous oxide layer with palladium increases the cathodic kinetics to those of a bare surface, which Kato, et al.,[9,10] pointed out could be due to either electrical shorting of an inner barrier layer or an increasing catalytic activity of the surface for oxygen reduction, but could not be due to low ionic conductivity in the corrosion product. Kato and Pickering[11] later proved that the catalytic effect was the correct explanation for this behavior. They resolved the confusion by developing a corrosion product film in sulfide-containing solutions.[12] This film shows high oxygen reduction rates, presumably because the sulfide is a good oxygen reduction catalyst, whereas the rate decreased greatly when the outer porous sulfide layer was removed. This is a convincing argument in favor of the catalytic mechanism.[11,12]

Individual Component Polarization Curves of Alloys

Alloy polarization curves demonstrate features that are important in understanding the corrosion of alloy systems. The typical polarization behavior for a binary (A-B) alloy, which has a difference in the standard potentials of its components on the order of $\triangle E^\circ >> RT/F$, is subject to unique polarization behavior not observed in pure metals. There is a limiting low current region that results from the dissolution of the less noble metal A, which is suppressed by a surface layer of the more noble metal B. As the potential becomes more oxidizing (more noble), a critical potential, E_C, is reached above which the current increases sharply with potential. In this region and in those cases when E_C is below the B equilibrium potential the current is due solely to the dissolution of the A component and the alloy surface undergoes a surface instability which results in fine-scale porosity. A recent review paper[30] discusses these characteristic features of alloy polarization curves in some detail.

Obtaining these individual component polarization curves is straightforward only for some of the noble metal alloys. Generally, special procedures are required because of the presence of reactions, with the exception of metal dissolution, occurring at the electrode surface in the potential range of interest, e.g., hydrogen or oxygen evolution. The absence of side reactions simplifies the experiment in that ammeter measurements may be used to determine the A component dissolution current. In the presence of side reactions, currents must be obtained by chemical analysis of the electrolyte and adherent reaction product layers. The most common side reaction is the evolution of H_2. This reaction is avoided when studying alloys that contain only noble metals; it must, however, be considered when working with base metal or base metal-noble metal alloys.

In the absence of side reactions, standard polarization techniques can be used. An example system is the Cu-rich half of the Cu-Au alloy system in acidified sulfate solution.[31] For this alloy/electrolyte system in the potential region free of side reactions, the measured current is due only to the dissolution of Cu. Certain limitations exist. For example, at $E<E_C$ it takes several hours to reach a steady-state current. Because of this, automated scanning equipment may be used only at very low scan rates. Since at potentials above E_C the surface instability occurs, the potential should be scanned in a more noble direction in order to observe the true current behavior for a planar surface below the critical potential.

When side reactions are present, the chemical analysis method must be very sensitive to the metal of interest since the current below E_C is usually quite low. Methods such as colorimetric analysis and atomic adsorption are used. To prevent the loss of metal ions via plating onto the counter electrode, the anodic and cathodic compartments must be separated. This can be accomplished with the use of frits between the two compartments. It is good practice to sample the cathodic compartment to ensure negligible transport of metal ions through the frits. Since it can take some time to reach a steady-state current, the electrolyte is generally drained after an initial dissolution period and the cell is then refilled with a known amount of fresh electrolyte. The sample is held for known lengths of time and the electrolyte sampled and analyzed. This procedure is then repeated at different potentials to obtain a polarization curve. The individual polarization curves of Cu and Zn for α, γ, and ϵ brass in a buffered (pH 5) Na_2SO_4 solution were obtained this way.[32] When performing such an experiment, care must be taken to ensure adequate mixing of the electrolyte to prevent the possibility of biased samples. The initial dissolution period must be long enough so that the fresh electrolyte is added after a steady-state dissolution rate is obtained.

Acknowledgment

The authors gratefully acknowledge the support of the National Science Foundation, Metallurgy Program, Division of Materials Research under Contract No. DMR80-10828.

References

1. D. D. Macdonald, J. Electrochem. Soc., Vol. 125, p. 1443, 1978.
2. C. Gabrimm, M. Keddam, H. Takenouti, Vu Quang Kinh, F. Bourlier, Electrochimica Acta, Vol. 24, p. 61, 1979.
3. H. Shih, H. W. Pickering, submitted to the J. Electrochem. Soc.
4. H. Sugawara, S. Shimodaira, J. Jap. Met. Soc., Vol. 30, p. 626, 1966.
5. H. Sugawara, S. Shimodaira, J. Jap. Met. Soc., Vol. 30, p. 635, 1966.
6. H. Sugawara, S. Shimodaira, J. Jap. Met. Soc., Vol. 30, p. 765, 1966.
7. D. C. Epler, J. E. Castle, Corrosion, Vol. 35, p. 451, 1979.
8. J. E. Castle, K. L. Bailey, D. C. Epler, A Cupric Aluminate Conversion Layer on Aluminum Brass, Annual Report, INCRA Project No. 254R, 1977.
9. C. Kato, B. G. Ateya, J. E. Castle, H. W. Pickering, J. Electrochem. Soc., Vol. 127, p. 1890, 1980.
10. C. Kato, B. G. Ateya, J. E. Castle, H. W. Pickering, J. Electrochem. Soc., Vol. 127, p. 1897, 1980.
11. C. Kato, H. W. Pickering, J. Electrochem. Soc., Vol. 131, p. 1219, 1984.
12. C. Kato, H. W. Pickering, J. Castle, J. Electrochem. Soc., Vol. 131, p. 1274, 1984.
13. R. F. North, M. J. Pryor, Corro. Sci., Vol. 10, p.

297, 1970.
14. F. P. Ijsseling, Corros. Sci., Vol. 14, p. 97, 1974.
15. F. L. LaQue, J. Am. Soc., Naval Engs., Vol. 29, p. 54, 1941.
16. H. H. Uhlig, Z. Elektrochem., Vol. 85, p. 307, 1944.
17. H. H. Uhlig, Z. Elektrochem., Vol. 62, p. 700, 1958.
18. P. F. Effertz, W. Fichte, P. Forchhammer, VGB Kraftwerkstechnik, Vol. 2, p. 54, 1974.
19. W. C. Steward, F. L. LaQue, Corrosion, Vol. 8, p. 259, 1952.
20. P. S. Keir, M. J. Pryor, P. R. Sperry, J. Electrochem. Soc., Vol. 114, p. 777, 1967.
21. S. R. J. Saunders, M. J. Pryor, J. Electrochem. Soc., Vol. 115, p. 1037, 1968.
22. R. F. North, M. J. Pryor, Corros. Sci., Vol. 8, p. 149, 1968.
23. U. R. Evans, The Corrosion and Oxidation of Metals, Edward Arnold Ltd., London, p. 134, 1926.
24. E. D. Verink, Jr., "Applications of Electrochemical Techniques in the Development of Alloys for Corrosive Environments," Electrochemical Techniques for Corrosion, R. Baboian, Ed., National Association of Corrosion Engineers, Houston, Texas, p. 43, 1977.
25. A. Bacarella, J. Greiss, Jr., J. Electrochem. Soc., Vol. 1210, No. 4, p. 457, 1973.
26. C. Bonfiglio, H. Albaya, O. Cobo, Corros. Sci., Vol. 13, p. 717, 1973.
27. R. North, M. Pryor, Corros. Sci., Vol. 10, p. 297, 1970.
28. J. Popplewell, R. Hart, J. Ford, Corros. Sci., Vol. 13, p. 295, 1973.
29. O. Kubschenski, B. Hopkins, Oxidation of Metals and Alloys, Buttersworth, London, p. 21, 1967.
30. H. W. Pickering, Corros. Sci., Vol. 23, p. 1107, 1983.
31. H. W. Pickering, P. J. Byrne, J. Electrochem. Soc., Vol. 118, p. 209, 1971.
32. H. W. Pickering, P. J. Byrne, J. Electrochem. Soc., Vol. 116, p. 1492, 1969.

Application of the Vibrating Probe to Localized Current Measurements

HUGH S. ISAACS* and YUICHI ISHIKAWA

Abstract

The principles, advantages, and applications of a scanning vibrating electrode technique are presented. The technique has been used to locate the position of localized corrosion and heterogeneities on surfaces and to determine their polarization characteristics. The technique has been applied to the study of aluminum pitting, crevice corrosion of stainless steel, and organic and metallic coatings on steels.

Introduction

Localized corrosion is one of the most destructive forms of corrosion found in aluminum alloys and stainless steels exposed to solutions containing chloride ions.[1] During localized corrosion the areas of electrochemical dissolution are generally well separated from the cathodic reactions, and potential gradients develop around anodic areas as a consequence of current flow. It is the purpose of this work to present corrosion-related applications where current density variations are measured. The variations are also important in the electroplating industry where uniform deposits are desirable, as well as during cathodic protection. At immersed offshore structures or pipelines, current flow patterns may become complex and the reliability of potential measurements less certain to ensure protection.[2] Other studies have shown that currents flow from one part of a single cell to another and during the growth or regeneration of bone or limbs.[3,4]

Figure 1 shows a schematic of the current flow in the electrolyte from a localized anode to the surrounding cathodic areas and equipotentials set up around the localized anode. The potential changes are most rapid over the anode; a potential maximum is observed over its center. A DC scanning reference electrode technique (SRET), where a fine capillary tip is scanned parallel and in close proximity to the corroding surfaces, has been used to measure the potential distribution, and to locate and monitor the corrosion sites. Applications of this

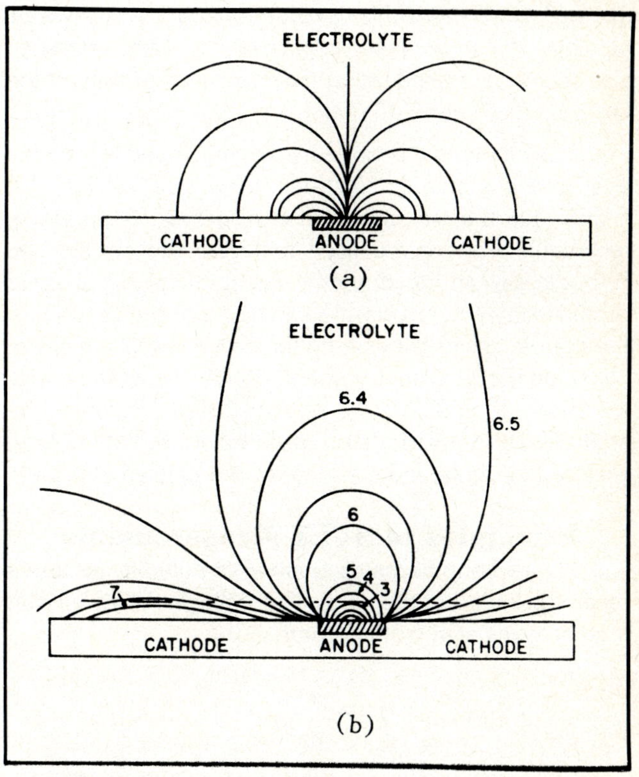

FIGURE 1 – Schematic of a local corrosion cell showing (a) current paths and (b) equipotential lines.[5]

*Brookhaven National Laboratory, Upton, New York.
**Mechanical Engineering Research Laboratory, Hitachi, Ltd., Tsuchiura, Japan.

method have been reviewed[5] and more recent studies are of interest.[6-8] These methods have generally been restricted to relatively high corrosion rates because of the limitations in the separation of small potential variation resulting from background noise. AC scanning, where an alternating current is applied in conjunction with scanning to overcome the noise problem, has also been investigated to determine variations in surface impedance.[9-12] Direct correlation with dissolution kinetics is complicated by the spread of the applied AC currents and the extensive analytical procedures required. However, with indirect calibration, the technique would become extremely sensitive for locating corrosion sites or coating defects but will have limited applications.[11]

The vibrating electrode is a modification of the DC technique and overcomes the problem arising from background noise. Vibrating the tip of a microelectrode in a potential gradient converts the DC potential variation associated with the current flow to an AC signal with a specific frequency. This allows better amplification of the signal, as well as filtering of background noise. This approach gives an increased current resolution of at least three orders of magnitude, and measured currents of 30 nA/cm^2 in 0.6 M NaCl have been reported.[3] Similar approaches have been applied to the monitoring of currents during the cathodic protection of underwater pipelines.[13-14] In the latter application, two zinc reference electrodes are rotated rather than oscillated to convert the DC potential to an AC signal.

The vibrating electrode has been adapted to develop a scanning vibrating electrode technique (SVET).[15-16] SVET, like the techniques discussed here, is an *in situ* technique that can be used to locate the anodic and cathodic sites and to study the electrochemical processes during localized corrosion. It can also be used to monitor the corrosion without altering the processes taking place, changing the local environment, or influencing the corrosion rates. In addition, the potential of the metal can be controlled and polarization measurement of the localized sites can be made independent of the applied current.

Principles of SVET Measurements

Figure 1 shows the distribution of potential around a relatively large anodic area. The voltage (V) at a point may be calculated from the equation:

$$V = \frac{I\rho}{2\pi r} \qquad (1)$$

where I is the current from the pit, ρ is the solution resistivity, and r is the distance from the pit.

A number of assumptions are made in using this equation. The pit is assumed to act as a point source with the current flowing radially from it into the solution above the metal surface. For this to occur the cathodic area around the pit must not modify the current flow. This is generally the case since the current density at the cathode surface is usually very small and forces the current from the pit to distribute itself over very large areas. The potential of the cathodic area is made equal to zero and the

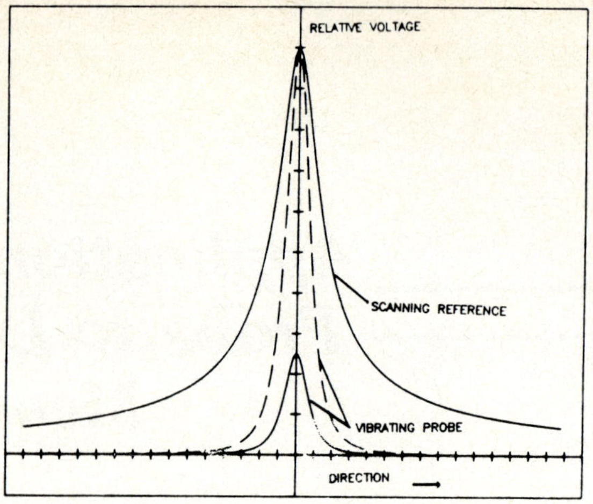

FIGURE 2 — Relative variations in measured voltages for the scanning reference electrode and the vibrating electrode techniques. The broken curve has been amplified to compare the widths of the curves. The X axis gradations are in units of probe height above the surface.

potential increases as the pit is approached. This convention is opposite to that normally chosen where the anode is negative to the cathode, so that the anodic areas graphically appear as peaks rather than troughs.

Equation (1) refers to the variations that would be expected for the SRET. For SVET, the voltage variations depend on the vibrations. The amplitude of the signal is given by the difference in the potential at the maximum and the minimum of the vibration. These variations have been calculated and are compared in Figure 2. The X-direction units were taken as multiples of the height of the probe above the surface. In the case of SVET, the magnitude of the vibration perpendicular to the surface was assumed to be 0.25 times the height for the purpose of the calculations.

The relative voltages show that the SRET gives a larger potential variation for equal heights above the surface and would appear at first sight to be a more sensitive method. However, because of the ability to analyze the AC signals from the vibrating probe and filter the noise using lock-in amplifiers, the larger voltages offer little advantage since the noisier DC SRET signals are often difficult to separate from the background noise.

Upon magnifying the vibrating probe signal, it is seen that the width of the peaks differed, being 3.46 and 1.53 times the height at half the maximum voltages. The wider SRET voltage variations would be observed more easily with larger step sizes, but a better spacial resolution of the anodic site can be obtained with SVET.

The voltage response of SVET also depends on the vibration direction. When the vibrations are perpendicular to the surface, a single peak is observed with a maximum directly above the site, while with vibrations parallel to the surface, the response shows a maximum and a minimum symmetrically positioned around the pit. The voltage response for different vibration angles is

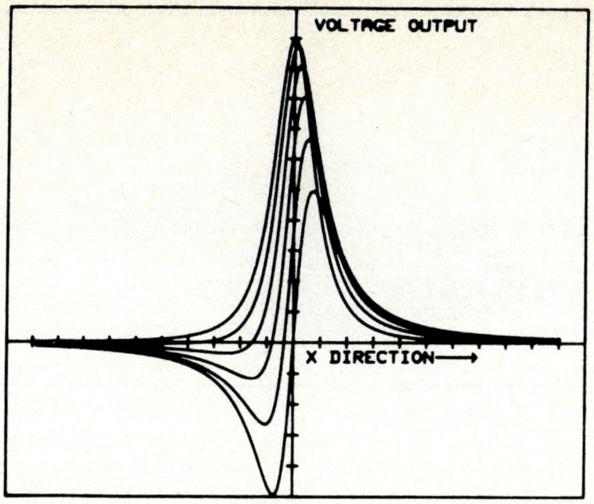

FIGURE 3 – Variations in output voltage for different vibration directions of 0, 18, 36, 45, 72 and 90° to the surface.

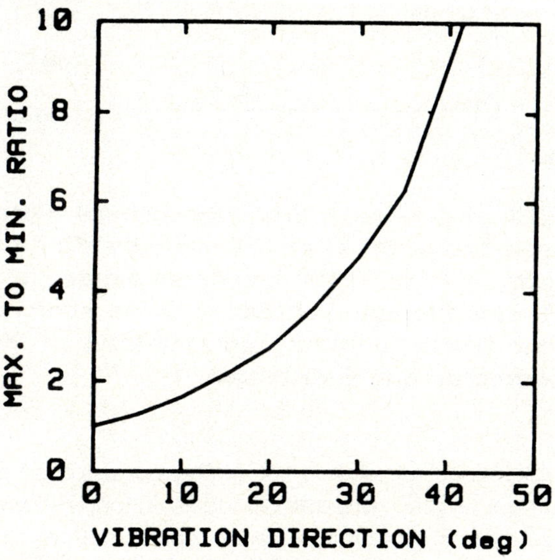

FIGURE 4 – Dependence of the ratio of the maximum to the minimum voltages on the angle of the vibration direction to the surface.

shown in Figure 3. The variations in the shape of the curves can also be used to determine the vibration direction and the height of the probe tip above the surface. The height of the probe can be calculated from the half width as mentioned above. There is a small error incurred in taking the height as 1/1.5 of the half width, but as can be seen from Figure 3, the error is only about 10% and generally within the accuracy of the voltage variations. The vibration direction can be determined from the shape of the voltage curve. The ratio of the maximum and the minimum can be related to the angle as shown in Figure 4 for angles to the surface of up to about 30°. The major error in this procedure arises from the difficulty in locating the average or zero voltage from which the maximum and minimum are measured above 35°. This error dominates the calculation and thus may limit this approach.

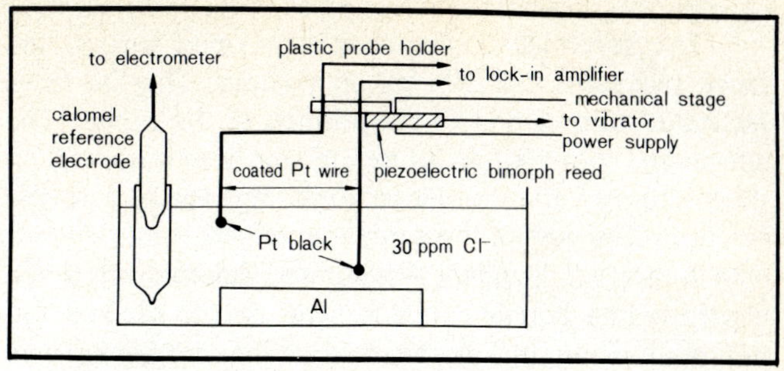

FIGURE 5 – Schematic of the scanning vibrating electrode configuration.[15]

It must be emphasized that the voltage variations observed during SVET are a direct measure of the component of the current density flowing parallel to the vibration direction of the vibrating electrode tip. They are a measure of the IR gradients in the solution rather than any potential at the metal/electrolyte interface. The gradients are proportional to the product of the solution resistivity and the current density component in the direction of the vibration of the electrode tip.

Experimental Technique

A schematic of the SVET apparatus is shown in Figure 5.[15] The vibrating electrode was a platinum wire 0.12 mm in diameter, coated with an insulating lacquer. The end of the wire was cut to expose its cross section and then electrocoated with platinum black to reduce its interfacial impedance. However, subsequent studies have been conducted without the platinum black.[17] The wire was scanned above the metal surface being studied at about a height of 0.1 mm and sensed the corrosion currents. A similar wire was positioned well above the surface where no significant currents were present and was used to reduce noise.

The wires were attached to a piezoelectric bimorph reed 5.4 × 1.6 × 22 mm in size and vibrated perpendicular to the specimen surface. The vibrator power supply activated the reed by a sinusoidal wave of 40 to 10 kilo-Hertz (kHz) with an output of 22 V peak-to-peak, and also supplied a reference signal for the lock-in amplifier. The frequency of vibration was adjusted to obtain the maximum vibration amplitude of the electrode in the vertical direction while limiting the horizontal vibration components. The optimum frequency was found to vary depending on the particular piezoelectric reed used, the length of the electrode, and the manner of attachment of the electrode to the leads. The frequency of vibration was generally 100 to 700 Hz and the vibration amplitude was about 0.03 mm.

The potentials at the two extremes of the vibrating electrode tip within the gradients generated during localized corrosion gave the peak-to-peak potentials of an AC signal at the special frequency. The signal was fed into the lock-in amplifier which amplified the signal associated with the vibration frequency and filtered out other signals associated with the noise. Thus, signals of the order of a few nV were measurable.

The specimen container was an open square flat-faced plexiglass cell 300 mL in volume. The flat faces facilitated observation of the position of the probe. The specimen, about 30 × 30 mm in size, was mounted on the bottom of the container, exposing one face to the solution. The rest of the sample and lead were covered with a stop-off lacquer. A saturated calomel reference electrode in a double container was used to monitor the potential using an electrometer. The container was placed on a finely adjustable X-Y-Z table. The table was leveled under a steroscopic microscopic to be parallel with the scanning probe, with the height of the vibrating electrode at 0.1 mm from the specimen surface during scanning.

The vibrating electrode was held on the mechanical stage driven by two stepping motors and scanned in both X and Y directions over the specimen surface. Scanning of the electrode was controlled with a computer, which was also used for data acquisition and analysis.

A single step was a multiple of 0.005 mm. The vibrating electrode was scanned over a predefined distance in the X direction from the specified origin in fixed steps. After data was acquired, the electrode returned to the starting point, and then moved a constant step distance toward Y and again scanned in the X direction. This operation was repeated to scan the desired area of the specimen. The data could also be acquired in the reverse X direction but this emphasized any hyteresis effects. Scanning time depended on the waiting time for data acquisition at the single step. At a standard waiting time of 200 ms, it took about 50 min to scan 100 steps over 100 times with a step distance of 0.3 mm in the X direction and 0.2 mm in the Y direction, i.e., a scanning area of 30 × 20 mm^2. After scanning, the data was graphically displayed either on the computer CRT display, a thermal printer, or an X-Y plotter, and also stored on a flexible disk.

There were five possible graphical representations: a three-dimensional plot, with or without perspective, in which the location in the X direction is represented on a coordinate and the location in the Y direction and the measured potential gradient are superimposed on an ordinate; a two-dimensional plot in which the location in the X or Y direction at a specified location in the Y or X direction is represented on a coordinate and the potential gradient is plotted on an ordinate, and finally a potential contour map in which equipotential lines are displayed.

Applications

Pitting Corrosion

The rate of corrosion of passivated metal surfaces is usually many orders of magnitude smaller than the high corrosion rate observed in the pits. It is generally accepted and has been demonstrated with stainless steels[17,18] that an anodic current flows from the pit through the electrolyte to the passive surface which behaves as a cathode. Thus, a local cell described in Figure 1 becomes effective and SVET can be applied to detect the location of pits and to monitor their growth.

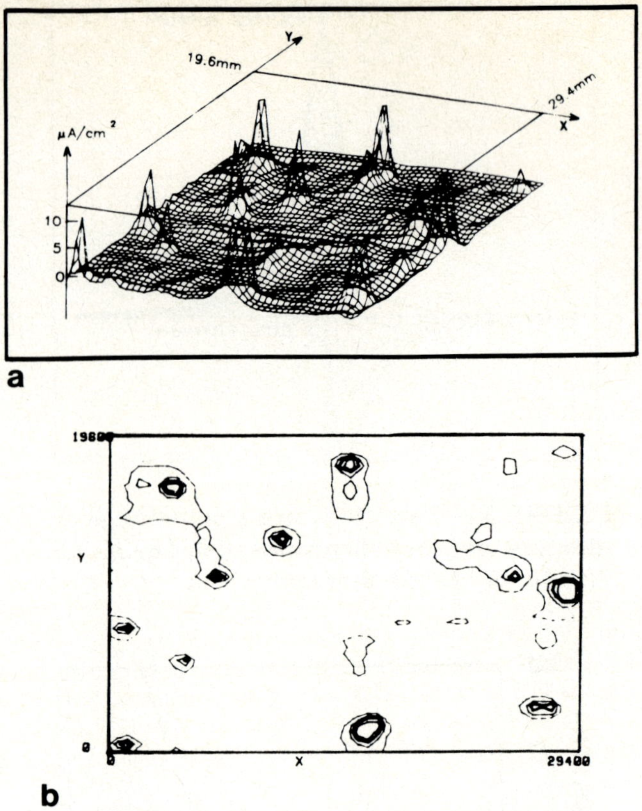

FIGURE 6 — (a) A three-dimensional representation of the current density distribution over a pitting A1100 aluminum surface in a 10 ppm chloride solution.[16] (b) An equicurrent density, contour, representation of the current density distribution.

Examples are shown for pitting corrosion of pure aluminum (A1100) in dilute chloride solutions.[15,16] Figure 6(a) shows a three-dimensional representation of the current density distribution over the aluminum surface in 10 ppm chloride ion solution. This figure was obtained by scanning 40, 0.6 mm steps in the X direction and 40, 0.4 mm steps in the Y direction. The voltages or rather voltage gradients at 2500 locations were converted to the current density variations. Each peak resulted from an active pit on the surface and the peak height was a function of its growth rate. The variable peak heights suggested a wide variation in the growth rate of the pits. Figure 6(b) shows equicurrent density lines of the pits shown in Figure 6(a). The exact location of the pits can be determined from this figure and superposition of such figures for different lengths of exposure provide information on initiation and repassivation of pits. Figure 7 shows a three-dimensional representation of current distribution over a relatively small area (0.2 mm steps, both in the X and Y directions). Of 7.8 × 4.8 mm, several pits are present. Figure 8 shows the results after a 10 ppm phosphate addition. Six pits out of nine shown in Figure 7 disappeared and two new pits initiated. The pit that continued to grow showed higher current densities after the phosphate addition.

By repeated scanning, it has been shown that pit ac-

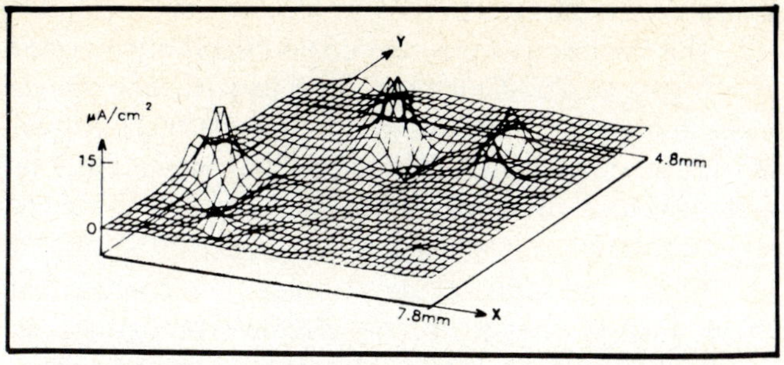

FIGURE 7 – Current density distribution over a relatively small area of A1100 aluminum in a 10 ppm chloride solution.

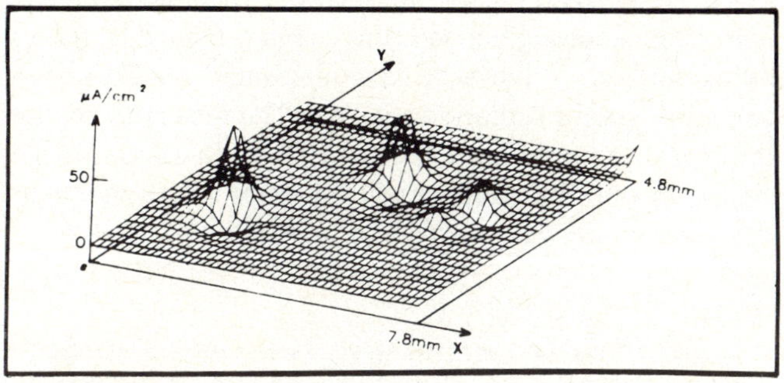

FIGURE 8 – Figure 7 after the addition of 10 ppm phosphate.

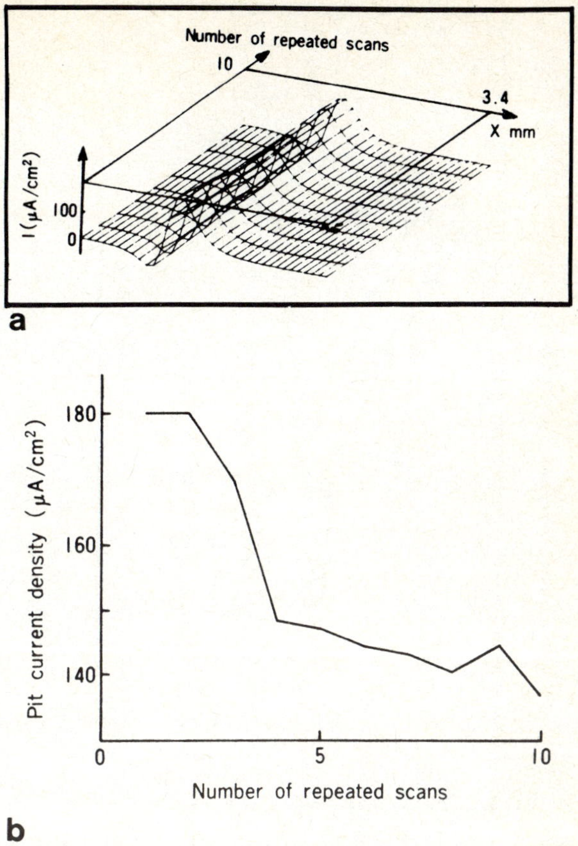

FIGURE 9 – (a) Repeated single line scans over a pit in A1100 aluminum in 30 ppm chloride. (b) Variations in the maximum current density above the pit at X = 1.4 mm as a function of repeated scans, i.e., time steps of 7.2 s, or period of each scan.[15]

tivity varies markedly with time, increasing and decreasing, or even disappearing and then reappearing. Figure 9(a) shows an example of a study of a single pit. Here a single line scan was repeated 10 times over a distance of 3.4 mm. The number of the scanning repetition is plotted in the Y axis. Figure 9(b) shows the variation of the activity of the pit with time (each scan took 7.2 s). This example showed relatively small changes with time for a well developed pit. Smaller pits showed very large variations with time but in general passivated after a number of reactivations at open circuit.

Similar measurements were made when potentials were applied to the specimen during this repetitive scanning, and enabled polarization curves of the pit to be obtained as shown in Figure 10(a). This figure was obtained by 100 repeated scans over a pit in 10 ppm sulfate ion and 10 ppm chloride ion in solution. The first two scans were at −0.50 V (vs SCE); the potential was then ramped to −0.35 V at a rate of 1 mV/s. The ramp direction changed again at −1.0 V and the scan ended at −0.75 V. The activity of the pit increased with increasing potential and decreased with decreasing potential. Anodic current disappeared at a potential at −0.8 V and below this potential cathodic current appeared.

A comparison of the potential dependence of the pit currents with the applied average current densities for the entire sample is shown in Figure 10(b). This figure shows that the polarization characteristics of the pit and the total surface are markedly different. It is also demonstrated that at the potential of zero current for the total surface, the pits are distinctly anodic indicating no significant cathodic activitiy within the pit at the corrosion potential.

Crevice Corrosion

In crevice corrosion the origin of the attack stems from the compositional differences developed in the solution inside the crevice. These differences lead to the formation of a local cell with the metal inside the crevice acting as the anode and the outer surface as the cathode. Thus, SVET can be used to measure the initiation and development of crevice corrosion.

In Figure 11, an example is shown of crevice corrosion of AISI 304 stainless steel in 0.25 M sodium chloride solution. Figure 11 shows a three-dimensional representation of current density distribution over the creviced stainless steel. This figure was obtained by scanning the vibrating electrode across the creviced area of the specimen by 34, 0.20 mm steps in the X direction and 41, 0.50 mm steps in the Y direction. The crevice gap was about 0.05 mm and its depth was 2.0 mm. The figure clearly shows a distinct anodic nature of the creviced area and cathodic sites are uniformly distributed over the passivated surface. The entire creviced area was not uniformly anodic and corrosion current density varied over the longitudinal location in the crevice. It is not clear at present whether this local variation of corrosion rate in the crevice reflects the characteristics of crevice corrosion or geometrical variations in the specimen.

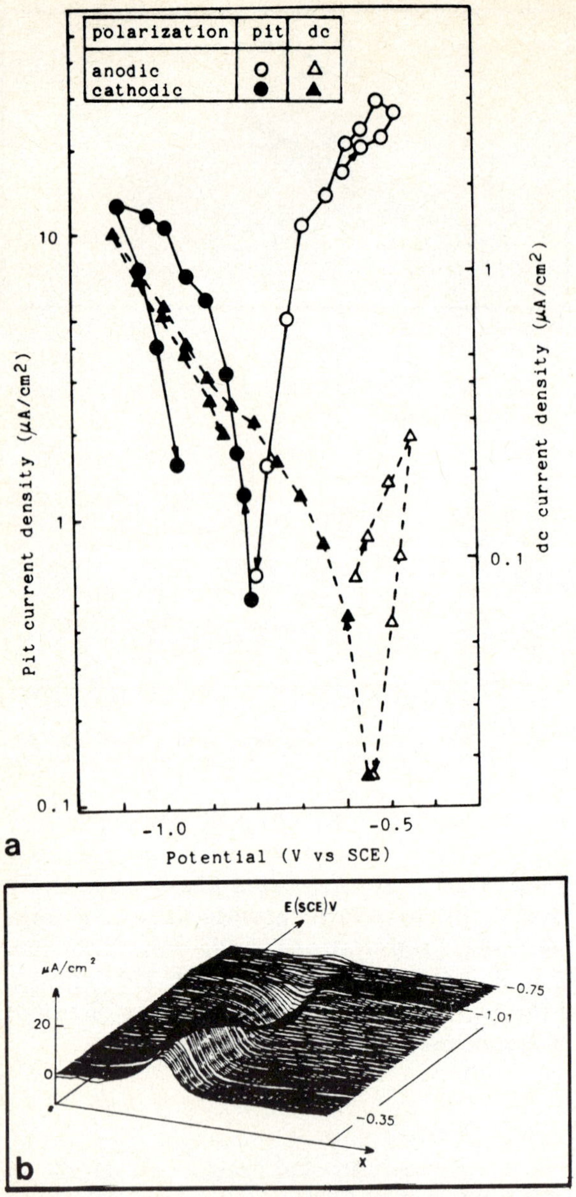

FIGURE 10 — (a) Repeated single line scans over a pit during potentiodynamic polarization of an A1100 aluminum sample in a 10 ppm chloride and 10 ppm sulfate solution. (b) Polarization curve of the pit showing the variation of the current density above the pit with potential.[16]

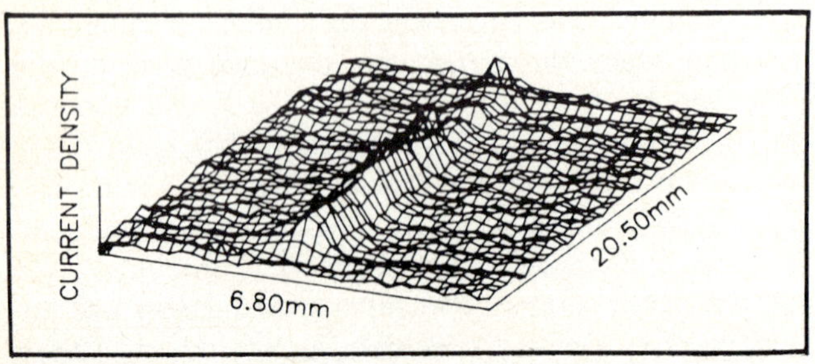

FIGURE 11 — The variation in current density above a creviced AISI 304 stainless steel sample in 0.25 M NaCl.

Organic Coating

The corrosion of polymer-coated steels often occurs in a highly localized manner. Scanning AC impedance measurements[9,10] have shown that the electrical properties of a coating are not uniform and that large reductions in the electrical properties occur locally when exposed to an electrolyte.

When corrosion occurs at areas of a coating that show a large change in the electrical properties, an anodic current flows from those areas to cathodic areas that might be elsewhere in the coating or present beneath the coating.

If a local cell is formed by breakdown of an area of the coating, SVET can measure the magnitude of the corrosion occurring beneath the coating. Figure 12 shows an example of a thin acryl lacquer-coated steel exposed to 0.25 M sodium chloride solution. The results show that there is one marked anodic area along the edge of the specimen and that surprisingly the edge where the coating was thin, also shows distinct cathodic behavior.

Inorganic Coating

The SVET can also be used as a sensitive device to detect the presence of defects or heterogenieties in coatings by taking into account differences in the electrochemical behavior of the coating and the substrate.[19] Figure 13 shows an example of differentiating between the defects in aluminum dip-coated steels. The tests were conducted in a neutral borate solution. The steel shows higher anodic and cathodic currents—the former because

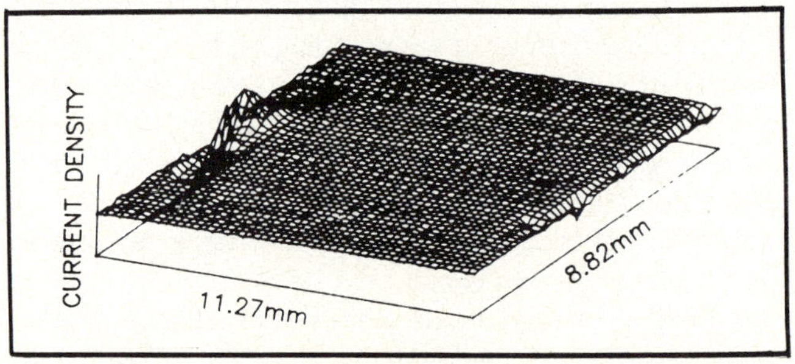

FIGURE 12 — The variation in current density above an acryl lacquer-coated steel surface.

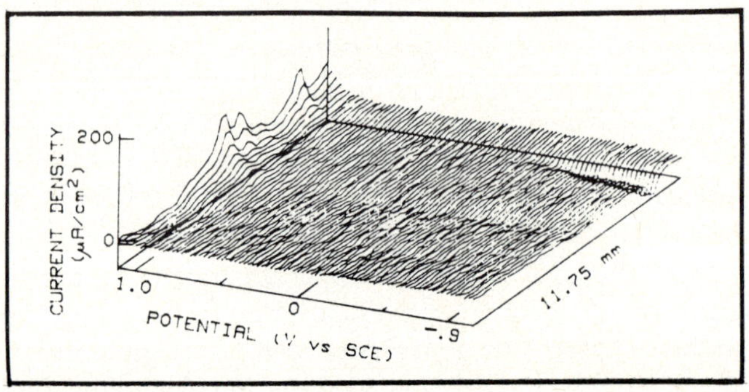

FIGURE 13 — Repeated single line scans over a dipped aluminum coated steel during potentiodynamic polarization.[19]

of higher rates of anodic dissolution and oxygen evolution and the latter because of the lower hydrogen overvoltage. Alloys of these materials would be expected to have an intermediate behavior. The figure shows the result of repeated scanning of a single line in the X direction, which crosses over active sites. The potential was decreased from 1.1 to −0.9 V at a rate of 1 mV/s. The result compares the polarization behavior of three specific sites in detail. The two adjacent high current density sites at 1.1 V show a decrease in current density with decreasing potential. At negative potentials, the cathodic currents remained low at the two adjacent sites while the cathodic currents observed from the third site increased. This result demonstrates that by scanning the surface at 1 V and then at −1 V it would be possible to separate two types of surface inclusions.

Conclusions

Vibrating an exposed tip of a thin electrode wire converts the DC potential variation associated with the current flow to an AC signal with a specific frequency. This enables the filtering of background noise and gives a current resolution orders of magnitude greater than the conventional DC scanning technique. A spatial resolution of 0.005 mm and potential resolution of 10 nV were obtained with an aid of a computer. Examples of the technique have been presented and give insight into the behavior of pitting corrosion, crevice corrosion, and coatings.

Acknowledgments

The authors wish to thank D. Becker for his experimental assistance and Y. Sanborn for producing the computer programs. This work was funded by the US Department of Energy, Office of Basic Energy Sciences, under contract No. DE-AC02-76CH00016.

References

1. Localized Corrosion, R. W. Staehle, B. F. Brown, J. Kruger, A. Agrawal, Eds., National Association of Corrosion Engineers, Houston, Texas, 1974.
2. T. Sydberger, Materials Performance, Vol. 22, No. 5, p. 57, 1983.
3. L. F. Jaffe, R. Nuccitelli, J. Cell Biology, Vol. 63, p. 614, 1974.
4. R. B. Borgens, J. W. Vanable, Jr., L. F. Jaffe, Proc. Natl. Acad. Sci., Vol. 74, p. 4528, 1977.
5. H. S. Isaacs, B. Vyas, Electrochemical Corrosion Testing, F. Mansfeld, U. Bertocci, Eds., STP 727, ASTM, p. 3, 1981.
6. L. J. Gainer, G. R. Wallwork, Corrosion, Vol. 35, No. 10, p. 435, 1979.
7. H. S. Isaacs, Y. Ishikawa, Proc. Intern. Congr. Metal. Corros., Vol. 2, p. 150, Toronto, June 1984.
7. G. Pallos, G. Wallwork, Vol. 38, p. 304, 1982.
8. R. J. O'Halloran, L. F. G. Williams, G. P. Lloyd, Corrosion, Vol. 40, No. 7, p. 344, 1984.
9. H. S. Isaacs, N. W. Kendig, Corrosion, Vol. 36, No. 6, p. 269, 1980.
10. J. V. Standish, H. Leidheiser, Jr., Corrosion, Vol. 36, No. 8, p. 390, 1980.
11. H. S. Isaacs, Novel NDE Methods for Materials, B. B. Rath, Ed., American Institute of Mining, Metallurgical, and Petroleum Engineers, p. 63, 1983.
12. S. Ito, M. Yabumoto, T. Murata, Proceedings of the International Meeting on Chemical Sensors, Kodansha, p. 753, 1983.
13. R. Strommen, A. Roland, "Computerized Numerical Techniques Applied in Design of Offshore Cathodic Protection Systems," CORROSION/80, Paper No. 241, National Association of Corrosion Engineers, Houston, Texas, 1980.
14. T. G. Egger, R. Strommen, E. Bardal, Proc. 8th Scandinavian Corros. Congr., p. 33, Helsinki, 1978.
15. Y. Ishikawa, H. S. Isaacs, Boshoku Gijutsu, Vol. 33, p. 147, 1984.
16. Y. Ishikawa, H. S. Isaacs, Conf. on Corrosion and Exploitation of Aluminum Alloys, Cranfield, England, April 1983.
17. H. S. Isaacs, "Potential Scanning of Stainless Steel During Pitting Corrosion," Localized Corrosion, R. W. Staehle, B. F. Brown, J. Kruger, A. Agrawal, Eds., National Association of Corrosion Engineers, Houston, Texas, p. 158, 1974.
18. H. S. Isaacs, Y. Ishikawa, Marcel Pourbaix Eightieth Birthday Symposium, Electrochemical Society, October 1984.
19. H. S. Isaacs, R. Jackson, in "Fundamental Aspects of Corrosion Protection by Surface Modification," E. McCafferty, C. R. Clayton, Eds., Electrochemical Society, p. 339, 1984.

New Approaches to the Study of Localized Corrosion

J. KRUGER*

Abstract

This paper focuses on some of the newer approaches to obtaining a more valid understanding of localized corrosion processes. The issues that are considered are: (1) relevant environments, (2) relevant structures, (3) relevant basic processes, and (4) relevant protective films.

Introduction

A number of approaches have been developed over the years to study localized corrosion. This paper will single out some of the newer approaches that attempt to focus on issues that should be considered if a valid understanding of local corrosion processes is desired. It will not review all of the relevant issues, nor will it be an all-inclusive description of the various new ways in which localized corrosion can be studied. Instead, it will concentrate primarily on the results of some experimental approaches that have influenced the way in which localized corrosion has been viewed in the past. It is not intended as a comprehensive look but rather a more narrow, personal, and, perhaps, opinionated enumeration of important issues to be considered when studying localized corrosion.

Relevant Environments

One of the primary reasons why many localized corrosion studies have failed to reveal important insights into the controlling processes has been the failure to consider relevant environments. This is because localized corrosion usually occurs within the restricted confines of a crack, pit, or crevice. Because ion diffusion into or out of the confines of this restricted space (called by B. F. Brown an *occluded cell*) is difficult, concentrations of these ions are either built up or decreased. Thus, the environment within a pit is not the same as that of the environment of the solution that exists outside the pit. Unfortunately, it is the environment outside the pit in which many workers choose to conduct experiments. For example, an important reaction occurring in an occluded cell is the hydrolysis reaction that can lower the pH below that of the surrounding environment in the following way:

$$M^{+X} + XH_2O \rightarrow M(OH)_X + XH^+ \qquad (1)$$

The relevant environment to be used in a study or to be considered in the interpretation of an experiment would be the one that takes the pH of the environment produced by Reaction (1) into consideration.

While it has been recognized for a number of years that the environment trapped in the interior of restricted regions on a metal (cracks, pits, and crevices) has a different composition than the bulk environment surrounding the metal, it has been only in recent years that such knowledge was considered important, and serious attempts made to measure the composition of the occluded environment in a crack. (An important exception is the use of ferric chloride solutions of low pH to conduct screening tests of the resistance of alloys to pitting.) Brown and co-workers,[1-3] using the insights provided by Pourbaix[4] that systematized the role of pH and potential in all corrosion processes, attempted to measure these parameters in cracks. Their initial approach was to immobilize the corrodent within stress corrosion cracks by freezing it in liquid nitrogen, breaking open the crack mechanically, and then analyzing the thawed solution

*Materials Science and Engineering Dept., The Johns Hopkins University, Baltimore, Maryland.

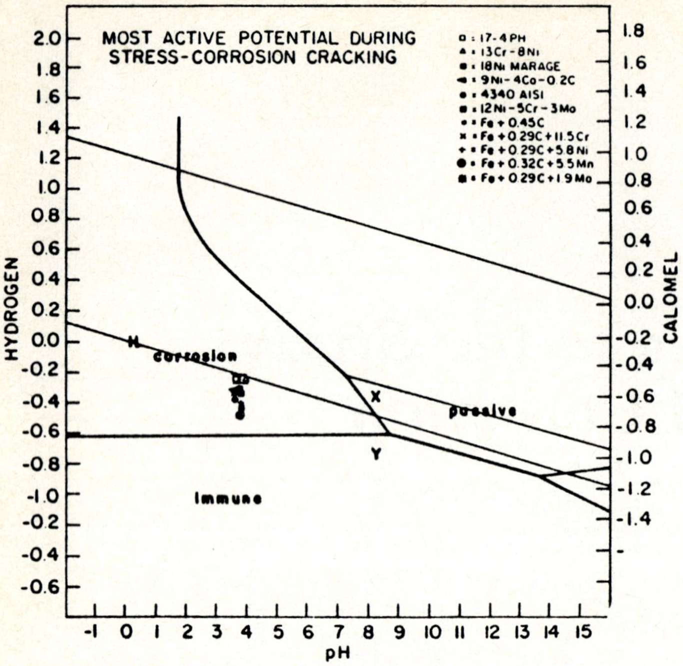

FIGURE 1 – Potential-pH diagram showing the potential-pH values in the crack tip environments for a number of high strength steels.[53]

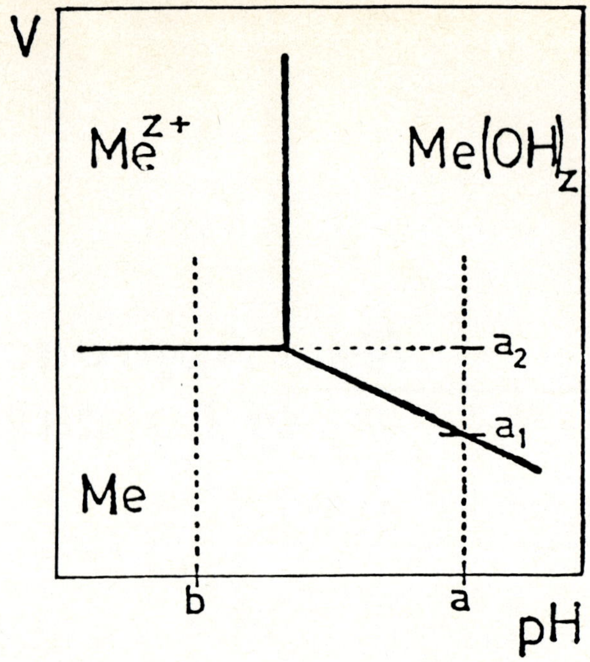

FIGURE 2 – Schematic potential-pH diagram, a – pH of the bulk solution (a_1 – a_2): passive zone; a_2 – pitting potential; b – pH of the local acidified zone.

removed from the crack. A later development involved the use of microelectrodes to determine the pH and potential in the crack's environment. Other developments in this field have included the use of a capillary to sample the crack solution[5] and the combination of a capillary technique with thin layer chromatography[6] to examine the solution in crevices.

The value of these crack, pit, or crevice chemistry techniques becomes evident when Figure 1 is inspected. For a large number of steels of different compositions in environments of either high or low pH, this figure shows that the potential and pH inside the crack falls in a region of the pH potential diagram (Pourbaix diagram) where one expects both corrosion and, because the points on Figure 1 are below the line marked H, the presence of hydrogen. Thus, any failure prevention measures using bulk composition or potentials could be based on unrealistic conditions and thus lead to an improper choice.

Galvele[7] used the ideas of Pourbaix to develop a local acidification mechanism of pitting. He assumed that a transient breaking of the passive layer occurs through mechanical or electrochemical means; hydrolysis then occurs during the reaction between the exposed metal and the environment. This reaction produces localized acidification at the site of the cracked film. In this model, the induction time is the time needed to achieve a critical pH in a broken passive film. The pitting potential is that potential needed to cause the current flowing to the initiating pit to reach a value where a critical pH is reached that renders repassivation impossible. Galvele has used a simplified Pourbaix diagram (Figure 2) to describe events that take place at the beginning of the propagation stage. In Figure 2, "a" represents the pH of the bulk environment and "b," the pH in the starting pit where conditions for the metal dissolution reaction

$$Me = Me^{z+} + Ze^- \quad (2)$$

becomes thermodynamically favorable.

The hydrolysis reaction:

$$Me + ZH_2O = Me(OH)_Z + ZH^+ + Ze^- \quad (3)$$

only takes place between a_1 and a_2 in Figure 3. When the potential is above a_2, the localized acidity produced by Reaction (3) will promote dissolution. Potentials between a_1 and a_2 will promote repassivation.

Relevant Alloy Structure and Composition

In many cases of localized corrosion, differential rates of attack for the different phases present in an alloy are observed. For example, the potential difference between the phase making up a grain boundary (or grain boundary region) and the interior of a grain may be the driving force for intergranular corrosion. A similar situation can exist when pitting or stress corrosion is observed. Thus, it becomes important, when one wishes to disentangle the origins of localized attack, to study in detail the process occurring on the individual phases. A fruitful approach is to study, for example, the electrochemistry of alloys that are identical or similar to the relevant phases that constitute the different regions of a real alloy whose localized corrosion is being studied.

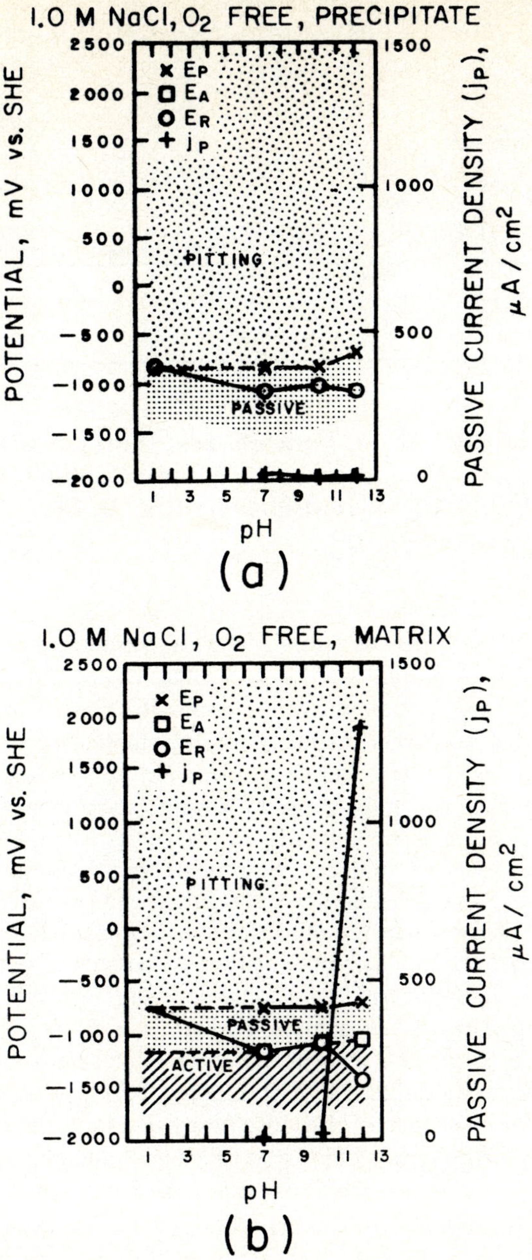

FIGURE 3 – Domains of pitting, passive and active behavior of: (a) the precipitate ($MgZn_2$) and (b) the matrix (Mg 2.05%, Zn 6.40%, balance Al) in 1.0 M NaCl, oxygen free.[8]

When a form of localized corrosion is considered where structure plays an important role, intergranular corrosion, the microstructure, or the nature of the phase of an alloy is of great importance. To observe the effect of the phases in an alloy on intergranular corrosion, Ugiansky[8] produced alloys that simulated the different phases making up an aluminum-zinc-magnesium alloy. He developed an alloy that was similar to the precipitate ($MgZn_2$), one with the composition of the precipitate-free zones, and one that simulated the matrix of the grains. He made a plot of pH vs potential (Figure 3). Using experimental data, he could show various parameters as a function of pH and potential. By doing so, as Figure 3

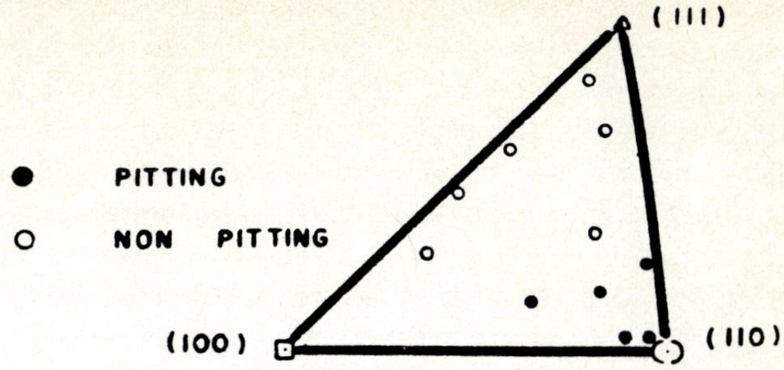

FIGURE 4 – Stereographic triangle showing pitting on grains of iron having different crystallographic orientations.[9]

shows, domains of pitting, passive and active behavior for each phase, or part of the Al-Zn-Mg alloy studied could be identified. Thus, by observing specific parts of an alloy, one can obtain a great deal of insight into the highly important localized corrosion process which is intergranular corrosion.

Not only are different phases or compositions important, but also structural variations, such as different crystallographic orientations of the grains of a metal or the presence of variations, for example, dislocations, within the same orientation, can be a factor underlying localized attack. For example, studies by Kruger[9] on single crystal iron surfaces showed that pitting was more predominant on surfaces whose orientations were near the {110} orientation than on those of other orientations (Figure 4).

Relevant Critical Potentials

There are vast numbers of experiments and discussions[7,10-14] in the literature on the factors or underlying causes of the existence of a critical potential for pitting and for stress corrosion. This is so because the concept of a critical potential, E_C, below which localized attack does not occur has immediate practical consequences. Obviously, if E_C for a given alloy and environment is known, one then has an incentive to control the potential of that system so that it never goes above E_C and thereby, presumably, never suffers localized attack. One does, however, have to know the relevant E_C. The concept of a critical potential, however, is held in disrepute by some workers,[10] because the relevant E_C, measured incorrectly, is not always known. Even if a relevant E_C can be measured, it is by no means established that all pitting or stress corrosion will be stopped by keeping a system below E_C. Nevertheless, one can never establish the practical validity of the concept until it is assumed that the correct, that is, relevant value of E_C is used.

The determination of the relevant E_C depends on what process or phenomenon determines the value of E_C. Unfortunately, this has not been established although a number of suggestions have been made.[14] The different explanations of the origin of E_C can be listed as follows:

1. E_C Determined by Potential of Zero Charge. Almost all mechanisms of localized corrosion require the adsorption of damaging anions as their first step. Roughly, the potential above which such adsorption takes place is the potential of zero charge (PZC). Thus, since some mechanisms for pitting require that damaging anions adsorb and thereby displace the ions forming the passive film, such an explanation of E_C would be reasonable, and the effect of such variables as the nature of the metal, concentration of anions, and temperature all affect E_C and PZC.[15,16] Because it is difficult to measure the PZC for solid electrodes,[15] especially when adsorption may coincide with the initiation of breakdown processes, there have been no good experimental verifications that E_C is related to the PZC.

2. E_C is Determined by Film Pressure to Fracture Film. Sato[17] has identified E_C with the critical potential above which the film pressure exceeds the critical compressive stress of film mechanical breakdown. The film pressure depends on both electrostriction and interfacial effects.

3. E_C is Determined by the Potential of Formation of an Unprotective Film. One possible factor determining E_C that may not apply to all cases may be that it is the potential above which an unprotective film may form or deposit. Ambrose and Kruger[18] suggest that this may be the case for iron.

4. E_C is Determined by Repassivation Kinetics. This concept of E_C is based on kinetic considerations. Some studies[19-22] have suggested that competition between repassivation and breakdown of films is important in determining initiation of localized attack. Videm[19] has proposed that the value of E_C for Al is determined by the relative rates of these two processes. He based this on his experiments that showed a narrow potential region where repair and breakdown were competitive. Below this region, repair predominated and above it, breakdown was the major process. Because potential can affect repassivation kinetics[23,24] and because a number of workers[20-22,25,26] suggest that initiation of localized attack depends on a dynamic balance between breakdown and repair, the determination of E_C by repassivation kinetics is a possibility that should be considered. (More will be mentioned about repassivation in the section on Relevant Basic Processes.)

Thus far, it is explanation No. 4 (above) that has led to new approaches for determining E_C. These show some promise of yielding more relevant values. Notable among these approaches is that of Pessall and Liu.[13] Their technique involves scratching the alloy surface whose E_C is to be determined at various potentials set by a potentiostat in the environment under examination.[(1)] When the potential is below E_C, the current observed upon scratching rapidly returns to the original values observed before scratching. When the potential is above E_C, no or very little repassivation occurs and the current either remains near the value found after scratching or it increases. Figure 5 shows the kind of current transients observed above and below E_C. A major advantage of this approach is that it eliminates the induction time for localized corrosion to be initiated. The usual methods for determining

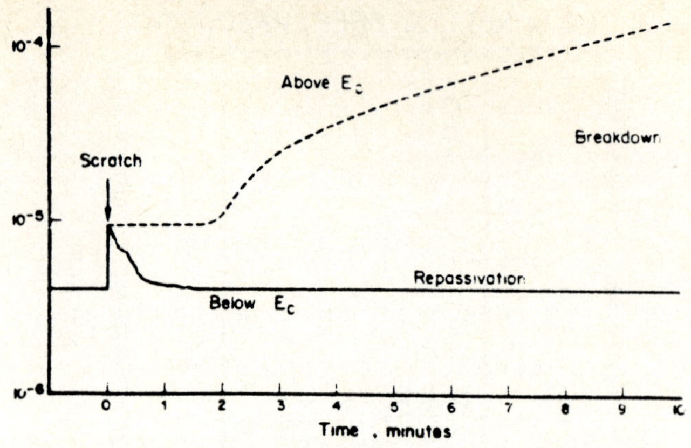

FIGURE 5 — Scratch test current-time curves for specimen held potentiostatically above E_C (breakdown) and below E_C (repassivation).[13]

E_C do not have this virtue. With the common methods, one may either scan a potential region potentiodynamically or potentiostatically sample a series of potentials in succession staying at each potential for varying amounts of time. When a value for the potential is reached at which the current measured starts to increase, this value is identified as E_C. However, it is known that there exists an induction time for localized corrosion attack. Therefore, unless one remains at all potentials for a time greater than the induction time, one can never be assured that a lower potential would not have been E_C. Since the induction time can be very long depending on the history of the sample studied (see Ambrose and Kruger[18]), it is no wonder that the E_C determined by the usual methods has been considered by some to be of questionable value.[10] Because Pessal and Liu's technique eliminates the problem of time of initiation, it appears that the E_C determined by its use is a more accurate value.

Similarly, the potential identified by Pourbaix[11] as the protection potential, E_P, may also be more relevant. This potential, which is more active than E_C, is the potential below which localized corrosion once started can be stopped. Pessal and Liu's work indicates that when E_C is determined by their technique, a way that eliminates the time for initiation of attack to occur, E_C then equals E_P.

Relevant Basic Processes

In the preceding section, a technique for determining E_C was considered to give relevant values because it was based on the supposition that repassivation was a process strongly affecting localized attack. A number of workers[19-21] have suggested that the rates of two major basic processes, namely, breakdown of passivity and repassivation, are of great importance in determining whether localized attack will occur. By focusing on these two basic processes, some valuable approaches have been developed for studying localized corrosion.

One technique that has been suggested as actually looking at breakdown/repair events is that concerned with examining the small fluctuations of potential or current occurring on a corroding surface, *corrosion noise*.

[(1)]Actually a more relevant environment would be that environment one would expect to develop in a pit, crack, or crevice when the surface is exposed to the ambient environment.

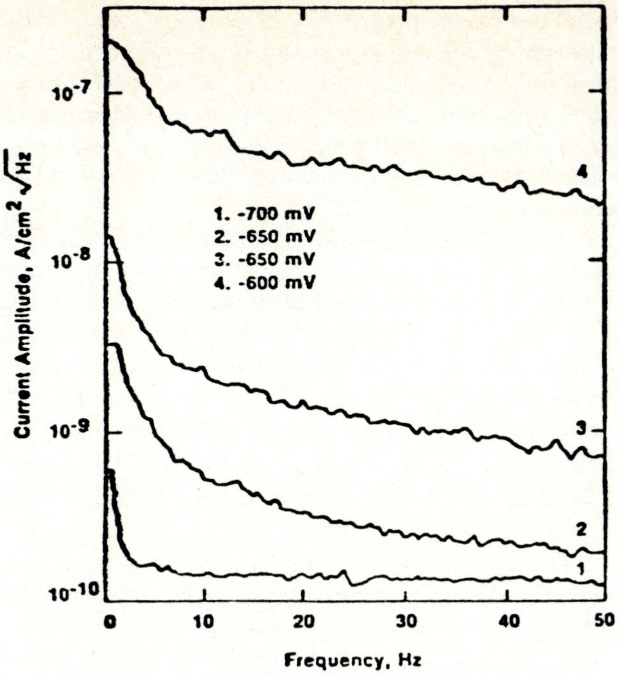

FIGURE 6 — Noise current spectra of AA (Aluminum Association) Al 6061 in borate buffer + 0.01 M NaCl, under potentiostatic conditions. (Reference—SCE.) Averages over 64 spectra: (1) −700 mV, (2) −650 mV, (3) −650 mV after 10 min, and (4) −600 mV.[29]

Iverson[27] demonstrated that the frequency and intensity of such noise could be markedly diminished by the addition of an inhibitor to a corroding system. Videm[19] and Okamoto, et al.,[28] have suggested that the noise represents breakdown and repassivation events and that the dynamic competition between these two basic processes determines whether localized attack such as pitting will be initiated.

Bertocci[29] developed a more quantitative way to look at corrosion noise, which he and others called by the more exact name of *electrochemical noise*. Bertocci plotted the noise current amplitude ($A/cm^2 \sqrt{Hz}$) vs the frequency of the noise events, i.e., current fluctuations. Figure 6 shows such a plot of noise current spectra for an Al alloy in a borate buffer solution containing Cl^- ions at potentials above and below E_C. It can be seen that as the potential goes from values below E_C to values above, the noise current amplitude can change by three orders of magnitude.

Electrochemical noise has been observed during both the initiation (induction period) and propagation stages of pitting. Whether the current fluctuations, or noise, correspond to breakdown and repassivation processes has not been established. Nevertheless, they do reflect some sort of dynamic processes that clearly and strongly affect the initiation and propagation of localized corrosion. The three order-of-magnitude differences between pitting and nonpitting conditions strongly emphasize such an assertion. For example, the function of the chloride ion in pitting is to tilt the scale toward breakdown and corrosion by slowing down or stopping effective repassivation.[19,21]

If repassivation kinetics is, therefore, an important basic process, attention must be focused on it and ways developed to separate its measurement from the breakdown process that always occurs simultaneously during localized corrosion. Scully[30,31] was one of the earliest workers to emphasize the importance of repassivation kinetics. It has become an important element, especially in the film rupture metal-dissolution mechanism of stress corrosion cracking (SCC), because combined with the concept of slip-step dissolution, it explains why restricted lateral dissolution occurs. The slower the repassivation, the more of the metal slip-step that has emerged and ruptured the film can be dissolved before new film forms. If too little dissolves, no SCC occurs; if too much dissolves, the dissolved areas are not sufficiently restricted, and the crack is too blunt to propagate. Only when the environment and alloy provide a system where the repassivation kinetics produce a highly restricted region for dissolution is SCC found. In addition to Scully, Staehle[32] has given a detailed discussion of the significance of repassivation kinetics.

These considerations have resulted in a number of techniques aimed at determining the repassivation kinetics.[33-38] These techniques produce a bare surface either by straining an alloy wire or by scratching or abrading a surface. In most of these techniques, the specimens are either strained or scratched while maintaining their potentials at a fixed value and measuring the current transient that takes place during the dissolution/repassivation event that occurs subsequent to exposure of bare metal. The rate of current decay is related to the rate of repassivation. A useful technique for measuring susceptibility to SCC by looking at repassivation kinetics is that developed by Galvele, et al.[39] This technique determines anodic and cathodic polarization curves for both an unstrained and strained electrode. The ratio between the corrosion current for the unstrained electrode to that of the strained can then be used to assess susceptibility.

In such primarily electrochemical techniques, the current involved in dissolution cannot be separated from that involved in passive film formation. Using a technique developed by Ambrose and Kruger,[23] however, it is possible to measure the film regrowth transient separately by optical means (ellipsometry) and to compare it to the overall dissolution/repassivation current. With this technique, called *tribo-ellipsometry*, the protective film on a metal in an environment is removed by means of a polishing wheel; the film growth kinetics and the current are then both recorded.

Tribo-ellipsometry shows that the film growth rate on a titanium alloy is considerably slower in the susceptible chloride environment than it is in the unsusceptible nitrate solution. With tribo-ellipsometry, the ratio of the total current to that involved only in repassivation can be determined. This ratio, R_p^*, when the exposed surface area is known is given by the expression:

$$R_p^* = 1 + \frac{Q_d}{Q_r} \qquad (2)$$

where Q_d = charge involved in dissolution and Q_r = charge involved in repassivation.

R_p^* is a measure of the effectiveness of the repassivation process. Work on the cracking of low carbon steel[23] has shown that SCC does not occur when the effectiveness of repassivation is either too high or too low. Thus, for low carbon steel, at the time required to form a complete film, R_p^* is 2.8 in 1 N $NaNO_2$ (25 C) where it is insusceptible; R_p^* is 26 in 4 N $NaNO_3$ (90 C) where it does undergo SCC; and R_p^* is > 75 in 1 N NaCl (25 C) where it undergoes widespread pitting attack but not SCC. Thus, a delicate balance between repassivation and dissolution must be maintained to lead to SCC. The determination of this balance is dependent on the environment, the alloy, and finally the metallurgy and mechanics of the system that control the rate at which bare metal sites (generally slip-steps) are produced, as well as their density and size.

It is on the rate at which bare metal sites are produced that another new approach to localized corrosion (primarily SCC) is based. This approach uses the constant strain rate technique which gives some information on the beneficial breakdown processes occurring in SCC. This technique, which produces experimental data as depicted in Figure 7, can be applied to mechanistic studies as well as be used for routine testing. The mechanistic importance of the technique is that it ties the metallurgically controlled rate of bare metal exposure, resulting from film rupture by slip-step emergence, to the chemical process of repassivation of the exposed metal. This can be seen by examining the significance of the three regions shown in Figure 7. For Region A, where strain rates are too low to lead to SCC, the rate of repassivation is high enough to prevent any significant and damaging interaction between the environment and the bare metal exposed during straining. At the opposite extreme, Region C, the strain rate is so rapid that the interaction of the exposed metal with the environment is of no consequence because the rate that the metal fails by ductile fracture exceeds the rate by which the environment can affect fracture via any of the possible SCC mechanisms (dissolution, hydrogen embrittlement, or adsorption of damaging species). It is only in the intermediate range of strain rates (Region B), where the rate of production of bare metal sites has a value sufficiently high that the rate of repassivation does not prevent environmental access. However, the strain rate is sufficiently low that the environment interaction and not ductile pulling apart produces fracture. Under these conditions, SCC takes place. If SCC is not possible for the system studied, the dip shown in Region B will not occur.

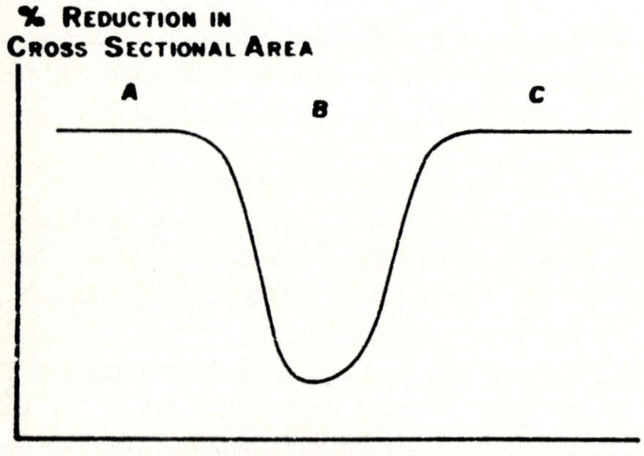

FIGURE 7 — A schematic representation of the results obtained in constant strain rate tests. Region A—repassivation is rapid enough and strain rate slow enough to prevent damaging interaction with the environment. Region B—strain rate is high enough so that the repassivation does not prevent environment access. Region C—strain rate is so rapid that ductile pulling apart takes place.

The question then arises whether this test, whose development and increasing use was stimulated for mechanistic reasons and was guided mainly by the implications of film rupture and repassivation, provides a valid routine material-environmental evaluation test. Parkins[40] has pointed out that there are certain features of the constant strain rate technique that make it a useful routine laboratory test. These features are as follows: (1) It is a relatively severe test so that it will promote laboratory SCC failures where other tests on smooth specimens will not unless inordinately long testing times are used; (2) it always produces fracture by SCC or some other mechanism and therefore is a positive test; and (3) the time of testing is relatively short. These three somewhat interrelated attributes, severity, positiveness, and rapidity, are especially valuable in comparative testing where intercomparison of materials and environments is desired.

The promise that the constant strain rate approach offers for laboratory testing is somewhat tempered by the lack of good ways to quantify the test results. For example, how does one relate the strain rate in Region B where SCC susceptibility is the greatest to crack growth rate? Moreover, it is not known what relation, if any, this strain rate bears to the strain rate that exists at the tip of a growing crack. Parkins[40] has made some valuable qualitative connections between this value and the threshold stress for SCC or the threshold value of the stress intensity factor (an important parameter in the fracture mechanics approach to SCC testing[41]), but the quantitative relationships require new theoretical and experimental advances. Despite these problems, the constant strain rate tests have provided evidence of SCC in systems where other tests have been difficult or not possible to apply.[42-45]

Relevant Films

The protective films that form on alloy surfaces in a given environment are of crucial importance in controlling localized attack because their ability to withstand the breakdown processes determines whether such attack will occur. In attempting to determine how these films affect localized attack, one must ascertain that certain relevant considerations with respect to the films' ability to affect localized corrosion are considered.

A major property of the passive film that enables its control of the initiation of localized corrosion is its structure. Hoar[46] has proposed that a good passive film is one that has a noncrystalline or glassy structure. Revesz and Kruger[47] examined in greater detail the question, what is the nature of a good passive film and what determines whether the film will be resistant to the initiation of localized corrosion? They proposed that one of the important characteristics of a more effective protective layer was that it contained elements whose oxides have the possibility of existing as different structures that have the same energy or, more exactly, *enthalpy*.

The oxides of chromium provide a good example of this. One can have a chromium III oxide (Cr_2O_3) or a chromium VI (CrO_3) and they both have essentially the same enthalpy. This leads to infinitely adapted structures, a concept developed by Anderson[49] in 1973. Such structures are similar to giant inorganic polymeric molecules; they have good short range order but not the long range order of crystalline solids. A protective layer with such a glassy structure is very effective. For example, stainless steel forms a better passive film than low carbon steel because the stainless contains chromium, an element that promotes the formation of noncrystalline, glassy structures. Of course, stainless steels break down also, but electron diffraction studies[49] show that when chromium is added to iron, the passive film goes from a well-oriented spinel to a noncrystalline amorphous or glassy structure as more chromium is added. Alloys bearing such films are more resistant to the initiation of localized corrosion. They are more resistant to initiation because they break down less rapidly and they repassivate more effectively. Optical studies[50] comparing the breakdown of Fe vs that of an Fe-Cr alloy showed that the addition of Cr made the passive film considerably more resistant to breakdown. Similarly, as Figure 8 shows, the passive film on stainless steel is repaired more effectively than that on low carbon steel.

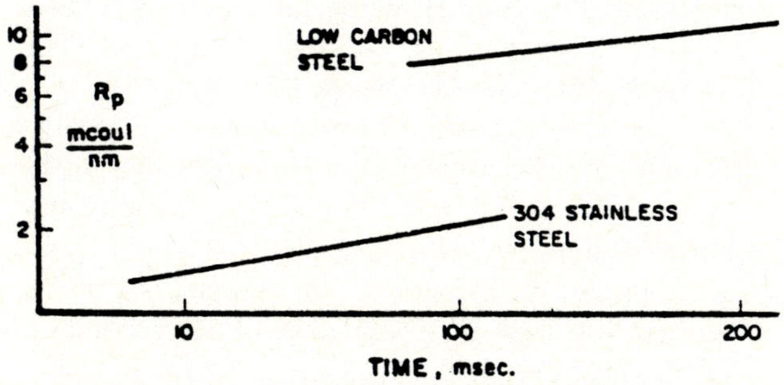

FIGURE 8 — Comparison between the variation in R_p in 1 N NaCl for AISI 304 stainless steel and low carbon steel. R_p is the repassivation ratio determined by tribo-ellipsometry. It is the charge involved in metal dissolution and passive film formation during repassivation divided by the charge involved in only film formation. R_p is a measure of the effectiveness of the repassivation process, the smaller R_p the more effective the repassivation.[48]

Another element, molybdenum, also has a beneficial effect on pitting but has not been found to exist in significant concentrations in the outer parts of the passive film of stainless steel.[51] Thus, it is unlikely that it can affect the structure of the passive film on stainless steels in the beneficial way just described for stainless steel.

Kodama and Ambrose[52] sought to understand the role of Mo by using Pourbaix diagrams. They used a combination of the iron and the molybdenum diagrams to provide insights into the nature of the protective layers that stop pit propagation in ferrous alloys containing molybdenum. They showed that the region of stability for $FeMoO_4$ corresponds to the pH and potential conditions found in initiated pits for molybdenum containing ferrous alloys. Therefore, they concluded that $FeMoO_4$ is a feasible constituent of the protective layer that chokes off pit propagation. It is, therefore, in the production of a protective layer during propagation rather than during initiation that Mo plays its role in combating pitting. Whether or not it also plays a role in initiation has not yet been established. Nevertheless, it can be said that the relevant protective film affecting localized corrosion is not the original film formed by passivation of the alloy surface, but a film whose origin is a precipitation reaction occurring in solution between ions that were formed during the initiation stage.

Conclusion

As can be determined from the paper just presented, a fruitful study of localized corrosion requires the separation of relevant factors from those that are irrelevant. It is only after doing so that the localized corrosion process can be fully understood.

References

1. B. F. Brown, C. T. Fujii, E. P. Dahlberg, J. Electrochem. Soc., Vol. 116, p. 218, 1969.
2. G. Sandoz, C. T. Fujii, B. F. Brown, Corros. Sci., Vol. 10, p. 839, 1970.
3. J. A. Smith, M. H. Peterson, B. F. Brown, Corrosion, Vol. 26, No. 6, p. 539, 1970.
4. M. Pourbaix, Theory of Stress Corrosion Cracking of Alloys, J. C. Scully, Ed., NATO, Brussels, p. 17, 1971.
5. H. Leidheiser, Jr., R. Kissinger, Corrosion, Vol. 28, No. 6, p. 218, 1972.
6. F. D. Bogar, C. T. Fujii, Naval Research Laboratory Report 7690, Washington, D.C., March, 1974.
7. J. R. Galvele, Passivity of Metals, R. P. Frankenthal, J. Kruger, Eds., Electrochemical Soc., Pennington, New Jersey, p. 285, 1978.
8. G. M. Ugiansky, J. Kruger, R. W. Staehle, Proc. Fourth Int. Cong. on Metallic Corrosion, ABRACO, Rio de Janeiro, p. 605, 1979.
9. J. Kruger, J. Electrochem. Soc., Vol. 106, p. 736, 1959.
10. M. J. Pryor, "The Influence of the Defect Structure of Aluminum Oxide Films on the Pitting of Aluminum

in Chloride Solution," Localized Corrosion, R. W. Staehle, B. F. Brown, J. Kruger, A. Agrawal, Eds., National Association of Corrosion Engineers, Houston, Texas, p. 2, 1974.
11. M. Pourbaix, "The Electrochemical Basis for Localized Corrosion," Localized Corrosion, R. W. Staehle, B. F. Brown, J. Kruger, A. Agrawal, Eds., National Association of Corrosion Engineers, Houston, Texas, p. 12, 1974.
12. M. Janik-Czachor, J. Electrochem. Soc., Vol. 128, p. 513C, 1981.
13. N. Pessall, C. Liu, Electrochim. Acta, Vol. 16, p. 1987, 1971.
14. J. Kruger, K. Rhyne, Nuclear and Chem. Waste Management, Vol. 3, p. 205, 1982.
15. J. O'M. Bockris, A. K. N. Reddy, Modern Electrochemistry, Vol. 2, Plenum Press, New York, New York, p. 706, 1970.
16. S. D. Argade, E. Gileadi, Electrodeposition, E. Gileadi, Ed., Plenum Press, New York, New York, p. 87, 1967.
17. N. Sato, Electrochim. Acta, Vol. 19, p. 1683, 1971.
18. J. R. Ambrose, J. Kruger, "Breakdown of Passive Films on Iron by Chloride Ion," Proc. Fourth Int. Cong. on Metallic Corrosion, National Association of Corrosion Engineers, Houston, Texas, p. 698, 1972.
19. K. Videm, Kjeller Report KR-149, Institutt for Atomenergi, Kjeller, Norway, 1974.
20. J. Kruger, J. R. Ambrose, NBS Report NBSIR 74-583, Washington, D.C., September, 1974.
21. J. Zahavi, M. Metzger, "Breakdown of Films and Initiation of Pits on Aluminum During Anodizing," Localized Corrosion, R. W. Staehle, B. F. Brown, J. Kruger, A. Agrawal, Eds., National Association of Corrosion Engineers, Houston, Texas, p. 547, 1974.
22. G. Okamoto, Corros. Sci., Vol. 13, p. 471, 1973.
23. J. R. Ambrose, J. Kruger, "Tribo-Ellipsometric Studies of the Relationship Between Repassivation Kinetics and Stress Corrosion of Low Carbon Steel," Proc. Fifth Int. Cong. on Metallic Corrosion, National Association of Corrosion Engineers, Houston, Texas, p. 406, 1974.
24. J. R. Ambrose, J. Kruger, J. Electrochem. Soc., Vol. 121, p. 599, 1974.
25. J. R. Galvele, S. M. deMicheli, I. L. Muller, S. B. deWexier, I. L. Alanis, "Critical Potentials for Localized Corrosion of Aluminum Alloys," Localized Corrosion, R. W. Staehle, B. F. Brown, J. Kruger, A. Agrawal, Eds., National Association of Corrosion Engineers, Houston, Texas, p. 580, 1974.
26. K. J. Vetter, H. H. Strehblow, "Pitting Corrosion in an Early Stage and Its Theoretical Implications," Localized Corrosion, R. W. Staehle, B. F. Brown, J. Kruger, A. Agrawal, Eds., National Association of Corrosion Engineers, Houston, Texas, p. 240, 1974.
27. W. P. Iverson, J. Electrochem. Soc., Vol. 115, p. 617, 1968.
28. G. Okamoto, K. Tachibana, S. Nishiyama, T. Sugita, "The Analysis of Passive Current Noise of Stainless Steels Under Potentiostatic Conditions With and Without Chloride Ions," Passivity and Its Breakdown of Iron and Iron Base Alloys, R. W. Staehle, H. Okada, Eds., National Association of Corrosion Engineers, Houston, Texas, p. 106, 1976.
29. U. Bertocci, J. Kruger, Surface Sci., Vol. 101, p. 608, 1980.
30. J. C. Scully, Corros. Sci., Vol. 7, p. 197, 1967.
31. J. C. Scully, Corros. Sci., Vol. 8, p. 513, 1968.
32. R. W. Staehle, Theory of Stress Corrosion Cracking in Alloys, J. C. Scully, Ed., NATO, Brussels, p. 223, 1971.
33. T. P. Hoar, J. M. West, Proc. Roy. Soc., Vol. 268A, p. 304, 1962.
34. H. Leidheiser, Jr., E. Kellerman, Corrosion, Vol. 26, No. 3, p. 99, 1970.
35. T. Murata, R. W. Staehle, "Theory and Application of Straining Electrode," Proc. Fifth Int. Cong. on Metallic Corrosion, National Association of Corrosion Engineers, Houston, Texas, p. 573, 1974.
36. N. D. Tomashov, L. P. Vershinina, Electrochim. Acta, Vol. 15, p. 501, 1970.
37. J. R. Ambrose, J. Kruger, Corrosion, Vol. 28, No. 1, p. 30, 1972.
38. R. B. Diegle, D. A. Vermilyea, J. Electrochem. Soc., Vol. 122, p. 180, 1975.
39. J. R. Galvele, S. B. deWexler, I. Gardiazabal, Corrosion, Vol. 31, No. 10, p. 352, 1975.
40. R. N. Parkins, Stress Corrosion Cracking—The Slow Strain Rate Technique, G. M. Ugiansky, J. H. Payer, Eds., ASTM STP 665, ASTM, Philadelphia, Pennsylvania, p. 5, 1979.
41. B. F. Brown, Ed., Stress Corrosion Cracking in High Strength Steels and in Titanium and Aluminum Alloys, Naval Research Laboratory, US Government Printing Office, Washington, D.C., 1972.
42. C. D. Kim, B. E. Wilde, Stress Corrosion Cracking—The Slow Strain Rate Technique, G. M. Ugiansky, J. H. Payer, Eds., ASTM STP 665, ASTM, Philadelphia, Pennsylvania, p. 97, 1979.
43. J. C. Scully, Stress Corrosion Cracking—The Slow Strain Rate Technique, G. M. Ugiansky, J. H. Payer, Eds., ASTM STP 665, ASTM, Philadelphia, Pennsylvania, p. 237, 1979.
44. J. H. Payer, W. E. Berry, R. N. Parkins, Stress Corrosion Cracking—The Slow Strain Rate Technique, G. M. Ugiansky, J. H. Payer, Eds., ASTM STP 665, ASTM, Philadelphia, Pennsylvania, p. 222, 1979.
45. W. H. Clark, M. J. Povich, T. M. Devine, Stress Corrosion Cracking—The Slow Strain Rate Technique, G. M. Ugiansky, J. H. Payer, Eds., ASTM STP 665, ASTM, Philadelphia, Pennsylvania, p. 132, 1979.
46. T. P. Hoar, J. Electrochem. Soc., Vol. 117, p. 17C, 1970.
47. A. G. Revesz, J. Kruger, Passivity of Metals, R. P. Frankenthal, J. Kruger, Eds., Electrochemical Soc., Pennington, New Jersey, p. 137, 1978.

48. J. S. Anderson, J. Chem. Soc. Dalton Trans., p. 1107, 1973.
49. C. L. McBee, J. Kruger, Electrochim. Acta, Vol. 17, p. 1337, 1972.
50. C. L. McBee, J. Kruger, "Ellipsometric Spectroscopic Study of Breakdown of Fe-Cn and Fe-Cr-Mo Alloys," Passivity and Its Breakdown of Iron and Iron Base Alloys, R. W. Staehle, H. Okada, Eds., National Association of Corrosion Engineers, Houston, Texas, p. 131, 1976.
51. J B. Lumsden, R. W. Staehle, Scripta Metal., Vol. 6, p. 1205, 1972.
52. T. Kodama, J. R. Ambrose, Corrosion, Vol. 33, p. 155, 1977.
53. B. F. Brown, Theory of Stress Corrosion Cracking in Alloys, J. C. Scully, Ed., NATO, Brussels, p. 197, 1971.

In Situ Studies of the Transpassive Behavior of Steel Anodes in Neutral Electrolytes

J. P. HOARE*

Abstract

Cyclic voltammetric studies were conducted on mild steel anodes in 4 M NaCl, 4 M $NaClO_3$, 4 M $NaClO_4$, and 2 M $Na_2Cr_2O_7$ electrolytes to determine the passive-transpassive behavior of steel in these solutions. It was found that the passive oxide film could be formed either by the direct oxidation of the metal surface or by the oxidation of a salt film deposited in the metal/solution interface. An ion-exchange model for the dissolution of the passive film in the transpassive region is proposed. The suitability of these electrolytes for the machining of steel by electrochemical machining (ECM) was discussed.

Introduction

Voltammetry is one of the best *in situ* methods of studying the metal/solution interface, since the potential of the electrode surface is very sensitive to what is present on the metal surface. Other methods such as ellipsometry, reflectance, Mossbauer spectroscopy, and enhanced Raman spectroscopy, are valuable, *in situ*, investigative tools, but all suffer from limitations to their application. *Ex situ* spectroscopy methods such as Auger (AES), photoelectron (XPS), electron energy loss (EELS), secondary ion mass (SIMS) spectroscopy, and low energy electron diffraction (LEED) require a hard vacuum. Consequently, the potential must be removed from the system and the system must be dehydrated. In such cases, the *ex situ* and the *in situ* systems are most likely to be different since many species present on the electrode surface are stable only in a given potential region or under a certain degree of hydration or both.

An example of this problem is encountered in studying the industrial operation known as electrochemical machining (ECM),[1,2] which may be considered a high rate, controlled, anodic-corrosion process. It is generally held[3-6] that the ECM process on iron and steel occurs in passivating electrolytes in the potential range corresponding to the transpassive region of the polarization curve where high metal removal rates take place.

It is the purpose of this report to demonstrate how considerable information may be obtained concerning the nature of the passive and transpassive anodic films present on mild steel surfaces in high-concentration, neutral solutions of NaCl, $NaClO_3$, $NaClO_4$, and $Na_2Cr_2O_7$ from steady-state polarization studies at constant potential, as well as from transient constant current stripping and cyclic voltammetric studies.

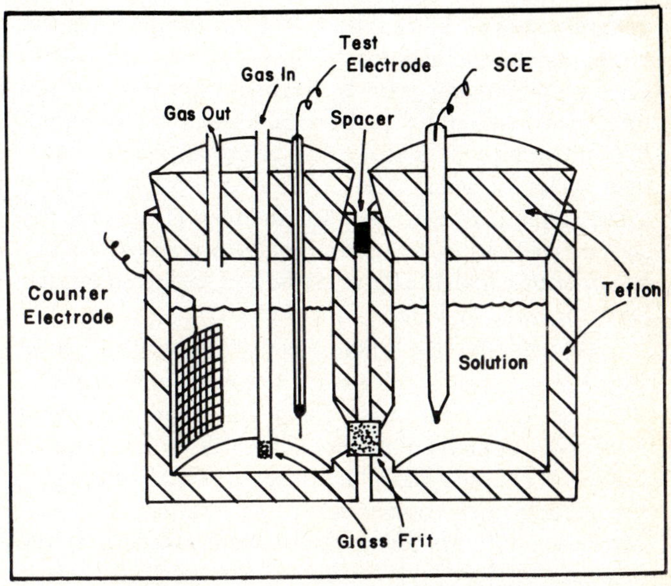

FIGURE 1 – Experimental PTFE cell.

*Electrochemistry Dept., General Motors Research Laboratories, Warren, Michigan.

Experimental

Test electrodes were made from mild steel (AISI 1020) wire (0.02 cm in diam) that had been abraided with fine emery, degreased in trichlorethylene, rinsed in acetone, and washed in triply distilled water. These wires were potted in polyethylene which was removed from the end of the wire to expose about 0.3 cm ($\sim$0.2 cm^2 in area).

The polytetrafluoroethylene (PTFE) cell consists of two compartments shown in Figure 1 connected by a glass frit. In the left or test compartment, a Pt gauze counter electrode is mounted, and in the right or reference compartment is mounted the saturated calomel reference electrode (SCE). After the cell had been filled with the electrolyte of interest, the solution was pre-electrolyzed for 48 h against Pt wire pre-electrolysis electrodes which were removed at the end of this period to break the pre-electrolyzing circuit. Three test electrodes (for duplicate runs) were mounted in the test compartment and the solution was saturated with purified N_2 gas bubbling through the cell at rates between 200 and 300 cm^3/min to minimize concentration polarization effects.

Once the solution is saturated with N_2 (30 min), the potential of the test electrode was cycled between two chosen values at 100 mV/s until a steady-state cyclic voltammogram was recorded. In some cases, up to 50 cycles were required to reach the steady state. It is believed that the steady state is reached when the surface roughness, caused by the periodic formation and reduction of oxide films, does not change significantly with further cycles of polarization. Three types of families of cyclic voltammograms were recorded by: (1) varying the sweep rate from 17 to 100 mV/s, (2) decreasing the upper or anodic limit in 50 mV steps, and (3) increasing the lower or cathodic limit in 50 mV steps.

All solutions were made with triply distilled water. Experiments were conducted in 4 M NaCl, 4 M $NaClO_3$, 4 M $NaClO_4$, and 2 M $Na_2Cr_2O_7$. All experiments were conducted at ambient temperatures (25 ± 1 C) and all potentials are recorded with respect to SCE. The data for the steady-state, potentiostatic polarization, and the constant-current stripping curves were taken from the literature.[7,8]

Results and Discussion

The steady-state polarization curve in Figure 2(a) and the plot of the charge associated with each point on the polarization curve in Figure 2(b) for mild steel in 4 M $NaClO_3$ solution show that in the active region (up to 400 mV) iron goes into solution and no oxide film is present on the electrode surface. Above 400 mV, the current falls to very low values in the passive region as the oxide layer builds up to a fairly steady value between 800 and 1400 mV. The thickness of the oxide layer decreases and the current for metal dissolution increases once more in the transpassive region. Mass balance studies[9,10] have shown that the transpassive current is consumed in metal removal (virtually 100%).

With these data at hand, it is possible to interpret the cyclic voltammograms recorded in Figure 3 for mild steel

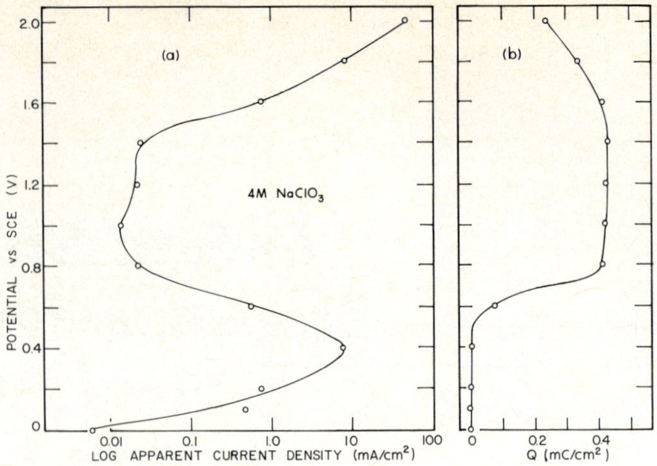

FIGURE 2 — (a) Steady-state potentiostatic polarization curve, and (b) the quantity of charge, Q, associated with the film formed on the anode surface for mild steel in O_2-stirred 4 M $NaClO_3$ solution.

anodes in 4 M $NaClO_3$. Figure 3(a) contains the data as a function of the sweep rate. Below 500 mV, the steel is active and iron goes into solution. The nonlinear dependence of the current peak in this region gives evidence of the electrochemical removal of metal from the anode surface into the solution. Above 600 mV, the current falls to very low values because the electrode surface is passivated by the formation of an anodic oxide layer which dissolves above 1300 mV in the transpassive region.

At 1500 mV, the potential sweep is reversed, and between 1200 and 1300 mV, the current becomes negative through the reduction of the thin oxide film remaining on the anode surface in the transpassive region. Here, the current peak depends linearly on the sweep rate, which indicates that the electrode process responsible for the negative current is a surface reaction and does not involve diffusion of the electroactive species to or from solution. With the formation of the passive film between 600 and 1200 mV, the current becomes very low again. This observation gives evidence that the nature of the thick oxide film in the passive region may be somewhat different than that of the thin oxide film present in the transpassive region.

It is noted in Figure 3(b) that for the first three cycles there is little or no change in the shape of the traces in the active region. Similarly in Figure 3(c), there is little or no change in the shape of the transpassive trace with the first 3 or 4 cycles of polarization. Apparently, the voltammograms for mild steel in $NaClO_3$ solution in the transpassive and active regions are independent of one another.

In contrast, the steady-state polarization curve [Figure 4(a)] and the charge associated with the passive film [Figure 4(b)] for mild steel in 4 M $NaClO_4$ exhibit a more gradual transition from the active to the passive state with a gradual increase in the thickness of the passive film. When the transpassive region is reached at about 1600

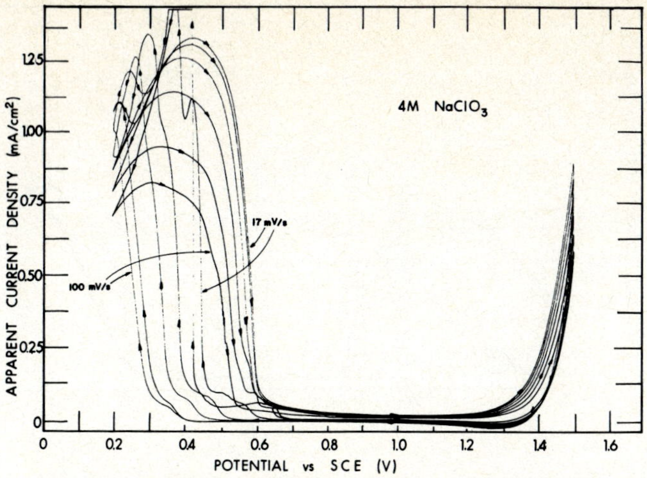

FIGURE 3(a) – Family of cyclic voltammograms as a function of the sweep rate (17, 33, 50, 67, 83, 100 mV/s) obtained on mild steel in N_2-stirred 4 M $NaClO_3$ scanned between 200 and 1500 mV; arrows indicate direction of potential sweep.

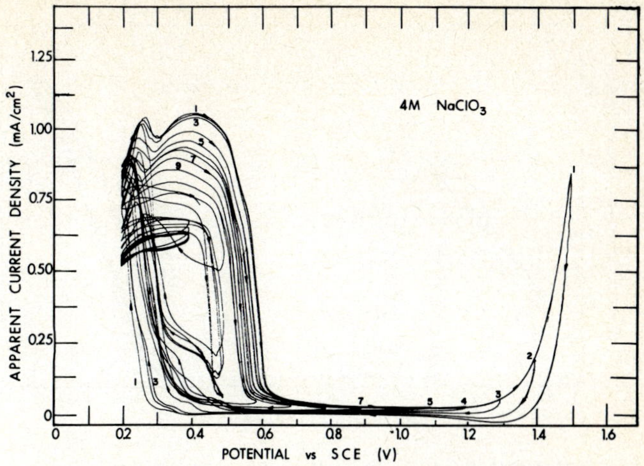

FIGURE 3(b) – Family of cyclic voltammograms obtained by decreasing the anodic limit in 50 mV steps on mild steel in N_2-stirred 4 M $NaClO_3$; sweep rate is 100 mV/s; numerals indicate cycle number.

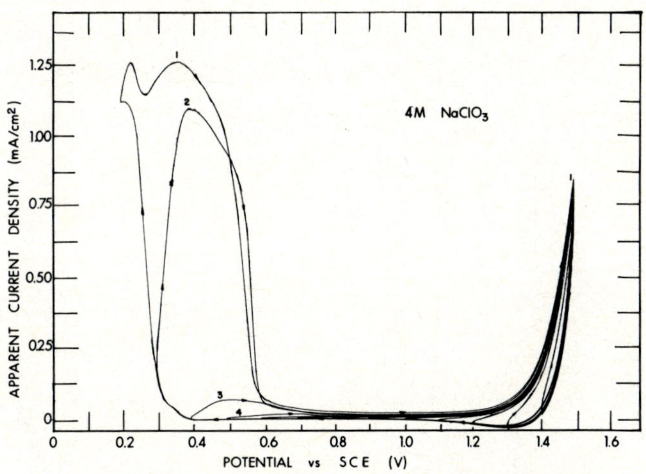

FIGURE 3(c) – Family of cyclic voltammograms obtained by increasing the cathodic limit in 50 mV steps on mild steel in N_2-stirred 4 M $NaClO_3$; sweep rate is 100 mV/s.

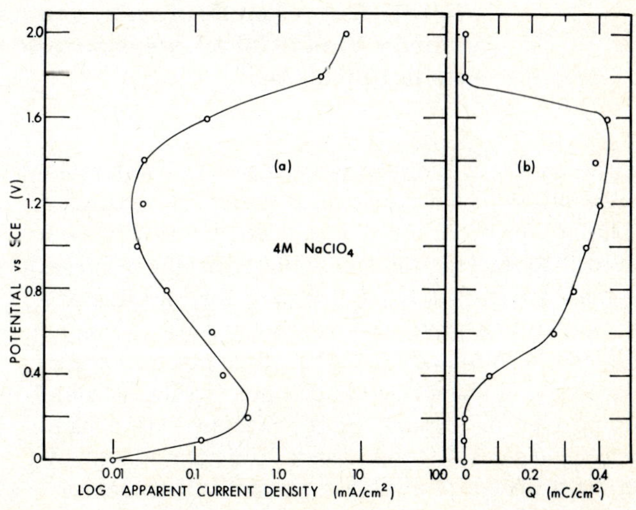

FIGURE 4 – (a) Steady-state polarization curves and (b) charge vs potential obtained on mild steel anodes in O_2-stirred 4 M $NaClO_4$ solution.

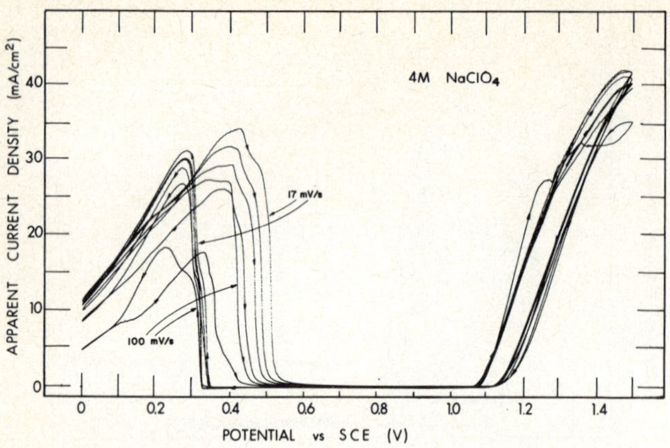

FIGURE 5(a) — Family of cyclic voltammograms as a function of sweep rate (17, 33, 50, 67, 83, 100 mV/s) obtained on mild steel in N_2-stirred 4 M $NaClO_4$ scanned between 0 and 1500 mV; arrows indicate direction of potential sweep.

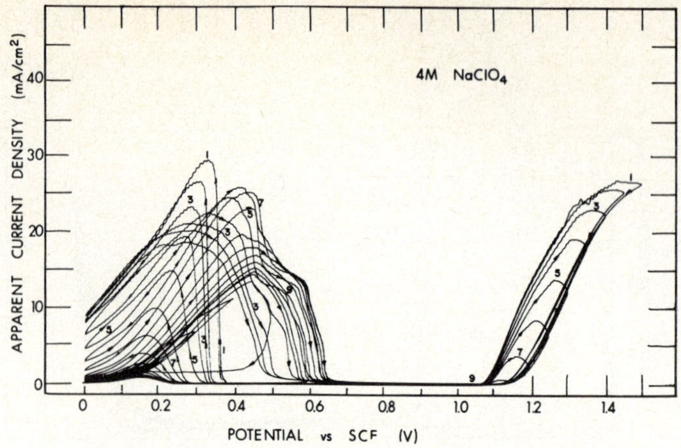

FIGURE 5(b) — Family of cyclic voltammograms obtained by decreasing the anodic limit in 50 mV steps on mild steel in N_2-stirred 4 M $NaClO_4$; sweep rate is 100 mV/s; numerals indicate cycle number.

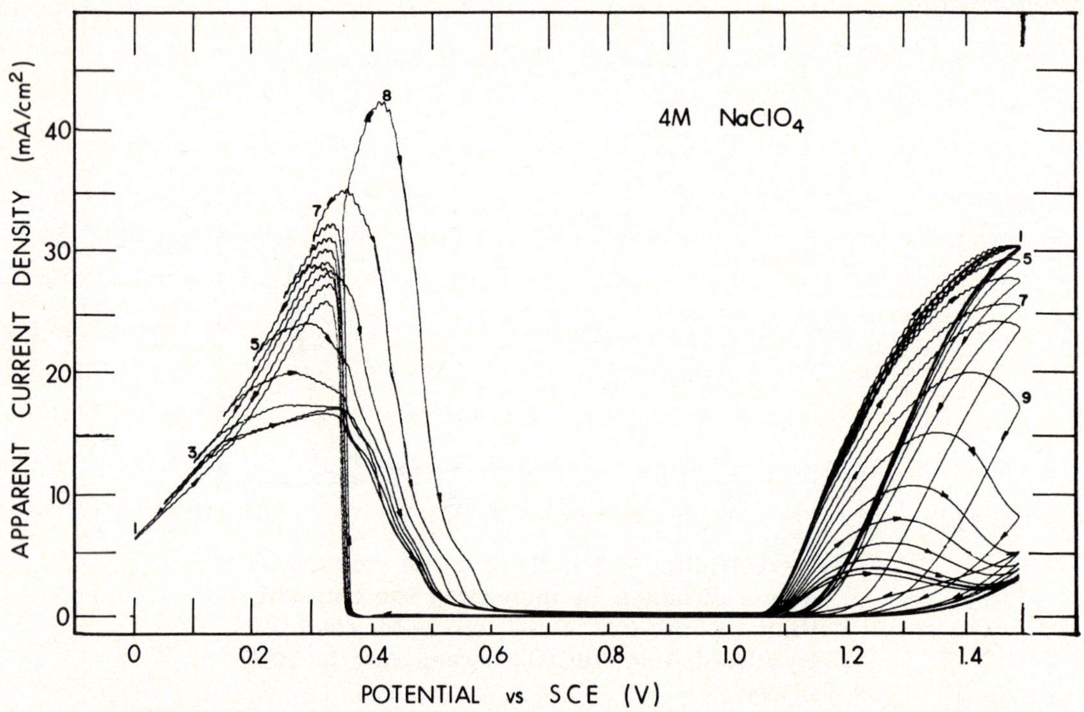

FIGURE 5(c) — Family of cyclic voltammograms obtained by increasing the cathodic limit in 50 mV steps on mild steel in N_2-stirred 4 M $NaClO_4$; sweep rate is 100 mV/s.

mV, the passive film disappears completely. From mass balance studies,[11] the transpassive current is accounted for virtually 100% in metal dissolution. It was observed from rotating steel hemisphere studies[12] that the polarization curve for steel in $NaClO_3$ [Figure 2(a)] is independent of the rotation rate, whereas that in $NaClO_4$ [Figure 4(a)] varies with rotation rate (the critical potential in the active region is shifted to more anodic values with increasing rates of rotation). It is concluded[13] that such behavior indicates that the passive oxide film is formed from the oxidation of a salt film present at the steel surface in $NaClO_4$.

The family of cyclic voltammograms obtained on mild steel in 4 M $NaClO_4$ as a function of the sweep rate is presented in Figure 5(a). Note that the current densities in Figure 5 are much larger than in Figure 3. In the active region between 0 and 400 mV, the high rate of metal dissolution produces a salt film at the metal/solution interface. As the potential becomes more anodic on the anodic-going sweep, the salt film is oxidized to an oxide film on the steel surface, which then passivates the electrode between 500 and 1100 mV. At this active-passive transition, the traces vary linearly with the sweep rate which indicates that the electrochemical process responsible for the current in this region is a surface reaction (oxidation of the salt film).

At the beginning of the transpassive region (above 1100 mV), the anodic film begins to dissolve where the salt film thickness is low. Once again the steel surface is open to attack and the metal dissolution current increases rapidly. This high current regenerates the salt film causing the increase in current to decelerate and the anodic-going trace to flatten out. When the sweep is reversed at 1500 mV, the anodic film is formed from the thick salt film at potentials above 1200 mV, producing a prominent hysteresis in the voltammograms in the transpassive region (1100 to 1500 mV). Evidence for the presence of an oxide film formed from a salt film in $NaClO_4$ arises from the observations in Figures 5(b) and (c) that the shape of the traces in the active and passive regions are interdependent on the current limits set for the potential sweep. For an oxide film formed by the direct oxidation of the surface, this is not observed [Figures 3(b) and (c)].

As ECM electrolytes, $NaClO_3$ provides better control of geometry and dimensions than $NaClO_4$ because the passive-transpassive transition is sharper in the curves of Figure 3 than in those of Figure 4. For a plunge-cut operation,[1,2] the current path at the bottom of the cut opposite the cutting face of the cathode tool is short with a low IR-drop. Consequently, the effective potential lies in the transpassive region with high metal removal rate. Along the sides of the cut, the current path is much longer with a larger IR drop. In this case, the effective potential lies in the passive region where metal removal is very low. In this way, tapering of the plunge cut is greatly reduced. According to the polarization curves, tapering in $NaClO_3$ should be less than in $NaClO_4$, although both are good ECM electrolytes. This conclusion was supported by the results obtained on the ECM test rigs.[1,2]

Electropolishing of a metal surface occurs according to the model proposed by Hoar and co-workers,[14,15] when a thin porous oxide film lies on the metal surface below a thick salt film. The thick salt film produces macroscopic leveling and the thin oxide film produces microscopic brightening. In $NaClO_3$ solutions, a thin porous uniform oxide film, which is formed by the oxidizing power of the ClO_3^- ion as quickly as it is dissolved in the transpassive region, lies on the steel surface beneath a thick salt film (generated by high metal dissolution rates). This situation produces electropolishing of the electrochemically machined steel surface.

In the case of the symmetrical ClO_4^- ion, its oxidizing power is so low[16] that a thin oxide film cannot be formed on the steel surface in the transpassive region. Under certain hydrodynamic and electrochemical conditions, a uniform porous compact salt film may be formed that can produce electropolishing of the machined surface.[17] It would, however, be expected that the electropolishing would depend on the electrolyte flow rate which would influence the properties of the salt film. This conclusion was confirmed by the results of the machining of steel bushings in an ECM through-hole test rig.[13] As expected, the electropolishing of steel in $NaClO_3$ was not affected

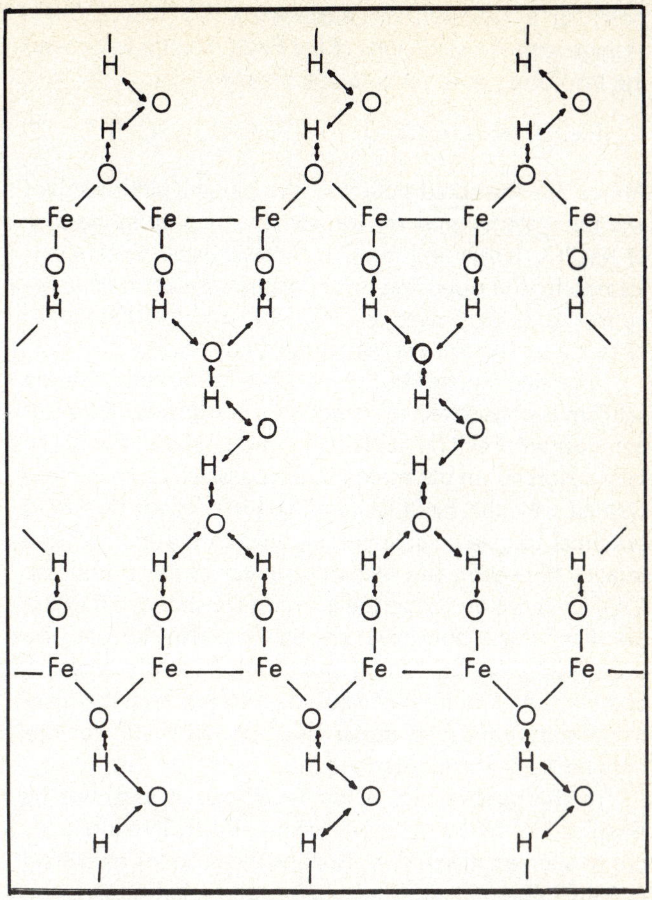

FIGURE 6 — Model of the passive film formed on a steel anode by Fe_2O_3 units forming puckered chains of FeOOH units, which hydrogen bond together through water molecules to form a three-dimensional matrix; hydrogen bonds are indicated by double headed arrows.

by flow rate as a result of the oxidized surface by ClO_3^- ions.

To account for the abrupt dissolution of the anodic oxide film on steel in $NaClO_3$ in the transpassive region, an ion exchange model was proposed.[18,19] From electron probe, vacuum fusion, and electron diffraction studies[20] of steel surfaces machined in $NaClO_3$, the oxide film present is composed of γ-Fe_2O_3. It is suggested then, that the passive oxide film on steel anodes consists of hydrated Fe_2O_3 or FeOOH units in the form of chains, which can hydrogen bond together with water molecules to give a three-dimensional matrix, sketched in Figure 6 in two dimensions. If it is assumed in the passive region that four water molecules are associated with each FeOOH unit, then as the potential is made more anodic ClO_3^- ions may replace water molecules,

$$FeOOH \cdot 4H_2O + ClO_3^- \rightarrow FeOOH \cdot 3H_2O \cdot ClO_3^- + H_2O \quad (1)$$

At high enough potentials, three waters will be replaced to give $[FeOOH \cdot H_2O \cdot 3ClO_3]^{-3}$. When a threshold potential is reached, the adsorbed ClO_3^- ions can exchange with the oxide ions of the FeOOH units to convert the insoluble oxide units to soluble chlorates.

$$[FeOOH \cdot H_2O \cdot 3ClO_3]^{-3} \rightarrow [Fe(ClO_3)_3 \cdot 3OH]^{-3} \quad (2)$$

Above this threshold potential, the passive film dissolves and the system enters the transpassive region. In the case of $NaClO_3$, a thin uniform film of steady-state thickness is formed by the opposing forces of the formation of oxide by the oxidizing power of ClO_3^- ions and the dissolution of oxide by the ion exchange mechanism.

For steel in $NaClO_4$, the oxide film formed from the salt film is considered to be a matrix composed of hydrogen-bonded FeOOH units with water. Here, the ClO_4^- ions adsorbed on the matrix in the passive region can exchange with the FeOOH units above a given threshold potential to give soluble perchlorates in the transpassive region. However, the oxidizing power of the symmetric ClO_4^- ion cannot oxidize the iron surface so that a thin oxide film does not exist on steel in $NaClO_4$ in the transpassive region. Under sufficiently high potentials, a compact salt film may be laid down to produce the electropolishing observed under ECM conditions,[13] but this brightening is stirring-dependent.

If the passive film is too tenacious, it may not be dissolved in the transpassive region and ECM cannot occur in such an electrolyte. Such is the case for mild steel in $Na_2Cr_2O_7$ solution. The steady-state polarization curve for mild steel in 2 M $Na_2Cr_2O_7$ is plotted in Figure 7(a) and the charge associated with the oxide film in Figure 7(b). It is observed that the steel surface is passive from 0 to ~1200 mV and a relatively thick passive film is present at all potentials studied. The mass balance results indicated negligible metal removal over the complete range of potential[21] and the current was consumed in O_2 evolution. In this case, $Cr_2O_7^=$ ions may adsorb on the hydrated FeOOH matrix in the passive region, but the threshold potential for exchange of $Cr_2O_7^=$ ions with FeOOH is so high that O_2 is evolved on the conducting, Cr-contaminated, iron oxide film[22] before exchange can take place.

The cyclic voltammograms obtained on steel in $Na_2Cr_2O_7$ solution are quite simple as noted in Figure 8. Figure 8(a) contains the voltammetry as a function of the sweep rate. The surface is passive over the entire potential range swept (as noted by the very low current densities in Figure 8), since the steel surface is covered with a conducting oxide film. Above about 800 mV, oxygen is adsorbed on the electrode surface and the system goes into O_2 evolution above 1500 mV on the anodic-going trace. When the potential sweep is reversed, the adsorbed oxygen is reduced below 600 mV; below 100 mV

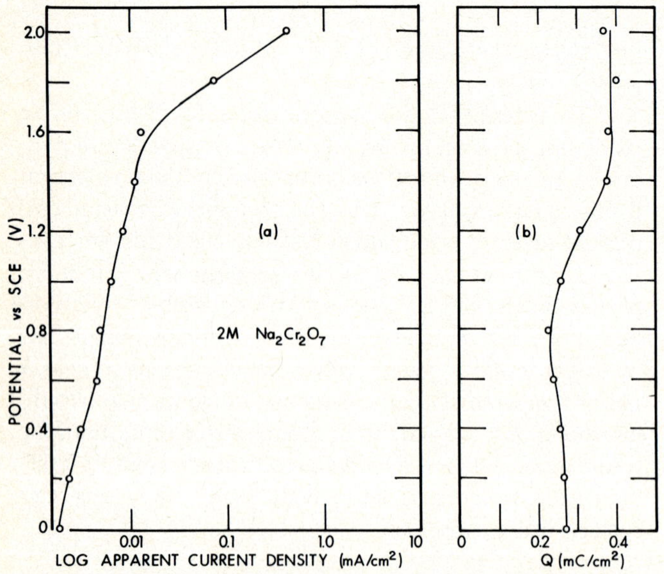

FIGURE 7 — (a) Steady-state polarization curve and (b) charge vs potential obtained on mild steel anodes in O_2-stirred 2 M $Na_2Cr_2O_7$ solution.

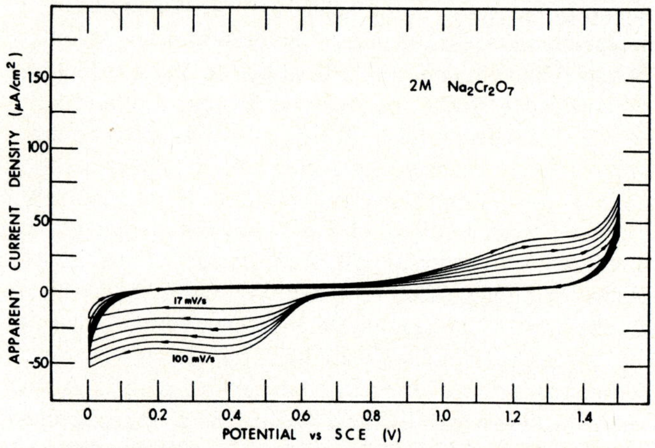

FIGURE 8(a) — Family of cyclic voltammograms as a function of the sweep rate (17, 33, 50, 67, 83, 100 mV/s) obtained on mild steel in N_2-stirred 2 M $Na_2Cr_2O_7$, scanned between 0 and 1500 mV; arrows indicate direction of potential sweep.

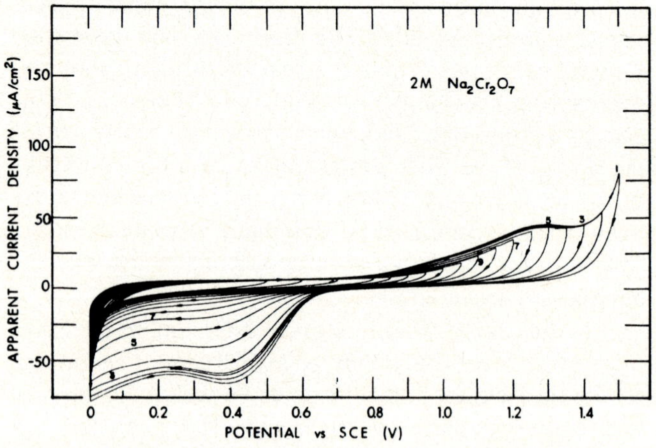

FIGURE 8(b) — Family of cyclic voltammograms obtained by decreasing the anodic limit to 50 mV steps on mild steel in N_2-stirred 2 M $Na_2Cr_2O_7$; sweep rate is 100 mV/s; numerals indicate cycle number.

(Figure 8 continued)

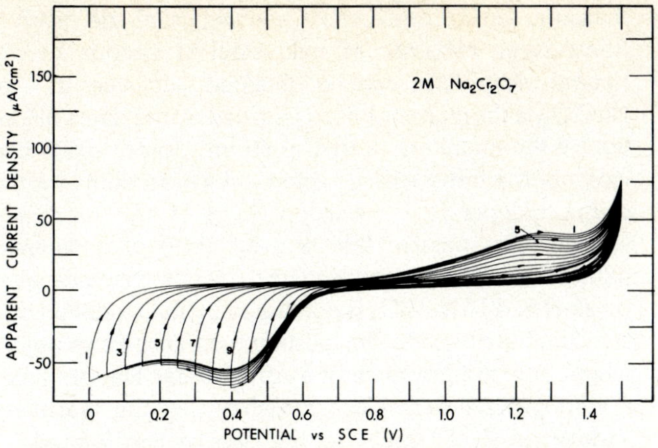

FIGURE 8(c) – Family of cyclic voltammograms obtained by increasing the cathodic limit in 50 mV steps on mild steel in N_2-stirred 2 M $Na_2Cr_2O_7$; sweep rate is 100 mV/s.

the system goes to H_2 evolution. The linear dependence of the oxygen adsorption and desorption peaks indicate that these processes are surface reactions.

The data in Figure 8(a) show that as the anodic limit is lowered, less oxygen is adsorbed above 800 mV, to be desorbed below 600 mV. If the oxygen adsorbed above 800 mV is not desorbed because the cathodic limit is made more anodic, more oxygen cannot be adsorbed on the return trace, as noted in Figure 8(c). Obviously, $Na_2Cr_2O_7$ is not a good electrolyte for ECM.

When mild steel is polarized in 4 M NaCl solutions, the system is active over the entire range of potentials investigated as noted by the polarization curve in Figure 9(a). Oxide films could not be detected as noted in Figure 9(b). The current efficiency for metal removal is nearly 100% over the potential range studied from mass balance data.[11] In the presence of the Cl^- ion, the FeOOH matrix cannot be laid down on the steel surface and only salt films are present.

The dependence of the cyclic voltammogram obtained on steel in 4 M NaCl on sweep rate is complex [Figure 10(a)], indicating that the salt films are not adsorbed on the metal surface to generate oxide films but merely impede the dissolved iron ions in their diffusion to the bulk solution. On the anodic-going trace, the metal removal rate increased between −400 and 0 mV. Above 0 mV, the concentration of ferrous chloride becomes high enough to precipitate a salt layer in the metal/solution interface causing the current to fall rapidly. This salt film is not protective since the current remains high over the entire potential range scanned. The data in Figures 10(b) and (c) emphasize the fact that salt films are present on steel surfaces anodized in NaCl. Here, Cl^- ions complex with the Fe^{+2} ions before a FeOOH matrix can be formed.

Since the metal dissolution rate is high at all anodic potentials, metal will be removed on sides of a plunge-cut during ECM of steel in NaCl. This behavior results in significant tapering of cuts with a serious loss of control of

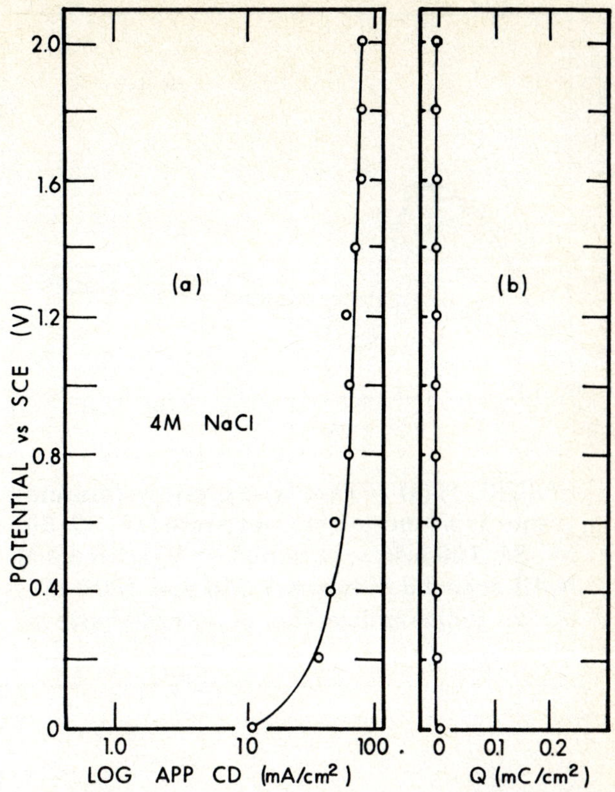

FIGURE 9 – (a) Steady-state polarization curve and (b) charge vs potential obtained on mild steel anodes in O_2-stirred 4 M NaCl solution.

geometry and dimensions. The NaCl electrolyte is a good electrolyte for rapid removal of metal such as deburring but is not good for precise work requiring close tolerances.

Conclusions

The best *in situ* method for an overall study of the electrochemical processes taking place at the metal/solution interface is voltammetry. Although salt films cannot be detected since electrical contact is not made with the electrode surface, secondary effects on the potential (dependence on flow rate, oxidation to an oxide film) can be observed.

A good ECM electrolyte is one that produces a protective anodic film on the given metal to be machined. In addition, this film breaks down in the transpassive region, probably by an ion exchange mechanism. The sharper the transition between the passive and transpassive region, the better will be the control of geometry during the machining by ECM of a given metal in a given electrolyte; this is because the sides of the cut are protected from metal removal by the passive film. The passive film may be generated either by direct oxidation of the metal surface or by the oxidation of a salt film precipitated in the metal/solution interface.

From the results of the three *in situ* voltammetry techniques (steady-state, potentiostatic, polarization studies; constant-current, stripping determinations; and cyclic voltammetric tracings), together with mass balance

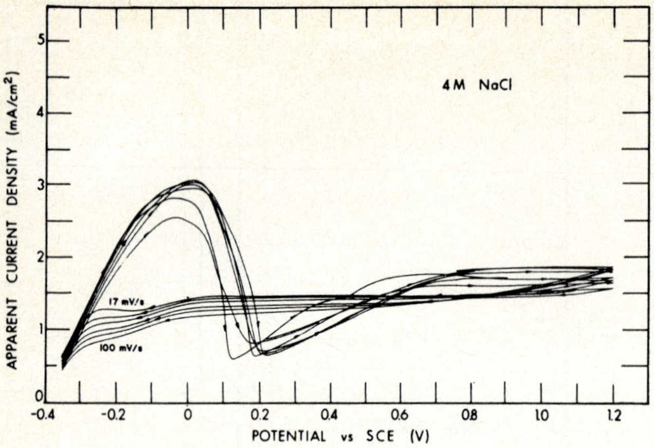

FIGURE 10(a) – Family of cyclic voltammograms as a function of sweep rate (17, 33, 50, 67, 83, 100 mV/s) obtained on N_2-stirred 4 M NaCl scanned between −350 and 1200 mV; arrows indicate direction of potential sweep.

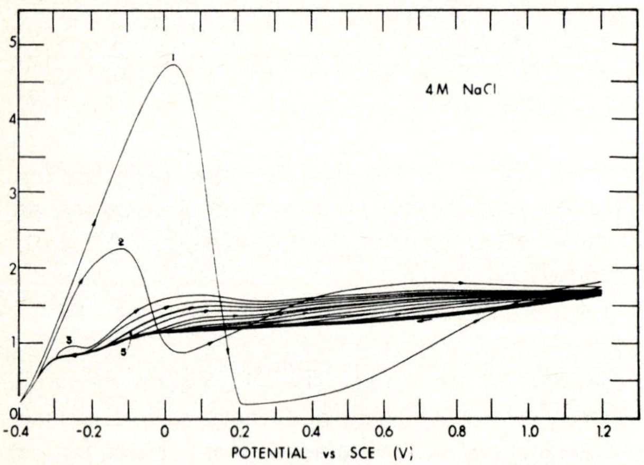

FIGURE 10(b) – Family of cyclic voltammograms obtained by decreasing the anodic limit in 50 mV steps on mild steel in N_2-stirred 4 M NaCl; sweep rate is 100 mVs; numerals indicate cycle number.

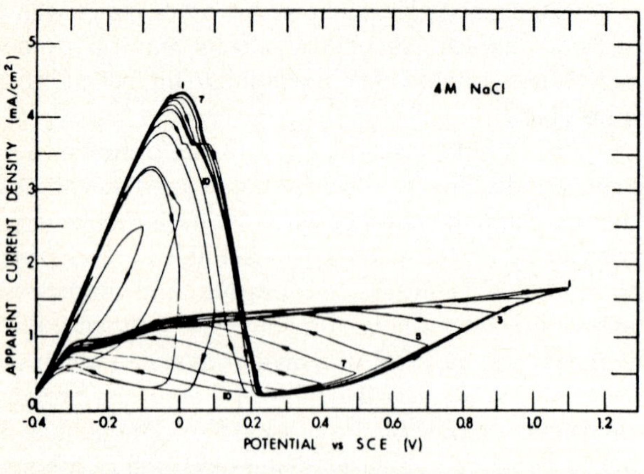

FIGURE 10(c) – Family of cyclic voltammograms obtained by increasing the cathodic limit in 50 mV steps on mild steel in N_2-stirred 4 M NaCl; sweep rate is 100 mV/s.

measurements, considerable knowledge of the passive-transpassive behavior of mild steel in various strong, neutral electrolytes can be obtained. On steel in 4 M $NaClO_3$, a thick passive film is formed by the direct oxidation of the metal and a thin, uniform, porous oxide film exists in the transpassive region, which accounts for the stirring-independent electropolishing of the machined surface. The passive film is most likely of a different nature than the transpassive film. The passive oxide layer on steel in 4 M $NaClO_4$ is generated by the oxidation of a salt film, but no oxide film exists in the transpassive region where only the presence of a compact salt film produces a stirring-dependent electropolishing of the machined surface. In 2 M $Na_2Cr_2O_7$, the passive film formed on steel anodes is so tenacious that in the transpassive region, O_2 is evolved on the passive film before a potential can be reached where the film will dissolve. Machining of steel by ECM cannot take place in $Na_2Cr_2O_7$ electrolytes. Ferrous ions complex with Cl^- ion so rapidly that a passive film cannot be deposited on the steel anode surface. Since the salt film formed is nonprotective, the steel is active over the entire potential range studied. Consequently, during the machining of steel by ECM in 4 M NaCl, the sides of the cut are not protected and significant metal removal takes place resulting in serious loss of geometry.

The results of *ex situ* spectroscopy analyses give only a hint of the species present in the metal/solution interface and should be used with extreme caution in a description of the anodic processes taking place.

References

1. J. P. Hoare, M. A. LaBoda, in Comprehensive Treatise of Electrochemistry, Vol. 2, p. 399, 1981.
2. J. P. Hoare, M. A. LaBoda, in Techniques of Electrochemistry, Vol. 3, p. 48, 1978.
3. J. P. Hoare, Nature, Vol. 219, p. 1034, 1968.
4. K. Chikamori, S. Ito, Danki Kagaku Oyobi Butsuri Kagaku, Vol. 37, p. 602, 1969.
5. D. Landalt, R. M. Muller, C. W. Tobias, J. Electrochem. Soc., Vol. 118, p. 36, 1971.
6. A. D. Davydov, R. A. Mirzoev, V. D. Kashchuv, B. N. Kabanov, Elektrokimya, Vol. 8, p. 1468, 1972.
7. J. P. Hoare, J. Electrochem. Soc., Vol. 117, p. 142, 1970.
8. J. P. Hoare, K-W. Mao, A. J. Wallace, Corrosion, Vol. 27, No. 5, p. 211, 1971.
9. K-W. Mao, D-T. Chin, J. Electrochem. Soc., Vol. 121, p. 191, 1974.
10. J. P. Hoare, C. R. Wiese, Corros. Sci., Vol. 15, p. 435, 1975.
11. K-W. Mao, J. Electrochem. Soc., Vol. 118, p. 1876, 1971.
12. D-T. Chin, J. Electrochem. Soc., Vol. 119, p. 1043, 1972.
13. J. P. Hoare, M. A. LaBoda, "Effects of Solution Flow Rate on the Stability of Anodic Films on Steel," Electrochemical Techniques for Corrosion, R. Baboian, Ed., National Association of Corrosion

Engineers, Houston, Texas, p. 106, 1977.
14. T. P. Hoar, J. A. S. Mowat, Nature, Vol. 165, p. 64, 1950.
15. T. P. Hoar, T. W. Farthing, Nature, Vol. 169, p. 324, 1952.
16. T. Moeller, Inorganic Chemistry, p. 443, John Wiley & Sons, Inc., New York, New York, 1952.
17. T. P. Hoar, private communication.
18. K-W. Mao, M. A. LaBoda, J. P. Hoare, J. Electrochem. Soc., Vol. 119, p. 419, 1972.
19. J. P. Hoare, K-W. Mao, J. Electrochem. Soc., Vol. 120, p. 1452, 1973.
20. J. P. Hoare, M. A. LaBoda, M. L. McMillan, A. J. Wallace, J. Electrochem. Soc., Vol. 116, p. 199, 1969.
21. M. A. LaBoda, J. P. Hoare, in Fundamentals of Electrochemical Machining, C. L. Foust, Ed., Electrochemical Society, Pennington, New Jersey, p. 227, 1971.
22. T. P. Hoar, D. C. Mears, G. P. Rothwell, Corros. Sci., Vol. 5, p. 297, 1965.

The Error in Polarization Resistance and Capacitance Measurements Resulting from Nonuniform Ohmic Potential Drop to Flush-Mounted Anodes

KEMAL NISANCIOGLU

Abstract

Systematic errors encountered in polarization resistance measurements resulting from the nonuniform ohmic-potential drop to flush-mounted electrodes is investigated. The error is introduced even though the ohmic-potential drop is taken into account by use of the interrupter technique or by measurement of AC impedance at high frequencies. The error is significant when the ohmic-potential drop is large, as, for example, in the presence of high corrosion rates and in solutions of low conductivity. Correction factors are calculated for a disk electrode immersed in a bulk electrolyte or covered by a thin film of conducting solution.

Introduction

The two most common methods used today in measuring corrosion rates by electrochemical means are linear polarization and AC impedance. Both methods are based on the estimation of the polarization resistance R_p, which is inversely proportional to the corrosion rate. The AC impedance technique has received considerable interest lately because it has several advantages over the linear polarization method.[1] An important advantage is the fact that the AC impedance technique allows a direct determination of ohmic resistance in the solution. Therefore, the method appears to be suitable for measuring high corrosion rates as well as monitoring corrosion in solutions of low conductivity.

For certain electrode designs, however, the correction of experimental data for the ohmic resistance may not be straightforward. Flush-mounted linear polarization and AC impedance probes are often designed so that the working electrode and the surrounding insulating material lie in the same plane.[2] Such designs may create a source of error in R_p measurements, originating from a nonuniform potential drop between the electrolyte and surface of the electrode. This is due to an edge effect that distorts the primary current distribution,[3] which, in turn, determines the magnitude of ohmic potential drop as it is measured by current interruption or AC impedance.[4]

This type of error is well-known in connection with the measurement of kinetic parameters of electrochemical reactions in electroanalytical chemistry.[5] In many practical applications of the polarization resistance concept in corrosion monitoring, the error is probably negligible because the corrosion rates are often low and the solutions have a sufficiently high conductivity. However, the problem may become significant in special situations involving corrosion, e.g., in low conductivity media,[6] in crevices, and under thin electrolyte films.[7]

This work is intended to provide a rough estimate of the magnitude of such error as a function of electrode dimensions, solution conductivity, the corrosion rate, and the position of reference electrode in the solution. The case of an electrode covered by a thin film of electrolyte is also investigated. Much of the theoretical work and numerical analysis were performed earlier for different purposes.[5,8-13] Although a complete theoretical treatment is given, many of the numerical results are presently limited to an application of earlier calculations for the specific problems at hand.

Mathematical Model

Basic Assumptions

A. The working electrode is a smooth disk em-

*Division of Metallurgy and Corrosion Center, SINTEF, Dept. of Industrial Electrochemistry, Norwegian Institute of Technology, Trondheim, Norway.

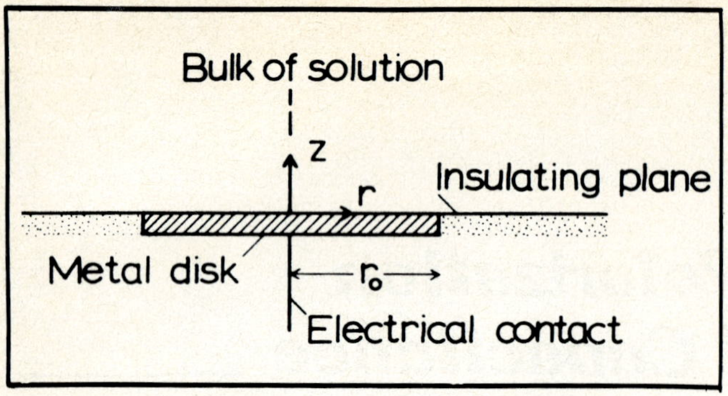

FIGURE 1 – Disk electrode.

bedded in an insulating plane as sketched in Figure 1.

B. The effect of concentration gradients in the electrolytic solution and the effect of adsorption and desorption in the double layer are neglected. The passage of charge across metal/solution interface is therefore due purely to charging of the double-layer capacity and faradaic reactions.

C. The electrode surface is homogeneous and active.

D. The potential dependence of anodic and cathodic reaction rates can be linearized in the vicinity of the corrosion potential.

Although the last three assumptions are not always easy to justify, they are intrinsic to the concept of employing polarization resistance as a quantitative means of estimating corrosion rates. Concentration gradients and adsorption effects are often present, and as a result, the capacitive and faradaic effects may not be separable *a priori* in a manner suggested by assumption B.[11] Moreover, a corroding surface can never be regarded as entirely homogeneous and active. Research seems to be underway to improve the available models for electrochemical impedance,[13-15] so that some of these factors can be treated in a rigorous fashion. However, experimental validation of the polarization resistance method as a quantitative tool for estimating corrosion rates has also been given,[1] thereby indicating that these assumptions are, in fact, fair approximations of surface phenomena in a number of situations characterized by simple corrosion kinetics.

Model for Disk Exposed to a Large Volume of Electrolyte

In light of the foregoing assumptions, the potential in the solution can be expressed by Laplace's equation:

$$\nabla^2\phi = 0 \tag{1}$$

On the insulating surface surrounding the disk, the current density and, therefore, the normal component of the potential gradient is zero:

$$\partial\phi/\partial z = 0$$

at

$$z = 0, r > r_o \tag{2}$$

The potential at infinity is also taken to be zero:

$$\phi \rightarrow 0$$

as

$$r^2 + z^2 \rightarrow \infty \tag{3}$$

Furthermore, the potential along the axis of the disk must be well-behaved.

The net current density at the disk surface is given by a linear combination of faradaic and capacitive contributions:

$$i = i_{cap} + i_f \tag{4}$$

The potential dependence of the faradaic term in the vicinity of corrosion potential E_{corr} can be expressed as:

$$i_f = \left(\frac{\partial i}{\partial E}\right)_{i=0} (E - E_{corr}) \tag{5}$$

The electrode potential E can be written in the form:

$$E = V - \phi_o \tag{6}$$

where V is the potential of the metal and ϕ_o, the potential in the solution adjacent to the metal surface. By definition, the polarization resistance R_p is given by:

$$\pi r_o^2 R_p = 1/(\partial i_f/\partial E)_{i=0} \tag{7}$$

where r_o is the disk radius. The capacitive part of the current density can be expressed in the form:

$$i_{cap} = C\frac{\partial(E - E_{corr})}{\partial t} \tag{8}$$

Thus, the net current density is given by:

$$i = C\frac{\partial(\nu - \phi_o)}{\partial t} + \frac{1}{\pi r_o^2 R_p}(\nu - \phi_o)$$

$$= -\varkappa\frac{\partial\phi}{\partial z}\bigg|_{z=0,\ r \le r_o} \tag{9}$$

where $\nu = V - E_{corr}$, and $\varkappa$ is the conductivity of solution, and C is the double-layer capacitance.

The parameters C and R_p are, respectively, the *true* double-layer capacitance and *true* polarization resistance of the disk electrode. The *measured* quantities C_{eff} and R_{eff} are, however, defined by the equivalent circuit depicted in Figure 2. The AC impedance of this circuit is given by:

$$Z = R_{ohm} + \frac{R_{eff}}{1 + j\omega C_{eff}\pi r_o^2 R_{eff}} \tag{10}$$

where ω is the angular frequency, and $j = \sqrt{-1}$. The im-

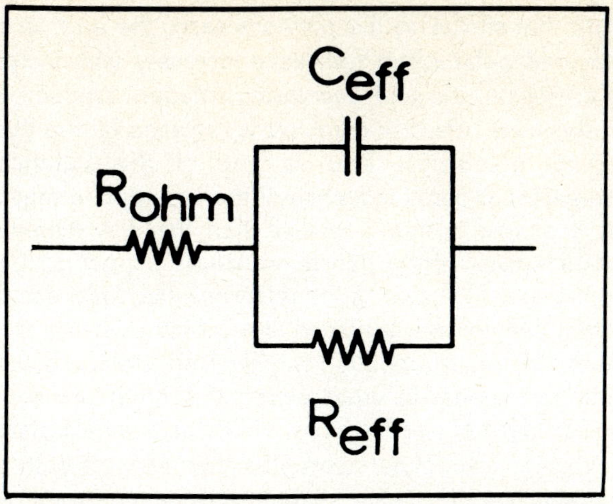

FIGURE 2 – Equivalent circuit for the working electrode. R_{eff} is the effective polarization resistance as measured by linear polarization or AC methods. C_{eff} is the effective double-layer capacitance as measured by an AC technique. R_{ohm} is the ohmic resistance in the solution determined by solution conductivity and the geometry of the working electrode; it can be measured by the interrupter technique or by AC impedance method at high frequency.

pedance of the disk electrode is, by definition:

$$Z = V/I \qquad (11)$$

where the total current I is given by:

$$I = \int_0^{r_o} 2\pi r i dr \qquad (12)$$

The current distribution on the electrode can be obtained by solving Equations (1), (2), (3), and (9). Such solutions are available in the literature,[5,8-11] and these are reproduced in Appendices A and B, for DC and AC situations, respectively. By comparing the impedance calculated from these solutions and that obtained from Equation (10), the error in the measured C_{eff} and R_{eff} relative to the true parameters C and R_p can be estimated.

Model for Disk Coated with a Thin Film of Electrolyte

The geometrical model for this case is shown in Figure 3. If h is small compared to the disk radius, then the current distribution across the film thickness is nearly uniform, and the current density in the radial direction can be expressed as:

$$i_r = -\varkappa \frac{\partial \phi}{\partial r} \qquad (13)$$

The conservation of charge criterion within the film requires for any r that:

$$2\pi r h i_r = 2\pi \int_0^r i r dr \qquad (14)$$

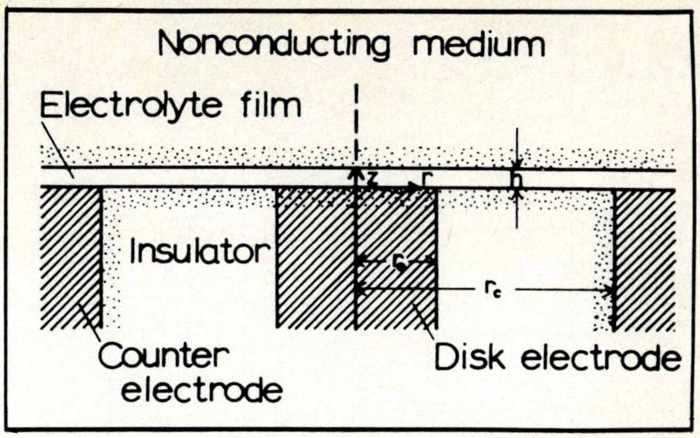

FIGURE 3 – Disk electrode wetted by a thin layer of electrolytic solution.

where i is the current density on the disk. After differentiating and combining the two equations, we obtain:

$$\frac{1}{r}\frac{\partial}{\partial r}\left(r\frac{\partial \phi}{\partial r}\right) = -\frac{i}{h\varkappa} \qquad (15)$$

For linearized reaction kinetics at the disk surface, the current density is given by Equation (9). Moreover, the current density is zero on the insulating surface around the disk. Thus, we can write:

$$i = \begin{cases} C\dfrac{\partial(\nu-\phi)}{\partial t} + \dfrac{1}{\pi r_o^2 R_p}(\nu-\phi) & (0 \le r \le r_o) \\ 0 & (r > r_o) \end{cases} \qquad (16)$$

The potential is well-behaved at the center of disk. Therefore, we have to satisfy the condition:

$$\frac{\partial \phi}{\partial r} = 0 \text{ at } r = 0 \qquad (17)$$

It is assumed that the potential in the film can be measured at a convenient location; specifically:

$$\phi = \phi_c \text{ at } r = r_c \qquad (18)$$

where r_c is the radius of the insulating plane around the disk. The steady-state (DC) solution for Equations (15) through (18) is given in Appendix C.

Numerical Results

Error in R_p Measurements

If the ohmic potential drop is large, the measured value of polarization resistance depends on the position of the reference electrode in the solution. The reference electrode is often placed as close to the surface as possible to minimize the ohmic drop. Depending on the design of measuring system, the reference electrode may be positioned close to the center or edge of the working elec-

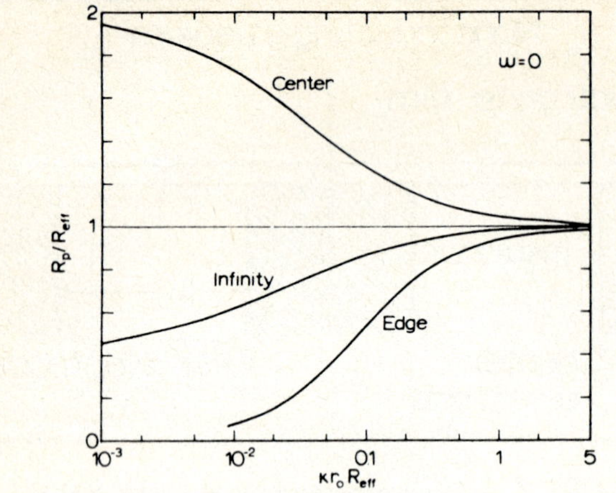

FIGURE 4 — The calculated error in polarization resistance measured by linear polarization or AC impedance on a disk electrode.

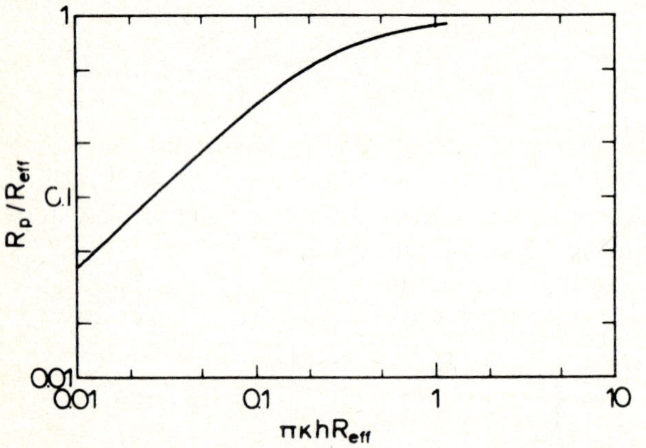

FIGURE 5 — The calculated error in polarization resistance measured by linear polarization or AC impedance on a disk coated by a thin film of electrolyte.

trode. The latter position is sometimes found more practical in the design of flush-mounted electrode systems employed in corrosion monitoring.[2] If the reference electrode cannot be positioned sufficiently close to the surface, the ohmic potential drop is measured by an appropriate technique, such as the interrupter method or the AC impedance method employed at a high frequency.

The calculated error (Appendix A) in R_p measurements for the three typical positions (center, edge, and far away) of the reference electrode with respect to a disk electrode is depicted in Figure 4. These results indicate that the measured polarization resistance, although corrected for the ohmic resistance, is expected to deviate from the true value with increasing corrosion rate, decreasing solution conductivity, and decreasing disk diameter. The discrepancy is largest for the two cases where the reference electrode is adjacent to the disk surface, especially for the case near the edge.

Analogous results for a disk electrode covered by a thin film of electrolyte are shown in Figure 5. These indicate that similar to the previous case, the error in the measured polarization resistance increases with increasing corrosion rate and decreasing solution conductivity. For this case, the disk diameter is replaced by the thickness of electrolyte film as one of the significant parameters as a consequence of the assumption that the film thickness is much smaller than the disk diameter. Another result of the thin film analysis (Appendix C) is that the exact location of the reference electrode is unimportant as long as it is placed at a position where $r > r_o$ and the ohmic contribution is taken into account by conventional means. In other words, the boundary condition 18 does not need to be as specific as originally stated. For thicker electrolyte layers, the diameters of both the disk and the insulating plane become important. An analysis of this more general case is given elsewhere.[12]

Frequency Dispersion of Capacitance and Polarization Resistance

The analysis of the nonsteady-state problem for an externally imposed AC current (Appendix B) suggests that the numerical results can conveniently be represented in terms of a dimensionless corrosion current density J and a dimensionless frequency Ω, defined by[9]:

$$J = 1/\pi r_o \kappa R_p \qquad (19)$$

$$\Omega = \omega C r_o/\kappa \qquad (20)$$

The results are expressed in Figures 6 and 7 in terms of the ratios C/C_{eff} and R_p/R_{eff}, respectively, as a function of dimensionless frequency for various J values. The reference electrode is assumed to be positioned far from the disk. The measured values R_{eff} and C_{eff} diverge from the true values R_p and C with increasing frequency and corrosion rate. Asymptotic expressions can be derived for these ratios at high frequencies.[9] In particular, C/C_{eff} was shown to vary linearly with ln Ω, and this relationship can also be visualized in Figure 7. It may also be noted that R_p/R_{eff} and C/C_{eff} attain asymptotic values as the frequency approaches zero. As expected, the asymptotic values for R_p/R_{eff} are identical to the results presented in Figure 4. A cross-plot of the frequency-independent C/C_{eff} values is shown in Figure 8 to reiterate the dependence of this ratio on the solution conductivity, electrode size, and corrosion rate.

The results of Figures 6 and 7 can also be presented in the form of Nyquist diagrams. The effective (measured) impedance of the disk for given C and R_p may be compared to the impedance of the circuit depicted in Figure 2, in which the capacitor and resistor are assumed to take, respectively, the true values $\pi r_o^2 C$ and R_p. This type of representation is shown in Figure 9 for different J values. With increasing J, an enlargement of the "semicircles" for the disk relative to those for the idealized equivalent circuit is visualized. Moreover, the data for the disk are slightly distorted toward higher frequencies[16]; they are distorted in such a way that these points cannot be fitted exactly to semicircles as is sometimes done in quantitative evaluation of impedance data.[18]

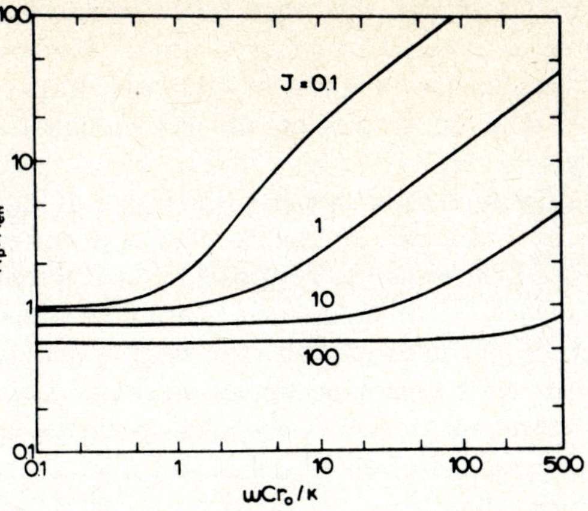

FIGURE 6 – Deviation of the measured polarization resistance from the true value with increasing frequency of applied AC current at various J values. The reference electrode is positioned far from the disk. Note that the zero frequency limits give the corresponding curve on Figure 4.

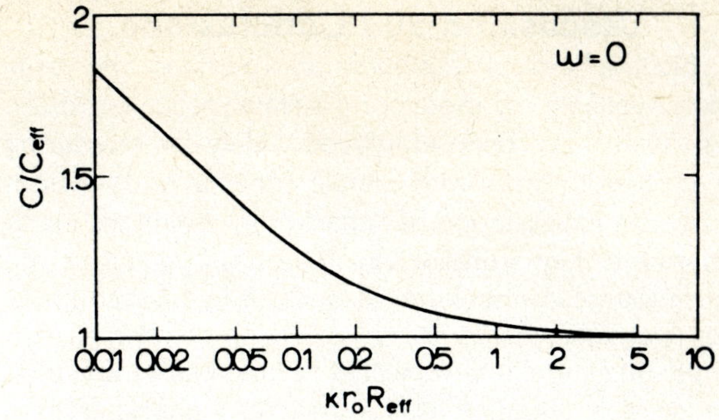

FIGURE 8 – Zero frequency limit of measured parallel capacitance as a function of measured parallel resistance (polarization resistance). The reference electrode is positioned far from the disk.

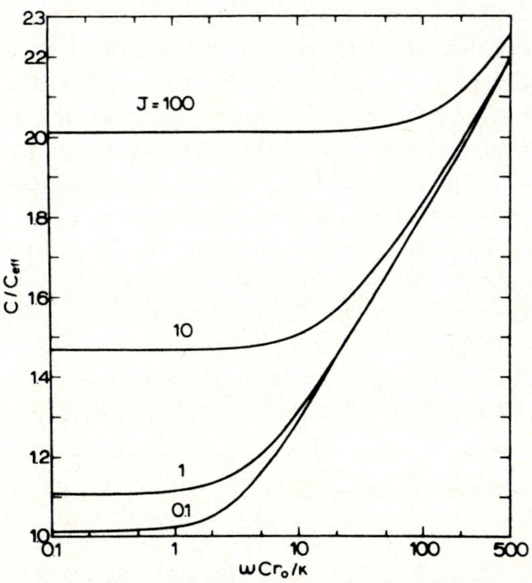

FIGURE 7 – Deviation of the measured capacitance from true double-layer capacity with increasing frequency at various J values. The reference electrode is positioned far from the disk.

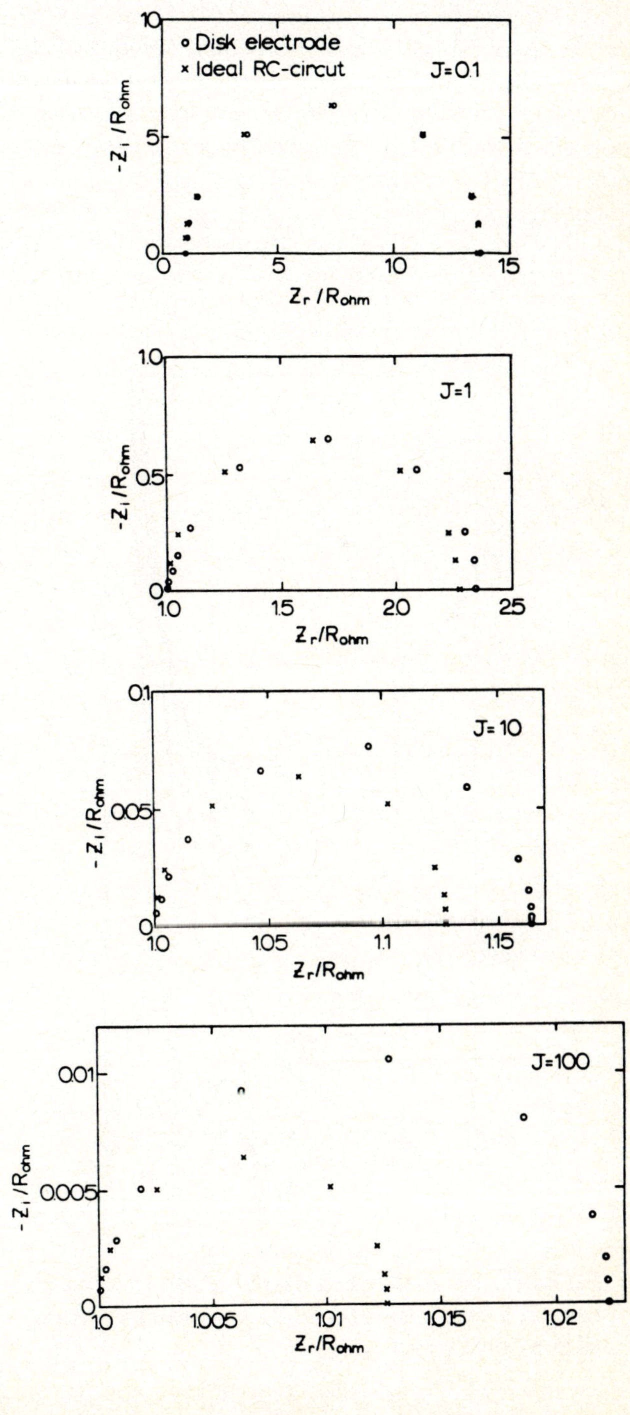

FIGURE 9 – Impedance of a disk electrode represented in the form of complex-plane Nyquist plots for different J values. The reference electrode is assumed to be positioned far from the disk. Impedance plots of an ideal equivalent circuit, given by a parallel combination of R_p and $\pi r_o^2 C$, are also shown. ➡

Discussion and Conclusions

Electrochemical methods are perhaps the only means available for monitoring instantaneous corrosion rates. They have been employed widely for measuring low corrosion rates with simple kinetics and without significant localization of attack, in solutions with moderate to high conductivity. The need to extend the applicability of electrochemical methods to more complex cases invariably raises the question as to how experimental data should be subjected to quantitative analysis. One possible situation was investigated arising from a combination of factors such as electrode geometry, low solution conductivity, and high corrosion rates, and causing a nonuniform ohmic potential drop that is not easy to correct by conventional means.

The extrapolation of high-frequency data by curve fitting and integration methods has been shown to be a useful approach in obtaining the polarization resistance from AC impedance measurements.[18] Normally, the double-layer capacitance is also obtained from high frequency data.[19] Such methods of data analysis have to be employed with care in the presence of frequency dispersion effects of the type described presently. The factors that render R_{eff} different than R_p can be compounded if the polarization resistance is estimated by means of fitting the high-frequency data to a depressed semicircle. The seriousness of the problem lies in the fact that the measured corrosion rate often becomes lower than the true rate as a result of incurred errors. Moreover, the capacitance cannot be determined reliably from high frequency data by employing the conventional analysis methods.[19]

Under such circumstances, the C_{eff} and R_{eff} values estimated from low frequency data and corrected by means of plots similar to Figures 4, 5, and 8 may prove more reliable if any frequency-independent regions for C_{eff} and R_{eff} can be located. The safest way of avoiding such errors is to design electrodes with a uniform primary current distribution. A few possibilities along this line have been discussed elsewhere.[5] If flush-mounted probes have to be used, the likelihood of an error can be estimated from Figures 4 and 5 and checked experimentally by varying the position of reference electrode in the solution. If, in addition, the ohmic drop can be measured properly, the error from positioning the reference electrode far from the working electrode may, in fact, be smaller than from placing it adjacent to the surface. In practice, the spacing between the two electrodes does not need to be more than a few working-electrode diameters.

The AC-impedance method is a relatively new tool in corrosion monitoring, and not all aspects of the theory behind the method are yet resolved. However, the real power of this method in comparison to DC techniques probably lies in the possibility of obtaining valuable information about the morphology of the corroding surface, as well as the corrosion mechanism, by utilizing the observed frequency dispersion effects. Interpretation of deviations from ideal RC-circuit behavior requires mathematical modeling. Such models are often based on equivalent circuits somewhat more complicated than the ideal case depicted in Figure 2. However, cases exist where circuits composed of a limited number of resistors and capacitors may not be sufficient for an adequate description of a corroding metal/solution interface. Therefore, a more rigorous mathematical modeling, starting from the fundamental principles, is required alongside the equivalent circuit representations.

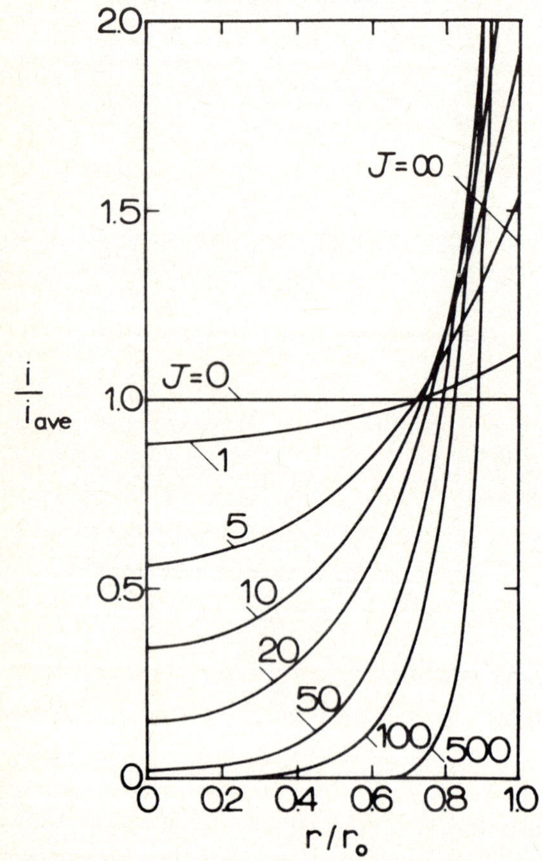

FIGURE 10 — Current distribution for linear kinetics on a disk electrode covered by a thin film of electrolyte.

Appendix A—Calculation of R_p/R_{eff} for a Disk Electrode

In rotational elliptic coordinates, Laplace's equation can be expressed as:[17]

$$\frac{\partial}{\partial \xi}\left[(1+\xi^2)\frac{\partial \phi}{\partial \xi}\right] + \frac{\partial}{\partial \eta}\left[(1-\eta^2)\frac{\partial \phi}{\partial \eta}\right] = 0 \qquad \text{(A.1)}$$

where:

$$z = r_o \xi \eta$$

$$r = r_o[(1+\xi^2)(1-\eta^2)]^{1/2} \qquad \text{(A.2)}$$

The boundary conditions 2 and 3 can be written in the form:

$$\partial\phi/\partial\eta = 0 \text{ at } \eta = 0 \text{ (insulating plane)}$$

$$\phi = 0 \text{ as } \xi \to 0 \text{ (far from disk)} \quad (A.3)$$

In addition, the condition that ϕ is well-behaved must be satisfied at $\eta = 1$ (axis of disk). The surface condition 9 can be expressed at steady-state as:

$$i_f r_o/\varkappa = -\frac{1}{\eta}\frac{\partial\phi}{\partial\xi}\Big|_{\xi=0} = J(\nu - \phi_o) \quad (A.4)$$

where:

$$J = 1/\pi r_o \varkappa R_p \quad (A.5)$$

is the dimensionless corrosion-current density.

The solution of Equations (A.1) and (A.3) yields the following expression for the potential in the electrolyte:

$$4r_o\varkappa\phi/I = \sum_{n=0}^{\infty} B_n P_{2n}(\eta) M_{2n}(\xi) \quad (A.6)$$

where $P_{2n}(\eta)$ is the Legendre polynomial of order 2n and $M_{2n}(\xi)$, a Legendre function with known properties. Combining Equations (A.4) and (A.6) and employing the orthogonality properties of Legendre polynomials give[11]:

$$4r_o\varkappa\nu/I =$$

$$1 + 4/\pi J + 2\sum_{n=0}^{\infty} a_{o,n}B_n \quad (A.7)$$

and

$$\sum_{n=1}^{\infty}\left[a_{m,n} - 2a_{o,m}a_{o,n} - \delta_{m,n}\frac{M'_{2m}(0)}{4m+1}\right]B_n =$$

$$\frac{4}{\pi J}a_{o,m} \quad (m = 0, 1, 2 \ldots) \quad (A.8)$$

where:

$$a_{m,n} = \int_0^1 \eta P_{2n}(\eta) P_{2m}(\eta)\, d\eta \quad (A.9)$$

$$a_{o,m} = \int_0^1 \eta P_{2m}(\eta)\, d\eta =$$

$$-\frac{P_{2m}(0)}{2(2m-1)(m+1)} \quad (A.10)$$

$$\delta_{m,n} = \begin{cases} 0 \text{ if } m \neq n \\ 1 \text{ if } m = n \end{cases} \quad (A.11)$$

$$M'_{2m}(0) = -\frac{2}{\pi}\frac{(2^m m!)^4}{[(2m)!]^2} =$$

$$M'_{2(m-1)}(0)\left(\frac{2m}{2m-1}\right)^2 \quad (A.12)$$

Equation (A.6) is normalized such that $B_o = 1$. The coefficients for $n > 0$ can be calculated by solving the set of linear simultaneous equations defined by Equation (A.8).

The effective polarization resistance of the disk electrode, as measured by the linear polarization method or the AC impedance technique, is given by:

$$R^*_{eff} = \nu/I \quad (A.13)$$

R^*_{eff} contains both the ohmic and faradaic contributions:

$$R^*_{eff} = R_{ohm} + R_{eff} \quad (A.14)$$

If the reference electrode is placed far from the disk, the ohmic resistance is given by[17]:

$$R_{ohm} = 1/4r_o\varkappa$$

After correcting for the ohmic resistance, the experimentally measured polarization resistance R_{eff} can be derived from Equation (A.7) to read

$$4r_o\varkappa R_{eff} = \frac{4}{\pi J} + 2\sum_{n=1}^{\infty} a_{o,n}B_n \quad (A.16)$$

This can also be expressed in the form:

$$\frac{R_p}{R_{eff}} = 1 - \frac{1}{4r_o\varkappa R_{eff}}\sum_{n=1}^{\infty} a_{o,n}B_n \quad (A.17)$$

Equation (A.16) indicates that the ratio of true polarization resistance to measured polarization resistance deviates from unity with increasing rate of corrosion, low solution conductivity, and small disk diameter.

Similarly, if the reference electrode is placed at the center of the disk, the measured polarization resistance can be expressed as:

$$4r_o\varkappa R_{eff} = \frac{\nu - \phi_o(1)}{I/4r_o\varkappa}$$

$$= \frac{4}{\pi J} + \sum_{n=1}^{\infty}(2a_{o,n} - 1)B_n$$

(center of disk) (A.18)

For the case where the reference electrode is placed near the edge of disk, Equation (A.19) can similarly be derived:

$$4r_o\varkappa R_{eff} = \frac{\nu - \phi_o(0)}{I/4r_o\varkappa}$$

$$= \frac{4}{\pi J} + \sum_{n=1}^{\infty}[2a_{o,n} - P_{2n}(0)]B_n$$

(edge of disk) (A.19)

Equations (A.15), (A.17), and (A.18) are the inverted forms of, respectively, Equations 12, 14, and 15 in Reference 5.

Appendix B—Frequency Dispersion of R_p/R_{eff} and C/C_{eff} for a Disk Electrode

For an alternating current, the potentials of the disk and bulk of the solution can be expressed, respectively, as[9]:

$$\nu = V_o e^{j\omega t} \tag{B.1}$$

$$\phi = V_o e^{j\omega t} U(r,z) \tag{B.2}$$

where V_o is the amplitude and ω (= $2\pi f$) is the angular frequency of the applied potential, and U (r,z) is a time-independent function. Substitution of Equation (B.2) into Equations (A.1) through (A.3) and solution of the resulting expressions give:

$$U = \sum_{n=0}^{\infty} B_n P_{2n}(\eta) M_{2n}(\xi) \tag{B.3}$$

where symbols have the same meaning as in Appendix A (see also Nomenclature, Appendix D) with the exception of the coefficients B_n, which are complex numbers. Substitution into Equation (9) and invoking the orthogonality properties of Legendre polynomials yield the set of linear equations:

$$\sum_{n=0}^{\infty} \left[a_{m,n} - \frac{\delta_{m,n}}{j\Omega + J} \frac{M'_{2m}(0)}{4m+1} \right] B_n = a_{o,m}$$

$$(m = 0, 1, 2, ...) \tag{B.4}$$

where J is the dimensionless corrosion current density [Equation (A.5)], and

$$\Omega = \omega C r_o / \varkappa \tag{B.5}$$

is the dimensionless frequency. Equation (B.4) can be solved readily by conventional matrix inversion techniques to obtain the coefficients B_n.

The total current to the disk is given by:

$$I = \int_0^{r_o} 2\pi r i dr$$

$$= 2\pi r_o^2 \varkappa \int_0^1 \eta \left. \frac{\partial \phi}{\partial z} \right|_{z=0} d\eta$$

$$= -2\pi r_o \varkappa V_o e^{j\omega t} B_o M'_o(0)$$

$$= 4 r_o \varkappa V_o e^{j\omega t} B_o \tag{B.6}$$

The impedance is given by:

$$Z = \frac{\nu}{I} = \frac{1}{4 r_o \varkappa B_o} \tag{B.7}$$

Hence, the first coefficient in the series expression (B.3) is of practical importance; however, the entire set of Equations (B.4) has to solved to obtain it. After combining (B.7) with Equation (10), separating the real and imaginary parts and rearranging, we obtain:

$$\frac{R_p}{R_{eff}} = \frac{4}{\pi J} \left[\frac{B_{or} - (B_{or}^2 + B_{oi}^2)}{(1 - B_{or})^2 + B_{oi}^2} \right] \tag{B.8}$$

$$\frac{C}{C_{eff}} = \frac{\pi \Omega}{4} \left[\frac{(1 - B_{or})^2 + B_{oi}^2}{B_{oi}} \right] \tag{B.9}$$

Appendix C—Steady-State Current Distribution on a Disk Coated With a Thin Film of Electrolyte Linear Kinetics

By combining Equations (15) and (16) (at steady-state), integrating and applying condition (17), we obtain:

$$\frac{\nu - \phi}{\Theta - \phi_c} = C_1 I_o \left(\sqrt{J} \frac{r}{r_o}\right) \quad (0 \leq r \leq r_o) \tag{C.1}$$

where:

$$J = 1/\pi h \varkappa R_p \tag{C.2}$$

is the dimensionless corrosion current density for the present system. C_1 is a constant of integration, which has yet to be determined. Before proceeding further with the analysis, calculating the current distribution on the disk surface may prove useful.

Thus, a combination of Equation (17), (C.1), and

$$i_{ave} = \frac{2\pi}{\pi r_o^2} \int_0^{r_o} i r dr \tag{C.3}$$

give:

$$\frac{i}{i_{ave}} = \frac{\sqrt{J}}{2} \frac{I_o\left(\sqrt{J}\, r/r_o\right)}{I_1(\sqrt{J})} \quad (0 \leq r \leq r_o) \tag{C.4}$$

A plot of Equation (C.4) for various J values is depicted in Figure 9. The current distribution becomes increasingly nonuniform as the corrosion rate becomes large. At very high corrosion rates (in the limit as $J \rightarrow \infty$), most of the charge transfer occurs at the edge of disk, while the current density drops to nearly zero toward the center. At the same time, the potential distribution in the solution adjacent to the metal becomes increasingly uniform. Equation (C.1), in fact, indicates that:

$$\lim_{J \rightarrow \infty} \phi = \nu \tag{C.5}$$

Therefore, this condition corresponds to the primary distribution for this system. The primary current distribution for a disk coated by a thin film of electrolyte can thus be express as:

$$i = \begin{cases} 0 & \text{for } 0 \le r < r_o \\ \infty & \text{for } r = r_o \end{cases} \qquad \text{(C.6)}$$

Returning to the original problem, we integrate Equation (15) for $r > r_o$ and apply condition (18). This gives:

$$\frac{\nu - \phi}{\nu - \phi_c} = C_2 \ln \frac{r}{r_c} + 1 \; (r > r_o) \qquad \text{(C.7)}$$

The constants C_1 and C_2 are obtained by matching the solutions (C.1) and (C.7) at $r = r_o$, i.e.,

$$C_1 I_o(\sqrt{J}) = C_2 \ln \frac{r_o}{r_c} + 1 \qquad \text{(C.8)}$$

$$C_1 \sqrt{J}\, I_1(\sqrt{J}) = C_2 \qquad \text{(C.9)}$$

Solution of these equations gives:

$$C_1 = \frac{1}{I_o(\sqrt{J}) - \sqrt{J}\, I_1(\sqrt{J}) \ln(r_o/r_c)} \qquad \text{(C.10)}$$

The total current on the disk can be calculated from:

$$I = 2\pi \int_0^{r_o} i r dr$$

$$= 2\pi h \varkappa C_1 (\nu - \phi_c) \sqrt{J}\, I_1(\sqrt{J}) \qquad \text{(C.11)}$$

The resistance of the system is given by:

$$R = R_{eff} + R_{ohm} = \frac{\nu - \phi_c}{I}$$

$$= \frac{1}{2\pi\varkappa h} \left[\frac{I_o(\sqrt{J})}{\sqrt{J}\, I_1(\sqrt{J})} - \ln \frac{r_o}{r_c} \right] \qquad \text{(C.12)}$$

Thus, after subtracting the ohmic resistance, we obtain:

$$\pi \varkappa h R_{eff} = \frac{I_o \sqrt{J}}{2\sqrt{J}\, I_1(\sqrt{J})} \qquad \text{(C.13)}$$

or

$$\frac{R_p}{R_{eff}} = \frac{2 I_1(\sqrt{J})}{\sqrt{J}\, I_o(\sqrt{J})} \qquad \text{(C.14)}$$

Appendix D—Nomenclature

$a_{m,n}$	Equation (A.9)
$a_{o,n}$	Equation (A.10)
b	Tafel slope (V/decade)
B_n	coefficients in the series for the potential
C	double-layer capacity, farad/cm^2
E	electrode potential, V
h	thickness of electrolyte film covering the electrode surface, cm
i	current density, A/cm^2
I	current, A
I_o	modified Bessel function of the first kind, of order zero
I_1	modified Bessel function of the first kind, of order one
j	$\sqrt{-1}$
J	dimensionless corrosion current density [Equation (19)]
$\mathcal{J}$	dimensionless corrosion current density [Equation (C.2)]
M_{2n}	Legendre function of order 2n
P_{2n}	Legendre polynomial of order 2n
R	resistance, ohm
R_p	polarization resistance, ohm
r_o	radius of disk electrode, cm
r	radial coordinate, cm
t	time, s
U	complex potential in the solution, V
V	potential of the metal, V
ν	$V - E_{corr}$
z	axial coordinate, cm
Z	impedance, ohm
η	rotational elliptic coordinate
$\varkappa$	solution conductivity, $ohm^{-1}\, cm^{-1}$
ω	frequency, radians/s
Ω	dimensionless frequency
ϕ	potential in the solution, V
ξ	rotational elliptic coordinate

Subscripts:

ave	average
cap	capacitive
corr	corrosion
eff	effective (experimentally measured)
f	faradaic
i	imaginary
o	value at the electrode surface
ohm	ohmic
r	real
r	radial component

Acknowledgments

This work was supported by Norsk Hydro A/S and Saga Petroleum A/S.

References

1. W. J. Lorenz, F. Mansfeld, Corros. Sci., Vol. 21, p. 647, 1981.
2. Roe Strommen, "Corrosion Monitoring Probe Designs," presented at ICOSM 83, London, 1983.
3. John Newman, "Mass Transport and Potential Distribution in the Geometries of Localized Corrosion," Localized Corrosion, R. W. Staehle, B. F. Brown, J. Kruger, A. Agrawal, Eds., National Association of Corrosion Engineers, Houston, Texas, p. 45, 1974.

4. John Newman, J. Electrochem. Soc., Vol. 117, p. 507, 1970.
5. William H. Tiedemann, John Newman, Douglas N. Bennion, J. Electrochem. Soc., Vol. 120, p. 256, 1973.
6. F. Mansfeld, M. W. Kendig, S. Tsai, Corros. Sci., Vol. 22, p. 455, 1982.
7. C. Fiaud, R. Chahrouri, M. Keddam, G. Maurin, H. Takenouti, "Impedance Measurements in Thin Electrolyte Layers," presented at Eighth Int. Cong. Met. Corros., Mainz, 1981.
8. John Newman, J. Electrochem. Soc., Vol. 113, p. 1235, 1966.
9. John Newman, J. Electrochem. Soc., Vol. 117, p. 198, 1970.
10. Kemal Nisancioglu, John Newman, J. Electrochem. Soc., Vol. 121, p. 523, Appendix II, 1974.
11. Kemal Nisancioglu, Lawrence Berkeley Laboratory Report No. LBL-1880 (Dissertation), University of California, Berkeley, September, 1973.
12. Kemal Nisancioglu, SINTEF Report No. STF34 F79022, Trondheim.
13. B. Tribollet, J. Newman, Abstract 194, presented at the 164th Meeting of the Electrochemical Society, Washington, DC, 1983.
14. A. Caprani, C. Deslouis, M. Keddam, Ph. Morel, B. Tribollet, Abstract 197, presented at the 164th Meeting of the Electrochemical Society, Washington, DC, 1983.
15. C. Deslouis, M. Duprat, M. Keddam, F. Moran, B. Tribollet, Abstract 198, presented at the 164th Meeting of the Electrochemical Society, Washington, DC, 1983.
16. S. H. Glarum, J. Electrochem. Soc., Vol. 124, p. 518, 1977.
17. John Newman, J. Electrochem. Soc., Vol. 113, p. 501, 1966.
18. F. Mansfeld, M. W. Kendig, Werkst. Korros., Vol. 34, p. 397, 1983.
19. Florian Mansfeld, Corrosion, Vol. 36, No. 5, p. 301, 1981.

Section 2

Basic Techniques

The Potentiostatic Technique for Corrosion Studies

*HARVEY P. HACK**

Abstract

One of the most widely used electrochemical corrosion testing methods is the constant potential, or *potentiostatic,* technique. This paper is designed as an introduction to potentiostatic testing. It first presents the rationale behind potentiostatic testing, and then describes the considerations for conducting simple potentiostatic tests and analyzing the data. The distinctions between potentiostatic and other "constant" potential or current test methods are explained next. Finally, a variety of common applications of potentiostatic testing to prediction or evaluation of corrosion processes is presented.

Introduction

The rates of the anodic and cathodic reactions constituting the corrosion process are usually a function of the voltage or "potential" of the corroding material.

This potential can, in turn, be altered by applying current to the material from an external source. This is illustrated in Figure 1 for a simple case of one anodic (corrosion) and one cathodic reaction under ideal charge transfer control conditions. When an external current is applied, the rates of the anodic and cathodic reactions change so that the difference between them equals the externally applied current. This causes the potential to change from the freely corroding potential, where no external current is applied. Weight loss is caused by the anodic reaction of metal oxidation. Thus, information about the anodic and cathodic reactions can be obtained if externally applied current and weight loss are measured at one or more potentials differing from the corrosion potential. This is the basis of the utility of potentiostatic, or constant potential, testing. Additional detail on the electrochemical aspects of corrosion and the relationship between applied current and potential is contained in References 1 through 3.

*David Taylor Naval Ship R&D Center, Bethesda, Maryland.

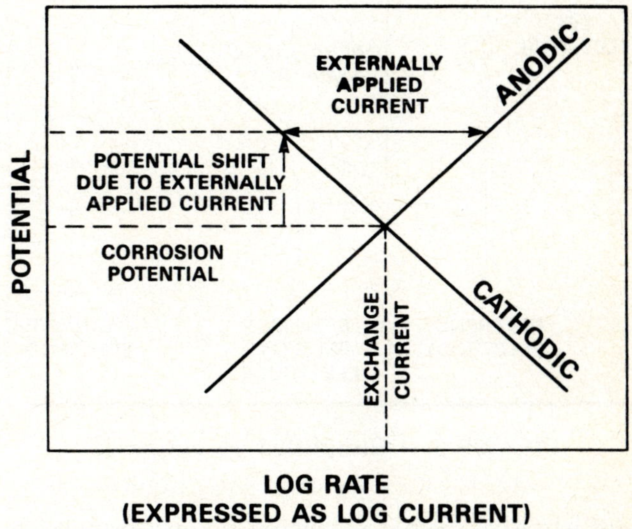

FIGURE 1 – Relationship between potential and externally applied current.

Test Approach

A typical potentiostatic test set-up is illustrated in Figure 2. The specimen, a reference electrode, and an inert counter electrode are immersed in the electrolyte of interest and connected to a potentiostat. Current output from the potentiostat is typically recorded on a strip chart recorder.

Potentiostat

The function of the potentiostat is to maintain a preset potential difference between the specimen and the reference electrode by applying a variable current between the specimen and the counter electrode. This is accomplished by comparing the measured potential difference with the preset value. The "error" signal thus generated causes a change in the applied specimen cur-

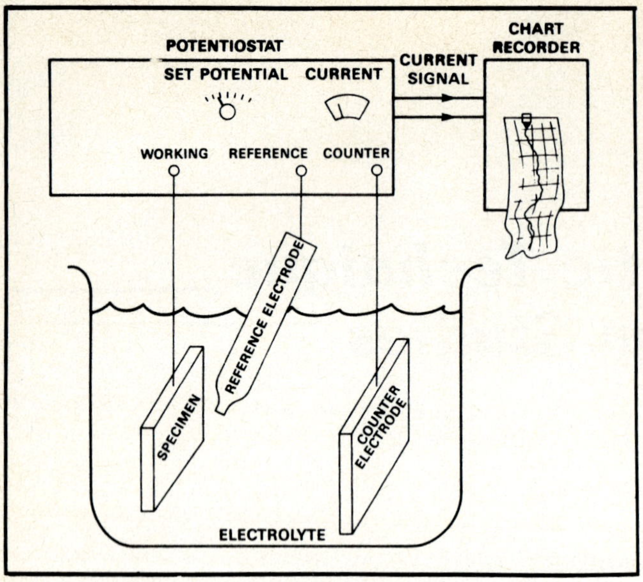

FIGURE 2 – Potentiostatic test set-up.

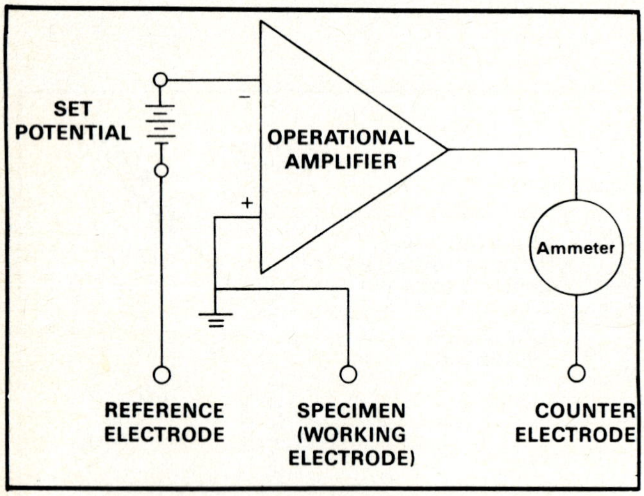

FIGURE 3 – Simplified potentiostat.

rent, which will result in a change in specimen potential in the desired direction. A schematic of a simple circuit that will accomplish this is presented in Figure 3.

A good potentiostat design has an extremely high input impedance between the specimen and reference electrode, usually 10^{12} to 10^{15} ohms. This is necessary so that the current drawn from the reference electrode is essentially zero, thus preventing problems of reference electrode polarization error. Good potentiostat design involves trade-offs between rapid response and some degree of damping on the output to prevent oscillation. Reference 4 (Chapter III, Section 1) and Reference 5 (Chapter 13) discuss operational amplifier theory and their application to potentiostat design. Although sophisticated and expensive potentiostats are available commercially, simple mini-potentiostats can be built for low-current applications very inexpensively. A discussion of design considerations for mini-potentiostats is contained in Reference 5.

Reference Electrode

The reference electrode is made of a special material/electrolyte combination which maintains a stable potential, and can therefore be used as a reference point for the potential measurements. Most reference electrodes are placed in a glass envelope so that the solution chemistry around the reference material can be controlled. Examples of these types of electrodes used for aqueous room-temperature tests are calomel (Hg/HgCl in KCl), silver/silver chloride (in KCl), and copper/copper sulfate. If the electrolyte path through the glass envelope, necessary for proper operation, is provided by a small hole or wick, intermixing of the electrolyte in the corrosion test with that of the reference electrode may occur. This causes contamination of the test electrolyte and may also cause an undesirable shift in the reference potential. To prevent this, some electrodes provide an electrolyte path by the use of semipermeable material such as Vycor.[1] Although this eliminates contamination, the semipermeable material may require special care. If the test electrolyte is well-characterized or small errors in potential are unimportant, then some materials can be used as reference electrodes without glass envelopes. Examples for seawater testing include zinc and silver/silver chloride. The considerations for selecting electrodes for a given application are too complex to enumerate here. Questions of thermal drifts, liquid junction potentials, etc., are discussed in detail in Reference 7.

Placement of the reference cell is important in electrolytes with a significant resistivity. In such solutions, the current between the specimen and counter electrodes will cause a significant voltage drop through the resistive electrolyte. A reference cell placed at some distance from the specimen will read a specimen potential that differs from the true potential by the IR drop or voltage drop in the solution between the specimen and the reference electrode. A rough idea of the maximum possible IR drop error can be calculated as follows:

Max IR drop (V) = Max specimen current density (A/cm^2) × electrolyte resistivity ($\Omega \cdot cm$) × distance of reference electrode from specimen (cm)

A more accurate calculation of the IR drop error must employ a technique that takes into account the exact geometry of the exposure vessel, specimen, and electrodes. Such methods include mathematical modeling, finite element analysis, finite difference analysis, surface-integral techniques. etc. Fortunately, it is frequently possible to measure IR drop error directly using several methods, including current interruption and alternating current (AC) impedance techniques. The IR drop error divided by the specimen current is usually referred to as the *uncompensated cell resistance*.

The current interruption technique involves use of a fast switch to quickly interrupt the current in the counter electrode lead. A very fast potential reading is taken just before and just after the interrupt, usually by means of a storage oscilloscope. The interrupt is kept short (1 to 20 μs) so that there is insufficient time for the specimen to depolarize. The current-off reading is the true specimen

[1]Registered trade name.

potential, while the current-on potential is affected by the uncompensated cell resistance. Thus, the difference between the two is the IR drop error. AC impedance measurement consists of applying a sinusoidal potential to the specimen and measuring the resultant current amplitude and phase shift from the original applied signal. By increasing frequency until no phase shift occurs, the uncompensated cell resistance becomes the ratio of the voltage amplitude to the current amplitude.[8] Other techniques of IR drop measurement include calibrated movement of the reference electrode, superposition of a square wave, and others.

If IR drop error is determined by one of the above techniques at one current level, it is first converted into uncompensated resistance. Then for each current level in the test, the product of that current and the uncompensated resistance yields the IR drop correction for that current. This can be a tedious process, and the net result is a potentiostatic test in which the true specimen potential is not constant. It is far better to prevent the IR drop error in the first place. This can be done instrumentally or, in some cases, by careful cell design.

Instrumental IR drop elimination is centered on the use of special potentiostats. Current interruption[9-11] is now a feature built into some potentiostats. The controlled potential is the current-off reading and IR drop error can thus be eliminated. Care should be exercised on some corroding systems with fast reactions, since the current-off phase must be short enough so that no significant change in the true specimen potential (depolarization) occurs. Other potentiostats use a positive feedback technique. This technique slows down potentiostat response time, and accurate settting of the amount of positive feedback is difficult, leading to under- or overcorrection. One popular technique involves increasing the extent of feedback until current oscillations occur, then backing off until they cease. It is, however, inaccurate because the feedback applied depends on the damping characteristics as much as on IR drop, and thus frequently leads to overcorrection.

IR drop error can also be minimized by proper cell design. The reference electrode may be placed very close to the specimen, thus minimizing the distance over which the IR gradient operates. To facilitate this, a salt bridge is commonly employed. This is a tube, filled with electrolyte, open at one end to a beaker containing the reference electrode and at the other end to the electrolyte near the specimen. Two typical arrangements are illustrated in Figure 4. (Certain arrangements of the salt bridge are sometimes referred to as a *Luggin probe*.) The tip of the salt bridge is usually positioned as close to the specimen as possible without blocking current flow to the specimen surface. The tip should be positioned about 1 to 2 times its diameter away from the specimen. Thus, small diameter tips are essential in high resistivity electrolytes. Since the reference electrode draws essentially no current because of the high impedance of the potentiostat, there will be no IR drop through the salt bridge, and thus the potential that the reference electrode measures will be the same as that at the tip of the salt bridge.

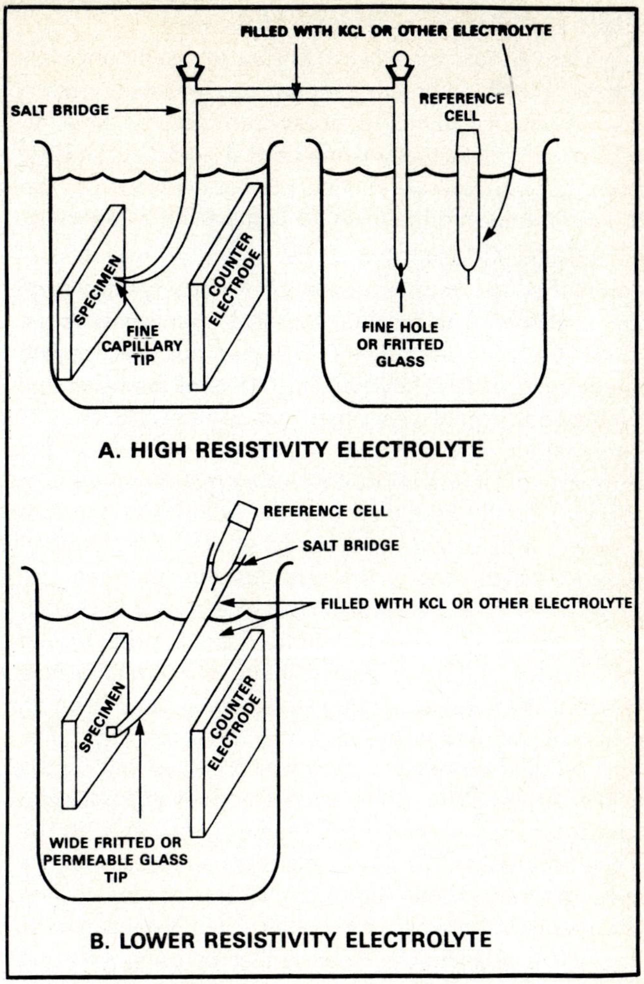

FIGURE 4 – Use of salt bridges.

Another test design method to minimize IR drop error is to add a supporting electrolyte. This is a substance that does not participate in the reactions being studied but which lowers the resistivity of the electrolyte, typically an ionic salt. Further detail on all aspects of IR drop error and its elimination is contained in Reference 12.

Counter Electrode

Counter electrodes are usually constructed of inert materials such as platinum or graphite. Platinum is expensive and is usually used as a coating. Graphite tends to be porous and may thus introduce cross-contamination if used in different electrolytes successively. Electrolyte contamination can also be introduced by corrosion or side reactions occurring at the counter electrode. In soil, scrap steel is sometimes used for a counter electrode since it is inexpensive and the contamination introduced by its deterioration never reaches the soil near the specimen. In cases where electrolyte contamination is critical, a counter electrode material identical to the specimen may be used, or the counter electrode may be placed in a separate container connected to the test vessel by means of a low-resistance fritted glass disk or semipermeable membrane. Care must be taken in the latter case to ensure that the IR drop through the frit or membrane is not so large that the overall cell voltage exceeds the potentiostat capacity.

Electrical Attachment and Mounting

Care should be taken to ensure that sufficient electrolyte is used. Although it is common to use an electrolyte volume of 250 mL/in^2 of exposed specimen area (40 mL/cm^2)[13], the actual volume used should be calculated from the assumed rate of generation or depletion of electrolyte species and the tolerable composition variance.

Electrical connection to the specimen may be accomplished in a number of ways.[14,16] Partially immersing the specimen enables the electrical connection to be made above the electrolyte. This is common practice with counter electrodes; however, the effect of the water line on the corrosion of a specimen should be considered. All connections made in the electrolyte introduce the possibility of crevice corrosion. A wire may be screwed or soldered to the specimen and the point of connection covered with a sealant impervious to the electrolyte. Threaded rods may be screwed into the specimen and protected from the electrolyte by a glass tube and O-ring. Special devices have also been used that expose a given area of the specimen face without creating an active crevice. These use an O-ring seal or gasket and have an electrical connection made to the dry back side. The method of attachment is not critical provided that (1) the active crevice site is not formed on anodically polarized specimens, (2) the connection remains dry, and (3) the electrical resistance is low. If electrical connection to the specimen must be through a significant resistance, then the specimen potential read by the potentiostat can be in error. This problem can be eliminated by using a second specimen lead, which carries no current but is used solely for the potential reading. Many commercial potentiostats have this feature available.

Data Analysis

The most common forms of data obtained from potentiostatic tests are specimen weight loss and applied current as a function of time. In addition, other data that may be useful under some circumstances are: corrosion morphology, corrosion product composition and morphology, amount and composition of gases generated, solution analyses for corrosion products, etc. Typically a potentiostatic test is conducted to determine weight loss or the value of the steady-state current, either of which is sometimes used to calculate corrosion rate as follows:

$$\text{corrosion rate } (\mu m/y) = \frac{\text{weight loss (g)} \times 3165\ (\mu m \cdot s/cm \cdot y)}{\text{area } (cm^2) \times \text{density } (g/cm^3) \times \text{time (s)}}$$

$$\text{weight loss (g)} = \text{current (C/s)} \cdot \text{time (s)} \cdot \text{equivalent weight (g/C)}$$

The weight loss calculation from current is usually accurate only when the specimen is at least 60 to 100 mV positive to its freely corroding value, since it is only under these conditions that the current generated by the cathodic reaction is negligible. The current-time product is frequently found by integration if the current varies significantly over the test period.

Comparison to Other Methods

Potentiostatic tests are easily contrasted to galvanostatic tests. In the latter, a constant current is supplied to the specimen. The setup is less complicated since a reference cell is not necessarily required and the electronics of a galvanostat are simpler. Constant current is desirable when the rate of a reaction must be held constant, such as for current integrations or for sacrificial anode efficiency calculations. However, when the reaction rate is held constant, the effect of other variables cannot be studied. Potentiostatic testing maintains a constant thermodynamic driving force, thus allowing the rate of a reaction to be studied as other parameters are changed. In addition, passivity of some active-passive materials cannot be maintained in a galvanostatic test, whereas a potentiostatic test easily accomplishes this. Thus, potentiostatic testing is preferred in many cases over galvanostatic testing.

A more difficult distinction exists between potentiostatic and potentiodynamic test methods. Potentiodynamic tests still use a controlled potential, but this potential is varied, usually linearly, as a function of time. In this way the currents required to hold a material at a whole range of potentials can be determined. Because this is a potential-controlled method, many authors refer to this swept-potential technique as potentiostatic, instead of using the correct term. The confusion increases, however, when similar information is generated using a stepped-potential scan. In this scan, the set potential of the potentiostat is held at a fixed value for a given time period, then stepped to a new value for the same time, and so on. This creates what is, in effect, a series of linked potentiostatic tests. Some investigators assume that the information so gained is equivalent to a series of individual potentiostatic tests, ignoring the fact that individual potentiostatic specimens experience a different potential history than stepped-potential specimens. In fact, it has been shown[17] in some systems that a stepped-potential test yields the same results as a potentiodynamic test, which scans the same potential range for the same time period. Thus, it may be reported that a series of potentiostatic tests were conducted when, in fact, a single stepped-potential test, equivalent to a slow-scan potentiodynamic test, was actually run.

Potentiodynamic and stepped-potential tests are useful for a large number of studies where performance of the material is needed over a range of potentials. In these tests, however, the effect of potential is confounded with the effect of exposure time. This is because the last potentials are applied to a specimen that has already been exposed for a length of time necessary to scan the first potentials. If a three-dimensional polarization curve of potential, current, and exposure were plotted, the scanned potential test actually gives a diagonal section as shown in Figure 5. Thus, potentiodynamic and stepped-potential tests are not accurate ways of studying the behavior of materials whose behavior changes significantly over the time frame of the test, and optimally a series of potentiostatic tests is required. Unfortunately, the latter is expensive and equipment-intensive, and is therefore usually not done.

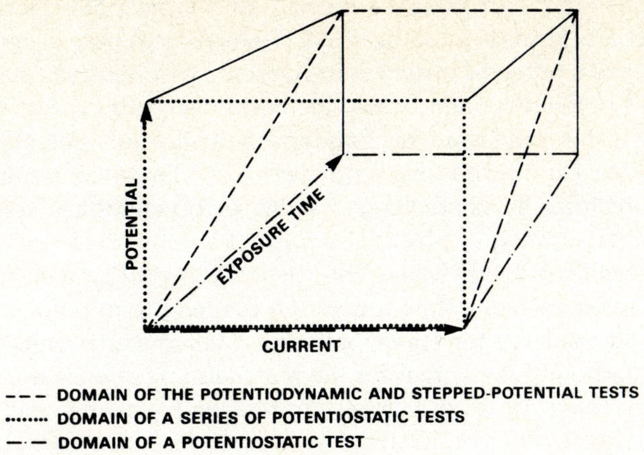

FIGURE 5 — Domain of controlled potential tests.

Applications

This section will deal almost exclusively with applications of the potentiostatic technique, and will not address the broader applications of potentiodynamic or stepped potential techniques.

General Corrosion

Potentiostatic anodic polarization can be used to artifically increase corrosion rate for some materials and thus can be considered as a form of "accelerated" corrosion test. The morphology of the resultant corrosion (uniform vs pitting vs dealloying) is frequently similar to that of naturally corroding material, if the potential is not too far anodic from the corrosion potential. Thus, anodic polarization can provide a quick, rough assessment of the expected corrosion morphology. Great care must be exercised, however, since (1) passivity may be artificially induced by the polarization, or (2) the nature of a protective corrosion product film may dramatically change if the film is grown at a different potential than the freely corroding potential.

A more well-accepted application of the potentiostatic technique is for single-point linear polarization measurements to obtain corrosion rate information rapidly.[18,19] This is the basis for several commercial corrosion rate meters. To obtain the corrosion rate, the specimen of interest is polarized either anodically or cathodically 5 to 10 mV away from its corrosion potential and the applied current is monitored until a steady state is achieved. This current, along with information on the Tafel slopes for the specimen materials and environment, is entered into the following equation to obtain the corrosion rate:

$$\text{corrosion rate } (\mu\text{m/y}) = \frac{B_A B_C}{2.3(B_A + B_C)} \times \frac{\text{steady state current (mA)} \cdot \text{equiv wt (g/C)}}{\text{potential offset (mV)} \cdot \text{area (cm}^2)} \times 3156\ (\mu\text{m} \cdot \text{s/cm-y})$$

Although this technique is usually fairly accurate, it suffers from several disadvantages[20]:

1. Tafel slopes must be known. Estimated values, frequently around 0.1 V, usually give reasonable corrosion rates, and trends can still be observed.
2. Mechanically enhanced corrosion such as erosion or severe intergranular corrosion will yield higher rates than predicted as a result of nonelectrochemical metal removal.
3. Specimens must be used that are similar to the structure of interest. Environment and flow conditions may differ between the specimens and the structure, causing errors.
4. Morphology of the attack is unknown. Small overall corrosion rates at localized areas can lead to failure before larger overall uniform corrosion. Cracking is not detected.

Detailed discussions of potentiodynamic linear polarization are contained in References 21 and 22.

Another corrosion rate measurement technique that can measure the anodic current directly is useful only for systems where the primary cathodic reaction is oxygen reduction. In this technique, the specimen is potentiostatically held at its corrosion potential in an aerated electrolyte. The electrolyte is then deaerated. The current to keep the specimen at its original potential is a direct measure of the rate of the corrosion reaction occurring under aerated conditions.[22]

Pitting and Crevice Corrosion

Potentiostatic testing can be used to verify protection potentials. Since those originally determined by potentiodynamic tests are frequently a function of scan rate, the true value is not known. The protection potential thus determined can be verified by running potentiostatic tests that bracket this value.[24-27] The specimens held below the true protection potential should not pit, and will experience a current that either falls with time or is stationary. The specimens held above the protection potential will pit and the current will begin to rise as soon as the pitting initiates. The time to initiation can thus also be determined. Since in some environments the initiation time can be quite large,[28] the length of the potentiostatic exposure should be adjusted accordingly.

Repassivation potentials can be verified similarly by using exposures both above and below the proposed repassivation potential. The specimens must first be activated by an anodic pulse or by scratching. Repassivation is indicated by a decrease in current and absence of pitting.[26-29] The decay rate is indicative of the rate of the repassivation and usually requires an oscilloscope to follow. Difficulties in measurement include the problem of establishing an exact time of the scratch and a scratch area for conversion to current densities. Despite these difficulties, a number of investigators do potentiostatic repassivation kinetics studies, using a variety of different scratching and abrading devices, including specimens rotating over needles, needles rotating on specimens, stretching wires, etc.[29-36] It is claimed that since potential affects repassivation, and since initiation of localized attack depends on a dynamic balance between breakdown and repair, then repassivation kinetics provides valuable information.[37]

In a pitting situation, pits are continually initiating and repassivating. Each of these events contributes a current spike to the specimen current. The sum of all spikes from all pitting events leads to a highly unstable current value. This electrochemical "noise" can be analyzed, frequency by frequency by Fourier transformation or other techniques to obtain information about initiation and repassivation kinetics. The presence of pitting can frequently be detected by the amount of noise, even before noticeable pits have formed.[38-42] Thus, the protection potential can be verified potentiostatically by analyzing noise in the specimen currents.

The propagation rate of pits has been determined potentiostatically by holding the specimen between the repassivation potential and the critical pitting potential.[43] The specimen is passivated and the current measured. An anodic pulse is then used to activate the specimen and the new current measured. The difference between these currents is the current at the initiated pit sites; this can be related to the rate of metal loss resulting from pitting.

Intergranular Corrosion

Most accelerated intergranular corrosion susceptibility tests are conducted galvanostatically.[44] The primary use of potentiostatic testing is to allow laboratory simulation of potentials of a particular service environment.

Dealloying

A number of investigators have used electrochemical techniques to study dealloying.[45-50] Acceleration of dealloying by the use of potentiostatic methods is possible, provided that the potential used has been carefully considered. Potentials too electronegative will give dealloying rates too slow to be useful, whereas the use of potentials too electropositive will result in nonselective dissolution. The appropriate potential may be chosen by reference to potential-pH diagrams or by experiment. In some systems sufficiently rapid dealloying cannot be achieved before the onset of nonselective dissolution, however. In addition, electrochemical acceleration of dealloying can bias the experimental results by masking the presence of diffusion-related selective-removal processes.[50] Dealloying can be verified by examination of the specimen or by analysis of the electrolyte. Once an appropriate potential is chosen, the effects of heat treatment, microstructure, minor alloying variations, etc., on the dealloying rate can be determined by measuring specimen current, weight loss, electrolyte composition, or by metallographic techniques.

Galvanic Corrosion

The extent of galvanic corrosion can be predicted by superposition of the polarization curves for the materials involved.[2,51] Many materials systems undergo changes in polarization characteristics with exposure time as a result of the buildup of corrosion product films, surface roughness changes, and even alloy surface enrichment.[52] Polarization curves generated potentiodynamically confound exposure time with potential. To predict long-term galvanic corrosion behavior, long-term polarization characteristics must be used. Galvanic couples experience a potential history that is closer to potentiostatic than to potentiodynamic; thus, the best polarization curves to use for prediction of long-term behavior are potentiostatically derived long-term curves.[53,54] These curves are obtained by conducting a series of potentiostatic tests over a range of potentials. Current is monitored until it reaches a steady value. This value and the set potential of that specimen define a point on the long-term potentiostatic polarization curve for the specimen material. Superposition of the curves for the materials and surface areas involved allows prediction of a coupled potential and a galvanic current. Galvanic corrosion rate can then be calculated from the galvanic current using Faraday's Law. Since total weight loss is available as a function of potential from the potentiostatic tests, the total corrosion rate of each material can also be predicted from the predicted coupled potential.

Although the above method is quite accurate, it is seldom used, primarily because of the cost and complexity of conducting multiple potentiostatic tests. Changes in flow rate, temperature, or electrolyte composition will change the polarization characteristics of the materials and thus necessitate retesting. Finally, steady-state behavior may indicate an acceptable corrosion rate while the initial transient behavior could result in unacceptable corrosion during the initial stages of exposure.

When full potentiostatic polarization curve generation is not used, potentiodynamic curves are usually employed. Potentiostatic tests can then be used to spot-check the validity of the potentiodynamic curves, providing an indication of the magnitude of error in galvanic corrosion prediction caused by using the potentiodynamic curves.

Cathodic Protection

The efficiency of sacrificial anodes can be determined potentiostatically; however, the usual practice is to perform these tests galvanostatically.[55-58] Similarly, cathodic or anodic protection requirements can be determined potentiostatically, but are usually found by using galvanostatic techniques. An exception to this is the determination of protection potential and current requirements for impressed current cathodic protection. A potentiostatic test will yield data on the current requirements to maintain that potential as well as the degree to which corrosion is reduced.[52-59] A series of such tests will indicate the least electronegative potential that will provide adequate reduction of corrosion.[59] The current at this potential can be scaled to the surface area of the structure of interest, and initial as well as steady-state current requirements can be ascertained.

Fretting Corrosion

The increased corrosion resulting from fretting can be detected using linear polarization techniques.[60] A test for fretting of surgical implant materials using linear polarization is currently under development by ASTM.[(2)]

[(2)] ASTM, Philadelphia, Pennsylvania.

Since single-point linear polarization, as described earlier, can be used in many cases instead of potentiodynamic linear polarization, it should thus be possible to detect fretting corrosion potentiostatically.

Corrosion Fatigue

Potentiostatic tests have been used to detect the effects of cathodic protection both on corrosion fatigue endurance limits[61] and on fatigue crack growth rates.[62] Since the effects of applied potential may be either beneficial or detrimental, depending on potential, material, environment, stress, etc., potentiostatic testing should be used as part of the procedure for establishing protection potentials for cyclically loaded structures.

Atmospheric Corrosion

The electrolyte in atmospheric corrosion is transient and of low volume and conductivity, consisting usually of rainwater or dew; thus, conventional potentiostatic tests cannot be conducted in this environment. Accelerated immersion potentiostatic tests to rate atmospheric corrosion performance have been developed, however. One such test is the electrolyte corrosion (EC) test,[63] which simulates extended atmospheric service on plated parts. This test is in acidic sodium nitrate at +0.300 V for a series of one-minute intervals, each interval being equivalent to 6 months service. The results of this test should be interpreted with caution when relating them to an atmospheric service condition, however, and the test may not prove useful for all alloy systems.

Environmentally Induced Cracking

Stress corrosion cracking and hydrogen embrittlement can be a function of applied potential, with more negative potentials promoting hydrogen evolution and hydrogen embrittlement. Thus, potentiostatic tests can be run on fracture specimens to yield the effect of potential on: time to initiation or failure, fracture toughness or energy-to-fracture parameters such as K_{IC} or J_{IC}, crack propagation rate, etc. Any of the cracking parameters can, in theory, be measured as a function of applied potential in a series of potentiostatic tests. This has been utilized on a range of specimen geometries from bent beam stress corrosion specimens[64] to compact wedge-opening loaded fracture specimens and slow-strain rate tests.[65-67] Details of fracture testing are beyond the scope of this paper. One interesting technique, however, is used for the determination of K_{ISCC}.[66] When the stretch zone typical of K_{ISCC} is formed, more specimen will be exposed to electrolyte and a larger current, either anodic or cathodic, will result.

Hydrogen generation is frequently accomplished galvanostatically to control the production rate. It is sometimes advantageous, however, to generate hydrogen potentiostatically since this maintains a constant fugacity.

Mechanism Studies

There are probably as many ways to use potentiostatic tests for corrosion mechanism studies as there are investigators. Some of the more "standard" uses and some novel approaches will be described below. (The following discussion is by no means comprehensive.)

Comparison of the integrated specimen current with weight loss for an anodically polarized specimen by the use of Faraday's Law leads to determination of the number of electrons passed in the anodic reaction. Examination of potentiostatically generated corrosion products can give clues about the change of mechanisms as a function of potential. Calcareous deposits on cathodically polarized specimens can be studied, as can their effect on protection currents. Current integration and solution analysis can be used to determine corrosion product type and growth kinetics.

Electron microscope studies of the surface of potentiostatically tested specimens can detect surface instability, indicative of reaction control by bulk diffusion in the alloy. Etching indicates activation control, while polishing is indicative of control by diffusion.

Kinetics of the reaction can be followed by following specimen current, and the effects of environmental variables and corrosion product films can be followed similarly. If current is followed after a potential step, information on surface coverage of adsorbed species,[3] double-layer capacitance, and uncompensated cell resistance[8] can be found. A technique used with a rotating disk electrode called *hydrodynamic modulation* can provide information about reaction kinetics.[5]

Conclusions

Potentiostatic testing is a common electrochemical technique for corrosion evaluation that is fairly simple and yet can yield a large amount of information about corrosion processes. It has a large number of applications in both corrosion science and technology and can be combined with apparatus such as fracture testing equipment or scratched electrodes to yield information unavailable to each technique alone. Modifications of the potentiostatic test, including the stepped potential and potentiodynamic techniques, increase the utility of the constant potential test to an even greater extent.

References

1. S. Barnartt, "Electrochemical Nature of Corrosion," Electrochemical Techniques for Corrosion, R. Baboian, Ed., National Association of Corrosion Engineers, Houston, Texas, p. 1, 1977.
2. M. G. Fontana, N. D. Greene, Corrosion Engineering, McGraw-Hill Book Co., New York, New York, 1978.
3. J. O'M. Bockris, A. K. N. Reddy, Modern Electrochemistry, Plenum Press, New York, New York, 1970.
4. E. Gileadi, E. Kirowa-Eisner, J. Penciner, Interfacial Electrochemistry—An Experimental Approach, Addison-Wesley, Reading, Massachusetts, 1975.
5. A. J. Bard, L. R. Faulkner, Electrochemical Methods, John Wiley and Sons, New York, New York, 1980.
6. H. A. Newborn, D. C. Bratlie, and C. R. Crowe, in Electrochemical Corrosion Testing, F. Mansfeld, U.

Bertocci, Eds., STP 727, ASTM, Philadelphia, Pennsylvania, p. 381, 1981.
7. D. J. G. Ives, G. J. Janz, Reference Electrodes-Theory and Practice, Academic Press, New York, New York, 1961.
8. F. Mansfeld, Corrosion, Vol. 36, No. 5, p. 301, 1981.
9. J. D. E. McIntyre, W. F. Peck, Jr., J. Electrochem. Soc., Vol. 117, No. 6, p. 747, 1970.
10. D. R. Flinn, M. Rosen, S. Schuldiner, J. Electrochem. Soc., Vol. 117, No. 1, p. 79, 1970.
11. J. Newman, J. Electrochem. Soc., Vol. 117, No. 4, p. 507, 1970.
12. D. Britz, J. Electroanal. Chem., Vol. 88, p. 309, 1978.
13. H. H. Uhlig, The Corrosion Handbook, John Wiley & Sons, Inc., New York, New York, p. 963, 1948.
14. N. D. Greene, W. D. France, Jr., B. E. Wilde, Corrosion, Vol. 21, No. 9, p. 275, 1965.
15. J. R. Myers, E. G. Gruenler, L. A. Smulczenski, Corrosion, Vol. 24, No. 10, p. 352, 1968.
16. E. A. Lizlovs, J. Electrochem. Soc., Vol. 117, No. 10, p. 1335, 1970.
17. N. D. Greene, R. B. Leonard, Electrochim. Acta, Vol. 9, p. 45, 1964.
18. H. Feitler, Materials Performance, Vol. 9, No. 10, p. 37, 1970.
19. J. C. Rowlands, M. N. Bentley, Br. Corros. J., Vol. 7, p. 42, 1972.
20. M. E. Indig, C. Groot, Corrosion, Vol. 25, No. 11, p. 455, 1969.
21. "Standard Practice for Conducting Potentiodynamic Polarization Resistance Measurements," G 59-78, 1984 Annual Book of ASTM Standards, Vol. 03.02, ASTM, Philadelphia, Pennsylvania, p. 331, 1984.
22. F. Mansfeld, "Polarization Resistance Measurements—Experimental Procedure and Evaluation of Test Data," Electrochemical Techniques for Corrosion, R. Baboian, Ed., National Association of Corrosion Engineers, Houston, Texas, p. 18, 1977.
23. J. Postlethwaite, in Electrochemical Corrosion Testing, F. Mansfeld, U. Bertocci, Eds., STP 727, ASTM, Philadelphia, Pennsylvania, p. 290, 1981.
24. S. Smialowska, M. Czachor, "The Analysis of Electrochemical Methods for the Determination of Characteristic Potentials of Pitting Corrosion," Localized Corrosion, R. W. Staehle, B. F. Brown, J. Kruger, A. Agrawal, Eds., National Association of Corrosion Engineers, Houston, Texas, p. 353, 1974.
25. J. R. Galvele, S. M. deMicheli, I. L. Muller, S. B. deWexler, I. L. Alanis, "Critical Potentials for Localized Corrosion of Aluminum Alloys," Localized Corrosion, R. W. Staehle, B. F. Brown, J. Kruger, A. Agrawal, Eds., National Association of Corrosion Engineers, Houston, Texas, p. 580, 1974.
26. Z. Szklarska-Smialowska, M. Janik-Czachor, Corros. Sci., Vol. 11, p. 901, 1971.
27. "Standard Test Methods for Pitting or Crevice Corrosion of Metallic Surgical Implant Materials," F 746-81, 1984 Annual Book of ASTM Standards, ASTM, Philadelphia, Pennsylvania, p. 79, 1984.
28. J. R. Ambrose, J. Kruger, "Breakdown of Passive Films on Iron by Chloride Ion," Proc. 4th Inter. Cong. on Metal. Corros., National Association of Corrosion Engineers, Houston, Texas, p. 698, 1972.
29. T. R. Beck, Electrochim. Acta, Vol. 18, p. 807, 1973.
30. T. P. Hoar, J. R. Galvele, Corros. Sci., Vol. 10, p. 211, 1970.
31. N. Pessall, C. Liu, Electrochim. Acta, Vol. 16, p. 1987, 1971.
32. J. R. Ambrose, J. Kruger, Corrosion, Vol. 28, No. 1, p. 30, 1972.
33. T. P. Hoar, J. M. West, Proc. Roy. Soc., Vol. 268A, p. 304, 1962.
34. T. Murata, R. W. Staehle, "Theory and Application of Straining Electrode," Proc. 5th Int. Cong. Met. Corros., National Association of Corrosion Engineers, Houston, Texas, p. 513, 1974.
35. N. D. Tomashov, L. P. Vershinina, Electrochim. Acta, Vol. 15, p. 501, 1970.
36. R. B. Diegle, D. A. Vermilyea, J. Electrochem. Soc., Vol. 122, p. 180, 1975.
37. J. Kruger, "New Approaches to the Study of Localized Corrosion," Electrochemical Techniques for Corrosion, R. Baboian, Ed., National Association of Corrosion Engineers, Houston, Texas, p. 35, 1974.
38. U. Bertocci, J. Kruger, Surface Science, Vol. 101, p. 608, 1980.
39. U. Bertocci, J. L. Mullen, Y-X, Ye, in Passivity of Metals and Semiconductors, M. Froment, Ed., Elsevier Science Publishers, B. V. Amsterdam, Netherlands, p. 229.
40. U. Bertocci, J. Electrochem. Soc., Vol. 127, No. 9, p. 1931, 1980.
41. U. Bertocci, J. Electrochem. Soc., Vol. 128, No. 3, p. 520, 1981.
42. U. Bertocci, Y-X Ye, J. Electrochem. Soc., Vol. 131, No. 5, p. 1011, 1984.
43. B. Syrett, "The Application of Electrochemical Techniques to the Study of Corrosion of Metallic Implant Materials," Electrochemical Techniques for Corrosion, R. Baboian, Ed., National Association of Corrosion Engineers, Houston, Texas, p. 93, 1974.
44. Intergranular Corrosion of Stainless Alloys, R. F. Steigerwald, Ed., STP 656, ASTM, Philadelphia, Pennsylvania, 1978.
45. E. E. Langenegger, F. P. A. Robinson, Corrosion, Vol. 24, No. 12, p, 411, 1968.
46. E. E. Langenegger, F. P. A. Robinson, Corrosion, Vol. 25, p. 59, 1969.
47. H. W. Pickering, C. Wagner, J. Electrochem. Soc., Vol. 114, p. 698, 1967.
48. H. W. Pickering, "Preferential Anode Dissolution in Binary Alloys," Proc. Conf. Fund. Aspects Stress Corros. Crack., R. W. Staehle, A. J. Forty, D. Van Rooyen, Eds., National Association of Corrosion

Engineers, Houston, Texas, p. 159, 1969.

49. F. P. A. Robinson, M. Shalit, Corrosion Technology, Vol. 11, p. 11, 1964.

50. R. H. Heidersbach, Jr., PhD Dissertation, University of Florida, Gainesville, Florida, 1971.

51. R. Baboian, in Galvanic and Pitting Corrosion-Field and Laboratory Studies, STP 576, ASTM, Philadelphia, Pennsylvania, p. 5, 1976.

52. H. P. Hack, "Exposure Time Effects on Current Densities of Polarized Marine Materials," CORROSION/83, Paper No. 210, National Association of Corrosion Engineers, Houston, Texas, 1983.

53. H. P. Hack, "Galvanic Corrosion Prediction Using Long-Term Potentiostatic Polarization Curves," CORROSION/83, Paper No. 73, National Association of Corrosion Engineers, Houston, Texas, 1983.

54. J. R. Scully, H. P. Hack, "Galvanic Corrosion Prediction Using Long and Short Term Polarization Curves," CORROSION/84, Paper No. 34, National Association of Corrosion Engineers, Houston, Texas, 1984.

55. O. Saetre, F. Jensen, U. Aspaas, "Electrochemical Testing of Anodes During Fabrication," CORROSION/84, Paper No. 337, National Association of Corrosion Engineers, Houston, Texas, 1984.

56. J. F. Brown, Jr., "Quality Control Testing of Aluminum Anodes, T-7L-2 Task Group Progress Report," CORROSION/84, Paper No. 346, National Association of Corrosion Engineers, Houston, Texas, 1984.

57. R.A.N. Cathodic Protection Manual, ABR 5023, Royal Australian Navy, June 1975.

58. Hydral 2 Sacrificial Anodes in Aluminum Alloy for Cathodic Protection of Marine Installations, Société de Vente de l' Aluminium Pechiney, Pans, France.

59. B. E. Miller, H. P. Hack, Hovering Craft and Hydrofoil, Vol. 13, No. 1, p. 26, 1973.

60. B. C. Syrett, S. S. Wing, Corrosion, Vol. 34, No. 11, p. 379, 1978.

61. B. F. Brown, Corrosion Fatigue Mechanics, Metallurgy, Electrochemistry, and Engineering, STP 801, ASTM, Philadelphia, Pennsylvania, p. 508, 1983.

62. T. W. Crooker, F. D. Bogar, W. P. Cares, Corrosion Fatigue Technology, H. L. Craig, Jr., T. W. Crooker, D. W. Hoeppner, Eds., STP 642, ASTM, Philadelphia, Pennsylvania, p. 189, 1978.

63. R. L. Saur, R. P. Basco, Plating, Vol. 53, p. 33, 1966.

64. D. C. Vreeland, Materials Protection, Vol. 9, No. 1, p. 32, 1970.

65. D. L. Dull, L. Raymond, Hydrogen Embrittlement Testing, STP 543, ASTM, Philadelphia, Pennsylvania, p. 20, 1974.

66. H. C. Chan, A. C. C. Tseung, A. I. Onuchukwu, Corros. Sci., Vol. 19, p. 209, 1979.

67. E. M. Hackett, "An Investigation of the Environmental Cracking Susceptibility of a High Strength Steel Using J Integral Techniques," Masters Dissertation, Johns Hopkins University, Baltimore, Maryland, October 1984.

Polarization Resistance Measurements — Today's Status

F. MANSFELD*

Abstract

The polarization resistance technique is a well-developed method for determining corrosion rates, and its theoretical background is well understood. ASTM G59[1] outlines the experimental approach for determining R_p from a potentiodynamic sweep in the vicinity of E_{corr}. The sources of error in R_p measurements include: (1) uncertainty in the parameter B, which is needed to calculate corrosion rates, (2) choice of a scan rate that is too fast, (3) neglect of the solution resistance, and (4) nonlinearity. The latter problem affects only linear polarization measurements. The electrochemical impedance technique has certain advantages, especially for measurements in low conductivity media and for systems with low corrosion rates. As with all electrochemical measurements of corrosion phenomena, care should be taken to ensure that the system to be studied obeys the simple kinetics that form the theoretical basis for the polarization resistance technique. An example is given where R_p measurements conducted in the usual manner give erroneous results.

Introduction

The experimental procedure and the evaluation of the test data that were collected by ASTM Subcommittee G01.11 as a basis for a standard practice have been discussed elsewhere.[2] The document referenced (ASTM G 59) contains standard polarization curves that should be reproduced by others who follow this standard reference method. It should be noted that the procedure in ASTM G 59 produces only the polarization resistance, R_p. The corrosion current density i_{corr} can be calculated only with a knowledge of the values of the anodic (b_a) and cathodic (b_c) Tafel slopes for the same experimental conditions according to:

$$R_p = \frac{B}{i_{corr}} \qquad (1)$$

with

$$B = \frac{b_a b_c}{2.3\,(b_a + b_c)}$$

The approach for measuring or calculating precise values of the parameter B has been discussed in a review article.[3] Various computer programs have been utilized for this purpose, among them CORFIT,[4] which was the first program to be published, and its improved version POLCURR,[5] and BETACRUNCH,[6] which has been critized[7] (see also rebuttal[7]). With the increased availability of computer programs that can be used to perform electrochemical experiments, it can be expected that very useful software will soon become available, which can be used to analyze experimental data in terms of R_p and B [Equation (1)] and to calculate precise values of i_{corr} and corrosion rates. At present, a value of B = 20 mV is often used for cases where Tafel slopes are unavailable or where only a qualitative measure of corrosion rates is needed.

The fact that the polarization resistance technique has been standardized shows that it is a well-developed technique, and its background is well understood. Therefore, it is unnecessary to discuss here the basic theoretical or experimental aspects of the polarization resistance technique. However, it is important to point out several considerations that are often neglected in the design and

*Rockwell International Science Center, Thousand Oaks, California.

analysis of polarization resistance measurements and can cause significant errors in the calculation of corrosion rates. These topics, which will be discussed below, include: (1) choice of the correct scan rate in potentiodynamic experiments, (2) neglect of solution resistance, and (3) nonlinearity in "linear polarization" experiments.

The electrochemical impedance technique is attracting a great deal of interest at present. Its theoretical background and experimental approach[8,9] and application to corrosion rate measurements[10] have been discussed recently. Examples for polarization resistance measurements in the presence of corrosion inhibitors are given elsewhere.[11] Certain aspects of the application of electrochemical impedance spectroscopy to polarization resistance measurements will be discussed below.

Finally, it must be pointed out that there can be corrosion systems where the polarization resistance technique cannot be applied in the usual manner. An example is the case for the system iron/H_2SO_4 with propargylic alcohol, which has been discussed by Lorenz and Mansfeld.[10]

Factors That Can Cause Errors in Polarization Resistance Meaurements

Choosing the Correct Scan Rate

Mansfeld and Kendig[12] have discussed the importance of choosing the correct scan rate for potentiodynamic measurements of the polarization resistance R_p, which is determined as the slope of the potential E-current I curve at the corrosion potential E_{corr}, where I = 0:

$$R_p = \left(\frac{\partial E}{\partial I}\right) I=0 \qquad (2)$$

An upper limit for the scan rate, s_{max}, was given as:

$$s_{max} < \frac{\pi \Delta E f_2}{10} \qquad (3)$$

where ΔE is the peak-to-peak amplitude of the scan (often 10 mV) and f_2 is the lower breakpoint in a Bode plot of the impedance spectrum.[8,9] The value of f_2 is given by:

$$f_2 = \frac{1}{4\pi R_\Omega C_{dl}} \left(1 - \frac{1}{R_p}\sqrt{R_p^2 - 4R_\Omega R_p - 4R_\Omega^2}\right) \qquad (4)$$

For a case where the solution resistance R_Ω = 1 ohm, R_p = 1000 ohm, and the electrode capacitance C = 100 μF, s_{max} = 20 mV/s, which is a much higher scan rate than is applied in most cases. For systems with fairly high corrosion rates [in this case i_{corr} ~20 mV/1000Ω = 20 μA (Equation (1)], the effect of the scan rate is not important. However, for well-inhibited systems such as the one shown in Figure 1, it is very important not to exceed the maximum scan rate s_{max}. In Figure 1, the log of the modulus of the impedance |Z| is plotted vs

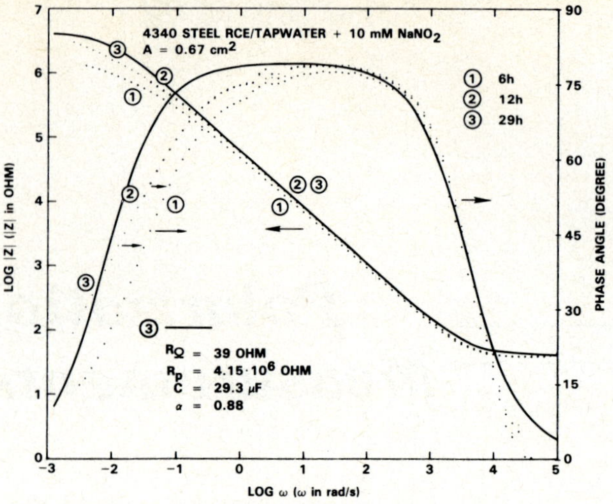

FIGURE 1 – Bode plot or AISI 4340 steel RCE in tapwater containing 10 mM $NaNO_2$.

the log of the frequency ω of the applied signal ($\omega = 2\pi f$), for an AISI 4340 steel rotating cylinder electrode (RCE) in tapwater containing 10 mM $NaNO_2$. The solid line is a fit of the experimental data to a theoretical model.[11,13] For the data in Figure 1, one obtains f_2 = 1.3 mHz and s_{max} = $4 \cdot 10^{-3}$ μV/s or 14.7 mV/h. This is a very low scan rate that cannot be exceeded if the correct value of R_p in a potentiodynamic experiment is desired. A scan rate of 0.2 mV/s, which is commonly used, corresponds to ω = 0.4 rad/s [Equation (3)]. If this scan rate were used to determine R_p for the example in Figure 1, an impedance value would be obtained in the capacitive part of the Bode plot corresponding to $1.7 \cdot 10^5$ ohm, which is considerably lower than the correct value of R_p = $4.5 \cdot 10^6$ ohm.

The case shown in Figure 1 might be considered exceptional, since it corresponds to a very low corrosion rate. It serves, however, very well to illustrate the factors that should be considered when the experiment is set up, and it shows the type of error that can occur. If the scan rate is too high, R_p will be determined too low and the calculated corrosion rate will be too high. Another factor that has to be considered is the stability of the corrosion potential during a slow scan. It is, therefore, necessary to perform both a forward and a backward scan to ensure that hysteresis has not occurred as a result of drift of the corrosion potential.

Neglect of Solution Resistance (Ohmic Drop)

In a rebuttal to the discussion by Mansfeld and Kendig[14] of several shortcomings of the BETACRUNCH computer program,[6] Greene and Gandhi state, "BETACRUNCH fits data to the Tafel equation, and therefore, cannot compensate for resistance errors. However, these errors are often small near the corrosion potential, especially if resistance is low. This is theoretically correct, widely accepted, and represents one of the major advantages of linear polarization analysis." The last two sent-

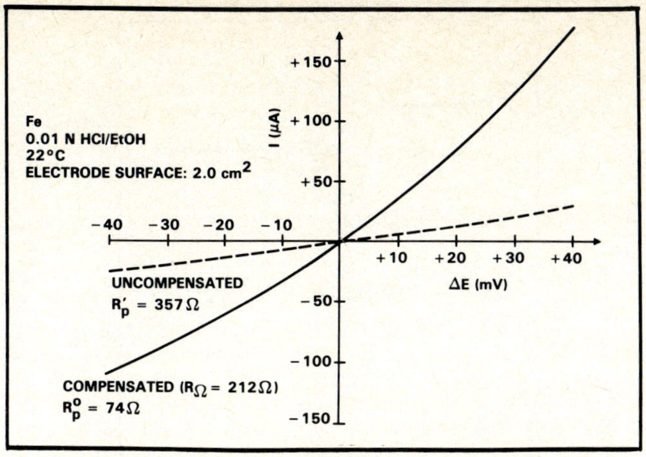

FIGURE 2 — Potentiodynamic sweeps with and without compensation of the ohmic drop for iron in ethanol containing 0.01 N HCl.

ences contain erroneous information which could be dangerous, especially if it were "widely accepted."

As pointed out several years ago,[15] it is a basic feature of electrochemical measurements of the polarization resistance that the experimental value R_p^1 is the sum of the true value R_p^0 and the ohmic resistance R_Ω, which is essentially the electrolyte resistance, but can also contain the resistance of surface films. Since:

$$R_p^1 = R_p^0 + R_\Omega, \qquad (5)$$

the relative experimental error ϵ:

$$\epsilon = \frac{R_p^1 - R_p^0}{R_p^0} = \frac{R_\Omega}{R_p^0} \qquad (6)$$

equals the ratio R_Ω/R_p^0. Even if R_Ω is low, the error can be appreciable if R_p^0 is also low, as found, for example, in systems with high corrosion rates. The same magnitude of error can occur for systems with low conductivity and low corrosion rates. It is, therefore, not a question of whether the absolute value of R_Ω is low, but whether the ratio R_Ω/R_p^0 is low.

An example of the type of error that can occur when R_Ω is not compensated during the measurement of R_p or subtracted from the experimental value R_p^1 after the measurement according to Equation (5) is given in Figure 2 for the case of iron in ethanol containing 0.01 N HCl.[2] Without compensation of the ohmic drop, R_p^1 = 357 ohm was obtained. Using positive feedback[2] for compensation of R_Ω, which was found to be 212 ohm, R_p^0 = 74 ohm was measured. The error $\epsilon = R_\Omega/R_p^0 = 2.86$ or almost a factor of three. The dangerous consequence of neglecting R_Ω is that corrosion rates will be underestimated.

At present, unfortunately, only very limited instrumentation is available commercially for routine compensation of the ohmic drop. Positive feedback techniques[3] are not very useful for corrosion studies since R_Ω will change with time as soluble corrosion products are formed and especially during anodic polarization. The interrupter technique (see Reference 16 and literature cited) is a useful technique, but at present is not generally available in commercial instrumentation. It is to be expected, however, that the major manufacturers will in the near future introduce potentiostats with an interrupter technique for automatic compensation of the ohmic drop.

Nonlinearity in "Linear Polarization" Experiments

The term *linear polarization* should be reserved for experiments where only one current ΔI-potential ΔE point is recorded as opposed to *polarization resistance* as defined by Equation (2). For linear polarization, R_p is defined as:

$$R_p = \frac{\Delta E}{\Delta I} \qquad (7)$$

It has been realized, of course, for several years that the polarization curve in the vicinity of E_{corr} is linear only for a very limited number of cases,[17] where the anodic and cathodic Tafel slopes are exactly equal. In fact, the nonlinearity of the polarization curve near E_{corr} is used in the various three- and four-point methods for simultaneous calculation of B and R_p.[18-21]

In linear polarization tests, usually an anodic and a cathodic $\Delta I, \Delta E$- pair are measured. Because of the nonlinearity of the polarization curve, a certain error occurs that is different in magnitude and sign for the anodic and the cathodic tests. The error from nonlinearity has been calculated[3,22] and a correction factor has been given. Unfortunately, the possibility of using this correction factor seems to have been ignored in most cases. For a theoretical case with b_a = 30 mV and b_c = 120 mV, the error is −38.5% for ΔE = +10 mV and +22.1% for ΔE = − 10 mV. Some of this error could be reduced by taking the average of these two readings.

The error from nonlinearity is perhaps one of the smaller contributions to the overall error in R_p measurements. However, it needs to be recognized to eliminate the belief occasionally voiced that there are different anodic and cathodic corrosion rates.

The AC Impedance Technique for Measuring the Polarization Resistance

Lorenz and Mansfeld[10] have discussed the application of the AC impedance technique to R_p measurements. The experimental approach for recording impedance data and methods for analysis have been discussed by Mansfeld and Kendig.[8,9,23,24] The polarization resistance R_p is defined as the limit of the impedance spectrum for $f \rightarrow 0$:

$$R_p = \lim_{f \to 0} R_e \{Zf\} \qquad (8)$$

where R_e (Z_f) is the real part of the complex faradaic impedance Z_f. In Figure 1 the frequency-independent part of the impedance modulus $|Z|$ at the lowest frequencies corresponds to R_p according to Equation (8). In a Nyquist plot,[8,9] R_p equals the diameter of the semicircle.

The primary advantages of the impedance techniques are that mechanistic information is obtained, that the measurement is independent of the solution conductivity (in fact R_Ω is obtained in the analysis), and that the electrode capacitance, which is affected by oxide film growth, adsorption of inhibitors, etc., is obtained together with R_p. A disadvantage can be seen in the fact that the Tafel slopes and the parameter B [Equation (1)] cannot be obtained from the impedance spectrum.

Mansfeld and Kendig[25] have compared R_p values obtained from potentiodynamic measurements according to ASTM G 59 (R_p^{dc}) and from impedance data (R_p^{ac}) for AISI 4340 RCE in 0.5 M Na_2SO_4 (Figure 3) and in the same solution containing 10 mM $NaNO_2$ (Figure 4). Complete agreement between both measurements was observed. Similar agreement was noted for tapwater with and without $NaNO_2$.[16] The impedance technique is discussed in more detail in the paper by Kendig, et al.[11]

Cases Where the Polarization Resistance Technique Cannot Be Used

The application of electrochemical techniques such as R_p measurements requires that the corroding metal follows simple corrosion kinetics. Lorenz and Mansfeld[10] have discussed the case of iron and steel in 1 N H_2SO_4 containing propargylic alcohol, 2-butyne-1,4 diol and hexynol as inhibitors, where desorption of the inhibitor and/or its decomposition products at or slightly positive to E_{corr} has made recording and analysis of electrochemical data very difficult. The desorption of the inhibitor produces pronounced hysteresis in the potential sweep in the vicinity of E_{corr} as shown in Figure 5, which leads to great uncertainty in the determination of R_p. Similar problems occur with impedance measurements at E_{corr}, where an inductive loop occurs, the extent of which depends on the magnitude of the applied signal indicating nonlinear behavior. Solution analysis by atomic absorption showed that R_p as determined by the extrapolation of the inductive loop for $f \rightarrow 0$ has no relation to the true corrosion rate. Aksut, et al.,[26] showed that non-steady-state pulse polarization and steady-state and quasi-steady-state cathodic measurements avoiding the nonlinearity of the system gave reasonable agreement with corrosion rates from solution analysis.

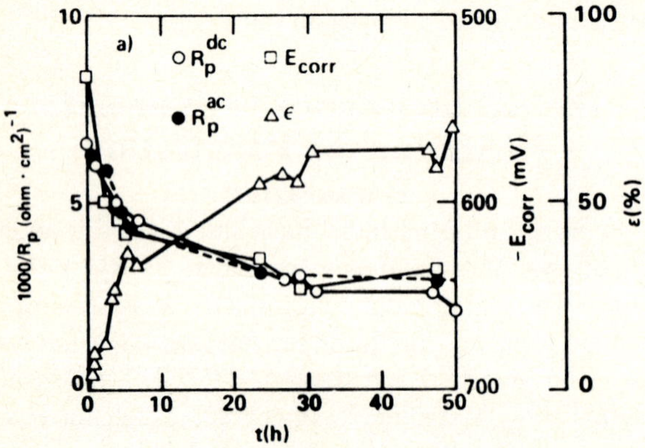

FIGURE 3 — Time dependence of R_p^{ac}, R_p^{dc}, and E_{corr} for AISI 4340 steel RCE in aerated 0.5 M Na_2SO_4.

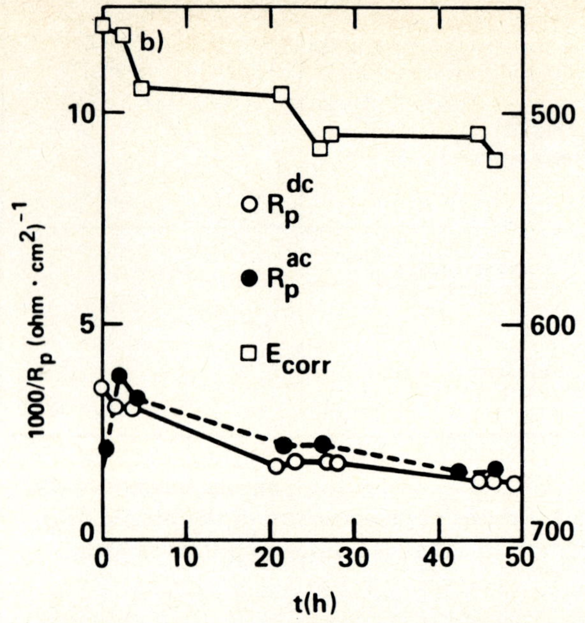

FIGURE 4 — Time dependence of R_p^{ac}, R_p^{dc}, and E_{corr} for AISI 4340 steel RCE in aerated 0.5 M Na_2SO_4 containing 10 mM $NaNO_2$.

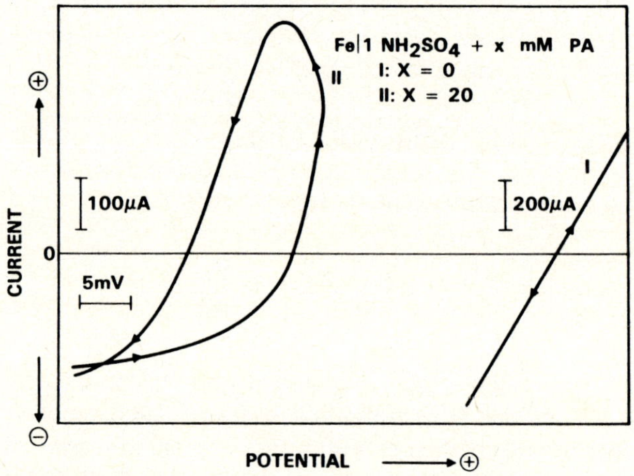

FIGURE 5 — Potentiodynamic sweeps for iron in aerated 1N H_2SO_4 without (curve I) and with 20 mM propargylic alcohol (curve II).

Summary and Conclusions

The polarization resistance technique is a well-developed technique for determining corrosion rates, and its theoretical background is well understood. ASTM G 59 outlines the experimental approach for determining R_p

from a potentiodynamic sweep in the vicinity of E_{corr}.

The sources of error in R_p measurements include uncertainty in the parameter B [Equation (2)], choice of a scan rate that is too fast, neglect of the solution resistance R_Ω [Equation (5)], and nonlinearity. The latter problem only affects linear polarization measurements.

The electrochemical impedance technique has certain advantages especially for measurements in low conductivity media and for systems with low corrosion rates.

As with all electrochemical measurements of corrosion phenomena, great care should be taken to ensure that the system to be studied obeys the simple kinetics that form the theoretical basis for the polarization resistance technique.[3] Examples have been given where R_p measurements carried out in the usual manner give erroneous results.

References

1. ASTM G 59, "Standard Practice for Conducting Potentiodynamic Polarization Resistance Measurements," ASTM, Philadelphia, Pennsylvania.
2. F. Mansfeld, "Polarization Resistance Measurements—Experimental Procedure and Evaluation of Test Data," Electrochemical Techniques for Corrosion, R. Baboian, Ed., National Association of Corrosion Engineers, Houston, Texas, pp. 18, 113, 1977.
3. F. Mansfeld, Adv. Corrosion Science and Technology, M. G. Fontana, R. W. Staehle, Eds., Vol. 6, p. 163, Plenum Press, 1976.
4. F. Mansfeld, Corrosion, Vol. 29, No. 10, p. 397, 1973.
5. S. M. Gerchakov, L. R. Udey, F. Mansfeld, Corrosion, Vol. 37, No. 12, p. 696, 1981.
6. N. D. Greene, R. H. Gandhi, Materials Performance, Vol. 21, No. 7, p. 34, 1982.
7. F. Mansfeld, M. W. Kendig, Materials Performance, Vol. 22, No. 12, p. 40, 1983.
8. F. Mansfeld, Corrosion, Vol. 37, No. 5, p. 301, 1981.
9. F. Mansfeld, M. W. Kendig, S. Tsai, Corrosion, Vol. 38, No. 11, p. 570, 1982.
10. W. J. Lorenz, F. Mansfeld, Corros. Sci., Vol. 21, p. 647, 1981.
11. M. W. Kendig, F. Mansfeld, A. T. Allen, this symposium, paper # ...
12. F. Mansfeld, M. W. Kendig, Corrosion, Vol. 37, No. 9, p. 545, 1981.
12. F. Mansfeld, M. W. Kendig, W. J. Lorenz, "Corrosion Inhibition in Neutral, Aerated Media," J. Electrochem. Soc. (in press).
13. F. Mansfeld, M. W. Kendig, W. J. Lorenz, "Corrosion Inhibition in Neutral, Aerated Media," J. Electrochem. Soc. (in press).
14. F. Mansfeld, M. W. Kendig, Materials Performance, Vol. 22, No. 12, p. 40, 1983.
15. F. Mansfeld, Corrosion, Vol. 32, No. 4, p. 143, 1976.
16. F. Mansfeld, M. W. Kendig, S. Tsai, Corros. Sci., Vol. 22, p. 455, 1981.
17. K. B. Oldham, F. Mansfeld, Corrosion, Vol. 27, No. 10, p. 434, 1971.
18. J. Jankowski, R. Judmiewicz, Corros. Sci., Vol. 20, p. 841, 1981.
19. R. Bandy, Corros. Sci., Vol. 20, p. 1017, 1980.
20. S. Barnartt, M. Donaldson, Corrosion, Vol. 39, No. 1, p. 33, 1983.
21. S. Barnartt, Electrochim. Acta, Vol. 15, p. 1313, 1970.
22. F. Mansfeld, Corrosion, Vol. 30, No. 3, p. 92, 1974.
23. M. W. Kendig, E. M. Meyer, G. Lindberg, F. Mansfeld, Corros. Sci., Vol. 23, p. 1007, 1983.
24. M. W. Kendig, F. Mansfeld, Corrosion, Vol. 39, No. 11, p. 466, 1983.
25. F. Mansfeld, M. W. Kendig, Werkst. und Korros., Vol. 34, p. 397, 1983.
26. A. A. Aksut, W. J. Lorenz, F. Mansfeld, Corros. Sci., Vol. 22, p. 611, 1982.

Primer on the AC Impedance Technique

D. C. SILVERMAN*

Abstract

The AC impedance technique (electrochemical impedance technique) is an emerging technology. The technology is becoming important to the corrosion engineer in accurately predicting corrosion behavior. In fact, the technology is now moving out of the laboratory. This primer is written to provide the background necessary for the corrosion engineer to become sensitized to the technique and the rather unique role it can play in corrosion prediction.

Introduction

Theory and applications of the AC impedance technique are currently so developed that the technique can aid in practical corrosion prediction and control. A gulf exists, however, between the practical corrosion engineer and the laboratory scientist in both the understanding of and the emphasis areas for AC impedance applications. A number of reviews exist on the AC impedance technique.[1-5] The primary problem with the majority of these reviews is that they either tend to be oriented only toward the laboratory without the introduction of "real world" factors or tend to assume a reasonably advanced background in either electronics or mathematics.

Two of the listed reviews that seem to require the least background for understanding are those by EG & G Princeton Applied Research[2] and by Hladky, et al.[5] Two of the reviews are oriented to particular products.[2,3] The purpose of this short overview is to serve as a simple primer for practical corrosion engineers so that they can understand the published literature more easily as well as determine how the AC impedance technique can best serve their needs. A reasonable understanding of the referenced literature is vital in making full use of the technique. Instrumentation is reviewed elsewhere[2-4] and is not discussed here.

Corrosion processes are electrochemical in nature. As such, they involve electron transfer, which facilitates the dissolution of metal. If a means can be found to model this process, then corrosion prediction and control might be facilitated. From a practical standpoint, this outcome is the goal of most industrial efforts to use electrochemical tools.

Technique

In principle, an electrochemical process may be modeled by electrical circuit elements such as resistors, capacitors, and inductors. For example, the corrosion reaction itself can often be modeled by one or more resistors. The ability to model a corrosion process in this manner gives rise to the practical utility of the AC impedance technique. The corrosion engineer can use established AC circuit theory to characterize the electrochemical corrosion process. Such characterization can facilitate understanding and lead to more accurate predictions of corrosion rates and overall corrosion behavior.

Before examining some of the different models that might be established, a brief theoretical overview is necessary. The purpose is to introduce the physical meaning and use of imaginary numbers, which are important to interpreting the data from and creating models for the AC impedance technique. Direct current can be viewed as alternating current in the limit of zero frequency. Under conditions of direct current, e.g., zero frequency, Ohm's Law can be written as:

$$E = IR^{(1)} \tag{1}$$

*Monsanto Chemical Company, St. Louis, Missouri.

(1)All symbols are defined at the end of the paper.

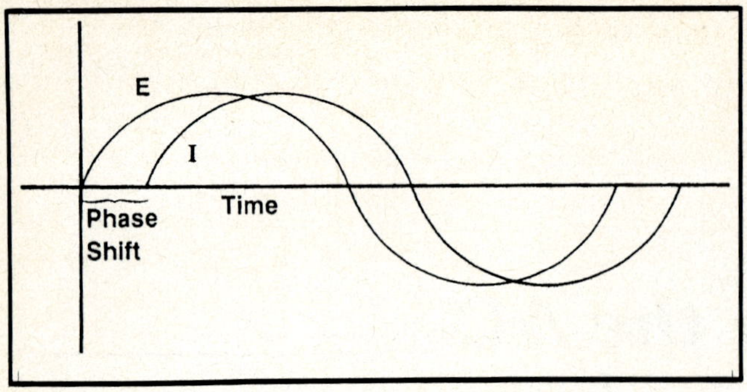

FIGURE 1 – Sinusoidal AC voltage and current signals.

In this case, the resistance is composed of only one or more actual resistors. When the frequency is not zero, as would occur from imposition of an alternating current, Ohm's Law becomes:

$$E = IZ \qquad (2)$$

Under these conditions, the resistance is caused by all circuit elements that can impede the flow of current, e.g., resistors, capacitors, and inductors. The value of the resistance created by capacitors and inductors depends on frequency. The value of the resistance created by a resistor is independent of frequency.

When an AC voltage sine or cosine wave is applied across a circuit composed only of a resistor, the resultant current is also a sine or cosine wave of the same frequency, with no phase shift but with a different amplitude. If the circuit consists of capacitors and inductors, the resulting current will not only differ in amplitude but will also be shifted in time, i.e., have a phase shift. This phenomenon is shown in Figure 1.

Use of sines and cosines are mathematically cumbersome. Vector analysis, however, provides a convenient method of describing the analogous circuit in mathematical terms. The relationship between such vector analysis and imaginary numbers provides the basis for the AC impedance analysis. A sinusoidal current or voltage can be pictured as a rotating vector as shown in Figure 2. In this figure, the current rotates at a constant angular frequency, f (hertz) or ω (radians-s^{-1} = $2\pi f$). For example, standard AC current can be pictured as a vector that rotates at 60 hertz or 377 radians-s^{-1}. In Figure 2, the x component defines the observed current. Therefore, it becomes the *real* component of the rotating vector. The y component is a contribution that is not observed; therefore, it is termed the *imaginary* component of the rotating vector. The mathematical description of the two components is written as:

$$\text{Real current} = I_x = |I|\cos(\omega t) \qquad [3(a)]$$

$$\text{Imaginary current} = I_y = |I|\sin(\omega t) \qquad [3(b)]$$

$$|I|^2 = |I_x|^2 + |I_y|^2 \qquad [3(c)]$$

The voltage can be pictured as a similar rotating vector

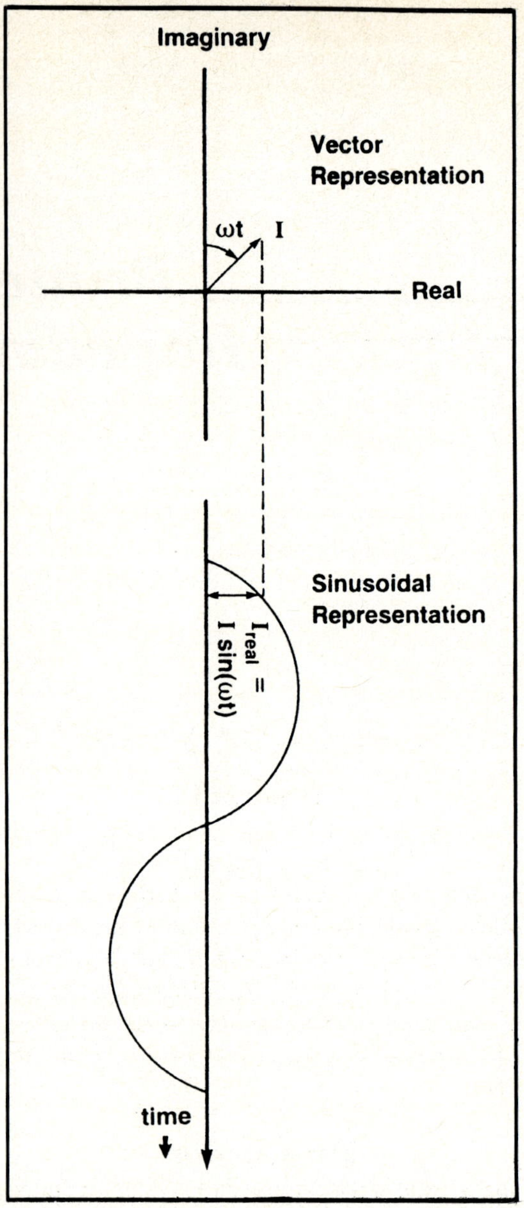

FIGURE 2 – Relationship between sinusoidal AC current and rotating vector representation.

with its own amplitude, E, and the same rotation speed, ω. As shown in Figure 3, when the current is in phase with the applied voltage, the two vectors are coincident and rotate together. This response is characteristic of a circuit containing only a resistor. When the current and voltage are out-of-phase, the two vectors rotate together at the same frequency, but are offset by an angle called the *phase angle shift*, θ. This response is characteristic of a circuit that contains capacitors and inductors in addition to resistors.

In the measurement of AC impedance, one vector is viewed using the others as a frame of reference. Thus, the reference point rotates and the time dependence of the signals (ωt) is not viewed. In addition, both the current and voltage vectors are referred to the same reference frame. The voltage vector is divided by the current vector to yield the final result in terms of the impedance (voltage divided by current), as shown in Figure 4.

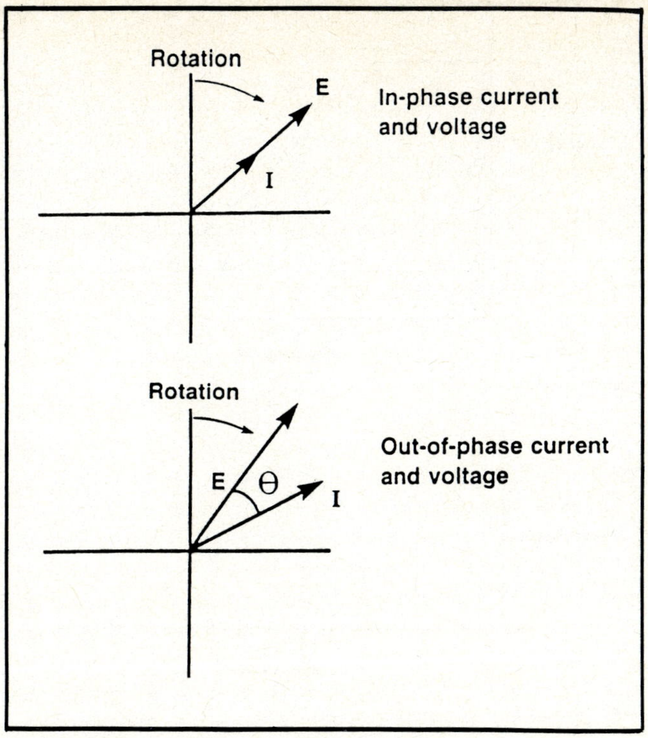

FIGURE 3 – In-phase and out-of-phase rotation of current and voltage vectors.

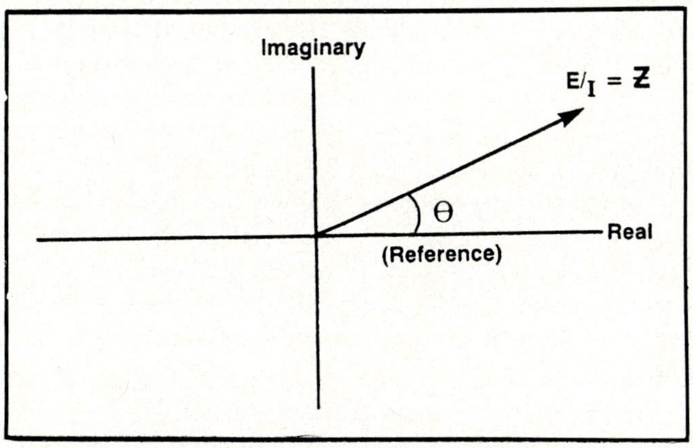

FIGURE 4 – Impedance vector.

The mathematical convention for separating the real (x) and imaginary (y) components is to multiply the magnitude of the imaginary contribution by j and report the real and imaginary values separately. The equations for AC impedance become:

$$E = E_{real} + E_{imaginary} = E' + jE'' \qquad [4(a)]$$

$$I = I_{real} + I_{imaginary} = I' + jI'' \qquad [4(b)]$$

$$Z = Z' + jZ'' = \frac{E' + jE''}{I' + jI''} \qquad [4(c)]$$

$$\tan\phi = \frac{Z''}{Z'} \qquad [4(d)]$$

TABLE 1 – AC Impedance Basic Circuit Elements

Element	AC Impedance Equation	
Resistor	$Z = R$	
Capacitor	$Z = -\frac{1}{j\omega C}$	$\omega = 2\pi f$
Inductor	$L = j\omega L$	$\omega = 2\pi f$

$$|Z| = \sqrt{(Z')^2 + (Z'')^2} \qquad [4(e)]$$

The goal of AC impedance is to measure the impedance Z as Z′ and Z′′ and then to use simple circuit analogues to model the response. If the amplitude of the voltage excitation is kept low (e.g., usually 10 mV or less), the measured response may often be assumed to be composed of a linear array of resistors, capacitors, and inductors. Then, the electrode response can, in principle, be modeled in terms of such an array. The three basic circuit elements can be written as shown in Table 1.

Table 1 shows that a resistor has only a real contribution. That is, the response of a resistor would be parallel to the real axis, independent of frequency. Both the capacitor and inductor have purely imaginary contributions. Their contributions would be parallel to the imaginary axis. The object of AC impedance is to model the corrosion process in terms of circuit elements such as those shown in Table 1 and from that model to be able to make conclusions about characteristics of the corrosion process. Since measurements can be taken at the corrosion potential, the characteristics are obtainable at the alloy state found in practice.

Simple Corrosion Process

The simplest type of corrosion process would be the combination of a corrosion reaction and a double layer. One example of this type of process is uniform corrosion involving one chemical reaction. For example, the corrosion of carbon steel in 1 molar (M) sulfuric acid falls into this category.[6] The reaction

$$Fe + 2H^{+} \rightarrow H_2 + Fe^{+2} \qquad (5)$$

describes the corrosion reaction. This reaction can be represented by a simple resistor. The double layer is created by the fact that the surface of a metal in contact with any oxidizing agent, such as water or oxygen, is itself an oxide or hydroxide. The latter component would tend to ionize in water so that the surface becomes charged.[7] Counter-ions would be present in the liquid layer adjacent to the metal surface to balance the surface charge. This structure responds to an AC voltage signal in a way analogous to a capacitor. If the real component Z′ and

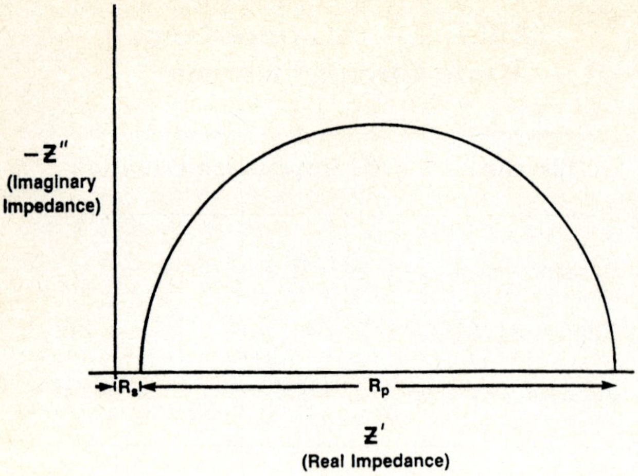

FIGURE 5 – Nyquist plot of simple charge transfer corrosion process.

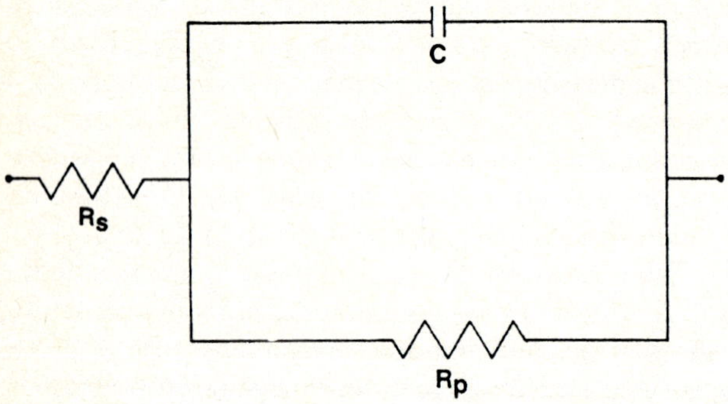

FIGURE 6 – Circuit that models impedance in Figure 5.

the negative of the imaginary component Z'', measured as a function of frequency, are plotted against each other for this type of simple corrosion process, the plot would appear as in Figure 5. This method of plotting is called the *Nyquist plot*.

For this simple process, the model circuit is that shown in Figure 6. The circuit is a resistor, R_p, in parallel with a capacitor, C. The entire parallel circuit is in series with another resistor, R_s. The utility of AC impedance lies in the fact that R_s equals the solution resistance uncompensated by the potentiostat and R_p equals the polarization resistance. For the case of carbon steel corrosion in 1 M sulfuric acid, for example, R_p is inversely proportional to the corrosion rate. This relationship is found for many processes. Use of R_p with Tafel slopes allows the corrosion rate to be estimated.[6] Thus, the AC impedance technique enables the corrosion rate to be estimated rapidly in the absence of uncompensated solution resistance and at the corrosion potential.

If Figure 6 adequately describes the corrosion process, the plot shown in Figure 5 can be described by Equation (6).

$$Z = R_s + \frac{R_p}{1 + \omega^2 R_p^2 C^2} + \frac{j\omega R_p^2 C}{1 + \omega^2 R_p^2 C^2} \quad (6)$$

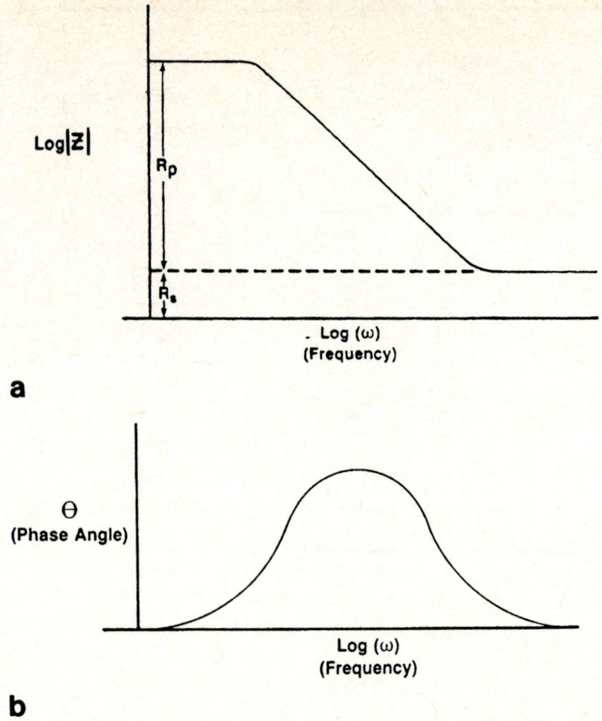

FIGURE 7 – (a) and (b): Bode type of plots of circuit in Figure 6.

Considerable attention has been paid to analyzing the circuit described by Equation (6).[1] By examining the impedance at appropriate frequency limits, accurate values of R_s, R_p, and C can be obtained. A full explanation of how Equation (6) can be used to understand the simple corrosion process shown in Figure 5 is given elsewhere.[1]

An alternative method of plotting the circuit in Figure 6 is by means of Bode plots. These plots, shown in Figures 7(a) and (b) are plots of impedance modulus, $|Z|$, versus frequency on log-log paper and the inverse of the phase angle shift versus frequency on semilogarithmic paper.

This method of plotting the data provides more information than that of Figure 5 because the frequency is explicit in Figures 7(a) and (b), but is implicit in Figure 5. Certain characteristics of these Bode plots can be used to obtain the circuit elements, R_s, R_p, and C. For example, R_s is the high frequency limiting value of Z. R_p is the difference between the low frequency limit and the high frequency limit. The double layer capacitance can be evaluated from the peak in the phase angle plot according to:

$$\omega_{\theta max} = \frac{1}{CR_p}\sqrt{1 + \frac{R_p}{R_s}} \quad (7)$$

where $\omega_{\theta max} = 2\pi f_{max}$. Further analysis can be made.[1] For a more complete discussion see Reference 1.

Unfortunately, a number of corrosion processes are not as simple as the case of one reaction dominating the process as represented by the circuit in Figure 6. These complications can arise from a number of sources.

Diffusion Control

The rate of a chemical reaction may be strongly in-

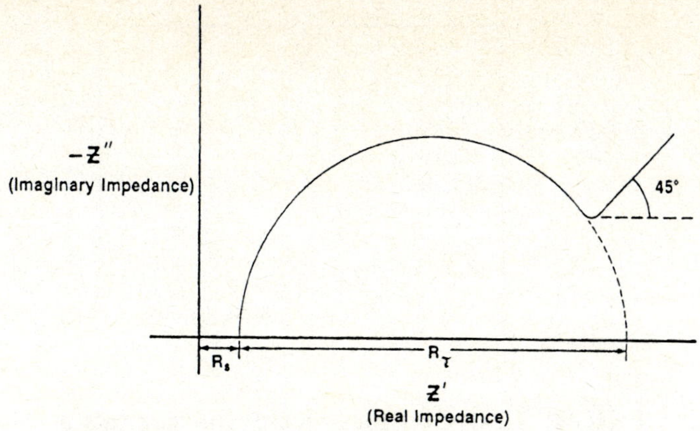

FIGURE 8 – Nyquist type of plot of simple charge transfer process in the presence of diffusion.

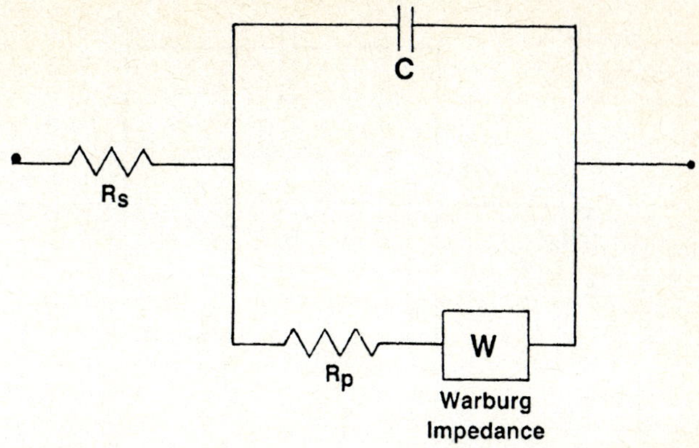

FIGURE 9 – Circuit that models impedance in Figure 8.

fluenced by the diffusion of one or more of the reactants or products to or from the surface. This situation can arise when diffusion through some type of surface film becomes the dominating process. This process may result in the surface being covered with reaction products or adsorbed solution components. An example of this type of corrosion process that has extreme practical importance is the corrosion of carbon steel in concentrated sulfuric acid. Such corrosion has been shown by Ellison and Schmeal to be controlled by the diffusion of $FeSO_4$ from a saturated film at the surface to the bulk fluid.[8]

The AC impedance response to a corrosion process dominated by diffusion control frequently has a unique characteristic known as the *Warburg impedance*. The Nyquist plot created by this phenomenon is shown in Figure 8. In the low frequency limit, the current is a constant 45 degrees out-of-phase with the potential excitation. The impedance response may ultimately deviate from this relationship and return to the real axis at very low frequencies.[9] The practical significance of this result is that if the diffusion control is in the fluid phase, the corrosion process might be sensitive to fluid flow.

The equivalent circuit is shown in Figure 9. The term W is the Warburg impedance. It creates the conditions so that the real and imaginary components of the impedance are equal at lower than a certain frequency. The value of Z for a diffusion controlled system varies inversely with the square root of the frequency. By appropriate manipulation of the data in Figure 8, the appropriate values of the circuit elements can be evaluated.[2,3] For example, for a reversible reaction under pure diffusion control, the Warburg impedance is given by:

$$Z_W = \frac{\sigma}{\sqrt{\omega}} - j\frac{\sigma}{\sqrt{\omega}} \qquad (8)$$

The Warburg coefficient, σ, can be used to calculate a diffusion coefficient. As indicated in Figure 8, the extrapolation of the high frequency semicircle back to the real axis yields R_T, a quantity that may be inversely proportional to the rate of the charge transfer corrosion process.

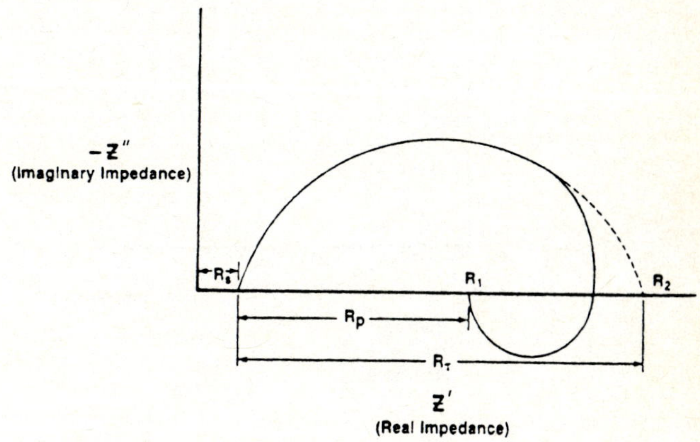

FIGURE 10 – Nyquist type of plot of simple charge transfer corrosion process in the presence of pseudo-inductance.

Inductance

The Nyquist plot will frequently have a low frequency portion that lies below the real axis. This behavior seems to arise from one of a number of sources,[3,10,11,12] for example, some types of adsorption processes. There is no uniform agreement as to the cause of inductive type of behavior. Indeed, this inductance may aptly be termed *pseudo-inductance* since the processes giving rise to inductance are probably not like those of a real inductor.[13] An example of an impedance response in the presence of inductance is shown in Figure 10.

The polarization resistance R_P is defined as R_1-R_S while the charge transfer resistance R_T is defined as R_2-R_S. There is controversy over which value is inversely proportional to the corrosion rate. Usually, R_1-R_S is used for corrosion rate calculations. Under some circumstances R_2-R_S seems to be more appropriate,[11] although this case may be unique.[10] In either case, evaluations of R_S, R_P, and R_T are important because either or both R_P and R_T may be related to the corrosion rate or corrosion process. Again, R_S is the solution resistance.

The polarization resistance R_P is defined as R_1-R_S while the charge transfer resistance R_T is defined as R_2-R_S. There is controversy over which value is inversely

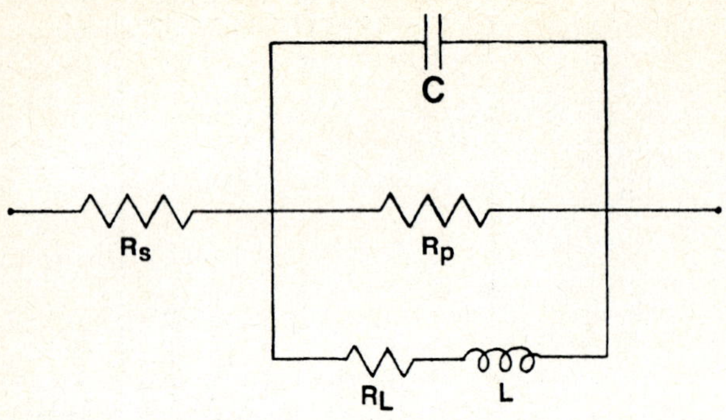

FIGURE 11 – Circuit that models impedance in Figure 10.

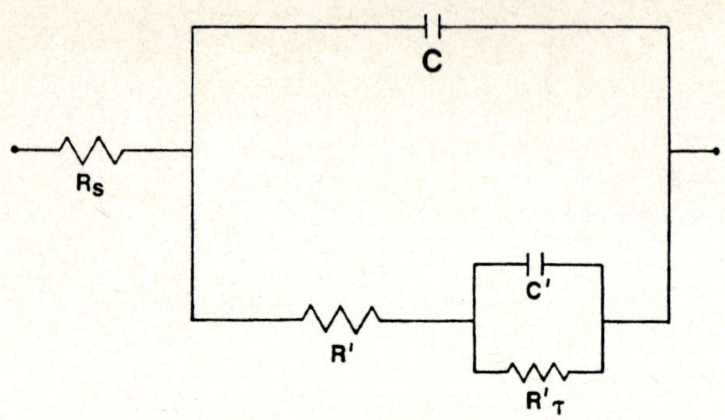

FIGURE 13 – Complex circuit that models a coated metal.

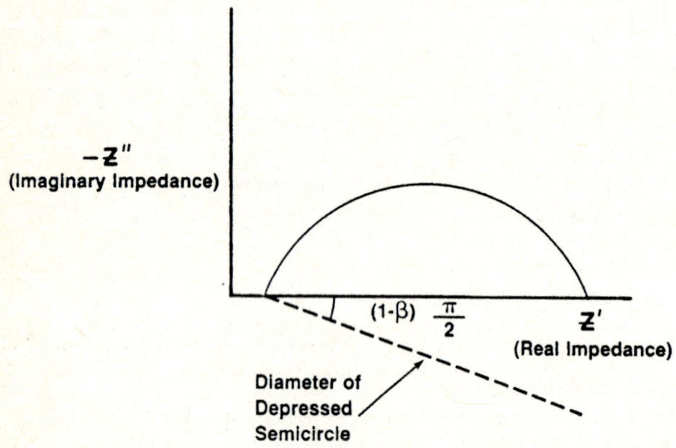

FIGURE 12 – Nyquist type of plot showing depression below the real axis.

proportional to the corrosion rate. Usually, R_1-R_S is used for corrosion rate calculations. Under some circumstances R_2-R_S seems to be more appropriate,[11] although this case may be unique.[10] In either case, evaluations of R_S, R_p, and R_T are important because either or both R_p and R_T may be related to the corrosion rate or corrosion process. Again, R_S is the solution resistance.

If there is one time constant, the circuit giving rise to the response in Figure 10 might be modeled as shown in Figure 11. Such a circuit may prove useful in the analysis of the response. This circuit can be solved as long as R_p can be estimated.[13] The accuracy of the values of R_p and R_L so calculated can be ascertained by comparing the calculated Nyquist and Bode plots with the measured Nyquist and Bode plots. Thus, the corrosion rate can be estimated in the presence of inductance.

Depression of Nyquist Semicircle

In real systems, the Nyquist type of semicircle for a simple corrosion process often exhibits some depression below the real axis; an example is shown in Figure 12. The cause of this behavior is believed to be a repartitioning of time constants around a central value.[3] Two explanations have been offered. The depression may be caused by increased surface roughness or by geometrical effects leading to a nonuniform repartitioning of the current density on the surface.[3] The practical significance of this depression of the semicircle is the fact that Figure 12 and not Figure 5 often represents the appearance of a real Nyquist plot of even a simple charge transfer process. Examples that can fit this characteristic include carbon steel in 1 M sulfuric acid and carbon steel in water.[14,15] Thus, the ability to extract the polarization resistance from this type of curve is important if the data are to be used for corrosion rate estimates. The circuit creating such depression is best described by:[14]

$$Z = R_S + \frac{R_p}{1 + (j\omega\tau)^{\beta}} \tag{9}$$

In Equation (9), the phenomenological term $(j\omega\tau)^{\beta}$ replaces the term $j\omega R_p C$ for the case when $\beta < 1$. In Figure 5, $\beta = 1$. The value of R_p can still be estimated by curve-fitting the semicircle and by allowing both the radius and origin to vary.[14,15] Thus, corrosion rates can still be estimated even in the presence of such depression.

More Complex Phenomena

When a metal is coated with a porous nonconducting film, the equivalent circuit must simultaneously take account of the polarization resistance caused by the corrosion process (R_p) and the pore resistance (R'). The circuit can appear as shown in Figure 13.[16] In Figure 13, corrosion is assumed to occur only under the coating. The circuit can be analyzed mathematically to retrieve the values of the various circuit elements.[14] In fact, Figure 13 could also represent corrosion in a simulated pit or crevice.[14]

Sometimes, two or more time constants can arise for a corrosion process.[17] The causes can be a multistep reaction in which both steps have comparable rates, an adsorbed intermediate, and electrocrystallization, among others. These phenomena require more complex equivalent circuits. However, the procedure for analyzing these complex responses still requires the proposal of an equivalent circuit and then the examination of the predicted versus measured frequency response to verify the model. Even in these cases, the value of the polarization resistance(s) can be estimated from an appropriate model so that the corrosion rate and behavior can be understood and corrosion predictions can be made.

Acknowledgment

H. P. Hack and A. S. Krisher are thanked by the author for their valuable comments on this manuscript.

Symbols

C	=	capacitance (farad $-cm^{-2}$)
E′	=	real component of voltage (volts)
E′′	=	imaginary component of voltage (volts
E	=	complex voltage (volts)
f	=	frequency (s^{-1})
I′	=	real component of current (amp-cm^{-2})
I′′	=	imaginary component of current (amp-cm^{-2})
I	=	complex current (amp-cm^{-2})
j	=	$\sqrt{-1}$
L	=	inductance (henry)
R_S	=	solution resistance (ohm-cm^2)
R_P	=	polarization resistance (ohm-cm^2)
R_T	=	charge transfer resistance (ohm-cm^2)
Z′	=	real component of impedance (ohm-cm^2)
Z′′	=	imaginary component of impedance (ohm-cm^2)
Z	=	complex impedance
β, τ	=	phenomenological coefficients caused by depression of the Nyquist plot below the real axis
θ	=	phase angle (deg)
ω	=	frequency (radians-s^{-1})

Subscripts

x	=	in-phase component
y	=	out-of-phase component

References

1. F. Mansfeld, Corrosion, Vol. 36, No. 5, p. 301, 1981.
2. "Basics of AC Impedance Measurements," EG&G Princeton Applied Research, Application Note AC-1, EG&G Princeton Applied Research, Princeton, New Jersey.
3. C. Gabrielli, "Identification of Electrochemical Proc- Instrumentation Group, 1980, Solartron, Farnborough, Hampshire, UK.
4. D. D. Macdonald, M. C. McKubre, "Impedance Measurements in Electrochemical Systems," Modern Aspects of Electrochemistry, J. O'M. Bockris, B. E. Conway, R. E. White, Eds., Vol. 14, Plenum Press, New York, New York, p. 61, 1982.
5. K. Hladky, L. M. Callow, J. L. Dawson, Brit. Corros. J., Vol. 15, No. 1, p. 20, 1980.
6. F. Mansfeld, Corrosion, Vol. 29, p. 397, 1973.
7. J. O'M. Bockris, A. K. N. Reddy, Modern Electrochemistry, Plenum Press, New York, New York, 1970.
8. B. T. Ellison, W. R. Schmeal, J. Electrochem. Soc., Vol. 125, p. 524, 1978.
9. J. L. Dawson, D. G. John, J. Electroanal. Chem., Vol. 110, p. 37, 1980.
10. W. J. Lorenz, F. Mansfeld, Corros. Sci., Vol. 21, No. 9, p. 647, 1981.
11. I. Epelboin, M. Keddam, H. Takenouti, J. Appl. Electrochem., Vol. 2, p. 71, 1972.
12. R. D. Armstrong, K. Edmondson, Electrochim. Acta., Vol. 18, p. 937, 1973.
13. D. D. Macdonald, J. Electrochem. Soc., Vol. 125, No. 12, p. 2062, 1978.
14. M. W. Kendig, E. M. Meyer, G. Lindberg, F. Mansfeld, Corros. Sci., Vol. 23, No. 9, p. 1007, 1983.
15. D. C. Silverman, unpublished results, 1984.
16. F. Mansfeld, M. W. Kendig, S. Tsai, Corrosion, Vol. 38, p. 9, 1982.
17. R. D. Armstrong, M. F. Bell, A. A. Metcalfe, "The AC Impedance of Complex Electrochemical Reactions," Electrochemistry, Vol. 6, p. 98, The Chemical Society, Burlington House, London, England, 1976.

Application of the Rapid-Scan Potentiodynamic Polarization Technique for Corrosion Studies

O. W. SIEBERT*

Abstract

Electrochemical techniques have been used to study fundamental phenomenological corrosion reactions of metals in corrosive environments. The two methods used for the generation of anodic polarization curves for those studies differ in the rate of change of the applied electrical potential. Neither method is more correct to the exclusion of the other; while both techniques provide a commonality of understanding of the electrochemical processes that result in metallic corrosion, each supplies a unique insight into that evaluation. A review of the so-called rapid-scan potentiodynamic polarization technique is presented, along with examples of its use as a tool for the accurate assessment of actual corrosion behavior.

Introduction

In 1903 Whitney proposed that corrosion of metals in aqueous electrolytes was an electrochemical reaction.[1] Since then, many others have contributed to an understanding of the close relationship of electrochemistry to thermodynamics, chemical kinetics, and atomic and molecular structure; some of the major researchers are cited.[2-6] The application of electrochemistry to an understanding of the basic theories of corrosion as well as to the solution of many engineering problems has also been covered in the literature.[7,8]

In the study of the anode-cathode polarization behavior of a metal-environment system, the potentiodynamic polarization technique provides a plot of the relationship of current changes resulting from potential changes, most often presented as a plot of log current density vs potential, or Evans' diagram.[9] The two methods generally used for the generation of anodic polarization curves for corrosion studies differ in the rate of change of the applied electrical potential. Neither method is more correct to the exclusion of the other; while both techniques provide a commonality of understanding of the electrochemical processes that result in metallic corrosion, each method supplies a unique insight into that evaluation. In any electrochemical study, the freely corroding potential/open-circuit potential (E_{OCP} or E_{CORR}), is the only "window" into the real world of corrosion, regardless of the scan rate. All first-principle explanations cover only the polarization techniques, not the scan rates.

A review of the so-called rapid-scan potentiodynamic polarization technique is presented, with comments about the influence of the potential sweep rate.[8] The results of the studies cited in this paper illustrate the benefits of using this electrochemical method and demonstrate that they can provide an accurate assessment of actual behavior.

Background

Electrochemical Theory

If a metal is placed in an electrolyte, it may react with the electrolyte (corrode) to form metal ions. As the ions are ejected into the solution, electrons are left at the metal surface. At the interface, there is a separation of electrical charge. The potential, E, or corrosion potential, exhibited by a metal is a result of this separation. The metal behaves as a single electrode in an electrical cell; a potential exists between the metal and the electrolyte. This potential cannot be measured directly, but this charge separation can be indirectly compared to the charge separation of some reference, e.g., a standard saturated calomel electrode (SCE). A complete electric cell is estab-

*Siebert Materials Engineering, Inc.

lished, i.e., metal to electrolyte to standard electrode. The voltmeter measurement of the cell potential is used as a unique description of the interface condition, and is reported with respect to a SCE. This value is called the freely corroding potential and indicates the state of corrosion of that metal-electrolyte at that exact set of conditions.

For any given metal-electrolyte system, if an electrical potential exists, there must be a corresponding current. This current (as current density) is directly related to the magnitude of the corrosion of the system, and can be expressed as a corrosion rate by means of Faraday's law. This proposition applies to any point on the metal surface, but it cannot be measured directly. The overall current density is measured and averaged for the exposed area of the metal specimen.

Polarization Scan Rates

Sweeping a range of potentials in the anodic (more electropositive) direction of a potentiodynamic polarization curve at a high scan rate of about 60 V/h (high by the perspective of the corrosion scientist, slow by the reference of a physical chemist) is done to indicate regions where intense anodic activity is likely. Secondly, for otherwise identical conditions, sweeping at a relatively slow rate of potential change of about 1 V/h will indicate regions where relative inactivity is likely. The rapid sweep of the potential range minimized film formation, so that the currents observed relate to relatively film-free or thin-film conditions. The object of the slow sweep rate experiment is to allow time for filming to occur.[10] A zero scan rate provides the opportunity for maximum stability of the metal surface, but at high electropositive potentials, the environment could be affected or changed. A rapid-scan rate compromises the steady-state nature of the metal surface but helps maintain the stability of the environment. Whenever possible, corrosion tests should be conducted using as many of the techniques available (potentiodynamic polarization at various scan rates, crevice, stress, velocity, etc.) A holistic evaluation of these results can greatly reduce or temper their individual limitations.

Slow-Scan Technique. In the polarization procedure, ASTM Standard G 5, all oxygen in the test solution is purged with hydrogen for a minimum of one half hour before introducing the test specimen.[12] The test material is then allowed to reach a steady state of equilibrium (open circuit corrosion potential, E_{CORR}) with the test medium before the potential scan is conducted. Starting the evaluation of a basically passive alloy that is already in its "stable" condition precludes any detailed study of how the metal reaches that protected state (the normal intersection of the theoretical anodic and cathodic curves is recorded as a zero applied current on the ASTM potentiostatic potential vs applied diagram). These intersections between the anodic and cathodic polarization curves occur where the total oxidation rate equals the total reduction rate.[13]

Rapid-Scan Corrosion Behavior Diagram. When a rapid-scan polarization study is conducted by the Morris and Scarberry method, the combined plots are called a *corrosion behavior diagram* (CBD)[14] (Figure 1). This technique includes both an initial and a final cathodic polarization curve in the electropositive and electronegative directions, respectively. Basically, the same equipment used in the conductance of a slow-scan polarization study is used for rapid-scan CBDs (that is, a standard test cell, potentiostat, voltmeters, log converters, X-Y recorders, and electronic potential scanning devices). The differences are in technique; the slow-scan is run at a potential sweep rate of about 0.6 V/h, the rapid-scan CBDs at about 50 V/h.[14,15]

In the conductance of rapid-scan CBDs, the test specimen is mechanically and chemically cleaned immediately before immersion in the test cell. Without any further delay, a cathodic polarization scan is made at a sweep rate of about 50 V/h (compared to the 0.6 V/h for the ASTM method). This rapid cathodic charging (hydrogen evolution) further cleans the oxygen specimen so that the subsequent anodic polarization enables a full study of how the corrosion reaction progresses as the potential is changed potentiodynamically in the more noble (positive) direction without the interference of any pre-existing passive or corrosion product film or scale. The test is run open to the atmosphere without any purge gas, and the rapid-scan rate permits completion of the

Corrosion Behavior Diagram

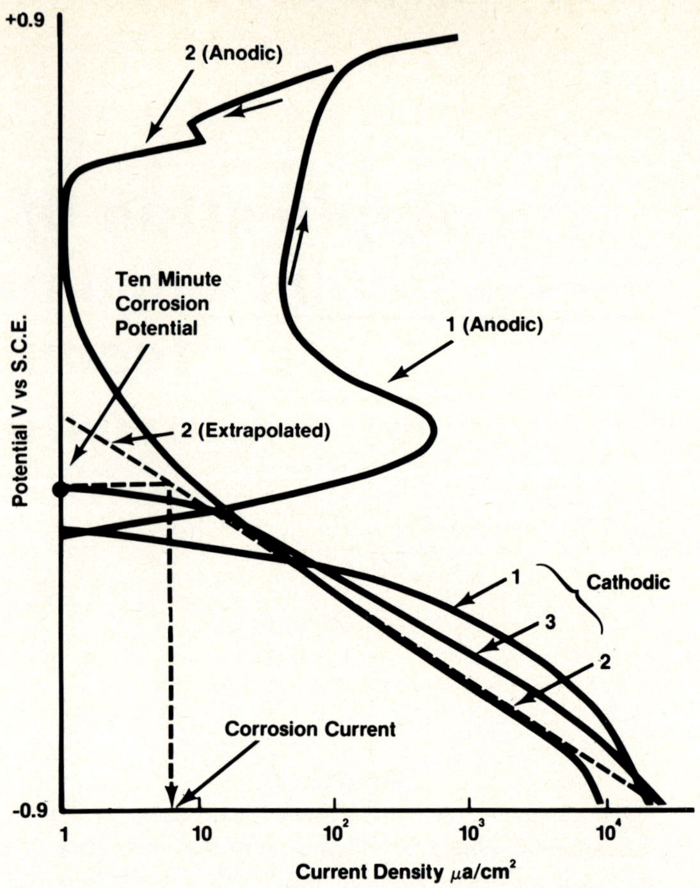

FIGURE 1 — Typical corrosion behavior diagram (CBD) for an active-passive alloy in a corrosive environment.

anodic curve before oxygen is replenished. This cathodic cleaning further activates the specimen and allows it to follow the pattern of its corrosion reaction(s) before it has an opportunity to spontaneously passivate or return to equilibrium with its environment.[11,14-16]

Different from the slow-scan technique, which is generally limited to Case I alloy-medium systems, the rapid-scan technique allows full anodic polarization study of alloys showing all Case I, II, and III behavior. In Case I, the single reversible corrosion potential (the anode-cathode intersection) is in the active region. A wide range of corrosion rates is possible. In Case II, the cathodic curve intersects the anodic curve at three potentials, one in the active region and two in the passive. Since the middle active-passive intersection is not stable, the intersections indicate the possibility of very high corrosion rates, depending on the environment or even slight changes to the exposure/environment system. In Case III, the curves intersect in the most stable, passive region; the alloys generally passivate spontaneously and exhibit low corrosion rates. Case II exhibits the most corrosion; it is difficult to study and presents the greatest risk for selection of materials of construction. Anything that can change the oxygen solubility of the oxidizing agent can alter the corrosion reaction. Environmental conditions seemingly identical from one day to another, from one laboratory or plant to another, and so forth, can give rise to widely differing results. Those desiring a more detailed review of the subject are directed to the works of Pourbaix, Edeleanu, and other works referenced therein.[17-20]

The conductance of a CBD is generally as follows: At a rapid scan rate of about 50 V/h, the clean sample is cathodically charged (polarized), starting at an electronegative position (with respect to SCE) that will produce a current density of 10^4 $\mu A/cm^2$ (nickel, chromium, and iron alloys in aggressive environments) and proceed in a noble (positive) direction. After completing the cathodic curve, proceed anodically until the current density reaches 10^3 $\mu A/cm^2$. The scan direction is reversed and potentials/currents are recorded back to the starting point. The potentiostat is then shut off. The freely corroding potential or open circuit potential (OPC) is recorded after 10 min (or when it is stable; that is, the potential value remains constant). A third cathodic polarization curve is then generated by scanning from the corrosion potential in the negative direction to i_{start}, again at a scan rate of 50 V/h. As noted earlier, this diagram can be produced in about 15 min.[15]

The CBD diagram can provide information about the performance of an alloy-medium system. The technique can be used for a direct calculation of the corrosion rate as well as for indicating the conditions of passivity and tendency of the metal to suffer local pitting and crevice attack.

Somewhat similar to the Tafel extrapolation method, the corrosion rate is found by intersecting the extrapolation of the linear portion of the second cathodic curve with the equilibrium stable corrosion potential. The intersection corrosion current is converted to a corrosion rate [mils per year (mpy)] by use of a conversion factor (based on Faraday's law—the electrochemical equivalent of the metal, its valence, and gram atomic weight). For 13 alloys, the conversion factor ranged from 0.42 for nickel to 0.67 for Hastelloy[(1)] B or C. For a quick determination, 0.5 was used in all of these studies. Generally, the accuracy of the corrosion rate calculation depends on the degree of linearity of the second cathodic curve; when it is less than 1/2 decade long, it becomes more difficult to extrapolate accurately. Obviously, the longer the better.[20,21]

An expected order of magnitude of corrosion may be identified by the CBD. Generally, the third cathodic curve will fall between the first two, or the first and third curves will coincide and fall positive to the second curve. These two curve configurations are the only valid relationships for the projection of accurate corrosion rates. When, however, the second or third curve falls positive to the other two curves, a very high corrosion rate may be indicated. Another possibility is that the second or third curve in this positive position is not controlling corrosion because of surface films, etc.[16,18]

The anodic nose of the first curve describes the primary passive potential (EPP) and critical anodic current density (the transition from active to passive corrosion). If the initial active to passive transition is 10^3 $\mu A/cm^2$ or less, the alloy will spontaneously passivate in the presence of oxygen or any strong oxidizing agent.[22]

The general literature for predicting pitting tendency with the slow-scan reviews the use of the reverse scan (if a hysteresis loop develops and returns to the repassivation potential below the FCP, E_{CORR}, the alloy will pit at crevices; when the value of the repassivation potential identified by the return of the hysteresis loop is above the FCP, E_{CORR}, the pits will tend to grow.[23] These rules of thumb have been expanded for rapid-scan. (1) If E_{CORR} is equal to or greater than the pitting potential, E_{PIT}, both pitting and crevice type attack are likely to occur on the specimen in an immersion test. (2) If E_{CORR} is less than E_{PIT}, pitting type attack of the specimen is unlikely in an immersion test. (3) If the repassivation potential, E_{REPASS}, is greater than E_{CORR}, crevice type attack is possible in an immersion test. (4) If E_{CORR} is equal to or larger than E_{REPASS}, crevice type attack is possible in an immersion test. (5) If no E_{PIT} can be determined but cathodic curves 1 and 2 cross at E_X forming a hysteresis loop, crevice type attack during the rapid-scan is indicated. (6) If E_X is equal to or greater than E_{PIT}, pitting type attack will be predominant during the rapid-scan test. (7) If E_{PIT} is equal to or greater than E_X + 100 mV, crevice type attack is indicated during the rapid-scan test. (8) If the plot of the No. 1 curve continues from E_{PIT} with a large increase in current density vs a small increase in voltage, the test specimen will suffer vigorous local attack during the rapid-scan test. (9) If the No. 1 curve, positive with respect to the value of E_{PIT}, is plotted with a small increase in current density per large increase in voltage, the test specimen in the rapid-scan test will suffer only mild local attack.[11,13,15,21]

With regard to the limits of CBD use, these very limitations often provide information, e.g., the corrosion

---

[(1)]Registered trade name.

rate will be too high, etc. Only in the case of interpretation of unstable active-passive conditions are the results of the rapid scan seen as a possible weakness of the technique (but it is not handled any better by other techniques). When the FCP is noble (positive) to the Epp, the alloy is passive; when the FCP is active (negative) to the Epp, the alloy will corrode in the active state; when the FCP is only slightly more noble (positive) than the Epp, the alloy could show active/passive tendencies. When the FCP oscillates and will not stabilize, the potential should be plotted against time. The alloy will probably not be resistant without anodic protection but in these latter two cases, the CBD does not indicate for certain.

In order to expedite the interpretation of the 50.0 V/h anodic polarization data, three logic sequence diagrams (LSD) were constructed.[11] These diagrams can be linked to each other to provide a more complete behavior analysis. These LSDs contain "benchmark" values that have been derived from conducting more than 10,000 potentiodynamic experiments at scan rates of either 50.0 or 60.0 V/h.

Experimental Procedure

Test Specimen Preparation

Rectangular immersion specimens of nominal 6 × 25 × 51 mm for laboratory and field exposures were prepared as outlined in the NACE Test Method TM-01-69.[24] Each specimen was manually abraded to a 240 grit surface finish with dry silicon carbide paper.

Cylindrical electrochemical specimens of nominal 6 × 32 mm were cut from 6-mm diameter rod. Each specimen was manually abraded to a 120 grit surface with dry silicon carbide paper just prior to testing at a scan rate of 50 V/h and to a 600 grit surface when tested at a scan rate of 1 V/h.

Solution Preparation

Various corrosion tests were conducted in individual process solutions of varying compositions, pH values, and temperatures. The solutions used in these studies were plant process fluids.

Apparatus

Laboratory, immersion tests were conducted in a 1-L, flat-bottom resin reaction kettle with a removable top, a reflux condenser with atmospheric seal, a thermowell and temperature regulating device, and a specimen support system.

Electrochemical tests were conducted in a similar kettle. The top contained openings that permitted a calomel reference electrode, or a standard Ag/AgCl reference electrode, a temperature probe, and a platinum (Pt) auxiliary electrode in the solution. A heating mantle was used to apply heat when required.

A potentiodyne analyzer was used to generate and simultaneously record the potential vs applied current curves. All potentials were measured vs saturated calomel electrodes (SCE), or a standard Ag/AgCl reference electrode. A 500 watt soldering iron was used as the heat source for all heat flux tests. A rotating cylinder within a cylinder electrode test unit was used in the standard laboratory reaction kettle noted above for electrochemical erosion-corrosion tests.

Experimental Technique

Immersion tests were conducted in the laboratory. Test methods, corrosion rate calculations, and evaluation of the immersion test results were as described elsewhere.[24]

Polarization curves were determined by one of two methods. Anodic polarization curves were determined by the ASTM G 5 test method except that the environment was not deaerated by purging and the potential sweep rate was 1 to 3 V/h. In the second procedure, cathodic and anodic polarization curves were determined in a manner described previously.[3] Potentials were scanned at a rate of 50 V/h from about −1.0 V (SCE) to about +1.0 V and back to the cathodic voltage starting point. After completion of the forward and return scans, the potentiostat was turned off to observe the test electrode corrosion potential in the absence of artificial polarization. The freely corroding potential, E_{CORR}, was allowed to reach a steady-state value that required time periods ranging from a few minutes to several hours.[11,25] A final cathodic polarization curve was plotted by scanning from the steady-state corrosion potential to the original starting point.

A modified disc-heat flux technique was followed by hot-wall effect immersion tests. By the Fisher method, the external soldering iron heats the disc specimen; the specimen is one side exposed to the process fluid by being clamped to a nozzle of the laboratory test vessel.[26,27]

The electrochemical erosion-corrosion tests were conducted under dynamic flow conditions using a rotating cylinder electrode by the method developed by Silverman.[28] The rotational speed was selected to match the shear stress of the process fluid at its maximum velocity.

Results and Discussion

Case No. 1

The first study was initiated to determine the effects of reducing the water content for a particular process solution. The proposed reduction in water would increase the sulfuric acid concentration from 3 to 4.5%, and would also raise the solid organic acid concentration from 12 to 17%. A concern was that this more concentrated acid solution would damage the existing titanium and AISI 304 stainless steel (SS) process equipment by action of corrosion and/or erosion-corrosion. Laboratory electrochemical, velocity, and immersion tests were conducted to resolve this concern.

Titanium Grade 2. The 50 V/h anodic polarization curves of titanium are plotted in Figures 2 and 3. The overall shape of the curves in these figures predict passive behavior for titanium in both the present and proposed solutions. Further interpretations of these tests were made by use of three LSDs, in a manner described

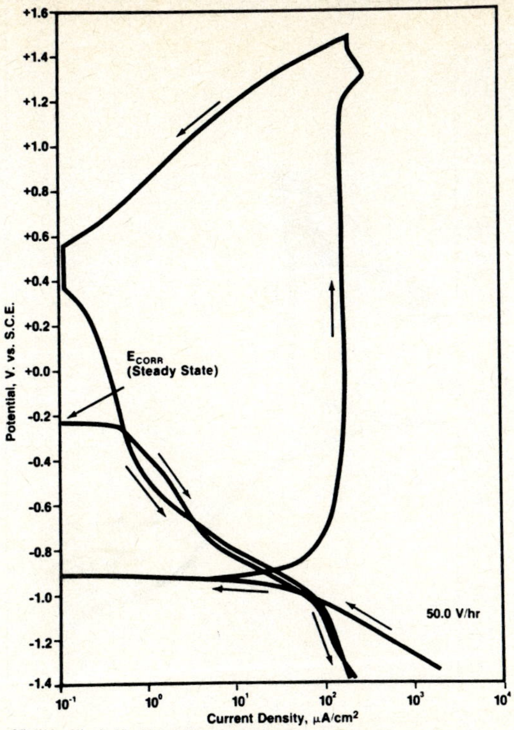

FIGURE 2 – A CBD for titanium (Grade 2) in a 3.0% sulfuric acid, 12% organic acid solution at 30 C.

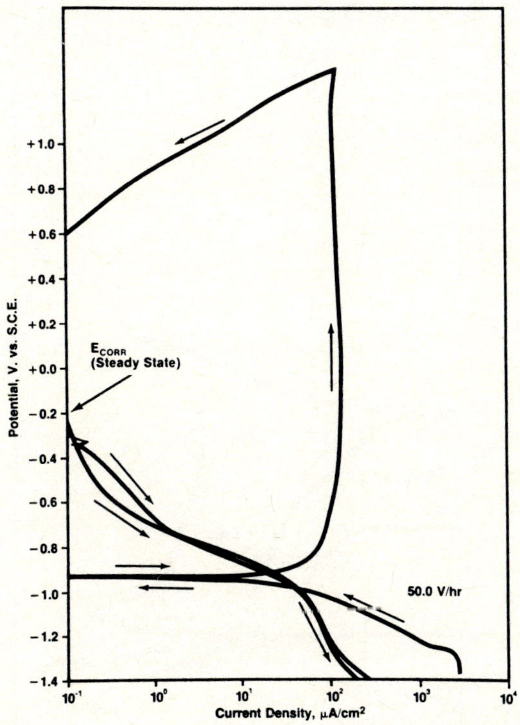

FIGURE 3 – A corrosion behavior diagram for titanium (Grade 2) in a 4.5% sulfuric acid, 17% organic acid solution at 30 C.

previously by the author.[11] Since these tests took less than one day each to conduct and evaluate, it was possible to predict very quickly, with a high degree of certainty, that the proposed process change would be free of any corrosion problems. It is the author's practice to also conduct polarization curves at scan-rates within the ASTM G 5 procedure. The fact that the 1.8 V/h scans shown in Figures 4 and 5 for both the existing and proposed solutions were almost exact overlays predicts that the metal will perform identically in both environments, for the same reasons.

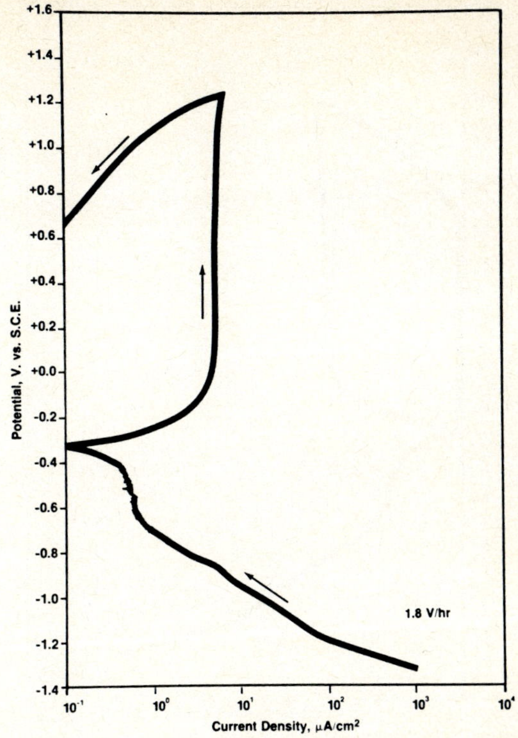

FIGURE 4 – The anodic polarization behavior of titanium (Grade 2) in a 3.0% sulfuric acid, 12% organic acid solution at 30 C.

Figures 6 and 7 show 1.8 V/h scans for both solutions under dynamic flow conditions. These electrochemical curves were produced using a rotating cylinder electrode that evaluates the metal's performance under fluid velocity conditions.[28] The 5000 rpm rotational speed of the cylinder was selected to match the shear stress of the process fluid at the maximum velocity of 36 m/s. Again, the curve in Figure 6 for the existing solution is almost an exact overlay of the curve in Figure 7 for the proposed solution. The similar match of the curves in Figures 6 and 7 with those in Figures 4 and 5, respectively, indicates that the performance of titanium is not affected in fluid velocities up to 3.6 m/s in these solutions. Immersion test results confirmed that titanium specimens did not exhibit any corrosion attack after 30 days' exposure in either the existing or proposed solutions.

AISI 304 SS. Figures 8 through 13 show results that are a direct copy of those electrochemical tests discussed for titanium. Again, these results were confirmed by the results of the 30-day immersion tests.

Conclusion. Results from all of the tests indicate that the proposed environment is no more corrosive or

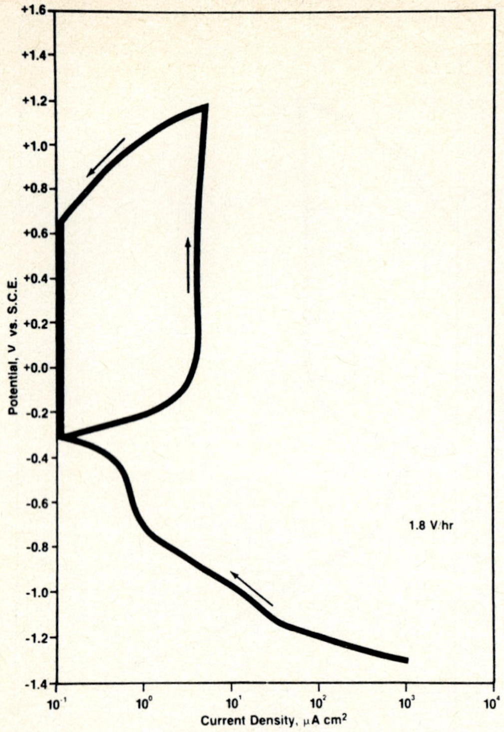

FIGURE 5 — The anodic polarization behavior of titanium (Grade 2) in a 4.5% sulfuric acid, 17% organic acid solution at 30 C.

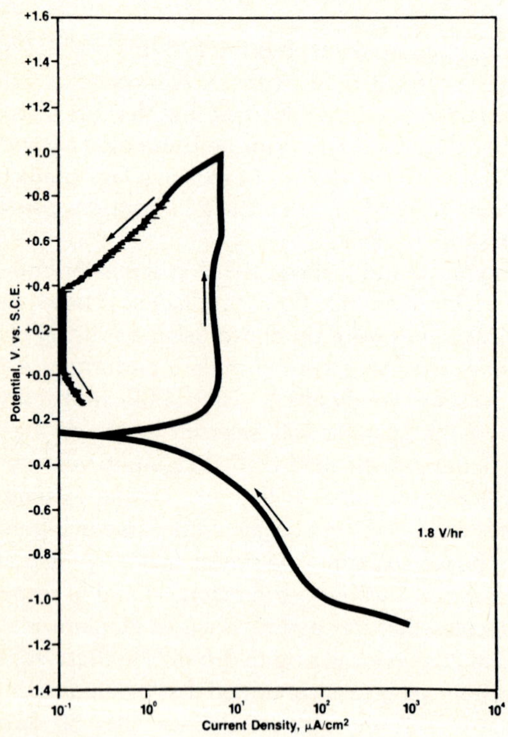

FIGURE 6 — The anodic polarization behavior of titanium (Grade 2) in a 3.0% sulfuric acid, 12% orgnanic acid solution at 30 C, under dynamic flow conditions of 3.6 m/s fluid velocity.

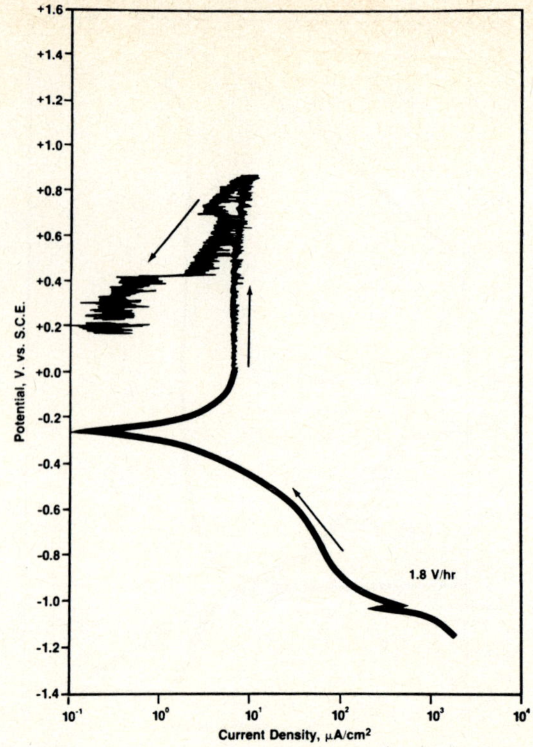

FIGURE 7 — The anodic polarization behavior of titanium (Grade 2) in a 4.5% sulfuric acid, 17% organic acid solution at 30 C, under dynamic flow conditions of 3.6 m/s fluid velocity.

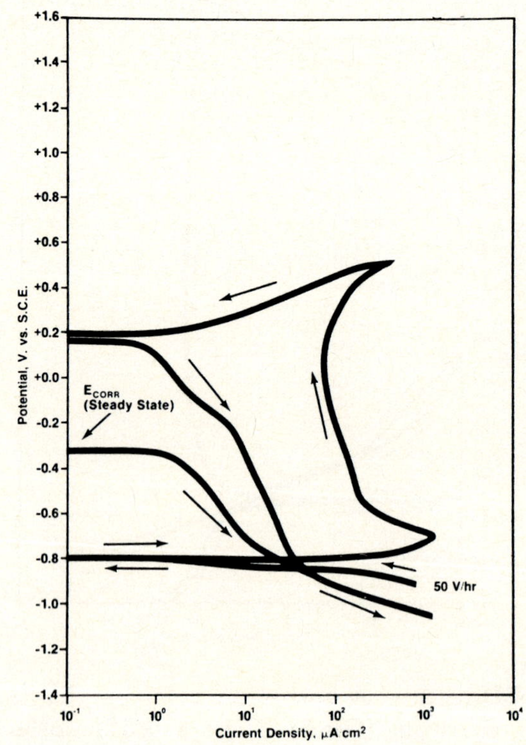

FIGURE 8 — A CBD for AISI 304 SS in a 3.0% sulfuric acid, 12% organic acid solution at 30 C.

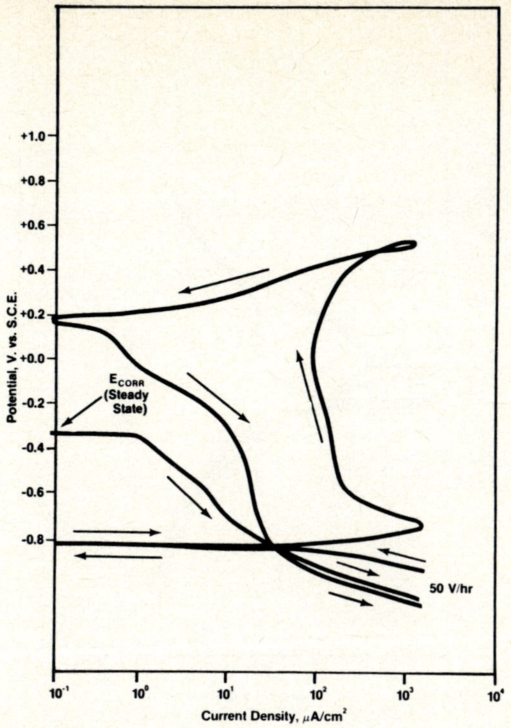

FIGURE 9 – A CBD for AISI 304 SS in a 4.5% sulfuric acid, 17% organic acid solution at 30 C.

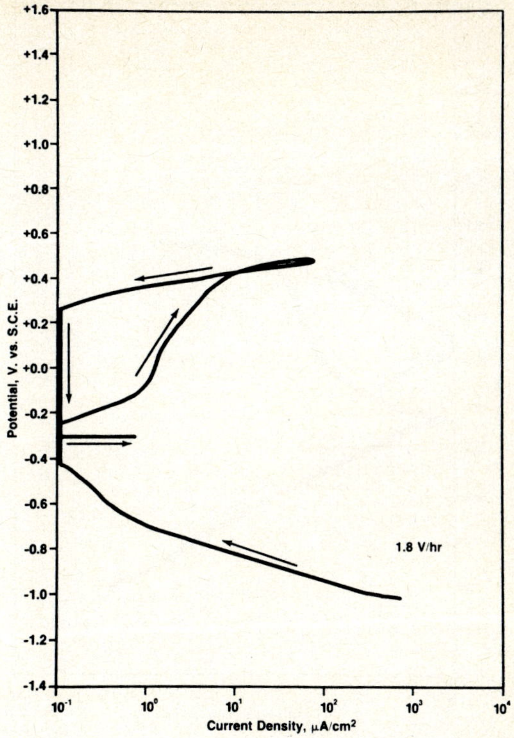

FIGURE 11 – The anodic polarization behavior of AISI 304 SS in a 4.5% sulfuric acid, 17% organic acid solution at 30 C.

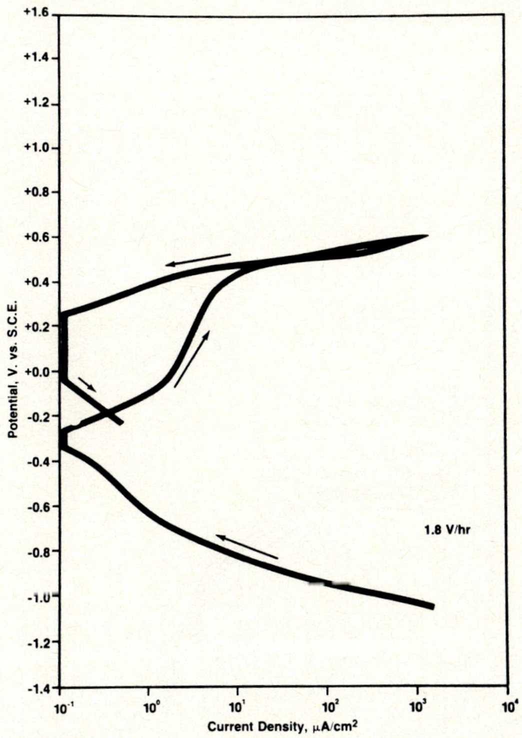

FIGURE 10 – The anodic polarization behavior of AISI 304 SS in a 3.0% sulfuric acid, 12% organic acid solution at 30 C.

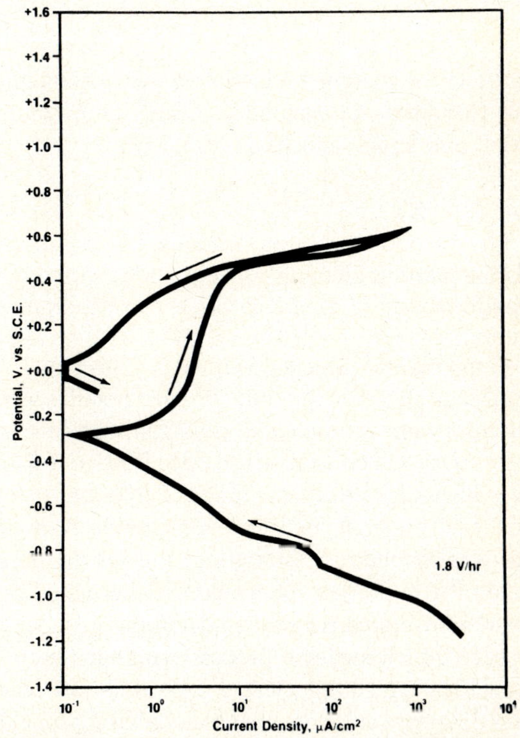

FIGURE 12 – The anodic polarization behavior of AISI 304 SS in a 3.0% sulfuric acid, 12% organic acid solution at 30 C, under dynamic flow conditions of 3.6 m/s fluid velocity.

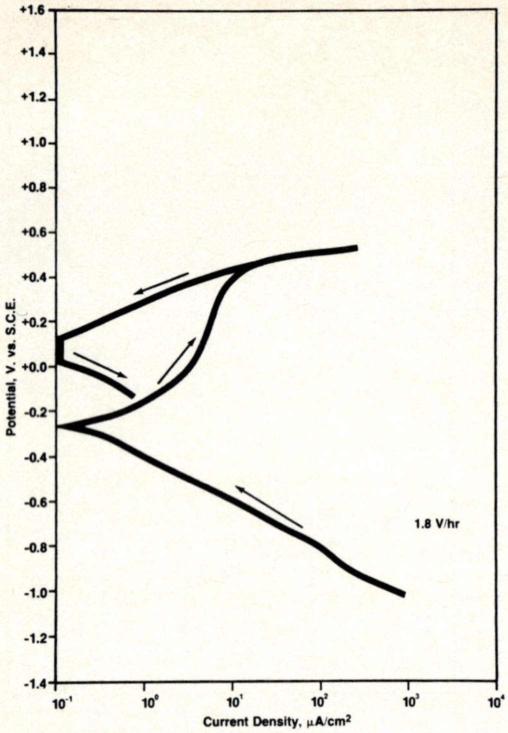

FIGURE 13 – The anodic polarization behavior of AISI 304 SS in a 4.5% sulfuric acid, 17% organic acid solution at 30 C, under dynamic flow conditions of 3.6 m/s fluid velocity.

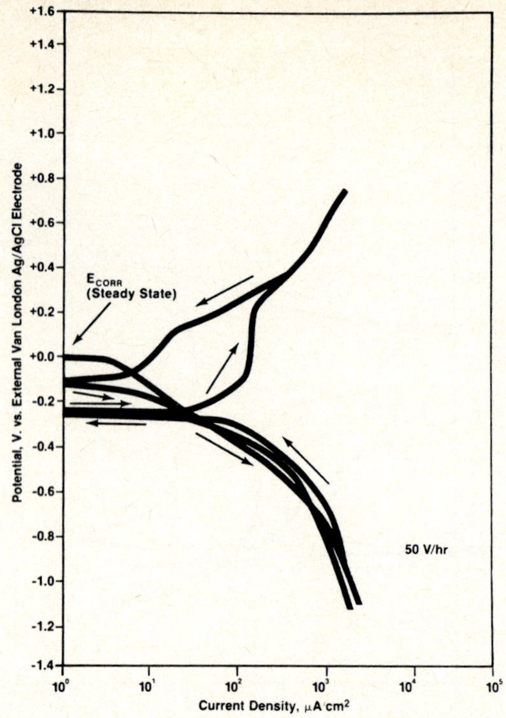

FIGURE 14 – A CBD for AISI 316 SS in a 60% organic acid solution at 98 C containing 336 ppm chlorides. Note that this test used an external Ag/AgCl reference electrode and electrically connected to the environment with a salt bridge.

abrasive to the materials tested than the existing environment. Conditions of evaluation include temperatures up to 30 C and liquid velocities up to 3.6 m/s.

Case No. 2

The second study was conducted to determine the effect that various chloride concentrations would have on the performance of AISI 316 SS in a 60% organic acid solution at 98 C.

In the manner described above in Case No. 1, laboratory electrochemical potentiodynamic polarization technique tests were conducted at potential scan-rates of 3.0 and 5.0 V/h. Chloride levels of 336, 500, and 800 ppm were evaluated, as this is the range of chlorides contained in the raw material feedstock evaluated. A hydrogen purge was maintained throughout the testing period to simulate the hydrogen subreaction of the actual process environment. Because some electrode instability had been seen during earlier in-process plant equipment tests, an initial test was conducted at each potential scan-rate for the 336 ppm chloride level material using an external type Ag/AgCl reference electrode (rather than the standard SCE). A salt bridge was used to connect the reference electrode electrically to the environment. The second comparison test used the internal reference electrode, which was immersed directly into the solution. Results indicated that both electrodes responded the same. Therefore, all tests used the internal (direct immersion) Ag/AgCl reference electrodes.

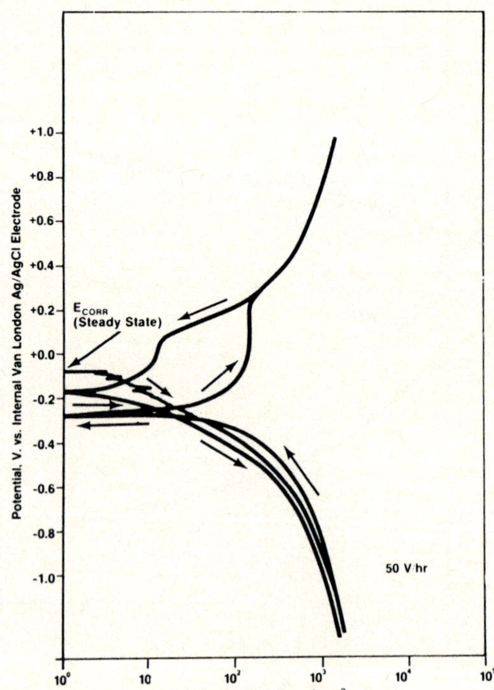

FIGURE 15 – A CBD for the same environment/test conditions as shown in Figure 14 except that a standard Ag/AgCl internal reference electrode was used.

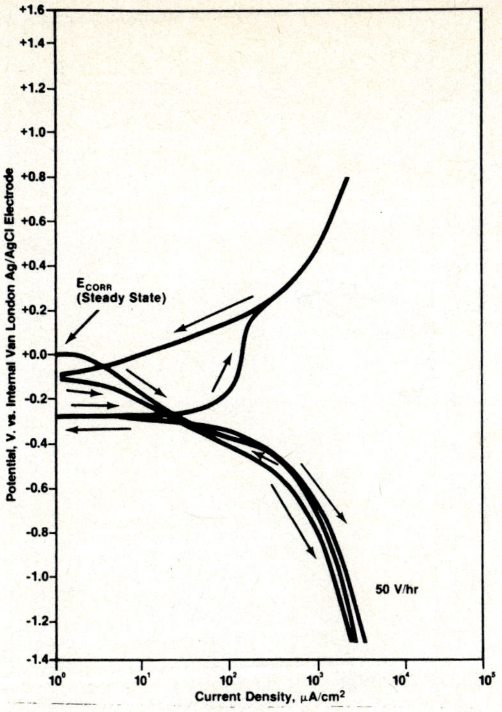

FIGURE 16 – A CBD for AISI 316 SS in a 60% organic acid solution at 98 C containing 500 ppm chlorides.

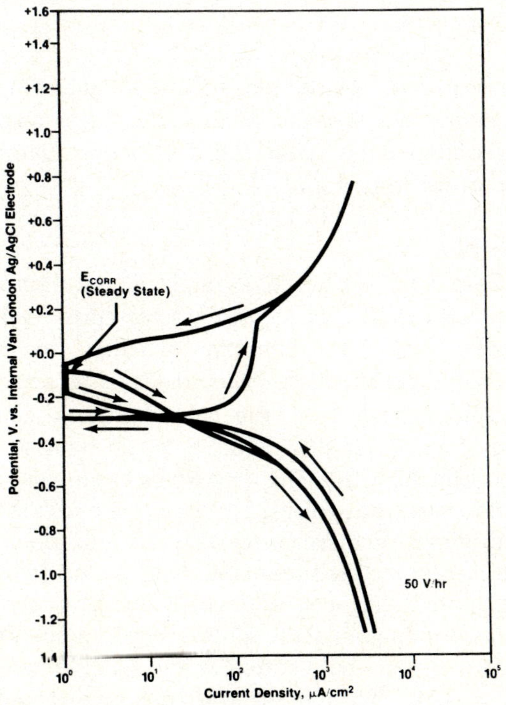

FIGURE 17 – A CBD for AISI 316 SS in a 60% organic acid solution at 98 C containing 800 ppm chlorides.

The electrochemical curves generated for this study graphically display the behavior of AISI 316 SS in this organic solution containing different concentrations of chlorides. Figures 14 through 17 are the CBD, generated at a scan-rate of 50 V/h. Interpretation of these diagrams indicated that AISI 316 SS behaves passively at the con-

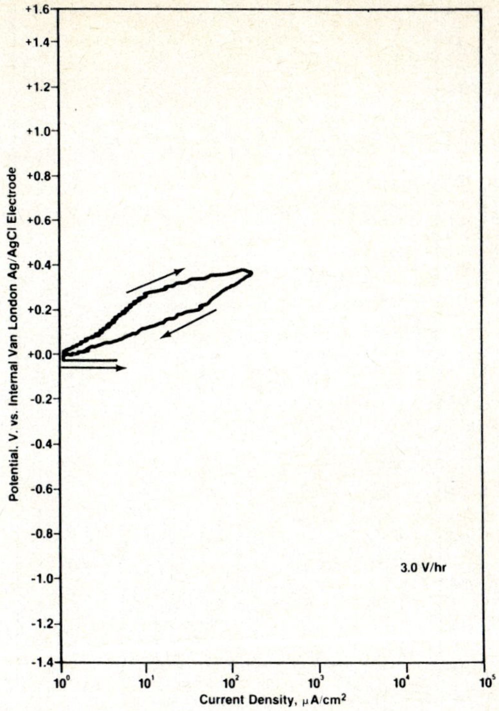

FIGURE 18 – The anodic polarization behavior of AISI 316 SS in a 60% organic acid solution at 98 C containing 336 ppm chlorides. Note that this test used an external Ag/AgCl reference electrode and is electrically connected to the environment with a salt bridge.

ditions tested. The curves also suggest a strong possibility that crevice corrosion can occur even though hysteresis is not present. This prediction is shown by E_{CORR} plotting positive to E_{REPASS} in these figures.[8,11]

The polarization curves shown in Figures 18 through 21 were generated at a scan rate of 3.0 V/h. These curves indicate that AISI 316 SS behaves passively, on a borderline basis. However, the probability of localized attack is indicated by the presence of hysteresis loops on these slow scan-rate generated curves.

Heat flux/immersion tests provide a graphic example of these electrochemical predictions. Immersion exposure of AISI 316 SS stressed specimens behaved passively, while heat flux specimens exposed in the same vessel suffered crevice corrosion and chloride stress corrosion cracking (CSCC). These latter types of localized attack occurring on the surface of the heat flux specimens are grouped into a category known as the *occluded corrosion cell.* Corrosion of this type occurs under conditions where the composition of the solution inside a restricted area is different from the composition of the bulk solution. Had the CSCC not been identified, the sensitivity to crevice corrosion could have been verified by using the test technique of Diegle.[29]

The results of these studies show that AISI 316 SS is susceptible to localized attack under all of the chloride conditions tested. As with the example cited in Case No. 1, the capability of corrosion predictions by the use of the

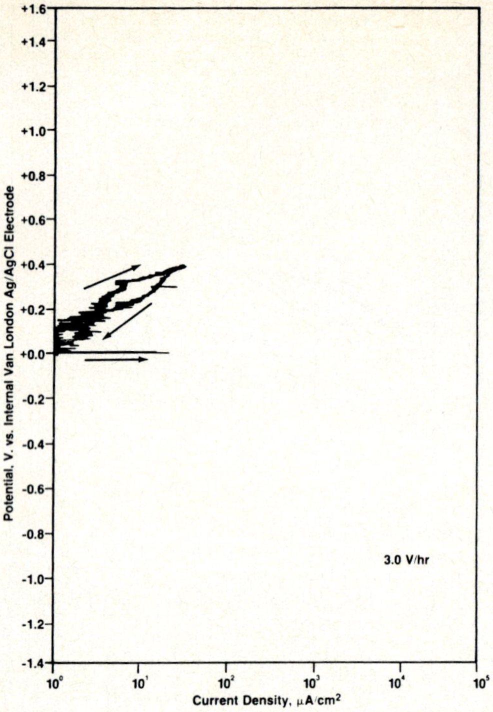

FIGURE 19 — The anodic polarization behavior of the same environment/test conditions as shown in Figure 18 except that a standard Ag/AgCl internal reference electrode was used.

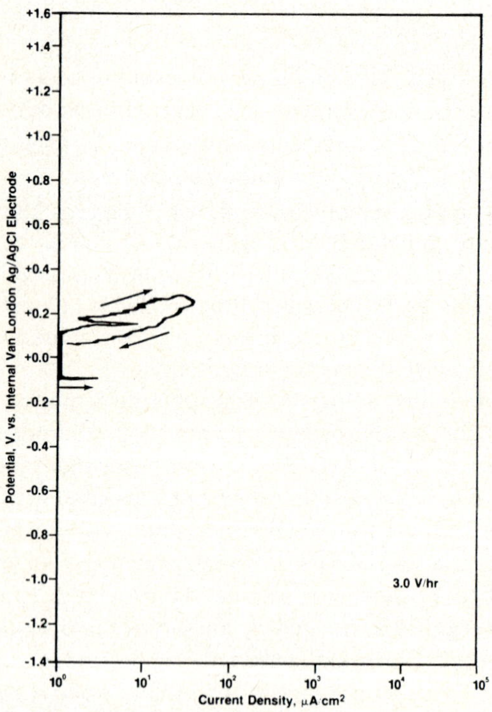

FIGURE 20 — The anodic polarization behavior of AISI 316 SS in a 60% organic acid solution at 98 C containing 500 ppm chlorides.

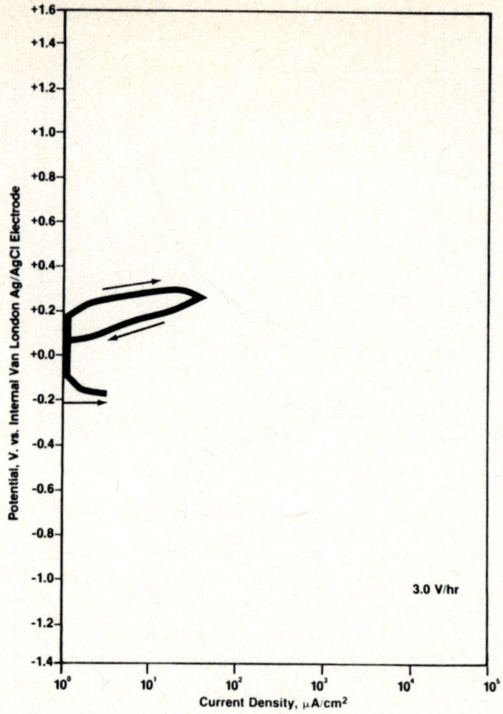

FIGURE 21 — The anodic polarization behavior of AISI 316 SS in a 60% organic acid solution at 98 C containing 800 ppm chlorides.

rapid scan-rate polarization technique, allows an early engineering evaluation to begin. Later, these results can be reconfirmed by other test techniques (which take much longer to perform).

Case No. 3

Case No. 3 is a typical example of how electrochemical studies can be used to rapidly screen proposed major process changes to determine whether the existing materials of construction will continue to be acceptable. Polarization studies of the initial process chemistry change found that AISI 316 SS would have to be replaced with Alloy 20. The economics of the project continued to favor the proposed process, even with this need to upgrade to the more costly Alloy 20. It was later found that a second recycle stream had to be added occasionally and more testing was required. These latter tests found that the introduction of the second recycle process stream made the system more corrosive and Alloy 20 would not be acceptable. To upgrade the equipment to an even higher cost alloy would eliminate the economic justification of the project.

While the economic screening concept is typical, the laboratory electrochemical testing for this project was anything but routine. The column feed had to be evaluated at 170 C, under 100 psi (0.6895 MPa). The same high temperature, high pressure reference electrode used in the Case No. 2 example was used.

The polarization curve generated at 50 V/h (Figure 22) shows that Alloy 20 was subject to severe local pitting

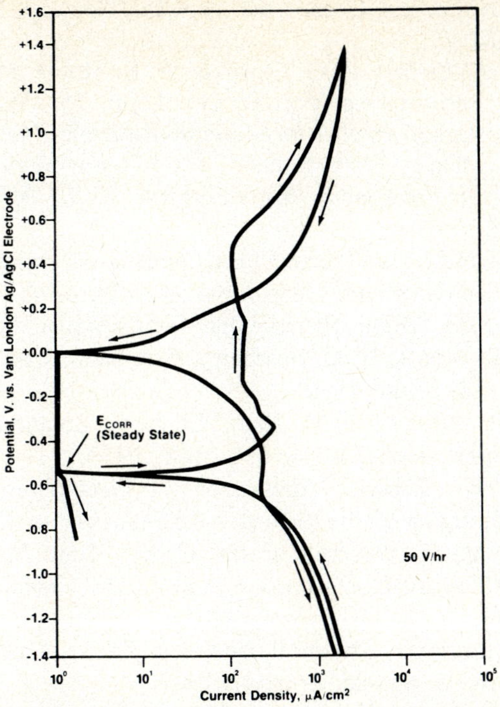

FIGURE 22 – A CBD for Alloy 20 in an organic acid recycle process stream, pH 1.5 at 170 C, under 100 psi (0.6895 MPa) pressure.

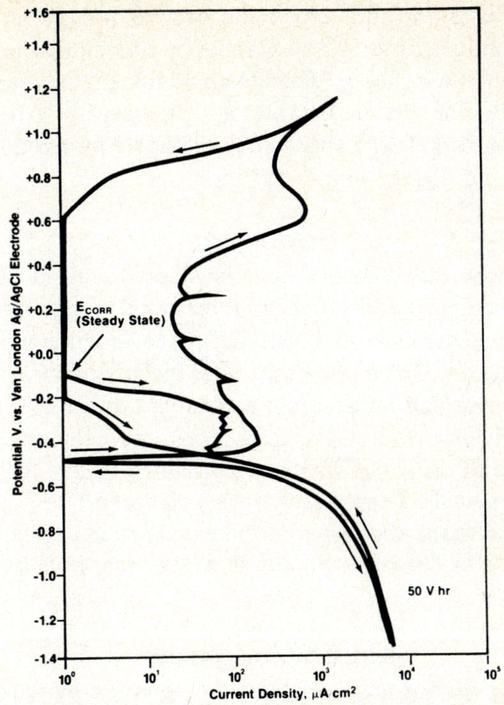

FIGURE 23 – A corrosion behavior diagram for Alloy 20 in the environment shown in Figure 22 after it had been preneutralized to pH 5.0.

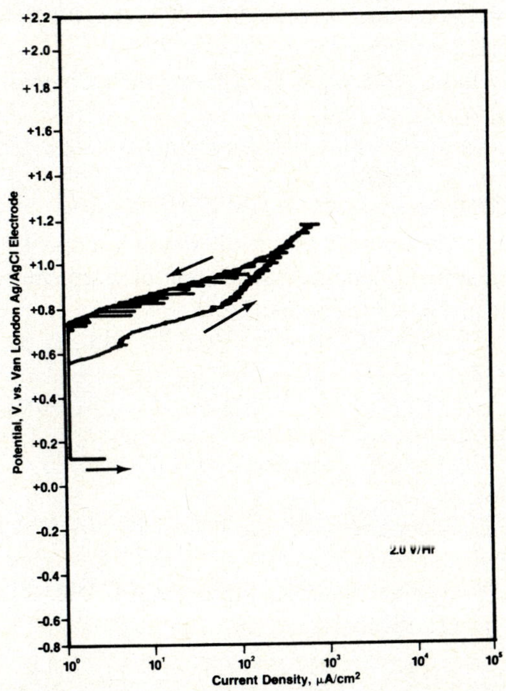

FIGURE 25 – The anodic polarization behavior of Alloy 20 in the environment shown in Figure 24 after it had been preneutralized to pH 5.0.

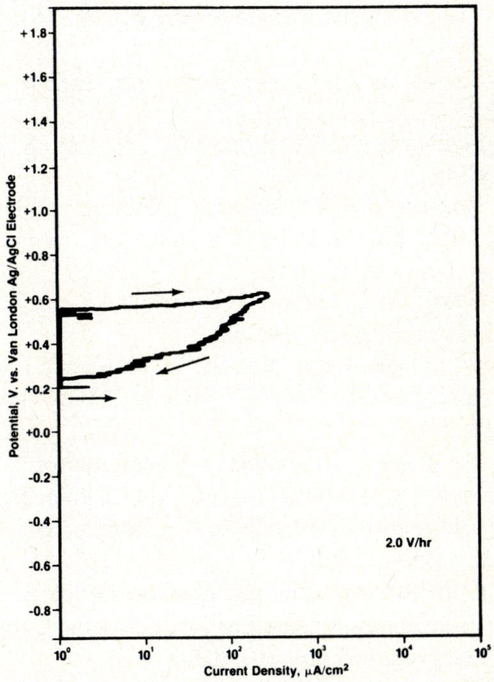

FIGURE 24 – The anodic polarization behavior of Alloy 20 in an organic acid recycle process stream, pH 1.5 at 170 C, under 100 psi (0.6895 MPa) pressure.

attack with the addition of the second recycle stream noted above. Figure 23 shows that by preneutralizing the second recycle slurry, Alloy 20 would remain the material of choice for this project. Curves generated at 2.0 V/h (Figures 24 and 25) confirmed the need for preneutralization as did 15-day immersion tests.

Summary

Three case histories have been presented in which electrochemical and immersion test results were used to investigate the corrosion resistance of several materials of construction. In each case, the electrochemical test results provided an accurate assessment of actual corrosion behavior. More specifically, for those alloy environments that displayed unstable passive behavior, the use of the so-called rapid-scan rate polarization technique provided more and faster information than either immersion tests or the conventional slow-scan rate polarization tests.

Acknowledgment

The author acknowledges the special contributions of J. E. Carrico, J. G. Stoecker, and M. E. Zerr in the conductance of the laboratory studies and/or the interpretation of the test results on this project.

References

1. W. R. Whitney, J. Amer. Chem. Soc., Vol. 25, No. 4, p. 394, 1903.
2. A. Hickling, Trans. Faraday Soc., Vol. 38, No. 1, p. 27, 1942.
3. C. Edeleanu, J.I.S.I. (London), Vol. 185, p. 482, 1957.
4. C. Edeleanu, J.I.S.I. (London), Vol. 188, p. 122, 1958.
5. G. Kortum, J. O'M. Bockris, Textbook of Electrochemistry, Elsevier Publ. Co., New York, New York, Vols. 1 and 2, 1951.
6. M. Stern, A. L. Geary, J. Electrochem. Soc., Vol. 104, No. 56, p. 559, 1957.
7. O. W. Siebert, Paper No. 66, 153rd Meeting of the Electrochemical Society, Seattle, Washington, 1978.
8. O. W. Siebert, "Laboratory Electrochemical Test Methods," presented at the ASTM G-1 Inter. Symp. on Laboratory Corrosion and Standards, Bal Harbour, Florida, 1983.
9. U. R. Evans, Zeitschift Fur Elecktrochemic Beriche Der Bunsenqesellschaft Fur Physikalische Chemie, Vol. 62, No. 6/7, p. 619, 1958.
10. R. N. Parkins, Corros. Sci., Vol. 20, No. 2, p. 147, 1980.
11. O. W. Siebert, J. G. Stoecker, P. E. Morris, Materials Performance, Vol. 22, No. 11, p. 13, 1983.
12. Polarization Practice for Standard Reference Method for Making Potentiostatic and Potentiodynamic Anodic Polarization Measurements, G 5, ASTM, 1982.
13. Recommended Practice for Conventions Applicable to Electrochemical Measurements in Corrosion Testing, G 3, ASTM, Philadelphia, Pennsylvania, 1981.
14. P. E. Morris, R. C. Scarberry, Corrosion, Vol. 28, No. 12, p. 44, 1972.
15. P. E. Morris, R. M. Kain, Chemical Engineering Progress, Vol. 73, No. 6, p. 103, 1977.
16. O. W. Siebert, "Application of Electrochemical Techniques to the Selection of Materials of Construction," presented at the Golden Gate Metals and Welding Conference, San Francisco, California, 1981.
17. M. Pourbaix, Thermodynamics of Dilute Aqueous Solutions, Arnold, London, 1949.
18. C. Edeleanu, Metallurgica, Vol. 50, No. 2, p. 113, 1954.
19. C. Edeleanu, Nature, Volume 173, No. 4407, p. 739, 1954.
20. C. Edeleanu, J. G. Gibson, Chemistry and Industry, Vol. 9, No. 3, p. 301, 1961.
21. P. E. Morris, Galvanic and Pitting Corrosion—Field and Laboratory Studies, STP 576, ASTM, Philadelphia, Pennsylvania, p. 261, 1976.
22. F. Flade, Zeitschgrift Fur Physikalisch Chemie, Vol. 76, No. 5, p. 513, 1911.
23. R. L. Martin, Materials Performance, Vol. 18, No. 3, p. 41, 1974.
24. NACE Standard TM-01-69 (1976 Revision), "Laboratory Corrosion Testing of Metals for the Process Industries," National Association of Corrosion Engineers, Houston, Texas, 1976.
25. B. E. Wilde, Corrosion, Vol. 28, No. 8, p. 283, 1972.
26. A. O. Fisher, F. L. Whitney, Jr., Corrosion, Vol. 15, No. 5, p. 257t, 1959.
27. A. O. Fisher, Corrosion, Vol. 17, No. 5, p. 93, 1961.
28. D. C. Silverman, Corrosion, Vol. 40, No. 5, p. 220, 1984.
29. R. B. Diegle, Materials Performance, Vol. 21, No. 3, p. 37, 1982.

Corrosion Rate Measurement Through Galvanostatic Polarization Transients

L. GIULIANI*

Abstract

When subjected to a step increase of current, a corroding electrode will react with a potential variation over time that depends on ohmic resistance, charge transfer, and diffusion overpotentials. By analyzing the potential variation over time, i.e., the polarization transient, it is possible to separate the three contributing factors by considering that the ohmic overpotential is established in a time of 10^{-9}s and that charge transfer and diffusion overpotentials, respectively, follow a time law such as:

$$Vt = Vo\left(1 - e^{\frac{-t}{RpC}}\right)$$

and

$$Vd = \frac{RT}{zF} \ln\left(1 + \sqrt{\frac{t}{\tau}}\right)$$

Using a dedicated microcomputer, the diffusion overpotential can be calculated and subtracted to obtain the charge transfer overpotential, from which the true Rp can be calculated. The advantages of the transient technique over the AC impedance are: (1) shorter measurement times, (2) simpler instrumentation, and (3) no assumption of equivalent electric circuits.

Introduction

Instantaneous corrosion rates of metals and alloys immersed in electrolytic environments can be calculated from electrochemical measurements. A number of methods are available in the laboratory, and the simplest one [generally known as *linear polarization resistance (LPR)]* has been widely accepted for field use, especially in water treatment control.

To overcome difficulties met when the LPR method is used in some systems, nonstationary methods (based primarily on electrode impedance measurements) are actively investigated. This paper describes a method based on the analysis of a galvanostatic polarization transient to extend the range of validity of the LPR measurement by removing ohmic and diffusion polarization from the total measured potential shift, thereby obtaining pure charge transfer polarization.

LPR Stationary Method

The original method, proposed by Stern and Geary, derives the corrosion current from a polarization resistance (Rp) measurement under the assumption that: (1) The potential at which the measurement is taken (i.e., the corrosion potential) is sufficiently far from the equilibrium potentials that it is not affected by the cathodic deposition of the metal and the anodic oxidation of the oxidant; and (2) the potential shift induced during the measurement is sufficiently small to allow a series expansion for the exponential function linking potential and partial currents and to limit such a series to the second term.

Actually, Stern and Geary assume that usual kinetic relationships can be written, which leads to further assumptions: (3) Both anodic and cathodic reactions exhibit Tafel slopes in a V-log I plot at the corrosion potential (or the cathodic reaction is under diffusion control and therefore has an infinite slope); and (4) only one anodic reaction occurs at the corrosion potential, i.e., the entire anodic current is due to metal dissolution.

*Atel, s.r.l., Rome, Italy.

Another practical assumption is that the corrosion system be in a steady state, i.e., that the potential variation over time is negligible, at least during the measurement time.

Prazak[1] has reviewed some of the problems that can be found when not all of such assumptions are met, while others[2] have analyzed the Stern and Geary equation primarily from a mathematical point of view, thereby defining limits of validity and possible errors. Practical experience indicates that generally, a very good agreement exists between electrochemical measurements and (properly determined) weight losses for a large number of corroding systems.

There are cases, however, for which the simple, stationary technique of applying a polarizing current and reading the potential shift (or vice versa) cannot be used because of experimental difficulties or because it would lead to large errors. In most practical environments, a corroding surface will be covered by corrosion products and/or some sort of deposit, which might increase overall ohmic resistance between working and reference electrodes to a value comparable to Rp. (Note that a high ohmic resistance between working and counter electrodes is only an instrumentation difficulty, which can be overcome by designing a high voltage output.) Many industrial systems involve very low conductivity solutions, such as demineralized or polished waters, or water/organic mixture, and can exhibit ohmic resistances even higher than Rp.

Very often a diffusion process controls either the arrival of reacting species to the electrode or removal of reacting species from it, especially when a redox reaction occurs at the corrosion potential with a rate comparable to that of the corrosion reaction itself: the result is a continuous increase of the polarized potential over time at any value of applied current.

In such instances, when measuring a polarization resistance, even with a three-electrode probe, part of the overall polarization will be attributable to ohmic resistance and/or to diffusion processes.

As a result, a corrosion rate calculated from total polarization will underestimate the real metal loss. Even worse will be the presence of diffusion-controlled redox reaction(s) as the corrosion potential will not be stable either before or after the polarization is applied. To overcome such difficulties, nonstationary methods of electrode polarization have been devised to separate ohmic, diffusion, and charge transfer polarizations.

Impedance Method

Most recent literature on nonstationary methods is dedicated to electrode impedance; it is based on the assumption that the corroding electrode can be represented by an equivalent electric circuit, including a condenser (representative of the double layer), in parallel with a resistance (Rp), with or without a series resistance to represent solution resistance (Rs).

Theoretically, the imaginary part of the measured impedance should describe a semicircle when plotted against the frequency (Nyquist diagram) and the two intercepts on the x-axis should give the Rs and Rs + Rp values.[3] The greatest drawback of this simplified impedance method is the difficulty in finding a second intercept at very low frequencies; therefore, the equivalent circuit is modified by adding a Warburg impedance component in series with Rp. Such a modification requires, in turn, a complexity in instrumentation and mathematical treatment of data; some simplifications are possible, however, for systems exhibiting a semicircle on a Nyquist diagram, with its center on the x-axis.[4]

One major objection that can be raised to the impedance method is the complexity of data interpretation, largely related to the rather factitious equivalence between a real corroding surface and an electrical circuit. Some difficulties of interpretation are schematically reported by K. Hladkly, et al.[3]

Galvanostatic Transient

One way of measuring Rp is to apply a step increase of current to a corroding electrode, wait until a stable polarization is reached, and divide such a polarization by the value of the applied current. When only charge transfer overvoltage is present, the electrode potential will vary with time, according to the equation describing the double-layer charging:

$$Vt = i\,Rp\left(1 - e^{\frac{-t}{RpC}}\right) \qquad (1)$$

where i is the applied current density and C is the double-layer capacitance.

Mathematically, the potential will reach 98% of the final asymptotic value at $t = 4\,RpC$: for a clean electrode with a capacitance of some hundreds of microfarads/cm^2 and a polarization resistance of some tens of kilo-ohms × cm^2, such a time should be a few seconds.

For polarization resistances in the order of mega-ohms × cm^2, the time to reach a stable value might be too long either for experimental needs or compared to physicochemical fluctuations of the system; it is nevertheless possible to obtain the asymptotic value by a trial and error method applied to the initial part of the curve Vt/t as demonstrated by Greene.[5] Electrodes exhibiting other types of overvoltages will yield different responses.

In the presence of ohmic overvoltage, as soon as the polarizing current is applied, there will be a step increase of polarization, which will be established in a time of the order of 10^{-9} s:

$$Vr = I\,Rs \qquad (2)$$

where Rs is the solution resistance between working and reference electrode.

In the presence of concentration overvoltage, the electrode potential will never reach a stable value but will continue to increase with time. This latter case has been observed in many systems, especially when corrosion currents are low or corrosion products are present on the electrode surface. An example of such behavior is given

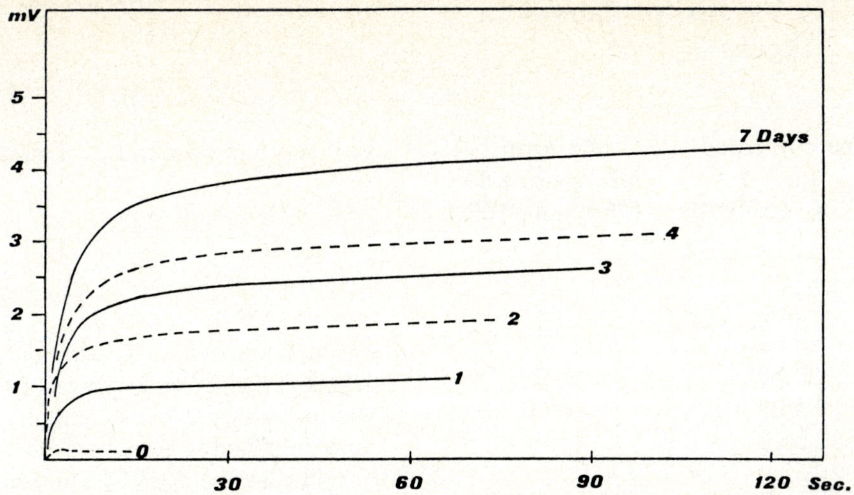

FIGURE 1 – Experimental polarization transients on Al-Brass in 3.5% NaCl solution (current density: 0.059 $\mu A/cm^2$) at different corrosion times.

by the AlBrass/3.5% NaCl system. Experimental curves obtained at different corrosion times (0 to 7 days) with a polarization current density step of 0.059 $\mu A/cm^2$ are reported in Figure 1.

An empirical method to derive the charge transfer contribution to total polarization has been devised by Azzerri,[6] who assumed that the V/t curve could be split into a Vt/t curve describing the charge transfer overvoltage and a V_{corr}/t curve having the same equation and a different set of Rp′ and C′ values.

When RpC<<Rp′C′, the two curves can be separated either by a mathematical curve fitting or by graphical extrapolation. A precise analysis of the polarization transient in the presence of charge transfer and diffusion overvoltages can be made under the assumption that the two overvoltages are additive as discussed by Vetter.[7] Whenever convection in the electrolytic solution can be neglected, it is possible to use the following equation to describe the time variation of diffusion overvoltage[8]:

$$Vd = \frac{RT}{zF}\left(\ln \; 1 + \sqrt{\frac{t}{\tau}}\right) \qquad (3)$$

where τ is a parameter of the diffusion process[9]:

$$\tau = \frac{\pi}{4} D \left(zF \frac{Co}{i}\right)^{2} \qquad (4)$$

where Co is the bulk concentration and D is the diffusion coefficient. For an electrode exhibiting ohmic, charge transfer, and diffusion overvoltages, a general equation for the transient curve will be:

$$V = Vr + Vt + Vd \qquad (5)$$

To obtain Rp, the ohmic and diffusion overvoltages should be removed from total polarization V.

The ohmic contribution is very easy to measure as it will be established immediately as the current is switched on. To split charge transfer and diffusion overvoltages, it is necessary to calculate τ. This can be done at times longer than 4 RpC when the charge transfer overvoltage has practically reached its final value. For t > 4 RpC, total polarization will be:

$$V = I\,Rs + i\,Rp + \frac{RT}{zF} \ln \left(1 + \sqrt{\frac{t}{\tau}}\right) \qquad (6)$$

Differentiating with respect to time:

$$\frac{dV}{dt} = \frac{RT}{zF} \frac{1}{1 + \sqrt{\frac{t}{\tau}}} \, 0.5 \frac{1}{\sqrt{\frac{t}{\tau}}} \qquad (7)$$

which, upon rearranging, gives:

$$\sqrt{\frac{t}{\tau}} = \frac{2t\left(\frac{dV}{dt}\right)}{\frac{RT}{zF} - 2t\left(\frac{dV}{dt}\right)} \qquad (8)$$

From Equation (8) it is possible to calculate τ and hence Vd:

$$Vd = \frac{RT}{zF} \ln \; 1 + \frac{2t\left(\frac{dV}{dt}\right)}{\frac{RT}{zF} - 2t\left(\frac{dV}{dt}\right)} \qquad (9)$$

Subtracting Vd and Vr from total polarization and dividing by the imposed current density, the correct value of Rp is obtained. Table 1 gives values of Rp calculated through Equation (9), together with values obtained from the trial and error method proposed by Greene (which does not take into account diffusion overvoltage) for the experimental curves of Figure 1.

TABLE 1 — Rp Values for Curves in Figure 1

Corrosion Time (days)	Rp corrected for Vd (ohm·cm·10^4)	Rp from Reference 5 (ohm·cm·10^4)
1	1.58	1.30
2	2.61	2.53
3	3.33	3.55
4	4.0	4.23
7	5.16	5.85

It can be seen that the longer the corrosion time, the higher the underestimation of Rp. Equation (9) gives the diffusion overvoltage provided that t be larger than 4 RpC. For smaller times, Equation (9) defines a different function Vd, as dV/dt is not equal to dVd/dt.

A numerical analysis of Equation (9), using different ratios Rp/τ and assuming C=400 microfarads/cm^2, yields curves of the type plotted in Figure 2. It can be seen that such curves show a well-defined maximum at t=Rp C; they then decrease to approach the true concentration overvoltage curves for t = 4 to 5 RpC, and finally merge with these latter.

Inspection of numerical results and of derivative of Equation (9) reveals that the position of the maximum is practically independent from τ, provided that Rp C < τ (a condition which is generally met in practical systems).

It is possible therefore to calculate Rp C from the value of t^* at which Vd is at a maximum and reduce Equation (1) to:

$$Vt = i\,Rp\left(1 - e^{\frac{-t}{t^*}}\right) \quad (10)$$

For values of t around t^*, Equation (3) can be reduced to:

$$Vd = \frac{RT}{zF}\sqrt{\frac{t}{\tau}} \quad (11)$$

From Equations (10) and (11) it becomes easy to calculate Rp. Calculations can be carried out in real time with a dedicated microcomputer and a program of the type described in Appendix A.

In case the condition that Rp C be small compared with τ is not considered valid, it is possible to calculate Rp through an iterative method as described in Appendix B.

Conclusions

It is possible to analyze the transient response of a corroding electrode to a galvanostatic step in a system where solution convection can be considered negligible. Using a dedicated microcomputer, charge transfer overvoltage can be calculated also in presence of ohmic and concentration overvoltages, thereby extending the applicability of the LPR method to systems involving high purity water (i.e., power plant primary loops) or water/organic mixtures and to those where a concomitant or subsequent redox reaction competes for the anodic current with the corrosion reaction.

The galvanostatic step technique is much easier to use then other nonstationary methods, from the point of view of both instrumentation and data interpretation, as its analysis is more closely related to electrochemical phenomena occurring at the electrode surface than to electrical simulation.

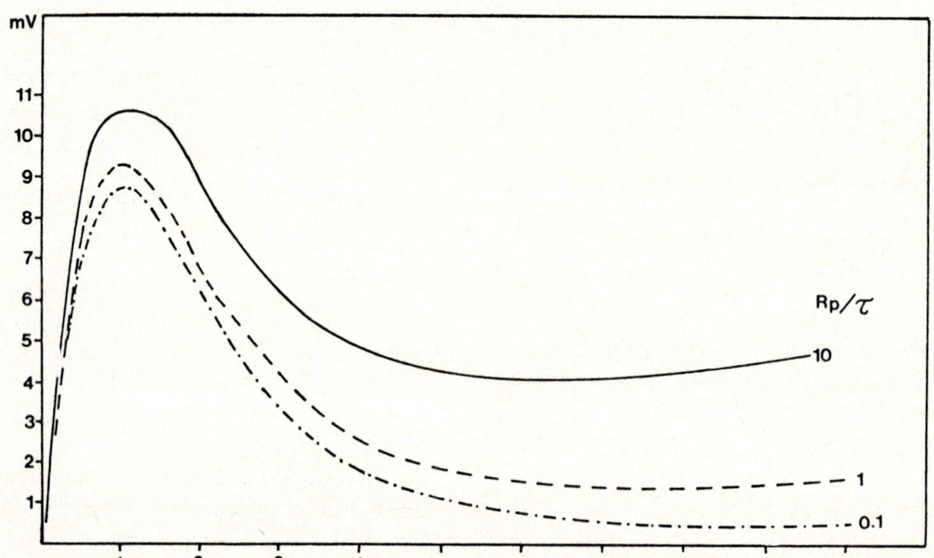

FIGURE 2 — Numerical analysis of Equation (9) for different ratios Rp/τ, showing maximum at Rp C = 1.

References

1. M. Prazak, Werk. Korros., Vol. 25, p. 104, 1977.
2. M. J. Danielson, Corrosion, Vol. 36, No. 4, p. 174, 1980; R. Bandy, D. A. Jones, Corrosion, Vol. 32, No. 4, p. 126, 1976; C. Chu-Nan, Corros. Sci., Vol. 22, p. 205, 1982.
3. K. Hladky, L. M. Callow, J. L. Dawson, Br. Corros. J., Vol. 15, p. 20, 1980.
4. J. L. Dawson, L. M. Callow, K. Hladky, ASTM Special Publication, ASTM, Philadelphia, Pennsylvania, 1979.
5. D. A. Jones, N. D. Greene, Corrosion, Vol. 22, No. 7, p. 198, 1966.
6. N. Azzerri, J. Applied Electrochem., No. 6, p. 139, 1976.
7. K. Vetter, Electrochemical Kinetics, Academic Press, New York, New York, p. 339, 1977.
8. K. Vetter, Electrochemical Kinetics, Academic Press, New York, New York, p. 209, 1977.
9. J. A. Butler, G. Armstrong, Proc. Royal Soc., Vol. 139A, p. 406, 1933.

APPENDIX A – Flow Chart of a Program Deriving Rp Under the Assumption that RpC $\ll \tau$

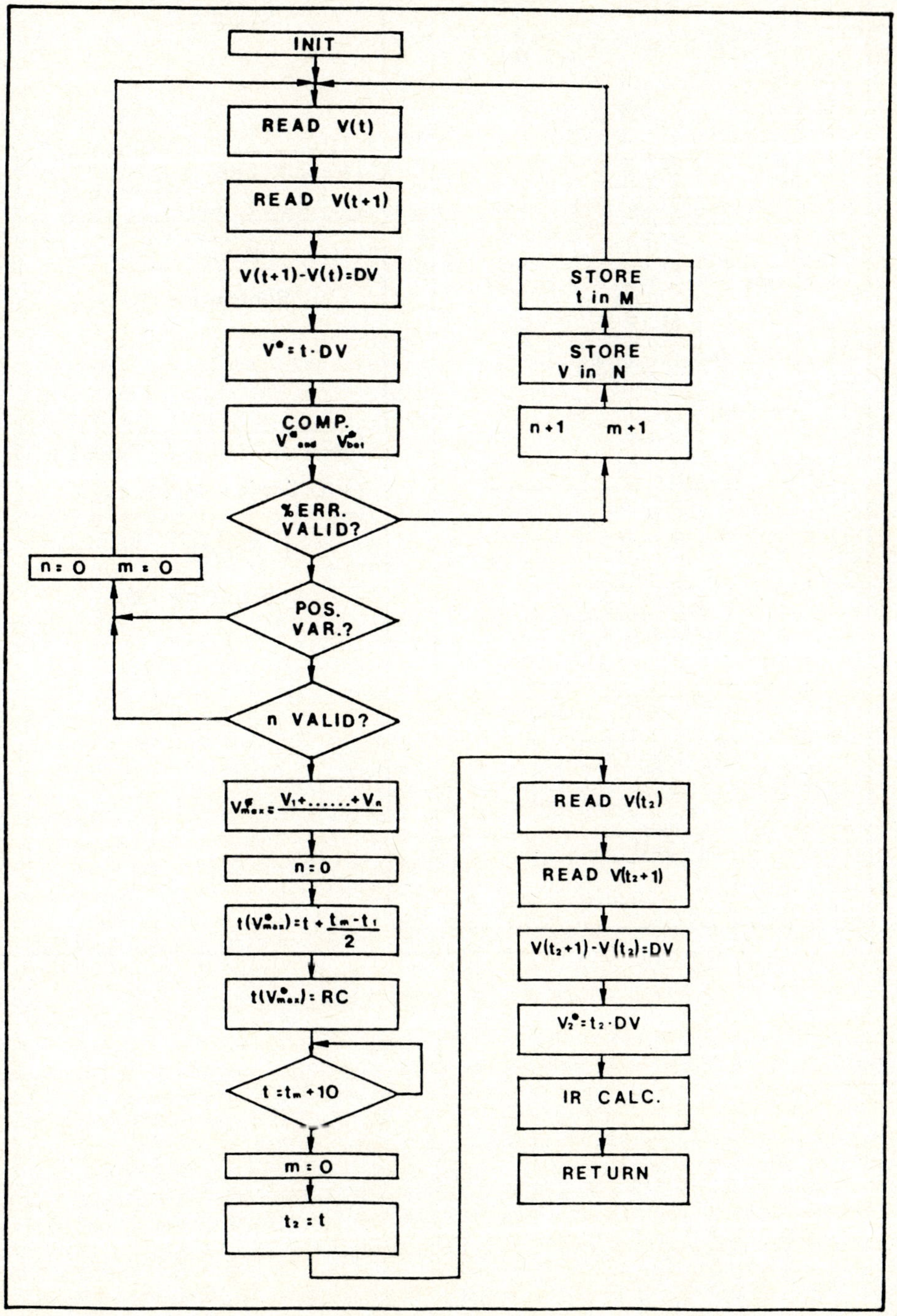

APPENDIX B — Flow Chart of a Program Deriving Rp With an Iterative Method

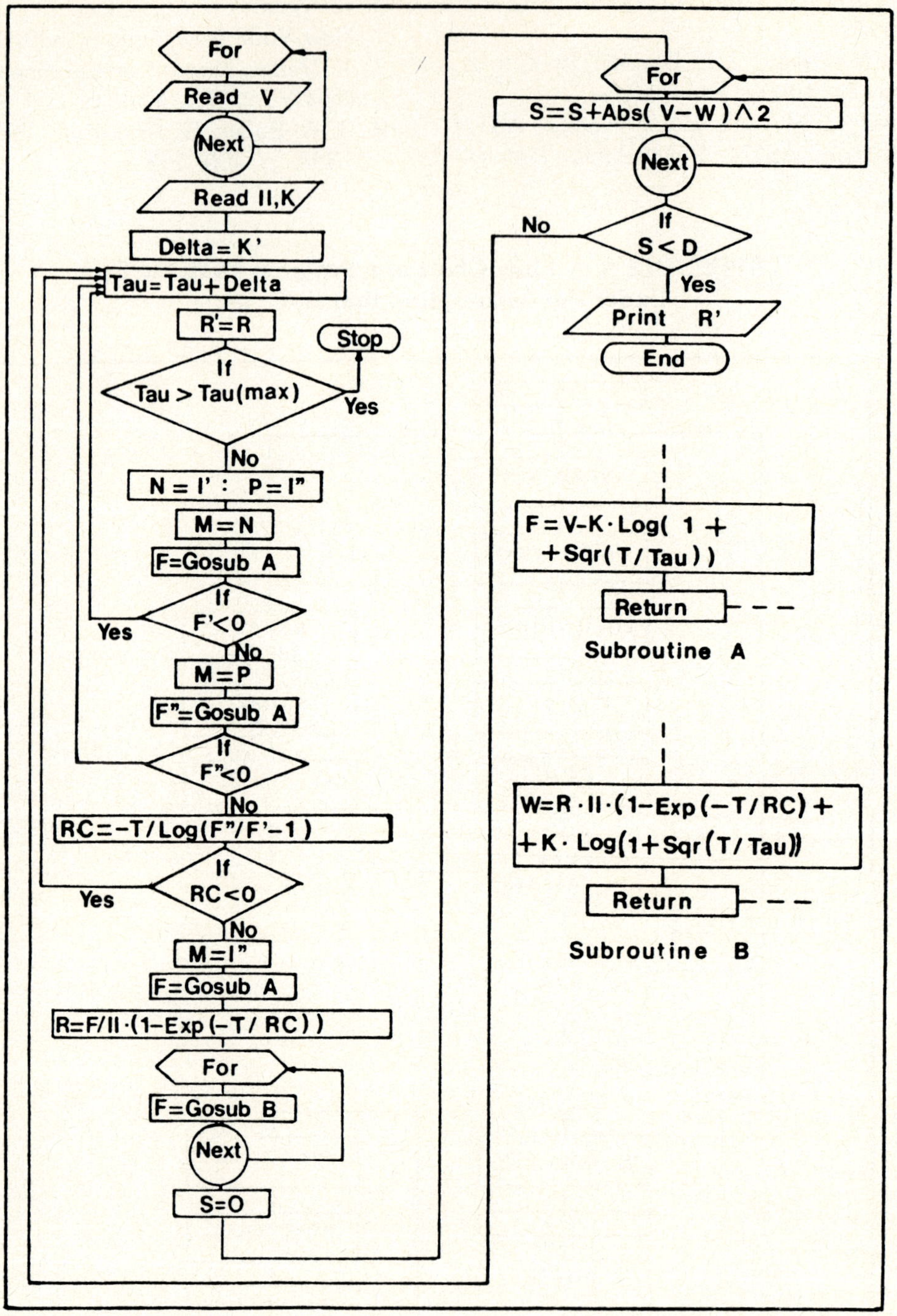

Section 3

Alloy Development

Application of Electrochemical Techniques in the Development of Alloys

E. D. VERINK, JR.*

Abstract

The combination of electrochemical methods with other analytical tools provides relatively rapid assessment of the corrosion behavior of metals and alloys in aqueous solutions. As such, they offer the opportunity to compare the influence of alloying additions on corrosion behavior of alloys within a single alloy family. They can be used to assess the tendency for crevice corrosion, pitting, dealloying, hydrogen embrittlement (HE), etc. Care must be exercised, however, to avoid overapplication of such results. Nonetheless, when properly used, electrochemical methods can greatly enhance the ability to predict aqueous corrosion behavior, particularly in increasing the cost effectiveness of research. Under no circumstances, however, should electrochemical tests be considered a substitute for doing the "critical experiment."

Introduction

As suggested by Morris Cohen,[1] "...when a metal is placed in contact with a corrosive solution, a very complex set of reactions occurs. It is the function of corrosion research to separate the various processes which take place and evaluate the effect of each on the overall corrosion reaction. A complete understanding of each individual process requires the solution of problems in mechanism (including chemical thermodynamics), kinetics, and because the reactions are heterogeneous, some knowledge of structural and topological factors. Although a single technique may yield information regarding more than one of these problems, in general the use of a number of techniques is required to obtain a complete understanding of the total corrosion process."

*Dept. of Materials Science and Engineering, University of Florida, Gainesville, Florida.

While this statement was made in 1969, it continues to be true today. In qualifying materials for corrosion service or in explaining the corrosion behavior of materials, it is necessary to use a combination of procedures that not only examine electrochemical behavior but also give detailed characterization of the chemical state and morphology of surfaces. Since the previous symposium on this topic,[2] there has been continuous progress in the development of sophisticated research techniques that can elucidate corrosion problems. Representative examples of some of the newer techniques are included herein, together with examples of the more traditional procedures used in alloy development.

Corrosion research programs at the University of Florida have been identified with the development of experimental potential-pH diagrams based on electrochemical measurements. Such experimental diagrams can overcome many of the shortcomings of the calculated (equilibrium) Pourbaix diagrams by providing a basis for the prediction of corrosion behavior of engineering alloys in aqueous media. This work has been directed at extending to alloys, techniques originally proposed by Pourbaix for pure metals. As a consequence, workers at the University of Florida have developed or refined techniques:

(1) Construction of three-dimensional Pourbaix diagrams (potential-pH-composition) for alloy systems, which are useful in alloy development.

(2) Inclusion of kinetic data on the experimental Pourbaix diagrams, which are useful in predicting corrosion and comparing the effects of alloy additions between members of alloy families.

(3) Evaluation of the influence of crevices and other occluded cells on the corrosion behavior of alloys in engi-

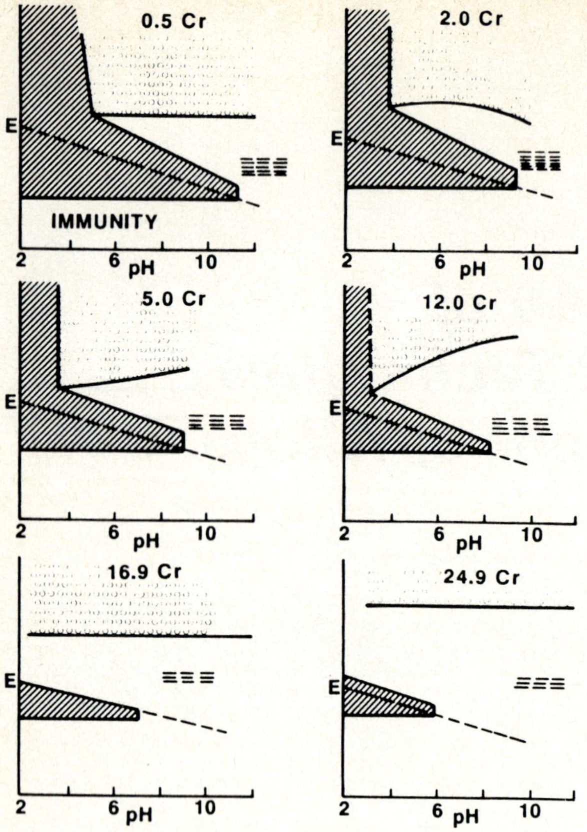

FIGURE 1 – Schematic experimental potential-pH diagrams for six binary Fe-Cr alloys presented for comparison to assess the influence of chromium content on corrosion behavior in solutions containing 0.1 M chloride ion.[3]

neering structures.

(4) Prediction of the tendency for dealloying as a function of potential and pH.

(5) Evaluation of the effect of environmental variables such as electrolyte composition, gas concentration, the presence of other ionic species, etc.

(6) Computerized calculation and plotting of equilibrium diagrams at various temperatures for electrolytes containing several ionic species in addition to water and metal ions.

The research in all of these areas continues and is enriched by the development of new sophisticated complimentary techniques, a number of which will be presented herein.

Experimental Potential Versus pH Diagrams for Alloys

Pourbaix first suggested a method for construction of experimental potential-pH diagrams using potentiokinetic polarization curves in connection with the electrochemical hysteresis method.[3] These methods were used to develop the schematic experimental potential-pH diagrams for six binary iron-chromium alloys exposed in solutions containing 0.1 molar (M) chloride ion (Figure 1). This figure illustrates the influence of chromium content on the corrosion behavior of binary iron-chromium alloys as well as delineating the locus of pitting potentials, protection potentials, etc. This data can be assembled into a three-dimensional diagram, potential vs pH vs composition, for the family of binary iron-chromium alloys by interconnecting the various features of each individual potential pH diagram as was done by Cusumono.[4] T. S. Lee[5] constructed a similar three-dimensional Pourbaix diagram for the copper-nickel binary system in 0.1 M chloride solutions. In the case of the binary copper-nickel system, it was possible to trace the change in passive behavior from "copper-like" behavior to "nickel-like" behavior as the compositions ranged from pure copper through a number of alloy compositions of increasing nickel content to pure nickel. This transition occurred in the vicinity of the 50% composition. A particular advantage of such three-dimensional diagrams is that they make it possible to compare quickly the results of solute additions on passive behavior. For example, in the copper-nickel system, it facilitated the assessment of the addition of solute materials such as iron and/or other alloying ingredients such as chromium on the passive behavior of individual copper-nickel alloys.

Kinetic Information

While no kinetic information is available from calculated, equilibrium potenital-pH diagrams, it is possible to incorporate kinetic information on experimental potential-pH diagrams. Such information is plotted in the form of current density contours. For example, for a given alloy, when the potentials are tabulated, the logarithms of the corresponding current densities taken from the potentiokinetic polarization curves are whole numbers for each pH. These current density data points are plotted at corresponding potentials and pH values on the experimental potential-pH diagrams and each point is labeled with its appropriate logarithmic value of current density (Figure 2). Isocorrosion contours result when lines are drawn to connect the data points having the same logarithmic index value. Since the absolute magnitudes of the measured corrosion current densities are strongly influenced by scan rate, it is necessary to "correct" these indicated corrosion rates by other methods, such as gravimetric tests, etc. Regardless, the corrosion contours are useful for comparing the *relative* behavior of alloys within a family provided that a constant scan rate is scrupulously maintained. Comparison of the relative corrosion rates among alloys of a family provides another rapid method for assessing the effect of alloy additions on passive behavior.

A powerful tool for distinguishing between soluble and insoluble reaction products of a corrosion process uses forced convection of the electrolyte to maintain a well-defined diffusion layer at the electrode interface and the transport of a soluble component to a concentric collector electrode where it can be electrochemically analyzed. The rotating ring disk electrode in its present form is due primarily to the efforts of Albery, Bruckenstein, and Miller.[6,7] Rotation defines the hydrodynamic diffusion layer, which is a function of the angular velocity

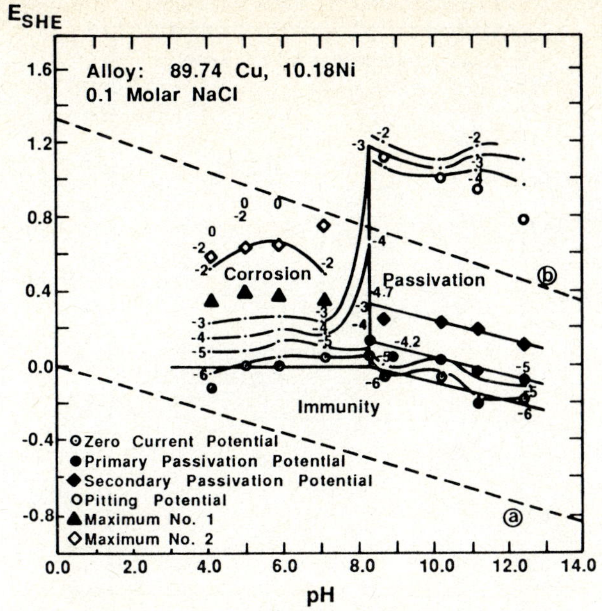

FIGURE 2 — Current density contour diagram for 90:10 cupro-nickel alloy.

of the electrode. At sufficiently high velocities, the diffusion layer assumes a constant minimum thickness such that anodic current density is limited by charge transfer kinetics alone. At lower values of angular velocity, even though the hydrodynamic diffusion layer is defined, the reaction kinetics are under mixed control, limited by charge transfer at short time intervals and by mass transfer at extended time intervals. By studying current-voltage relationships at various values of angular velocity, it is possible to characterize anodic reaction kinetics under various limiting conditions.

A fraction of the soluble reaction products generated at the disk may be detected on a ring electrode concentrically positioned outside the disk and under separate potential control, which is maintained at some value corresponding to the redox potential of a given product. Dividing the ring into discrete sections allows simultaneous detection of as many soluble reaction products as there are ring components and associated redox potentials. That fraction of reaction products transported to a given ring segment is determined by the electrode geometry. Accordingly, a given value of detected ring current can be translated into that fraction of the disk current associated with a particular electrochemical process. Thus, it should be possible to distinguish conceptually between soluble and insoluble reaction products generated at the disk. Rotating disk electrodes and rotating ring disk electrodes have been used to study metal dissolution kinetics of a number of materials including magnesium, manganese, zinc, iron, copper, nickel, lead, silver, gold, palladium, indium, a variety of steels and brasses, as well as copper-gold alloys. Accurate control of ring potential is essential since a true transport-limited ring redox current is seldom achieved. Ring and disk potentials as supplied may be coupled. Therefore, any variation in disk current will affect the applied ring potential in proportion to the electrical circuit and solution resistances. Such problems can normally be compensated for by suitable ohmic drop circuitry.

While the diffusion layer of a rotating disk electrode is reasonably well-defined, it is not uniform across the disk surface. When the disk current changes with time, as would be the case when disk coverage by a surface film occurs, the situation becomes more complex. The relationship between ring and disk current measurements is complicated by the fact that transport of electrolyte active species generated at a given point in time from different areas of the disk requires different transport times depending on the specific active area position with respect to the location of the ring. This is a particular complication when attempting to study the influence of electrode rotation velocity.

This problem can be overcome by either of two methods. McKubre, et al.,[8] has described the rotating cylinder electrode. This electrode "sandwiches" the working electrode between the two collector electrodes. The large ratio of the collector area to the working electrode area eliminates the transport problem. The other method involves replacing the disk with a ring. Thus, this method is referred to as the *rotating ring-ring electrode.* Reducing the working electrode's surface area increases the possibility of "edge" effects or uneven current density distributions. However, this factor should be constant and does not depend on the prior mechanical history of the material being studied.

These rotating electrode methods have been used extensively in the field of chemistry but only to a limited extent in corrosion studies. Their versatility and capability recommend them for considerably more widespread use in assessing the corrosion behavior of alloys.

Occluded Cells

Efird[9] developed a crevice corrosion cell that consists of a bulk solution container of approximately 1 L capacity with a small side arm separated from the larger volume by a fritted glass disk. Details of the cell are presented elsewhere.[2,3] The so called "bulk" sample was placed in the larger volume, whereas the so called "crevice sample" was placed in the smaller volume (i.e., the side arm). These samples were connected through suitable circuitry to permit monitoring of potential, pH, and chloride ion as a function of time, as well as current flow as the potential of the bulk sample could be controlled to any desired value. Using this apparatus, it was postulated that a crevice protection potential existed for 90:10 copper:nickel. It was also anticipated that dealloying might be expected in this system within the crevice under certain circumstances. Figure 3 presents the results. Consider first the points numbered 1 in Figure 3. The electrode potential in the bulk solutions is above (more noble than) the equilibrium potential for the Cu/CuCl reaction, $Cu + Cl^- = CuCl + e$.

The change in pH and electrode potential as a result of acidification caused by crevice corrosion, is represented by the phantom line connecting the points numbered 1. If this phantom line crosses the vertical

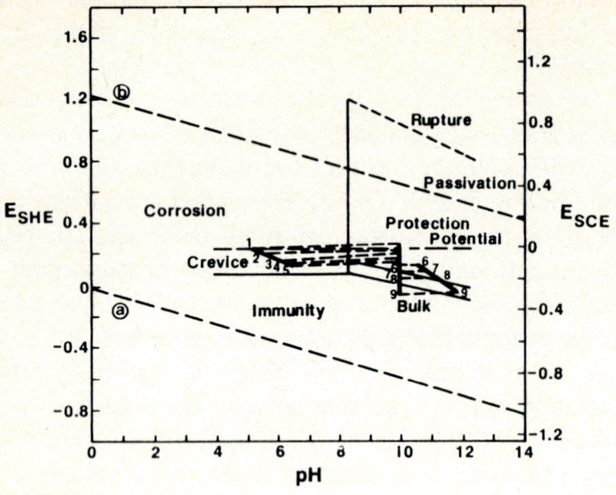

FIGURE 3 – The influence of external polarization on the potential-pH characteristics of a crevice on 90:10 copper:nickel with 1.46% iron. The dotted lines indicate simultaneous data for the crevice and the external surface.[9]

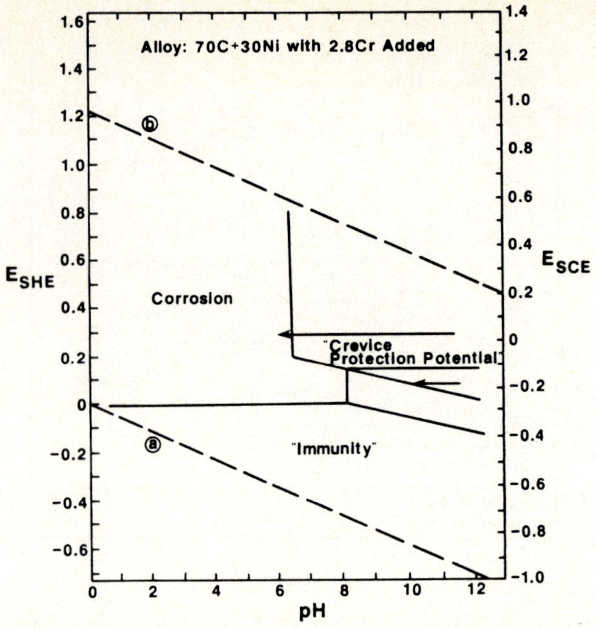

FIGURE 4 – Schematic potential-pH diagram for 70:30 cupro-nickel containing 2.8% chromium showing the "crevice protection potential" below which crevice corrosion does not occur.[10,11]

boundary (between passivation and corrosion) above 0.200 V_{SHE}, and the final electrode potential within the crevice is below 0.200 V_{SHE}, copper will be deposited in the crevice. (This phenomenon will be discussed later under Dealloying.) For data points numbered 6 in Figure 3, the final pH in the compartment containing the crevice specimen was more alkaline than that of the bulk solution. This phenomenon was ascribed to the existence of a crevice protection potential and is discussed more fully below.

Bowers,[10,11] using a modified specimen assembly consisting of a metal sample and a glass slide overlapped and held together with a rubber band, confirmed the concept of a crevice protection potential in chromium-modified 70:30 copper:nickel alloy containing 2.8% chromium. Referring to Figure 4, at a potential above (more noble than) the crevice protection potential (for example, line A on Figure 4), the composition and/or character of the corrosion reaction film does not change significantly as local acidification develops until the pH reaches the vertical line that forms the boundary between the passive and corrosion regions of the experimental potential pH diagram. When this vertical line is crossed, severe crevice corrosion occurs. If the initial electrode potential in the crevice is below the crevice protection potential (for example, line B), local acidification again results in a pH change. In this case, however, the path of pH change intersects the boundary of another passive region rather than the corrosion region on the experimental potential pH diagram. Chronoamperometric measurements have shown that the steady-state current necessary to maintain the passivating species on the more active side of this coexistence boundary is considerably less than necessary to maintain the passive species on the noble side of the boundary line. This indicates that the lower passive species is more protective. Hence, the passive film is reinforced and crevice corrosion is avoided since the pH does not move to the active (corrosion) region of the diagram. This work has important implications since it suggests that in saline solutions with pH values greater than about 8, crevice corrosion of chromium-modified 70:30 copper:nickel should not occur at potentials more active than the crevice protection potential.

Starr[11,12] also used a cell similar to that of Efird to measure variations in composition and pH in solutions surrounding the iron-12% chromium samples exposed in the crevice compartment. Figures 5(a) and (b) show the variation in pH and potential for an iron-12% chromium alloy within artificially occluded cells in hydrogen-saturated bulk solutions as compared to oxygen-saturated bulk solutions. In each case, the pH within the crevice goes to very acid values and the potentials are below the equilibrium hydrogen line. The fact that the final potential was below the equilibrium hydrogen line suggests the likelihood that hydrogen charging could occur, which in turn could result in hydrogen embrittlement. Because of the test chamber features, the occluded cell compartment was sufficiently large to make possible detailed analysis of the final artificial occluded cell solutions. Starr made such measurements for the iron-12% chromium alloy and the data are shown in Figure 6. The solid line in this figure was calculated ignoring carbide presence and assuming the iron-to-chromium concentration ratio was equal to that in the alloy. The dashed line was calculated taking into account the presence of the compound $Cr_{23}C_6$. The confidence limits shown were calculated assuming a possible experimental deviation from the observed value of ±2 ppm. The correspondence between the data points and the dotted line indicates that no macroscopic dealloying occurred in this case since the corrected ratio of alloy constituents in the alloy and in the solution were approximately the same. Figure 7 shows the results of chloride analysis compared in such a way as to test the assumption that Fe^{2+} and Cr^{3+} in the form of $FeCl_2 \cdot xH_2O$ and $CrCl_3 \cdot yH_2O$ are the predominant corrosion products in the occluded cells in which the iron-12% chromium binary alloy was exposed. The closeness of the data points to the calculated line

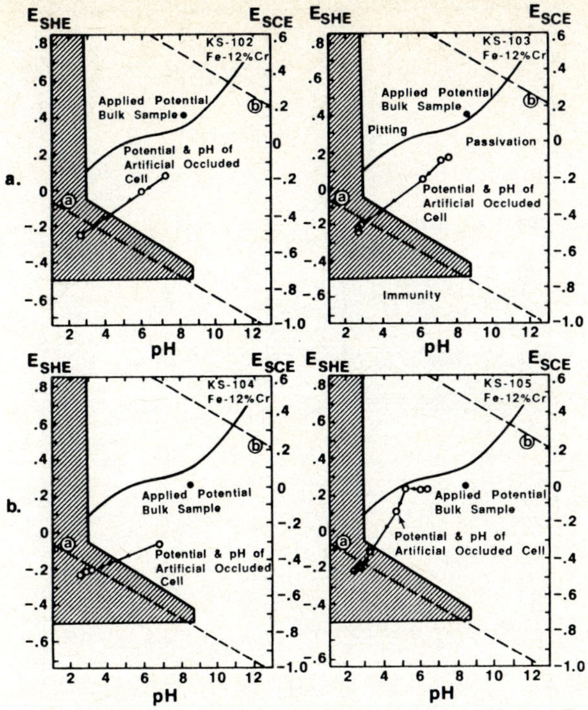

FIGURE 5 — (a) Progress of artificial occluded cell experiments for the Fe-12% Cr alloy in solutions initially 0.1 M NaCl + 0.01 M $NaHCO_3$, with H_2-saturated bulk solutions, and (b) progress of artificial occluded cell experiments for the Fe-12% Cr alloy in solutions initially 0.1 M NaCl + 0.01 M $NaHCO_3$, with O_2-saturated bulk solutions. The results are superimposed on the experimental potential vs pH diagram (0.1 M Cl^-, H_2-saturated solutions) for the same alloy.[11,12]

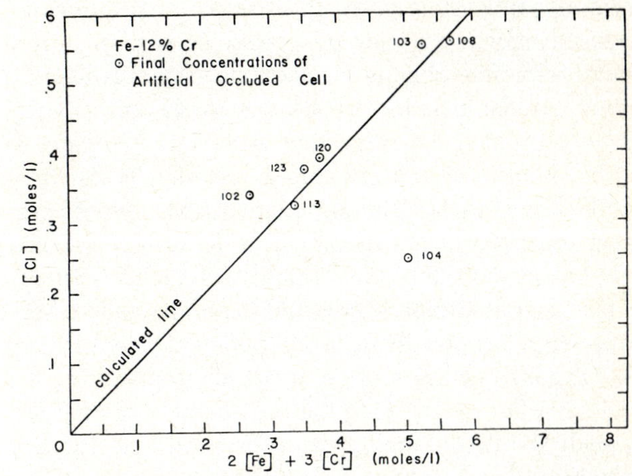

FIGURE 7 — Plot of chloride concentrations (determined by $AgNO_3$ titrations) vs linear combinations of the data from tests referred to in Figure 5. The solid line was calculated assuming the predominant corrosion products to be $FeCl_2 \cdot xH_2O$ and $CrCl_3 \cdot yH_2O$.[11,12]

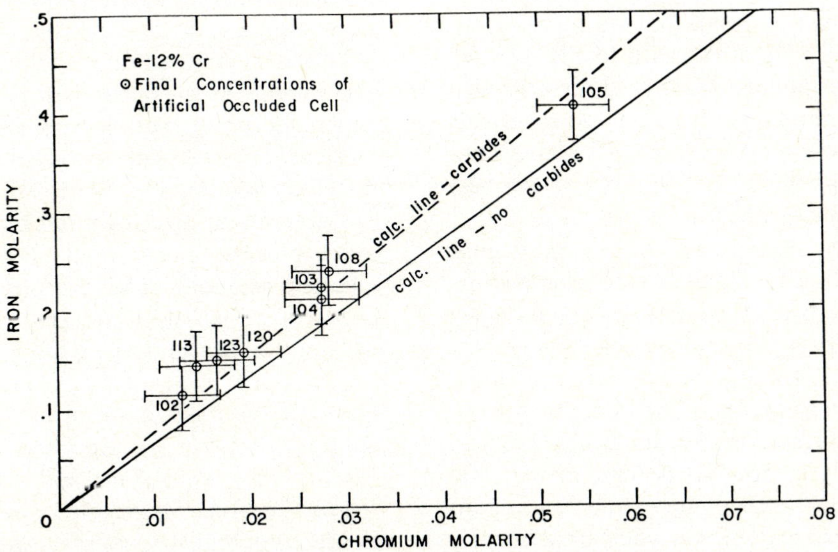

FIGURE 6 — Results of atomic absorption spectrophotometric analyses of the final artificial occluded cell solutions for the Fe-12% Cr alloy. The solid and dashed lines were calculated assuming no macroscopic dealloying occurred.[11,12]

lends weight to the argument that these species predominate.

Mechanistic Studies

For alloy systems that show a definite protection potential, it is possible to gain some insight into the mechanism by which corrosion ceases at the protection potential.[12] Tests with a series of iron-chromium binary alloys provided information regarding the mechanism by which the corrosion current was suppressed at the protection potential in chloride solutions. A deactivation mechanism is said to occur when, as a result of lowering its electrode potential sufficiently, the material is made thermodynamically immune to corrosion. Experimentally, deactivation is detected by a rapid increase in current density with little change in electrode potential when a second anodic polarization scan is made immediately after determining the protection potential. A repassivation mechanism occurs when upon lowering its electrode potential, the kinetics of thermodynamically favored corrosion reactions at the metal electrode surface become very slow as a result of the formation of a reaction product film. Repassivation is detected experimentally when, with a large increase in electrode potential, there is very little increase in current upon anodically repolarizing the specimen after determining the protection potential. Using such test procedures, it was observed that iron-2% chromium alloys exhibited deactivation, alloys containing 16.9% chromium or more exhibited repassivation, whereas the iron-12% chromium alloy showed either deactivation or repassivation depending on the pH of the solution.

Dealloying

The most prominent example of dealloying is dezincification. There has been a long and vigorous dialogue as to the exact mechanism of dezincification.[13,14] One group of researchers contends that the entire alloy is dissolved and that one of its constituents (the less active) is then replated from solution. Another group contends that one species (the more active) is dissolved selectively from the alloy, leaving a porous residue of the more noble species. Still others believe that both mechanisms take place more or less simultaneously.

Figure 8 shows the superposition of an experimental potential vs pH diagram for 70:30 copper:zinc alloy in 0.1 M chloride solutions over the copper-chloride-water equilibrium diagram for 0.1 M chloride solutions and the zinc-water equilibrium diagram. Superposition of these diagrams delineates regions that correspond to the experimentally observed results from a number of dezincification tests. At potentials more active than about 0.000 V_{SHE} (below the cross-hatched region in acid solutions in Figure 8), it is inferred that zinc ion would be selectively leached from an alloy containing zinc and copper. Atomic absorption spectroscopy confirms the presence of zinc ions at potentials in this range, but that copper was not present in detectable amounts (limit of detection, 0.25 ppm). In copper-free chloride solutions, specimens potentiostatted at potentials below about 0.000 V_{SHE}

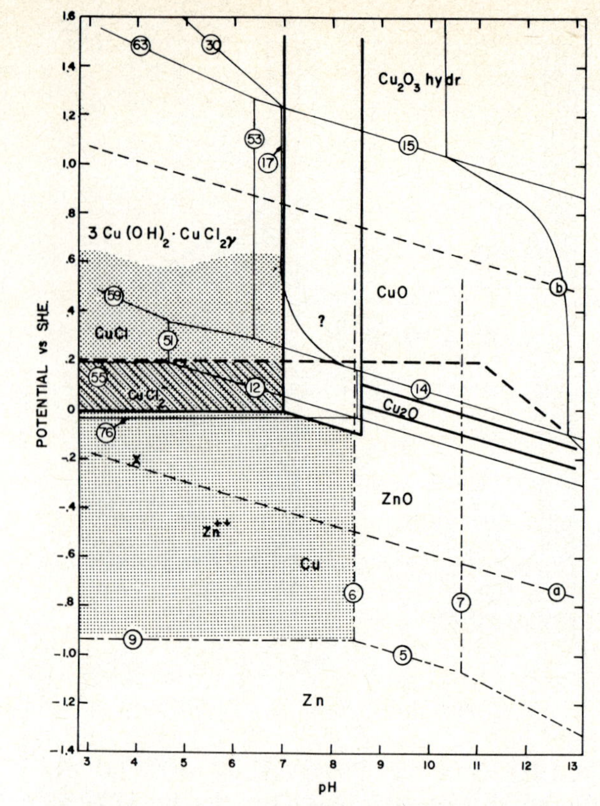

FIGURE 8 — Cu-30% Zn alloy in 0.1 M chloride solution. Superposition of the experimental potential diagram on the Cu-Cl-H_2O and the Zn-H_2O Pourbaix diagrams.[13,14]

would be expected to suffer dezincification by selective loss of zinc. This was confirmed by x-ray evidence. Within the cross-hatched region between the potentials of about 0.000 and +0.200 V_{SHE}, atomic absorption spectroscopy shows that both copper and zinc go into solution. With vigorous stirring, much of the copper may be scavenged on the auxiliary electrode. In quiescent solutions, a large amount of the copper is deposited on the sample as a copper film. This corresponds to the observation that metallic copper may be discovered in crevices or pits in copper-based alloy structures. Recall also the experiment of Efird discussed above (Figure 2).

If samples are potentiostatted above (at potentials more positive than) about +0.200 V_{SHE} in 0.1 M chloride solutions, and the potential is then dropped below (to potentials more active than) this value, copper will deposit, presumably by the reaction $CuCl + e \rightarrow Cu + Cl^-$. Fort[15] used auger electron spectroscopy (AEM) to elucidate the mechanism of the dezincification of 70:30 copper:zinc alloy. A specimen was potentiostatted at −0.450 V_{SCE} for four and one-half days. Figure 9 shows the auger composition profile. Depletion of the zinc content of the near surface layers is evidenced by initially low peak-to-peak heights for zinc which increase with increasing depth from the sample surface. The amount of zinc measured increases until a constant value is reached. The peak-to-peak heights measured for copper remain relatively constant after the initial cleaning period. The copper:zinc ratio curve indicates the level of

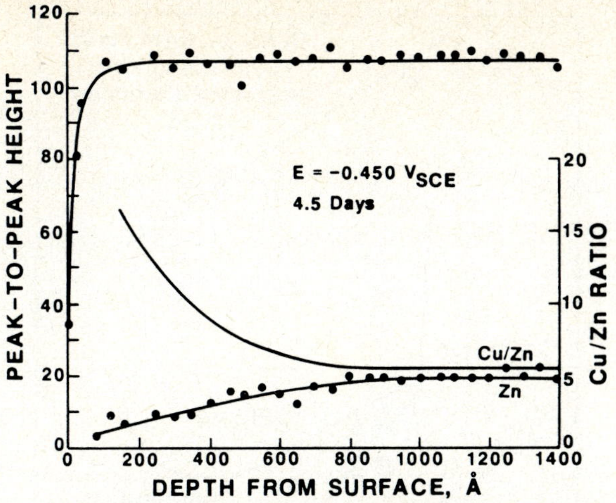

FIGURE 9 — Auger composition profile obtained from a Cu-30% Zn sample potentiostatted at -0.450 V_{SCE} for 4½ days.[15]

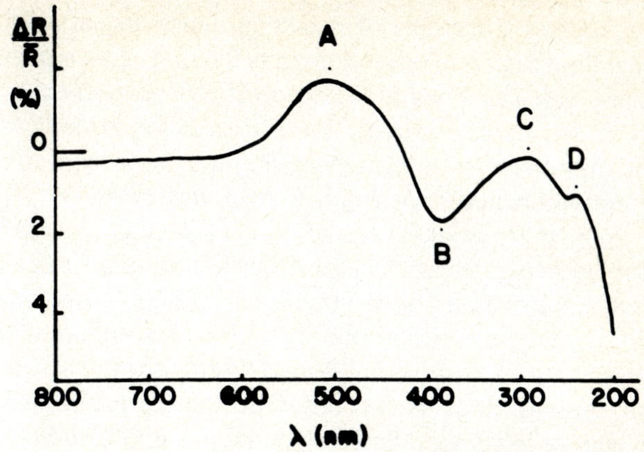

FIGURE 11 — Differential reflectogram of Cu-21.4% Zn after one hour exposure in pH ~4 and E = 0.105 V_{SHE}. (A "positive peak" A indicates a loss of zinc in the potentiostated half of the specimen.[18])

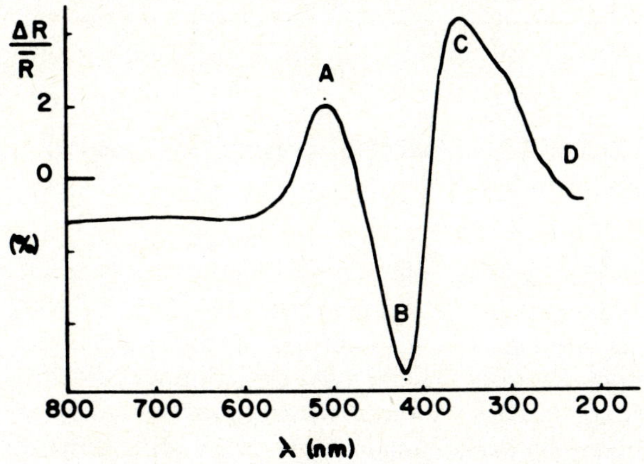

FIGURE 10 — Differential reflectogram (△R/R vs wavelength λ) of Cu-18.2 At% Zn vs Cu-19.01 At% Zn. Average composition 18.6 At% Zn (yellow brass).[16,17]

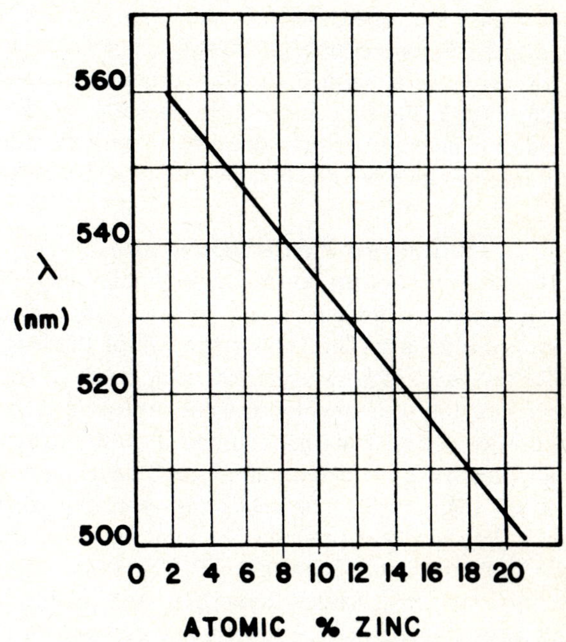

FIGURE 12 — Wavelengths of peak "A" vs average zinc concentration in copper-zinc alloys.[18]

dealloying at each point in the composition-vs-depth profile. Fort observed that comparison of such curves for several exposure times at -0.450 V_{SCE} shows that the level of zinc depletion at the surface increases with increasing time of exposure.

The differential reflectometer in combination with electrochemical techniques is a valuable tool for studying dealloying. The differential reflectometer is described elsewhere.[16,17] A typical differential reflectogram is shown in Figure 10. This reflectogram was generated under conditions wherein the two sample halves of a couple have different compositions, as is the case when dezincification occurs on one and not on the other. Figure 11 depicts a differential reflectogram of a copper-21.4 wt% zinc alloy, one-half of which was potentiostatted for one hour in a solution of pH = 4 at a potential of -0.105 V_{SHE}. The other half (the "reference half") was in the unpotentiostatted state (i.e., freely corroding). This figure shows that such electrochemical exposure leads to a change in zinc content, i.e., to compositional modulation. The wavelength of peak A decreases with increasing average zinc concentration as is shown in Figure 12. In Figure 11, peak A appears at 506 nanometers (nm) which, according to Figure 12, is equivalent to an average zinc concentration of 19.2% for the couple. By knowing the original composition of the unexposed specimen (in this case 21.4% zinc) and the measured alloy composition of the couple, the zinc content at the surface of the dealloyed specimen can be estimated by linear extrapolation. In this case, the dealloyed half was found to contain approximately 17% zinc which corresponds to a loss of about 4.4% zinc. The differential reflectometer is capable of detecting dealloying at a far earlier time than any other known technique. This is particularly valuable since the differential reflectometer permits *in situ* measurements, thereby avoiding complications that might arise from removing a sample from the test apparatus and exposing it to high vacuum.

Finnegan, et al.,[18] tested the frequently reported observation that "dezincification does not occur in brasses containing less than 15 atomic % zinc." Using a combination of electrochemical methods and the differential reflectometer, Finnegan confirmed that dezincification of such brasses does occur but that the dezincification behavior of red brass differs both in degree and kind from that observed on yellow brasses. Copper-21.4% zinc alloys (yellow brass) showed dezincification in less than 1 hour potentiostatting time at a pH of approximately 4 and a potential of $-0.105\ V_{SHE}$. Copper-9.2% zinc alloys (red brass) required approximately 16 hours under these experimental conditions to dealloy to the same extent as judged from instrumental signal intensity measurements. The loss of zinc at the surface was 4.4% for the copper-21.4% zinc alloy; 3.8% for a copper-9.2% zinc alloy; and 2.6% for a copper-5.3% zinc alloy. The presence of Cu_2O as a surface reaction product retarded dezincification in the red brasses. This reaction product also has a red color. Cuprous oxide was not prominant in the differential reflectograms obtained from yellow brass.

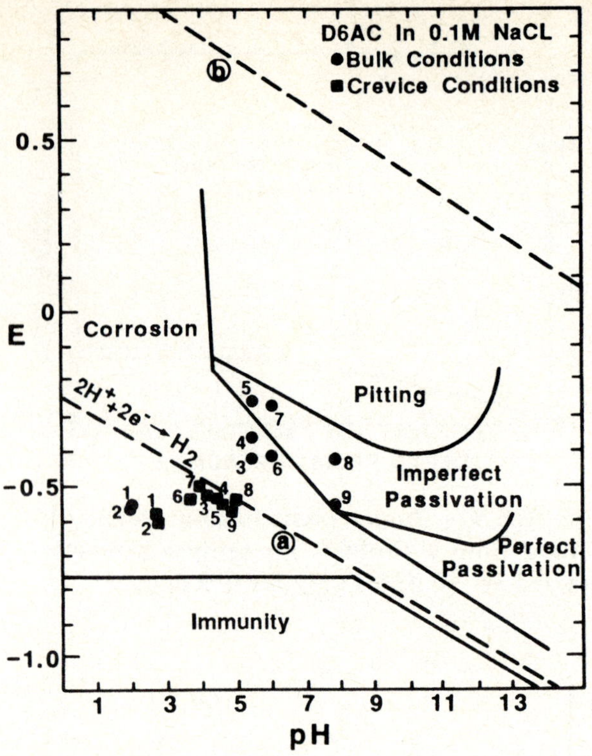

FIGURE 13 — Artificial crevice cell results, pH 2.1 to 8.1.[19,20]

Hydrogen Embrittlement

The prediction of the tendency for certain alloys to suffer hydrogen embrittlement can be made using electrochemical methods. The crevice test cell of Efird was used for tests with high strength, low alloy steel, D6AC, exposed in 0.1 M sodium chloride solution. Results of artificial crevice cell tests are superimposed on the experimental potential vs pH diagram for D6AC developed by Johnson,[19] (Figure 13). The circles represent the initial pH and the electrode potential in both the bulk and crevice compartments of the test cell. The square data points are the final, steady-state pH, and electrode potentials of the samples in the crevice compartment. Each test is numbered separately so the change in pH and potential for a particular test can be ascertained by selecting a circle and a square having the same index number.

For all experiments with this alloy in 0.1 M sodium chloride solutions above pH = 5, the final electrode potential within the crevice fell below the potential of the equilibrium hydrogen electrode regardless of the initial-potential, while the electrolyte in the occluded cell became more acid than the bulk solution. Parrish, et al.,[20,21] using compact fracture toughness specimens of D6AC exposed under conditions selected on the basis of these crevice tests, observed the effect of the occluded cells on K_{ISCC} and measured the hydrogen content. When the test conditions corresponded to those represented by the square data points in Figure 14, the hydrogen content of the fracture surface was substantially above the baseline hydrogen content of unexposed metal and K_{ISCC} was low. When oxidizing inhibitors (hydrazine for example) were added, the electrode potentials within the experimental occluded cells no longer fell below the potential of the equilibrium hydrogen electrode. Under these circumstances the fracture toughness tests showed increased values of K_{ISCC} and hydrogen concentrations were of the same order as unexposed metal.

Computerized Methods

Froning, et al.,[22] developed computerized methods employing mainframe computers to calculate complex Pourbaix diagrams. This program now has been adapted to personal computers[23] and greatly simplifies the task constructing equilibrium Pourbaix diagrams. Subprograms have been developed to construct diagrams for elevated temperatures using the Criss and Cobble approximation,[24] which appears to be acceptable up to about 200 C. The use of such programs permits the researcher to perform "thought experiments" which will refine the design of critical experiments and thereby make the research process considerably more cost effective.

Summary

The combination of electrochemical methods with other analytical tools provides relatively rapid assessment of the corrosion behavior of metals and alloys in aqueous solutions. As such, they offer the opportunity to compare the influence of alloying additions on corrosion behavior of alloys within an alloy family. They can be employed to assess the tendency for crevice corrosion, pitting, dealloying, hydrogen embrittlement, etc. However, care must be exercised to avoid over-application of such results. Nonetheless, when properly used, electrochemical methods may greatly enhance the ability to predict aqueous corrosion behavior, particularly from the standpoint of increasing the cost effectiveness of research. Under no circumstances, however, should electrochemical tests be considered a substitute for doing the "critical experiment."

References

1. M. Cohen, Modern Techniques in Corrosion Research, Basic and Applied Research, NACE, Houston, Texas (1969).
2. E. D. Verink, Jr., "Applications of Electrochemical Techniques in the Development of Alloys for Corrosive Service," Electrochemical Techniques for Corrosion, R. Baboian, Ed., National Association of Corrosion Engineers, Houston, Texas, p. 43, 1977.
3. E. D. Verink, M. Pourbaix, Corrosion, Vol. 27, No. 12, p. 495, 1971.
4. R. L. Cusumano, "The Establishment of a Three Dimensional (Potential-pH-Composition) Diagram for Binary Fe-Cr Alloys in 0.1 Molar Chloride Solutions," MS Thesis, University of Florida, Gainesville, Florida, 1971.
5. T. S. Lee, "The Use of Experimentally Determined Potential vs pH Diagrams to Elucidate the Influence of Solute Additions on the Passive Behavior of Binary Cu-Ni Alloys in Solutions With and Without Chloride," MS Thesis, University of Florida, Gainesville, Florida, 1971.
6. W. J. Albery, Trans. Faraday Soc., Vol. 62, p. 1915, 1966; W. J. Alberty, S. Bruckenstein, Trans. Faraday Soc., Vol. 62, p. 1920, 1966.
7. W. J. Albery, S. Bruckenstein, D. C. Johnson, Trans. Faraday Soc., Vol. 62, p. 1938, 1966.
8. M. C. McKubre, D. D. MacDonald, J. Electrochem. Soc., Vol. 127, p. 632, 1980.
9. K. D. Efird, "Crevice Corrosion of a Copper-Nickel Alloy and Its Relationship to the Experimental Pourbaix Diagram," MS Thesis, University of Florida, Gainesville, Florida, 1970.
10. J. M. Bowers, "Crevice Corrosion of a 70-30 Chromium-Modified Cupronickel and its Relationship to the Experimental Pourbaix Diagram," MS Thesis, University of Florida, Gainesville, Florida, 1973.
11. E. D. Verink, K. K. Starr, J. M. Bowers,, Corrosion, Vol. 23, No. 2, p. 60, 1976.
12. K. K. Starr, E. D. Verink, M. Pourbaix, Corrosion, Vol. 32, No. 2, p. 47, 1976.
13. E. D. Verink, R. H. Heidersbach, Evaluation of the Tendency for Dealloying in Metal Systems, ASTM Special Publication 526, pp. 303-322, ASTM, Philadelphia, Pennsylvania, 1972.
14. R. H. Heidersbach, E. D. Verink, Corrosion, Vol. 28, No. 11, p. 387, 1972.
15. W. C. Fort, E. D. Verink, "Elucidation of the Mechanism of Dezincification by Auger Electron Spectroscopy in Combination with Electrochemical Methods," Proc. Fourth Int. Cong. Marine Corros. and Foul., Antibes-Juan les Pins, France, 1977.
16. C. W. Shanley, R. E. Hummel, E. D. Verink, Corros. Sci., Vol. 20, p. 467, 1980.
17. C. W. Shanley, R. E. Hummel, E. D. Verink, Corros. Sci., Vol. 20, p. 481, 1980.
18. J. E. Finnegan, R. E. Hummel, E. D. Verink, Corrosion, Vol. 37, No. 5, p. 256, 1981.
19. J. M. Johnson, "Use of Experimental Potential-pH Diagram in D6AC Steel to Interpret Its Behavior in Aqueous Media," MS Thesis, University of Florida, Gainesville, Florida, 1974; published as Technical Report AFML-TR-76-8, Air Force Materials Laboratory WPAFB, Ohio, 1976.
20. P. A. Parrish, "The Retardation of Crack Propagation for High Strength, Low Alloy Steels in Aqueous Media by Addition of Oxidizing Inhibitors," PhD Dissertation, University of Florida, Gainesville, Florida, 1974.
21. P. A. Parrish, C. M. Chen, E. D. Verink, Retardation of Crack Propagation for D6AC High Strength, Low Alloy Steel in Aqueous Media by Addition of Oxidizing Inhibitors, ASTM Special Technical Publication 610, ASTM, Philadelphia, Pennsylvania, 1976.
22. M. H. Froning, M. E. Shanley, E. D. Verink, Corros. Sci., Vol. 16, p. 371, 1976.
23. E. D. Verink, Jr., "Adaptation of a Computer Program for Pourbaix Diagrams to Personal Computers," Proc. Int. Symp. Honoring M. Pourbaix on his 80th Birthday, R. P. Frankenthal, J. Kruger, Eds., The Electrochemical Soc., Inc., Pennington, New Jersey, 1984.
24. C. M. Criss, J. W. Cobble, Jr., J. Amer. Chem. Soc., Vol. 86, p. 5385, 1964; C. M. Criss, J. W. Cobble, Jr., J. Amer. Chem. Soc., Vol. 86, p. 5390, 1964.

Electrochemical Techniques for Determining Potential-pH Diagrams

J. BAKAMA and R. HEIDERSBACH*

Abstract

Potential-pH (Pourbaix) diagrams were developed as a means of presenting large amounts of thermodynamic data in a manner useful to practicing corrosion engineers and electrochemists. While thermodynamically based diagrams have received acceptance in many fields, they have several important limitations, including: (a) They cannot be calculated for alloys, (b) they cannot adequately predict which of several possible species is most likely to form, (c) they cannot predict the kinetics of electrode reactions, and (d) they cannot predict the effectiveness of passive films that form on electrode surfaces.

The above limitations led several research groups to seek methods of experimentally determining potential-pH diagrams for pure metals and alloys in aqueous environments. This paper discusses electrochemical and spectroscopic techniques that have been used for the experimental determination of potential-pH diagrams. Applications to corrosion research and limitations of present techniques are also discussed.

Introduction

Potential-pH diagrams were developed by Marcel Pourbaix as a means of presenting thermodynamic data in a manner useful to practicing corrosion engineers.

The potential-pH diagram is divided into three regions:[1-2]

1. **Immunity**—The pure metal is thermodynamically stable and dissolution or corrosion cannot occur.

2. **Corrosion**—Ions of the metal are thermodynamically stable and dissolution or corrosion will occur; and

3. **Passivity**—A compound reaction product of the metal with elements from the environment is thermodynamically stable.

*School of Chemical Engineering, Oklahoma State University, Stillwater, Oklahoma.

Passive reaction products will frequently form protective films on surfaces and prevent further reactions between metals and their environments. Thus, *passivity*, as defined by Pourbaix, is a region where corrosion may occur or may be prevented depending on the protectiveness of the passive films that are formed.

Figure 1 is a potential-pH diagram for iron.[3] It is one of the many potential-pH diagrams that have been published for iron in water. The slanted dashed lines in the middle of the diagram define the region of thermodynamic stability for water. Above the top dashed line, oxygen will be evolved; below the bottom dashed line, hydrogen will be evolved. Because much of the region between these lines shows passivity, it is obvious that the protectiveness of passive compounds will strongly influence the corrosion behavior of iron in water. This passivity dependence is true of all metals with the exception of copper and the noble metals: gold, silver, platinum, etc. All other metals are thermodynamically unstable in water and, thus, the corrosion resistance of most metals depends on the protective properties of passive films.

Unfortunately, thermodynamic calculations cannot be used to predict the protection afforded by passive films.[1] Other limitations of calculated potential-pH diagrams include the following: (a) they cannot be applied to alloys, and (b) they cannot predict the effect of minor environmental impurities, such as chlorides, on the behavior of metals. Nonetheless, these diagrams have been widely used in the corrosion community and seem to be justified by practical experience. As an example, carbon steel, which is similar to pure iron in its corrosion behavior, corrodes at slow rates in most bases and corrodes rapidly in many acids.

The development of the commercial potentiostat in the 1960s allowed researchers to attempt to determine experimental potential-pH diagrams for pure metals and alloys in a wide variety of environments.[4] Potentiostats

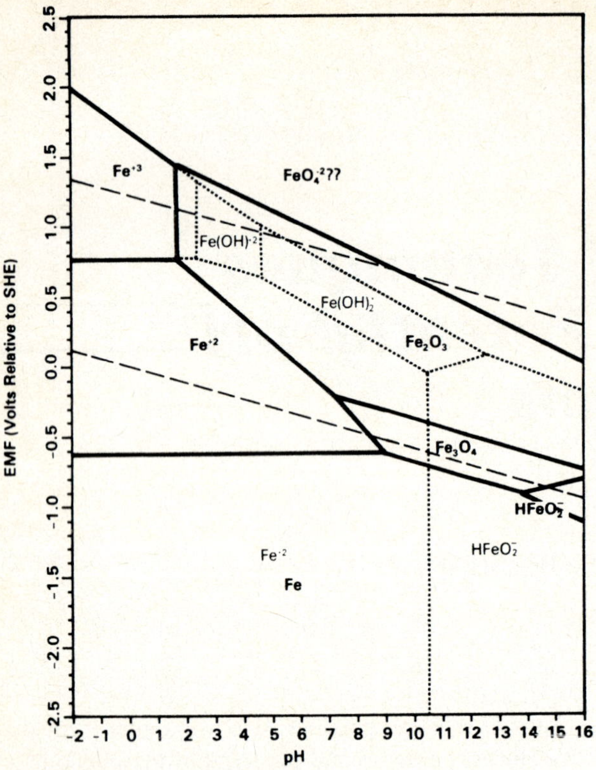

FIGURE 1 — Thermodynamically calculated potential-pH diagram for iron in water. Slanted dashed lines indicate thermodynamic stability region for water.[3]

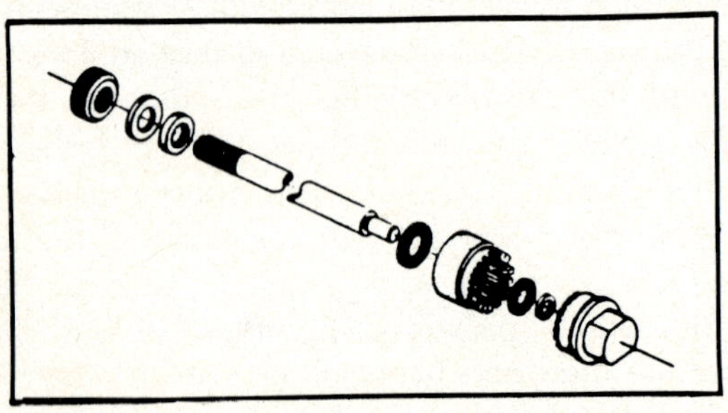

FIGURE 2 — Multiple crevice assembly working electrode holder.

also offered the possibility of using potential-pH concepts for anodic protection under conditions where passivity would either not occur or not remain stable.[5]

The first experimental potential-pH diagrams were based on the *electrochemical hysteresis technique* reported by Pourbaix, Verink, and co-workers.[6-9] Short-term polarization curves formed the experimental basis for the technique. Regions of passivity, transpassivity, general corrosion, and immunity from corrosion were identified from polarization curves of the alloy of interest in a series of buffered electrolytes having different pH values. A series of diagrams were developed for iron,[4,9-12] copper,[4,13-16] and aluminum[17-18] alloys.

The cyclic polarization techniques used by Pourbaix, Verink, and others were similar to those used in many laboratories for other electrochemical studies.[19] ASTM Standard G 61, "Cyclic Potentiodynamic Polarization Measurements,"[20] is based on these procedures. Many

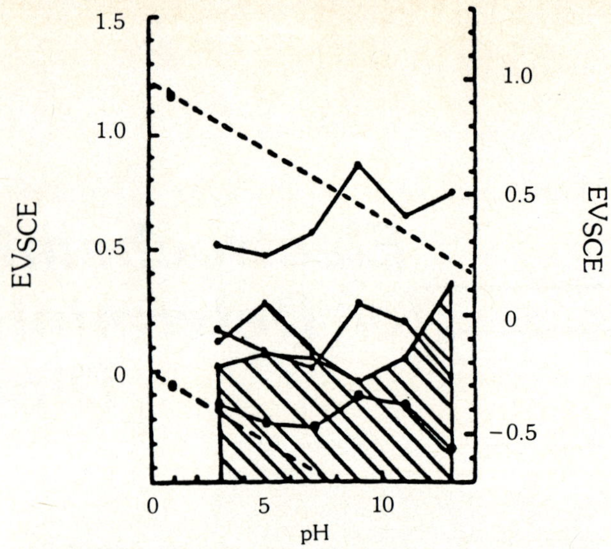

FIGURE 3 — Experimental potential-pH diagram for AISI 304 stainless steel at 20 C in 0.1 NCl^- solution. Cross-hatched region shows immunity from pitting and crevice corrosion.[26]

authorities have questioned the validity of these short-term test procedures. The criticism that occurs most frequently is that these tests may not adequately predict long-term corrosion behavior.[21-24]

Potentiostatic Tests

Long-term potentiostatic tests would seem an obvious method of confirming the cyclic polarization data developed using the techniques discussed above. Little work of this type has appeared, however. One of the obvious reasons for this lack of long-term testing is the expense and time necessary to develop long-term tests for a series of potentials and environments.

Figure 2 shows an electrode holder adapted from a multiple crevice assembly used for crevice corrosion testing.[25] This sample holder was used for long-term (two-week) potentiostatic tests which produced the potential-pH diagram for AISI 304 stainless steel (SS) shown in Figure 3.[26] The cross-hatched region shows regions where no corrosion (including pitting or crevice corrosion) is expected to occur. Several discrepancies between behavior predicted by short-term potentiodynamic tests and the longer potentiostatic exposures were noted during the development of this diagram.[26] These discrepancies confirm earlier criticisms of potentiodynamic tests by many authors.[21-24]

Passive Films

Figure 1 illustrated the importance of passive films to the performance of pure metals and alloys in aqueous environments. Several researchers have developed experimental potential-pH diagrams where the identify of passive films was determined experimentally.

Many techniques are available for the study of passive film composition and structure. Reviews by Vermilyea[27] and Leidheiser[28] discuss most of the important surface analysis techniques available for use in studying

TABLE 1 – Characteristics of Surface Analytical Techniques[32]

Analytical Method	Typical Background Pressure (Torr)	Analysis in Aqueous Solution	Minimum Film Thickness Observed (A)	Information Obtained
Raman spectroscopy	760	Yes	50	Compound ID
Infrared spectroscopy	760	No	10	Compound ID
X-ray diffraction	760	No	500	Compound ID
Ellipsometry	760	Yes	10	Film thickness
Low energy electron diffraction (LEED)	10^{-10}	No	Monolayer	Compound ID
Auger electron spectroscopy (AES)	10^{-9}-10^{-10}	No	Monolayer	Elemental comp.
Secondary ion mass spectrometry (SIMS)	10^{-10}	No	Monolayer	Elemental comp.
Electron microprobe analyzer	10^{-5}-10^{-10}	No	200	Elemental comp.
Scanning electron microscope (SEM)	10^{-4}-10^{-9}	No	50	Surface topography
Electron spectroscopy for chemical analysis (ESCA)	10^{-7}-10^{-10}	No	5	Elemental comp., electronic states
Differential reflectometry	760	Yes	5	Compound ID

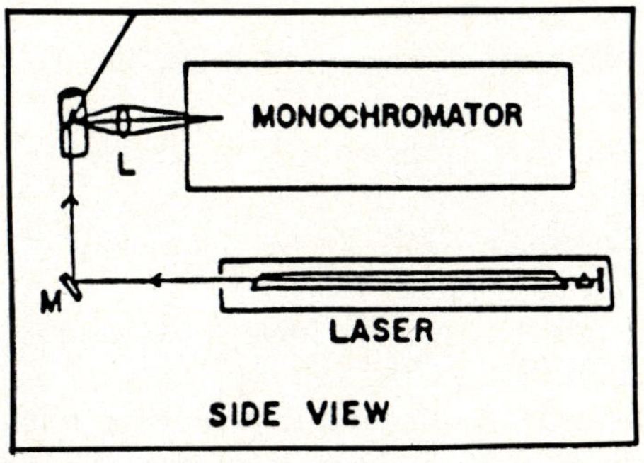

FIGURE 4 – Schematic of a laser Raman spectrometer.[32]

passive films. Table 1 summarizes the techniques used in corrosion studies.

The ideal surface analysis technique would be able to provide structural information, *in situ*, on a variety of substrates. It should require no vacuum, include amorphous as well as crystalline films, and be verifiable by other techniques. It should cover the entire atomic spectrum and not be limited, as is the case with Mossbauer spectroscopy, to certain elements.[29-30] Raman spectroscopy, with its capability for *in situ* analysis of passive films as they form in aqueous media, has several important advantages over other surface analysis techniques. The combination of infrared and Raman spectroscopy offers structural analysis possibilities that cannot be met by other surface analysis techniques with the possible exception of differential reflectometry.[31]

Figure 4 shows an experimental apparatus that can be used for *in situ* analysis of corrosion products on metal surfaces in aqueous environments. Figure 5 shows an electrochemical cell used for *in situ* Raman sectroscopy studies of electrode surfaces.

Figure 6 is a potential-pH diagram for lead. It compares the experimental results obtained by *in situ* Raman

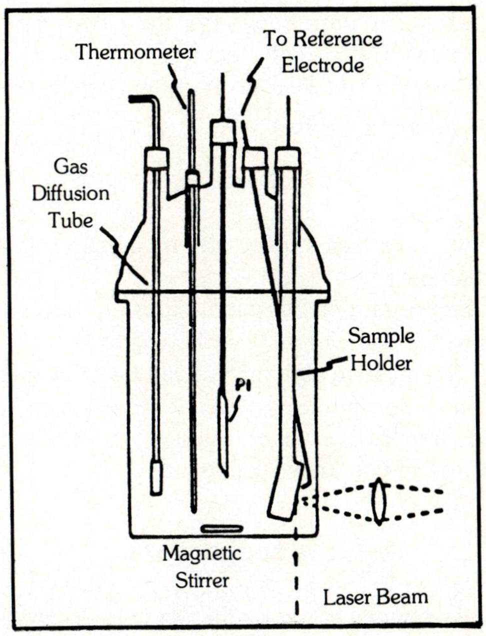

FIGURE 5 – Drawing of the cell used for *in situ* Raman spectroscopy of electrode surfaces.[32]

spectroscopy with those predicted using the latest available thermodynamic data.[32] Similar studies have been published showing the effects of chloride and sulfate anions on the species formed on lead in aqueous electrolytes.[32,34] Other reports describe how Raman spectroscopy can be used to develop experimental potential-pH diagrams for lead,[35] cobalt,[36] and nickel.[36]

Recent advances in Raman spectroscopy will allow analysis of thinner passive films as they form on other metals.[37]

Many studies of passive films assume that these films are homogeneous films of a single species or that they

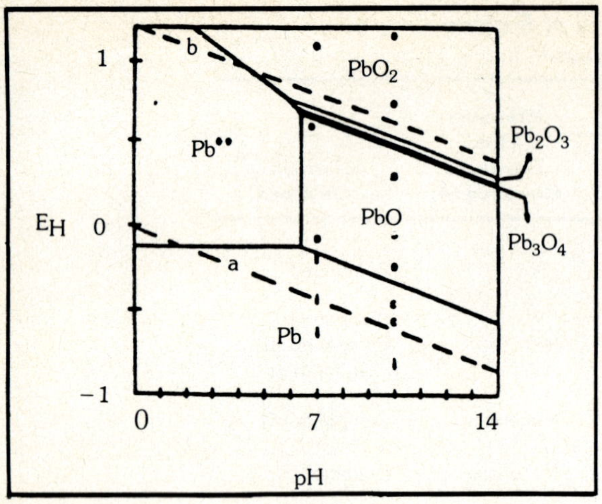

FIGURE 6 — Calculated Pb-H_2O potential-pH diagram using the most recent thermodynamic data. Experimental results: O = tetragonal PbO; i = apparent immunity; no spectrum observed; C = basic lead carbonate.[32]

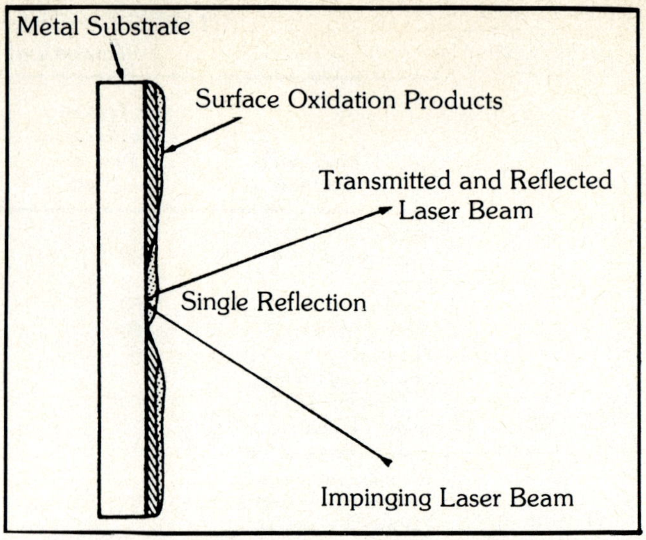

FIGURE 7 — Diagram showing complexity of some passive films and how spatially resolved Raman spectroscopy can be used to analyze complex passive films.[39]

consist of several layers of homogeneous species. However, careful examination of passive films using the scanning electron microscope shows that these films are often quite complex. The recently developed Raman microprobe[38] should prove useful in characterizing these complex films on metal surfaces.

Conclusions

1. Experimental potential-pH diagrams offer a number of advantages over thermodynamically calculated diagrams.

2. Long-term potentiostatic exposures are necessary to develop useful potential-pH diagrams.

3. *In situ* passive film analysis can be accomplished using Raman spectroscopy. Raman spectroscopy has several advantages over other techniques that have been used for passive film identification.

Acknowledgments

Portions of this research were sponsored by the Office of Naval Research and by the Materials Technology Institute of the Chemical Process Industries.

References

1. M. Pourbaix, Atlas of Electrochemical Equilibria in Aqueous Solutions, National Association of Corrosion Engineers, Houston, Texas, 1974.
2. P. Delaney, M. Pourbaix, P. van Rysselberghe, J. of Chem. Ed., p. 683, December 1950.
3. D. Silverman, Corrosion, Vol. 38, No. 8, p. 453, 1982.
4. E. Verink, Corrosion, Vol. 38, No. 6, p. 336, 1982.
5. O. Riggs, C. Locke, Anodic Protection, Plenum Press, New York, New York, 1981.
6. E. Verink, "Construction of Pourbaix Diagrams for Alloys with Special Application to the Binary Fe-Cr System," Report to the Advanced Research Projects Agency, University of Florida, Gainesville, Florida, June 1970.
7. M. Pourbaix, Corrosion, Vol. 26, No. 10, p. 431, 1970.
8. M. Pourbaix, Corrosion, Vol. 25, No. 6, p. 267, 1969.
9. R. Cusomano, MS Thesis, University of Florida, Gainesville, Florida, 1971.
10. E. Verink, K. Starr, J. Bowers, Corrosion, Vol. 32, No. 2, p. 60, 1976.
11. J. Johnson, Use of the Experimental Potential-pH Diagram in D6AC Steel to Interpret Its Corrosion Behavior in Aqueous Media, AFML-TR-15-8 Report, Air Force Materials Laboratory, WPAFB, February 1976.
12. E. Verink, M. Pourbaix, Corrosion, Vol. 27, No. 18, p. 495, 1971.
13. K. Efird, Corrosion, Vol. 31, No. 3, p. 77, 1975.
14. E. Verink, R. Heidersbach, ASTM STP 516, ASTM, Philadelphia, Pennsylvania, p. 303-322, 1972.
15. R. Heidersbach, E. Verink, Corrosion, Vol. 28, No. 11, p. 397, 1972.
16. W. Fort, E. Verink, Proceedings: 4th Int. Cong. Marine Corros. and Foul., Antibes—Juan les Pins, France, June 1977.
17. Ph. Gimenez, J. Rameau, M. Reboul, Corrosion, Vol. 37, No. 12, p. 673, 1981.
18.
19. J. Galvale, in Passivity of Metals, Electrochem. Soc., Princeton, New Jersey, pp. 285-327, 1978.
20. ASTM Standard G 61-78, Cyclic Potentiodynamic Polarization Measurements for Localized Corrosion, ASTM, Philadelphia, Pennsylvania.
21. B. Wilde, Corrosion, Vol. 28, No. 8, p. 283, 1972.
22. B. Wilde, E. Williams, J. Electrochem. Soc., Vol. 118, p. 1057, 1971.
23. F. LaQue, Materials Performance, Vol. 22, No. 8, p. 34, 1983.

24. J. Bates, Corrosion, Vol. 32, No. 1, p. 28, 1976.
25. D. Anderson, "Statistical Aspects of Crevice Corrosion in Sea Water," ASTM STP 576, ASTM, Philadelphia, Pennsylvania, p. 231, 1976.
26. M. Hubbell, J. Bakama, R. Heidersbach, "Methods for Determining Experimental Potential-pH Diagrams," in Equilibrium Diagrams—Localized Corrosion, ASTM STP 866, ASTM, Philadelphia, Pennsylvania, 1985.
27. D. A. Vermilyea, Physics Today, p. 23, September 1976.
28. H. Leidheiser, in "Passivity of Metals (Proceedings of the 4th Int. Symp. Passivity)," R. P. Frankenthal, J. Kruger, Eds., Electrochem. Soc., Princeton, New Jersey, p. 223, 1978.
29. Lieng-Huang Lee, in "Characterization of Metal and Polymer Surfaces," L. H. Lee, Ed., Vol. 1, Academic Press, New York, New York, p. 1, 1977.
30. U. Gonser, "Mossbauer Spectroscopy," Springer-Verlag, Berlin, 1975.
31. J. Finnegan, R. Hummel, E. Vernik, Corrosion, Vol. 37, No. 5, p. 256, 1981.
32. R. Thibeau, C. Brown, A. Golfarb, R. Heidersbach, J. Electrochem. Soc., Vol. 127, p. 37, 1980.
33. R. Thibeau, C. Brown, A. Golfarb, R. Heidersbach, J. Electrochem. Soc., Vo. 127, p. 1702, 1980.
34. R. Thibeau, C. Brown, A. Golfarb, R. Heidersbach, J. Electrochem. Soc., Vol. 127, p. 1913, 1980.
35. E. Reid, R. Cooney, P. Hendra, M. Fleischmann, J. Electroanal. Chem. Interfacial Electrochem., Vol. 80, p. 405, 1977.
36. C. Melendres, S. Xu, "*In situ* Laser Raman Spectroscopic Study of Anodic Corrosion Films on Nickel and Cobalt," submitted to J. Electrochemical Society.
37. J. Keiser, C. Brown, R. Heidersbach, American Laboratory, Vol. 4, p. 17, 1982.
38. R. Heidersbach, F. Purcell, "Analysis of Corrosion Products with the Use of the Raman Microprobe," in Microbeam Analysis—1984, A. Romig, J. Goldstein, Eds., San Francisco Press, San Francisco, California, p. 61, 1984.
39. P. Fabis, R. Heidersbach, C. Brown, T. Rockett, Corrosion, Vol. 37, No. 12, p. 700, 1981.

Derivation and Application of EMF-pH Diagrams

D. C. SILVERMAN*

Abstract

Thermodynamic electromotive force (emf)-pH diagrams calculated from physical property data are practical tools that aid in corrosion prediction. The implementation of these diagrams to aid in the design and execution of effective experimental programs and evaluation of the results has two thrusts. First an adequate data base must be provided. Very often, available data must be updated and supplemented with estimated values. Deficiencies in and methods of updating the present data base are discussed, using nickel and tantalum as examples. Second, the diagrams are used to provide guidance. They can show under what conditions corrosion may be possible or may be prevented. Thus, they can aid in the design of an experimental program before the first experiment is run. This thrust is discussed using an example involving titanium in a concentrated solution of HCl, NaCl, and an organic molecule, the thermodynamic properties of which are comparable to those of aspartic acid.

Introduction

One sometimes overlooked tool that can aid in corrosion prediction and corrosion control is the thermodynamic emf-pH diagram originally developed by Pourbaix.[1] The emf and pH coordinates in which these diagrams are plotted are useful to the corrosion engineer because pH, as a measure of acidity, and emf (potential), as a measure of oxidizing power, are important determinants of corrosion in electrolyte solutions. The diagrams are derived from thermodynamic data. The species shown at each (pH, emf) coordinate is that species that would have the minimum free energy for the system under those conditions. As long as all components important to corrosion are included in the calculation and as long as the data accurately represent the compounds that might be important to corrosion, the diagrams can show the conditions under which corrosion will not occur, e.g., immunity or pure metal.

The diagrams can show what conditions of pH (acidity) or potential (oxidizing power) might result in transformation of a metal into a metallic ion, a hydrolyzed species, or an oxidized solid. Thermodynamics provides no information on corrosion rates. Experiments are required to determine whether thermodynamic predictions of oxidized metal (dissolved or solid oxidized species) translate into reaction rates occurring in a rapid enough fashion that metal degradation is a problem. Therefore, the usefulness of the diagrams lies in their acting as "road maps" to provide direction before and during an experimental program and to provide guidance for interpretation and application of the results.

Scattered throughout the literature are examples of how these diagrams have been used to aid in data interpretation and corrosion prediction. Pourbaix has discussed a number of applications in which emf-pH diagrams have been used to aid in the interpretation of observed localized corrosion of, for example, copper and iron.[2,3] He showed that the theoretical diagrams can be useful for gaining mechanistic insight into the reactions that lead to and products that result from corrosion. In doing so, he demonstrated the important concept that electrochemical experimental results (kinetics) must not violate thermodynamics.

In the same vein, these diagrams have been used to aid in the interpretation of alloy corrosion in geothermal systems,[4,5] the effect of molybdate on the repassivation kinetics of iron,[6] the analysis of the corrosion of titanium carbide,[7] and the stress corrosion cracking (SCC) of low alloy steel.[8] Other examples are scattered throughout the literature. In all of these studies, the diagrams were used successfully to help to interpret the data primarily after the data were generated, not before the experiment was run.

*Monsanto Co., St. Louis, Missouri.

Using the emf-pH diagrams as practical aids in designing appropriate experiments or in conjunction with measured corrosion potential and pH to avoid excessive experimentation is an area that has received relatively little attention. Verink has recently discussed how these diagrams might be used in conjunction with simple experiments to aid in corrosion prediction.[9] The important limitation that the diagrams can be evaluated only for pure metals because of a lack of data for alloys was also discussed. Even considering this limitation, the diagrams were shown to be capable of explaining and predicting such phenomena as dealloying and passivation. More recent work has shown that a good indication of corrosion behavior can be obtained by interfacing the measured corrosion or controlled potential and measured or calculated pH with the theoretical emf-pH diagram.[10] Since the measured pH and steady-state corrosion or controlled potentials are not the same as the hydrogen ion activity and the equilibrium electrochemical potentials, this procedure is somewhat approximate. However, good agreement between experimental results and diagram predictions can be achieved.[10,11] This concept of using the diagrams to make *a priori* predictions and to provide guidance before and during experimentation coincides with the way that thermodynamics should be used in any experimental program involving reaction kinetics.

The first questions to be answered are whether corrosion of the metal can occur and, if so, what possible corrosion products would be stable. The answers to these questions relate to the states of the materials rather than the paths to form the materials. Thus, these questions relate to thermodynamics. The next questions to be answered are how fast and by what mechanism the corrosion reactions occur. The answers to these questions are path-dependent. Thus, they are not related to thermodynamics and must be found by experimentation. The thermodynamic answers should provide guidance for the experiments used to determine the kinetic answers. This guidance is important because it can decrease the number of required experiments and thus decrease the time and cost to make a materials evaluation.

There are two needs that must be fulfilled in order to make practical use of emf-pH diagrams. The first is an adequate computer program. As discussed by Linkson, et al.,[12] a number of algorithms exist for calculating the diagrams from thermodynamic data. Some are more efficient than others, although most of those that have been written are claimed to always give correct answers based on the data. Thus, as good algorithms are available, there is no need to write another one. Any of the ones that cannot fail can be chosen as a basis upon which a "user-friendly" software package can be developed to automatically calculate emf-pH diagrams for systems of interest.

The second need is an adequate data base with correct thermodynamic data and an appropriate method of estimating thermodynamic properties at elevated temperatures and pressures. There are at least three major data base compilations available for inorganic species.[1,13,14] The Russian compilation seems to be the most internally consistent.[14] This compilation also seems to be more consistent with the critical review of Baes and Mesmer[15] on the equilibria of hydrolyzed cations. However, even use of these sources leaves a number of deficiencies in the inorganic data base. As far as organic complexing agents, little free energy, entropy, or heat capacity data are available. Although compilations of equilibrium constants are available,[16,17] the method of translating these data into free energies of formation of the complexes in the standard state in solution for inclusion in emf-pH diagrams requires assumptions that have not been verified.[10] This discussion means that when emf-pH diagrams are to be applied to practical problems, care must be taken to ensure that the data are accurate, that the data include all necessary species, and that the limitations of any assumptions made in data estimation are understood.

The deficiencies in the data noted above have forced the application of emf-pH diagrams to have two parallel thrusts. The first thrust has been to ensure that the diagrams contain data derived from the best thermodynamic knowledge of the species. In addition, the data derived from different sources have been verified to be internally consistent. The second thrust is to apply these internally consistent data to real corrosion problems and to verify that the predictions coincide with reality. The purpose of this paper is to discuss implementation of these thrusts.

Calculation of emf-pH Diagrams

The algorithm used to calculate the emf-pH diagrams discussed here is that reported by Froning, Shanley, and Verink[18] as previously described.[10] This algorithm, while being somewhat inefficient, has been predicted to be infallible.[12] A subroutine was added to estimate free energies as a function of temperature. It uses the Criss-Cobble Correspondence Principle[19] for the ions. The entropy parameters contributing to the heat capacity of the ions have been curve-fit with respect to temperature by following the Taylor method.[20] This approach allows temperatures other than 60, 100, 150, and 200 C in the original correlation to be used with only minimal loss in accuracy. There is no attempt to correct for pressure. An additional subroutine was added to plot the diagrams. The program is menu-driven in an interactive mode. It contains a data base that is automatically accessed when the diagrams are generated. Species or properties not present in the data base can be added easily to the retrieved tabulation prior to generating the diagrams.

One point should be mentioned about the method used to predict the thermodynamic data at elevated temperature. The Correspondence Principle is applicable only to simple cations, simple anions, oxyanions, and acid oxyanions. Recent data suggest that many ions originally written as $HFeO_2^-$, or MO_q^{-n}, e.g., FeO_2^-, should be written as $M(OH)_x^{-n}$.[15] The question arises whether the values of parameters given by Criss and Cobble for the oxyanion and acid oxyanion apply to the hydroxy anion.

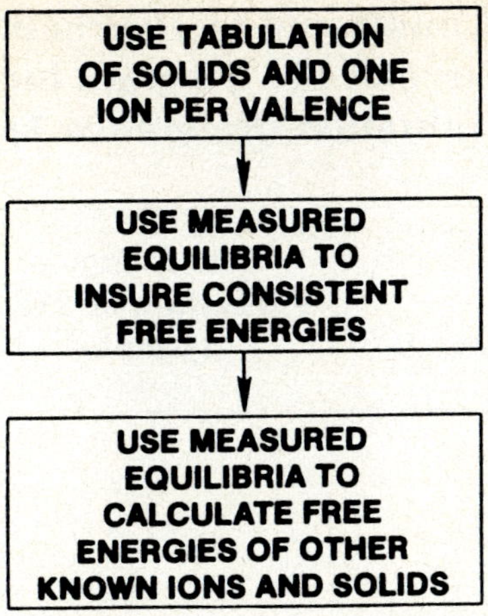

FIGURE 1 – Procedure to ensure thermodynamic internal self-consistency of standard free energies.

If not, new values are needed or new parameters must be defined. In addition, there are often no data on the heat capacities of neutral hydroxide solid and dissolved species, e.g., $Fe(OH)_2$ (dissolved), or on metallic complexes, e.g., Cd-EDTA^{-2}.[10] Thus, estimation of thermodynamic properties of these compounds in solution is approximate at best. The higher the temperature, the greater the error. Verified estimation procedures are lacking. Caution must be exercised when attempting to make predictions using emf-pH diagrams derived from these types of data. However, in spite of these deficiencies in the data, the diagrams are still useful for making practical corrosion predictions.

Development of a Consistent Data Base

Before the emf-pH diagrams could be applied to practical corrosion problems, an internally consistent data base had to be developed for each commonly used metal and its associated species. The methodology is shown in Figure 1. The procedure is an overall three-step process. The procedure ensures that all dissolved metallic species with a given metal valence and solids containing the metal at the same valence have internally consistent free energies of formation. The consistency of the free energy data for different valences is verified by agreement between measured and calculated electrochemical equilibrium potentials, e.g., between Fe^{+2} and Fe^{+3}. The tabulations of Naumov, et al.,[14] have been found to be consistent with the thermodynamic values calculated by following this procedure for inorganic species.

When ions such as Cl^- are to be included, the thermodynamic properties of known metallic complexes with the ion are estimated by including that ion and its complexes with the metal in the steps in Figure 1. Thermodynamic data for metal-organic ion species are estimated in a manner similar to that outlined previously.[10] In all cases, the data are only as good as the measured equilibrium constants for the hydrolysis[15] or complexing[16,17] reactions.

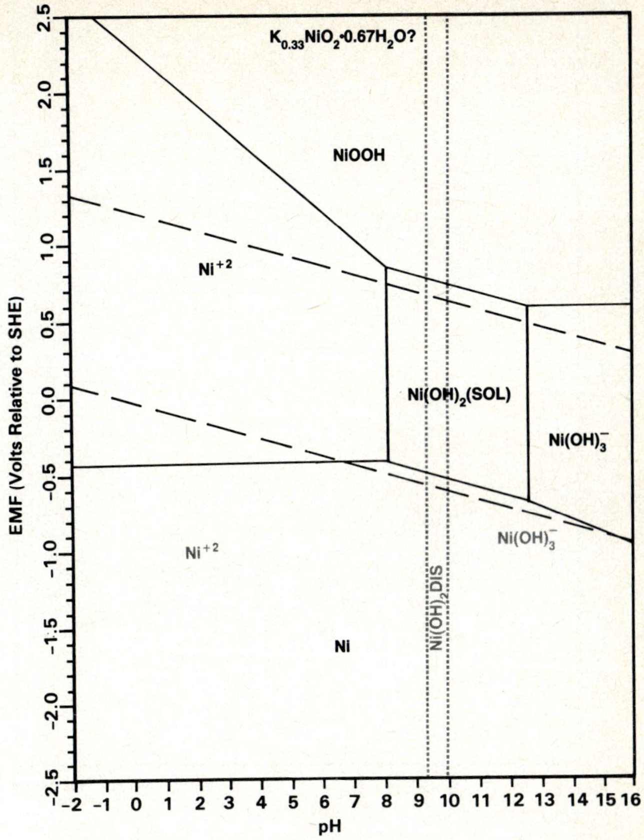

FIGURE 2 – Nickel at 25 C (emf-pH diagram). All ionic species are at an activity of 10^{-6}.

Revised emf-pH Diagrams (Nickel and Tantalum)

Improvements in and revisions to the nickel, iron, and chromium emf-pH diagrams that have been derived by following the procedure in Figure 1 have been discussed previously.[21-23] The revised nickel diagram requires some additional discussion since certain revised features have still been ignored in the recent literature and new data have been introduced since the original revision was made. The revised diagram is shown in Figure 2. One of the discrepancies found with the original emf-pH diagram is the fact that neither Ni_3O_4 or Ni_2O_3 or their hydrated forms are real compounds. Both electrochemical and x-ray diffraction data discussed previously lead to this conclusion.[21] This finding is important because investigators have interpreted observations in terms of Ni_3O_4 and Ni_2O_3 forming on the surface.[3,24] Unless these species can form on the surface and not as bulk compounds, these interpretations should be revised in terms of the solids NiO, (α,β) $Ni(OH)_2$, and (γ,β) NiOOH. Some reinterpretation has been presented.[21]

In addition, there have been more recent results that have suggested that nickel can exist at a valence state of IV in either NiOOH or in a compound of the form $K_{0.33}NiO_2 \cdot 0.67H_2O$.[25,26] These data suggest that an NiO_2-containing compound is possible but it is most likely not of the form $NiO_2 \cdot 2H_2O$, as originally proposed by

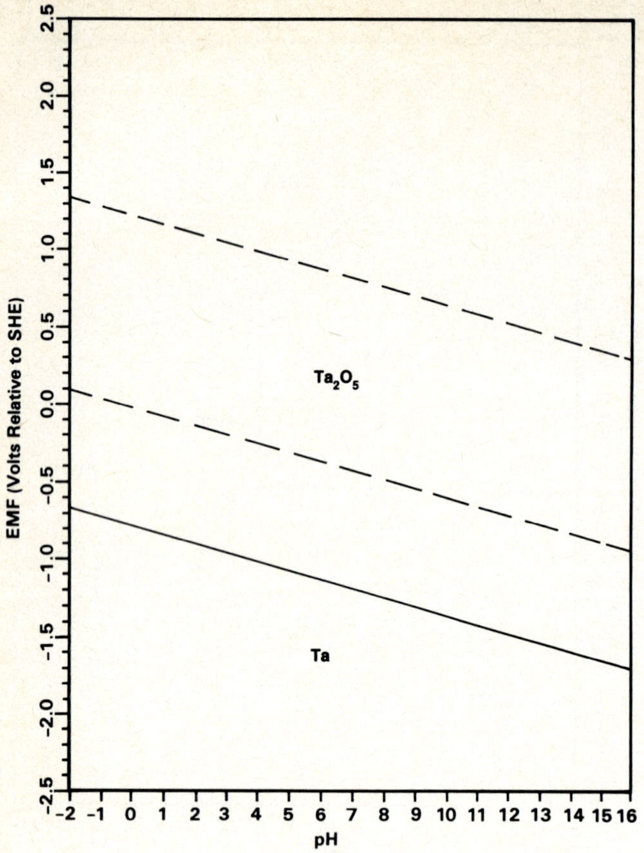

FIGURE 3 — Tantalum at 25 C (emf-pH diagram) using the data of Pourbaix.[1]

Pourbaix.[1] Unfortunately, no thermodynamic data exist for $K_{0.33}NiO_2{\cdot}0.67H_2O$. Therefore, the exact form and region of stability of a possible Ni(IV)-containing species cannot be placed on the emf-pH for nickel. The region most likely would exist at higher potentials. This observation is represented by including $K_{0.33}NiO_2{\cdot}0.67H_2O$ with a question mark at high positive potentials in Figure 2 but with no defined stability region. As electrochemical results for nickel oxidation have been interpreted in terms of an NiO_2 or Ni(IV)-containing species,[3,21,24] an effort is needed to verify the compounds and to determine thermodynamic data for them.

Recent work in deriving the emf-pH diagram for tantalum following the procedure outlined in Figure 1 has brought out another type of problem that can arise even when the emf-pH diagram is calculated from the best data. The diagram may still be imprecise because the data are imprecise. Figure 3 shows the emf-pH diagram for tantalum using the original data of Pourbaix as shown in Table 1.[1] As no ionic species were considered, the diagram would be applicable to all ionic activities.

The diagram predicts that the Ta_2O_5 would predominate for all hydrogen ion activities of 10^{-16} to 10^2. As a Ta_2O_5 surface layer might be expected to impart a low corrosion rate to tantalum, its presence suggests that tantalum would be resistant at all pH values in the stability region of water (between the two parallel dotted lines). There are data that indicate that tantalum is not resistant in strongly basic solutions.[27] The destruction seems to be by formation of successive layers of scale. This corrosion suggests that species in addition to Ta_2O_5 must account for the oxidation of tantalum at least at high pH.

TABLE 1 — Species Considered by Pourbaix for emf-pH Diagram for Tantalum (Reference 1)

Species	Free Energy of Formation at 298 K (j/mole)
Ta	0.0
Ta_2O_5	-1.91×10^6

TABLE 2 — Species Considered for Updated emf-pH Diagram for Tantalum

Species	Free Energy of Formation at 298 K (j/mole)	Reference
Ta	0.0	—
Ta_2O_5	-1.91×10^6	14
$Ta(OH)_4^+$	-1.29×10^6	15
$Ta(OH)_5$ Dissolved	-1.52×10^6	15
$Ta(OH)_6^-$	-1.72×10^6	15
$Ta(OH)_5$ Solid	-1.55×10^6	(see text)

More recent results suggest that a number of hydrolysis products exist for tantalum.[15,28] The compounds are shown in Table 2. As indicated by Baes and Mesmer, the hydrolysis equilibrium constants for formation of the dissolved hydrolysis species must be considered to be somewhat uncertain. This observation means that the free energies of formation calculated from them may be imprecise. For example, the relationship in free energies between freshly precipitated Ta_2O_5 [written as $Ta(OH)_5SOL$] and the oxide Ta_2O_5 is unknown. The value shown for $Ta(OH)_5SOL$ in Table 2 was estimated by assuming that the free energy of formation of $Ta(OH)_5SOL$ from $1/2\ Ta_2O_5 + 2.5H_2O$ is zero. Then, the free energy of formation of $Ta(OH)_5DIS$ was estimated by using the uncorrected equilibrium constant for the formation of dissolved $Ta(OH)_5$ [$Ta(OH)_5DIS$] from $Ta(OH)_5SOL$ as listed by Baes and Mesmer.[15] This procedure would most likely tend to overestimate the free energies of formation of both $Ta(OH)_5SOL$ and $Ta(OH)_5DIS$, since Ta_2O_5 might be expected to be more stable than its freshly precipitated analogue $Ta(OH)_5SOL$, especially when either is formed on the surface of tantalum. Then the hydrolysis constants[15] were used to estimate the free energies of formation of the dissolved, hydrolyzed ions. These hydrolysis constants are uncorrected for concentration as discussed by Baes and Mesmer.[15] The free energies of formation of all hydrolyzed species are most likely too negative because of the estimated Ta_2O_5-$Ta(OH)_5SOL$ relationship. That is, *calculated* stability of the dissolved species is most likely greater than their *actual* stability.

Hydrides of tantalum are known and phase diagrams have been derived for them.[29] They have not been included in Table 2. In addition, solid $K_8Ta_6O_{19} \cdot 6H_2O$ will dissolve in dilute KOH and will not be protonated in the pH range of 10 to 13.[30] The ion that forms is most likely $(Ta_6O_{19})^{-8}$. If this species behaves like its niobate analogue, hydrolysis species are possible.[15] However, as no thermodynamic data are available for these tantalum species, $K_8Ta_6O_{19} \cdot 6H_2O$ and its hydrolysis products have not been included in Table 2.

Figure 4 shows the emf-pH diagram derived from the data in Table 2. Ionic activities of 10^{-6} and 10^{-4} are included. The diagram shows that an ionic species has the lowest free energy at high pH. This result is consistent with the attack found in high-pH media. The soluble species $Ta(OH)_6^-$ could correspond to a dissolved corrosion product. The presence of the dissolved species $Ta(OH)_4^+$ at low pH and $Ta(OH)_5SOL$ as stable or even more stable than Ta_2O_5 is most likely inconsistent with the observation that tantalum is resistant to nitric acid and hydrochloric acid at all concentrations at 25 C.[27] The discrepancy is probably caused by overestimates of the free energies of formation of the dissolved species. A positive increase of 2 to 3% in the free energy estimates of the hydrolyzed species would eliminate $Ta(OH)_4^+$ from the diagram and make Ta_2O_5 most stable in that region. This change would leave $Ta(OH)_6^-$ most stable at a high pH. Also, note that the region of stability predicted for $Ta(OH)_4^+$ decreases greatly as ionic activity increases. However, resistance to corrosion and formation of $Ta(OH)_4^+$ at very slow rates may not be contradictory. Since thermodynamics gives no rate information, corrosion at microns per century (μm/100 y) would still have a corrosion product as the most stable state. More work is needed to resolve the possible inconsistencies at low pH.

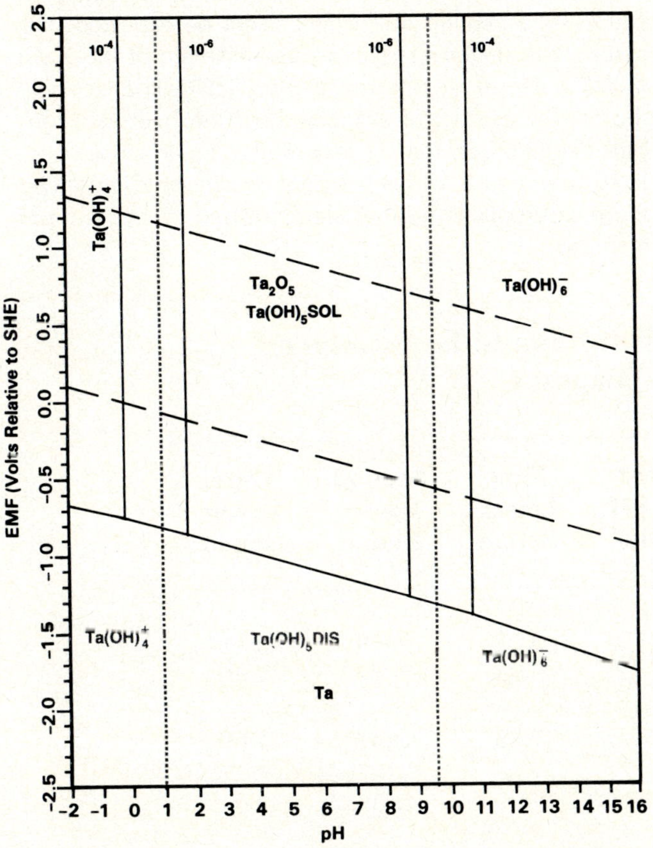

FIGURE 4 – Tantalum at 25 C (emf-pH diagram) using revised data. All ionic species are at activities of 10^{-6} or 10^{-4}.

Obviously, the boundaries between the regions of stability in Figure 4, especially for both $Ta(OH)_5$ species, are imprecise because of the inaccuracies in the data. This point means that although Figure 4 is more consistent with observation at high pH, more experimental work is still needed. Controlled potential experiments, solution and surface analyses, and new equilibrium measurements must be combined in order to properly determine the thermodynamic properties of and the emf-pH diagram for tantalum and the products formed from its interaction with water.

This discussion of the tantalum emf-pH diagram shows that attempts to derive an emf-pH diagram from best available data and by best estimation techniques may still result in an imprecise, though possibly usable diagram. This deficiency is not an isolated case because there are serious gaps in the information for a number of other metals. The controversy over the form of Ni(IV) mentioned previously and the controversy over the existence of Ti^{+2} as an aqueous ion[1,10] are two other examples. Inclusion of complexing agents also leads to approximate diagrams because of imprecise data. Alloy data is seriously lacking. Thus, the derivation of an accurate emf-pH diagram is an ongoing process and updating must be done on a continual basis. Obviously, much more experimental work is needed in developing a data base for these diagrams.

Application of emf-pH Diagrams

Even though there are omissions in the data base, the diagrams that are derived are very useful as aids to corrosion prediction. The relationship between the emf-pH diagram and corrosion prediction is analogous to the relationship between thermodynamics and kinetics in reaction engineering. The thermodynamics of a reaction is estimated first. If the reaction is favorable, e.g., the free energy change is negative, then the rate must be measured. If the free energy change is positive, the reaction is unfavorable and cannot proceed.

Similarly, if the pH as measured (if near neutrality) or as calculated (if far from neutrality) and the equilibrium potential (approximated by the steady-state corrosion potential if not far from equilibrium) of the system fall in an area of immunity, the corrosion reaction is thermodynamically unfavorable. If the coordinate lies in an area of oxidized species (dissolved ion or solid oxide, halide, etc.), the corrosion reaction is favorable to form corrosion products. The rate must then be measured. If the measured rate is low, then the product most likely helps to resist corrosion, e.g., TiO_2. If the measured rate is high, the emf-pH diagram might provide guidance as to how this rate can be suppressed. This procedure can work only if all important species are defined properly on the diagram and the thermodynamic properties of bulk and surface species are similar.

The above concepts are illustrated by another application involving the titanium-HCl system discussed previously.[10] The application is relevant because it shows how the simple experiments of measuring pH and corrosion potential and interfacing these results with emf-pH diagrams can lead to practical conclusions and decrease the number of required experiments.

Problem

The question arose as to the lowest pH that titanium withstand when exposed to a stream somewhat depleted in oxygen and containing NaCl, HCl, and an organic molecule whose acid-base equilibria and thermodynamic properties can be modeled by aspartic acid. The equipment involved was made from Grade 2 titanium; the temperature was to be 75 C; and the initial pH of the present operation as measured by a pH electrode was about 2.2 to 3.0. The question was open-ended because beyond the knowledge that a minimum pH must exist,[10] there was no estimate of its magnitude. With little *a priori* knowledge, many costly experiments might have been required. This predicament is an excellent example of the type of situation in which emf-pH diagrams can be most helpful. That is, before undertaking an involved experimental program, especially one for a metal that would either suffer a small or a catastrophic corrosion rate, emf-pH diagrams should be used. In this case, placement of the measured steady-state corrosion potential and the measured pH or calculated hydrogen ion activity on these diagrams could decrease the range of pH that needed to be considered experimentally.

Determination of pH

Proper use of emf-pH diagrams requires the inclusion of activities of all possible species important to the corrosion process. These species include the metal, products of its interaction with water, and any other ions in the solution. Activities, not concentrations, should be used because the activity is the natural thermodynamic variable that is related to the free energy of the component. This point extends to the hydrogen ion as well. Its activity, not its concentration, is the appropriate variable because the pH coordinate is actually the negative of the logarithm of the hydrogen ion activity.[1]

The pH was measured in solutions of different degrees of acidity. The solution compositions in molality are shown in Table 3 for the cases examined. Fisher reagent grade HCl was used to decrease the pH. The low pH values used suggest that the pH to be interfaced with the emf-pH diagram should be that calculated from molality and activity coefficients. The method is as follows: The NaCl concentration is at saturation. Its molality in solution shown in Table 3 was estimated from the correlation of Potter and Clynne.[31] The aspartic acid analogue was assumed not to affect the solubility of NaCl. As a "salting out" effect might be expected, this assumption might result in an overestimate of the NaCl molality. However, the molality of NaCl will affect only the activity coefficient calculated for the hydrogen ion. Since emf and pH may be regarded as log-log type of activity coordinates, errors in activity coefficients caused by these assumptions will be deemphasized on the emf-pH diagram.

TABLE 3 — Composition of Solutions to be Interfaced with emf-pH Diagram

Component	Original Solution (Molality)	Expected Lowest pH (Molality)	First Lower Bound	Second Lower Bound	Third Lower Bound
NaCl[(1)]	6.97	6.90	6.12	3.49	0.76
HCl	0.0	1.76	2.33	5.12	7.49
Aspartic Acid Analogue[(2)]	1.28	2.57	2.28	2.31	1.69
Na•Aspartic Acid Analogue	1.21	0.0	0.0	0.0	0.0

[(1)]Calculated from saturation conditions.[31]
[(2)]At approximate saturation.

TABLE 4 — Coordinates to be Placed on emf-pH Diagram

Solution	pH $= p(\gamma \pm m_{H+})$	E_{corr} (SHE)
Original	3.0	0.20
Expected Lowest pH	0.98	0.30
First Lower Bound	0.11	0.10
Second Lower Bound	−1.2	−0.35
Third Lower Bound	−1.6	−0.32

The data in Table 3 were used to estimate the activity coefficients of hydrogen ion. The estimation procedure was similar to that described previously.[10] The reaction:

$$C_4H_7O_4N{\bullet}H^+ \rightarrow C_4H_7O_4N + H^+ \qquad (1)$$

was assumed to be at equilibrium in the presence of free HCl. Reaction (1) has a pK of about 1.9[16,17] at 25 C. No data exist for the value at 75 C. The equation that relates activity to equilibrium constant defined by Reaction (1), the charge balance equation, and a mass balance equation were solved simultaneously for the hydrogen ion molality. Ideality was assumed for this calculation. The resulting hydrogen ion molality was then multiplied by the mean activity coefficient of HCl in solution as estimated by the methods of Kusik and Meissner[32] for the calculated molalities. The contribution of the organic ion to the activity coefficient was found by assuming that $C_4H_7O_4NH^+$ interacts in solution in a manner similar to NH_4^+ since the amine group of either ion creates the positive charge at low pH. Although the resulting hydrogen ion activity is an estimate, some of the error is depressed by the fact that its logarithm is used for the interfacing with the emf-pH diagram. Table 4 shows the estimated pH derived from the calculated molality and activity coefficient. In all cases, the calculated pH values and those measured by a standard pH electrode agreed within about 0.5 pH units. Note that the pH estimated from the pH electrode response would differ from the actual hydrogen ion activity, especially at the high acidity levels in these solutions.

Corrosion Potential

The steady-state corrosion potential of Grade 2 titanium was determined relative to a saturated calomel electrode (SCE) by means of a Keithley[(1)] 616 electrometer. The potential was monitored for about 15 hours. A slow, 30-cc/min purge of nitrogen saturated with water was maintained during the experiment. The nitrogen source had a small oxygen impurity. The purpose was to decrease the oxygen concentration and to minimize contamination of the solution from the outside atmosphere. The temperature was maintained at 75 C. Table 4 shows the final steady-state corrosion potentials. They have been corrected to the standard hydrogen scale by assuming that there are no liquid junction potentials or thermal liquid junction potentials.

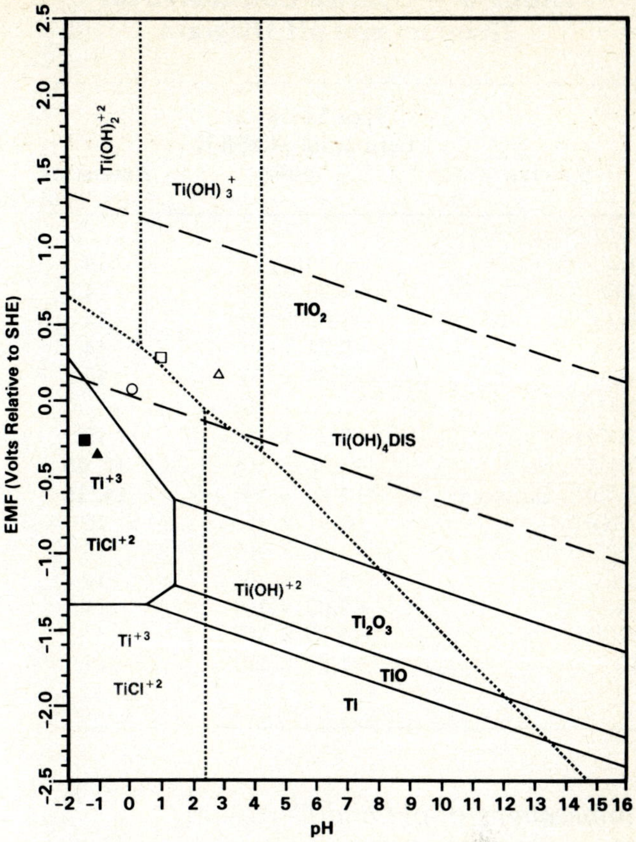

FIGURE 5 — Titanium at 75 C (emf-pH diagram). All titanium species are at an activity of 10^{-6}. Chloride ion is at an activity of 10: △ = original solution, □ = expected lower limit, ○ = first lower bound, ● = second lower bound, ■ = third lower bound.

Calculated emf-pH Diagram

Figure 5 shows the calculated emf-pH diagram for titanium at 75 C. Chloride ion was assumed to be at an activity of 10, the order of magnitude of the chloride ion activity expected. The thermodynamic data used to calculate the diagram are shown in Table 5. Note that titanium has been stated to assume a +3 or +4 valence state in water.[15] No evidence could be found for a titanium ion of the form Ti^{+2} as proposed by Pourbaix[1]; therefore, this ion was omitted. The free energies of formation of the other species were calculated as discussed previously.[10] The compound TiH_2 was omitted. Its inclusion would have resulted in an area of predominance of TiH_2 at low potential,[3] much more cathodic than the corrosion potentials measured in these experiments. Thus, the absence of TiH_2 has no effect on the present discussion.

[(1)]Registered trade name.

TABLE 5 — Species Considered for Titanium emf-pH Diagram

Species	Free Energy of Formation at 298 K (J/mole)	Reference
Ti	0	14
TiO	-4.90×10^5	14
Ti_2O_3	-1.43×10^6	14
Ti_3O_5	-2.32×10^6	14
TiO_2	-8.89×10^5	14
Ti^{+3}	-3.50×10^5	1
$Ti(OH)^{+2}$	-5.74×10^5	15
$Ti(OH)_4$ Solid	-1.26×10^6	14, 15
$Ti(OH)_4$ Dissolved	-1.24×10^6	14, 15
$Ti(OH)_2{}^{+2}$	-7.91×10^5	15
$Ti(OH)^{+3}$	-1.02×10^6	15
$TiCl_3$	-6.53×10^5	14
$TiCl_2$	-4.80×10^5	14
$TiCl^{+2}$	-4.79×10^5	33
Cl^-	-1.31×10^5	14

Implications of pH and Corrosion Potential Coordinates

The coordinates of pH and corrosion potential have been placed on Figure 5. The coordinates indicate that the stable state of titanium in these solutions is either an ion or the solid oxide TiO_2. The form of the ion with Ti(III) is very sensitive to the activity chosen for chloride ion. At a chloride ion activity of 10, both Ti^{+3} and $TiCl^{+2}$ are most likely present. The fact that a form of oxidized titanium is the most stable state of titanium indicates that a titanium reaction is possible with the environment in all solutions. However, the fact that for pH values above about 0, the minimum free energy state is TiO_2 suggests that in these solutions such reaction with the environment would proceed until metallic titanium is buried beneath the oxide. At that point, the corrosion rate would proceed only as rapidly as newly exposed titanium metal can react with the environment. The practical conclusion would be that titanium could withstand the environment for pH values above ~0.

At the lower pH values, the minimum free energy state of titanium (in the absence of oxygen) would be the dissolved ions Ti^{+3} or $TiCl^{+2}$. Such a state of the metal, if it formed, should impart no resistance to corrosion. Titanium corrosion might be expected to be rapid under these conditions. The practical conclusion would be that the resistance of titanium may be lost at these low pH values.

As mentioned previously, the emf-pH diagram can answer only the question "Is corrosion possible?" and can provide guidance as to where corrosion might be most rapid. The next question to be answered is "What is the rate of corrosion?" Experiments must be run to answer this question; the emf-pH diagram results indicate that these experiments should be concentrated around a pH of 0. The balance of the pH range can be eliminated from further consideration. The emf-pH diagram has greatly simplified the experimental program.

TABLE 6 — Results of Immersion Experiments

Measured Solution pH	Corrosion Rate g/cm²-s	mpy
2.4[1]	0	0
0.05	3.62×10^{-10}	1
−0.55	7.24×10^{-9}	20
−0.90	3.62×10^{-8}	100

[1]Original solution

Two-day static immersion experiments under a nitrogen purge were run using Grade 2 titanium that had been sanded with 120 grit SiC paper. The purpose is to determine whether the minimum pH suggested by the emf-pH diagram is reflected in the measured corrosion rates. The solutions had the same components at concentrations similar to those shown in Table 3. The pH values were adjusted by addition of Fisher reagent grade HCl. The pH values were determined at room temperature by means of a pH electrode. They are most likely within 0.5 pH of the values that could be calculated as discussed previously. The experiments were performed at pH values of about 0 and lower, the possible problem areas suggested by Figure 5.

Table 6 shows the resulting corrosion rates. The results indicate that the possibility of corrosion at a very low pH value is found as suggested by the emf-pH diagram. The minimum practical pH is about 0. A very small corrosion rate is seen at this pH. A finite, measureable corrosion rate in a region of stability of TiO_2 is not inconsistent. The compound TiO_2 is an oxidized solid which imparts passivity only through its ability to form a barrier. If the barrier can be affected, the corrosion resistance can be affected. On the emf-pH diagram, the coordinate corresponding to this pH shows movement toward the region of stability of the ions Ti^{+3} and $TiCl^{+2}$ and resides close to the boundary between H_2O and H_2. This reaction involving reduction of hydrogen ion could provide a driving force for the anodic reaction of corrosion of titanium metal to an oxidized state. One hypothesis would be that the minimum corrosion potential should be that potential anodic to the equilibrium potential of H^+ reacting to form H_2 (the lower parallel line corresponding to the breakdown of water to hydrogen). This hypothesis again suggests a minimum pH of ~0.

Conclusions

A practical opportunity to obtain information about corrosion before an experiment is conducted is offered by the emf-pH diagrams. The diagrams can provide guidance for the design and running of effective and efficient experimental programs. In order for these diagrams to become practical everyday tools, several areas must be addressed.

1. The data base must be expanded and verified to be thermodynamically self-consistent and all-inclusive. Thermodynamic data for complexing agents must be made available for inclusion in the diagrams when needed.

2. Methods of compensating for temperature and pressure must be developed for all types of dissolved and solid species. At present, temperature compensation can be made accurately only for some solids and four types of ions.[19]

3. Methods of estimating thermodynamic properties of inorganic species and metal-organic complexes must be modernized and shown to be internally self-consistent. Most methods now available are about thirty years old.

4. Methods of handling alloys in emf-pH coordinates must be developed, and thermodynamic data are needed. Preliminary results suggest that the interfacing of single metal emf-pH diagrams with actual alloy results can be done if the alloy surface can become essentially one component, e.g., chromium, as in such alloys as AISI 316 SS and Incoloy[(2)] 825.[34]

Acknowledgment

The author is grateful to J. E. Carrico for performing the experimental work in the titanium study.

References

1. M. Pourbaix, Atlas of Electrochemical Equilibria in Aqueous Solutions, National Association of Corrosion Engineers, Houston, Texas, 1974.
2. M. Pourbaix, Abh. Saechs. Akad. Wiss. Leipzig, Math-Naturwiss Kl., Vol. 53, No. 1, p. 9, 1974.
3. M. Pourbaix, "The Electrochemical Basis for Localized Corrosion," Localized Corrosion, R. W. Staehle, B. F. Brown, J. Kruger, A. Agrawal, Eds., National Association of Corrosion Engineers, Houston, Texas, p. 12, 1974.
4. D. D. Macdonald, B. C. Syrett, S. S. Wing, Corrosion, Vol. 35, No. 1, p. 1, 1979.
5. D. D. Macdonald, Rapp Tech. Cent. Belge Etude Corros., Vol. 142, No. 163, p. 235, 1982.
6. T. Kodama, J. R. Ambrose, Corrosion, Vol. 33, No. 5, p. 155, 1977.
7. M. Kh. Freid, S. A. Lilin, Zh. Prikl. Khim., Vol. 52, No. 7, p. 1541, 1979.
8. R. K. Melekhov, A. M. Krustan, M. V. Koval, Soviet Mater. Sci., Vol. 17, No. 6, p. 485, 1982.
9. E. D. Verink, Corrosion, Vol. 38, No. 6, p. 336, 1982.
10. D. C. Silverman, Corrosion, Vol. 38, No. 10, p. 541, 1982.
11. D. F. Taylor, C. A. Caramihas, J. Electrochem. Soc., Vol. 129, No. 11, p. 2458, 1982.
12. P. B. Linkson, B. D. Phillips, C. D. Rowles, Minerals Sci. Eng., Vol. 11, No. 2, p. 65, 1979.
13. D. D. Wagman, et al., NBS Technical Notes 270-3, 270-4, 270-5, 270-6, 270-7, 270-8, National Bureau of Standards (NBS), Gaithersburg, Maryland, 1971 through 1978.
14. G. B. Naumov, B. N. Ryzhenko, I. L. Khodakovsky, Handbook of Thermodynamic Data, USGS Translation, USGS-WRD-74-000, 1974.
15. C. F. Baes, R. E. Mesmer, Hydrolysis of Cations, John Wiley and Sons, New York, New York, 1976.
16. Stability Constants of Metal-Ion Complexes, L. G. Sillen, A. E. Martell, Eds., The Chemical Society, Burlington House, London, England, 1964.
17. Stability Constants of Metal-Ion Complexes, L. G. Sillen, A. E. Martell, Eds., The Chemical Society, Burlington House, London, England, Supplement No. 1, 1971.
18. M. H. Froning, M. E. Shanley, E. D. Verink, Corros. Sci., Vol. 16, p. 371, 1976.
19. C. M. Criss, J. W. Cobble, J. Amer. Chem. Soc., Vol. 86, p. 5385, 5390, 1964.
20. D. F. Taylor, J. Electrochem. Soc., Vol. 125, p. 808, 1978.
21. D. C. Silverman, Corrosion, Vol. 37, No. 9, p. 546, 1981.
22. D. C. Silverman, Corrosion, Vol. 38, No. 8, p. 453, 1982.
23. D. C. Silverman, Corrosion, Vol. 39, No. 12, p. 488, 1983.
24. R. M. Latanision, R. W. Staehle, Acta Met., Vol. 17, p. 307, 1969.
25. R. Barnard, C. F. Randell, F. L. Tye, J. Appl. Electrochem., Vol. 10, p. 109, 1980.
26. R. Barnard, C. F. Randell, J. Appl. Electrochem., Vol. 3, p. 97, 1983.
27. M. Schussler, "Corrosion Data Survey on Tantalum," Fansteel, Inc., Chicago, 1972.
28. A. K. Babko, V. V. Lukachina, B. I. Nabivanets, Russ. J. Phys. Chem., Vol. 8, No. 8, p. 957, 1963.
29. J. A. Pryde, I. S. T. Tong, Trans. Faraday Soc., Vol. 67, p. 297, 1971.
30. W. H. Nelson, R. S. Tobias, Inorg. Chem., Vol. 2, No. 5, p. 958, 1963.
31. R. W. Potter, M. A. J. Clynne, Chem. Eng. Data, Vol. 25, p. 50, 1980.
32. C. R. Kusik, H. P. Meissner, "Electrolyte Activity Coefficients in Inorganic Processing," in AIChE Symposium Series, Vol. 173, p. 14, 1978.
33. D. D. Feldman, V. M. Berdonikov, B. P. Matseevskii, Russ. J. Phys. Chem., Vol. 49, No. 12, p. 1858, 1975.
34. D. C. Silverman, unpublished results, 1983.

[(2)]Registered trade name.

Development of Corrosion-Resistant Alloys Using Electrochemical Techniques

*A. P. BOND, E. A. LIZLOVS, H. J. DUNDAS, and A. POZNANSKY**

Introduction

In the development of corrosion-resistant alloys, the first problem that must be considered is determining the best procedure for the evaluation of corrosion resistance. Normally, accelerated testing is used, since in many applications, it is expected that the alloys will provide quite long service. Many accelerated exposure type tests have been used, but the difficulties encountered include the need for a corrosive environment severe enough to cause some corrosion of all or most of the alloys involved, while not being so severe as to cause catastrophic corrosion of many of the alloys. Certainly these difficulties can be overcome by a judicious selection of corrosive environments, but for complete characterization of alloy systems, a large number of experiments are required.

Electrochemical polarization techniques using the potentiostat make it possible to vary the oxidation potential of an electrode-electrolyte system over the entire range of stability of the electrolyte quickly and easily. Thus, by choosing a few representative electrolytes and obtaining polarization data at a few different temperatures, it is possible to obtain an estimate of the corrosion behavior of an alloy under a wide range of corrosive conditions.

In an alloy development program for general-use alloys, it is not necessary to predict the exact corrosion behavior in a specific corrosive environment, at least at the outset of the program. Thus, electrochemical measurements that estimate relative corrosion resistance are very useful in sorting out the important variables and then arriving at compositions of greatest interest. This is especially true if existing alloys of known corrosion resistance are included in the electrochemical evaluations for purposes of calibration.

The work to be reviewed here was concerned with the development of new stainless steels as well as an evaluation of variants of existing alloys. Since stainless steels depend on achieving passivity to exhibit good corrosion resistance in most environments, only anodic polarization tests were used. For evaluating resistance to general or uniform corrosion, measurements of the current density required to produce passivity and the potential corresponding to this current were made. To study resistance to pitting, and to some extent crevice corrosion, the potential at which localized corrosion initiated was determined.

Of course, electrochemical techniques do not offer significant help in determining resistance to some forms of corrosion even though they may be very useful in mechanistic studies. Currently, there is no electrochemical test that directly measures susceptibility of stainless steel to stress corrosion cracking. Exposure tests are also simpler for determining resistance to intergranular corrosion, at least for most ferritic stainless steels.

Factors Affecting Uniform Corrosion

For studies of resistance to uniform corrosion in acids, polarization measurements to determine the critical current density for passivation (I_{cr}) and primary passivation potential (E_{PP}) were made in hydrochloric and sulfuric acids. In this way, the passivation tendencies in strong acids with or without the presence of aggressive

*AMAX Materials Research Center, Ann Arbor, Michigan.

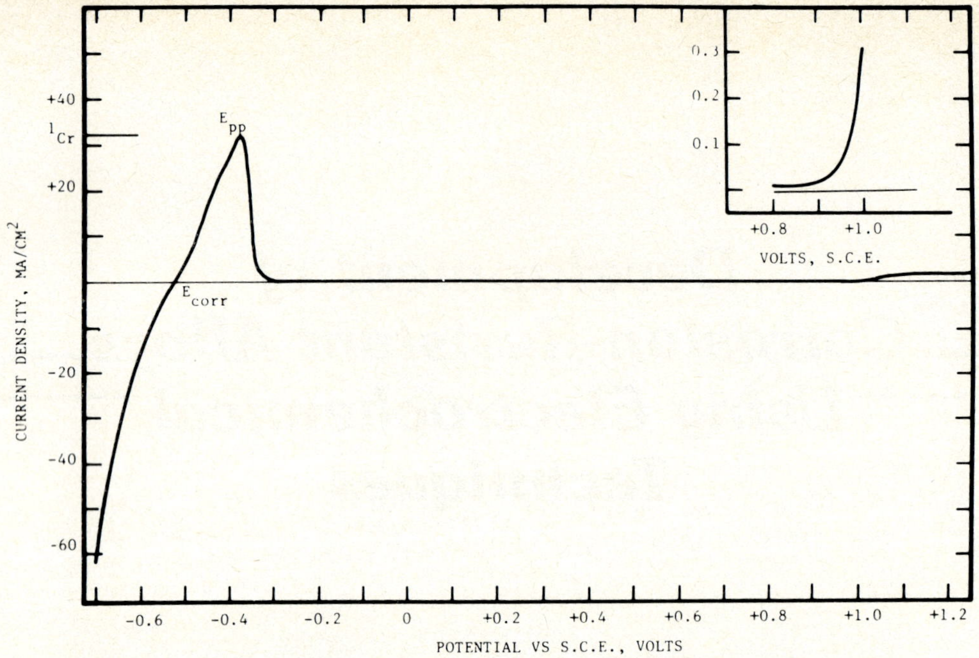

FIGURE 1 — Potentiodynamic polarization curve for a vacuum-melted 17% Cr stainless steel in 5% H_2SO_4 at 29.6 C. The insert shows the expanded view of the passive-transpassive transition.

TABLE 1 — Critical Current Densities for Various Ferritic Stainless Steels in Deaerated Sulfuric Acid and Hydrochloric Acid at 30 C

Nominal Composition	Interstitial Content	Critical Current Density (mA/cm) 1N H_2SO_4	0.1N HCl	1N HCl
17% Cr	Commercial	7.9	1.2	
17% Cr-1% Mo	Commercial	4.7	0.26	
18% Cr-2% Mo	Commercial	1.8	0.11	
17% Cr	Vacuum-melted	32	0.95	1.0
17% Cr-1% Mo	Vacuum-melted	8.8	0.48	46
17% Cr-3% Mo	Vacuum-melted	2.5	0.958	13
25% Cr	Commercial	7.4		30
25% Cr-1% Mo	Commercial	3.5		14
25% Cr-2.4% Mo	Commercial	1.1		9.6
25% Cr-3.5% Mo	Commercial	1.1		5.6
25% Cr-5% Mo	Commercial	0.62		3.2
25% Cr	Vacuum-melted	30	1.4	180
25% Cr-2% Mo	Vacuum-melted	3.0	0.24	16
25% Cr-3.5% Mo	Vacuum-melted	1.5		6.6
25% Cr-5% Mo	Vacuum-melted	0.25		2.1

chloride ions were determined. A typical result is shown in Figure 1. The normal scanning rate used was about 600 mV/h.

To evaluate the effects of chromium, molybdenum, and interstitial content on critical current density for passivity for ferritic stainless steels, the data shown in Table 1 were obtained.[1-3] It is clear that under all conditions studied, the greater the molybdenum content, the lower the critical current density, as would be expected. On the other hand, variation of chromium content from 17 to 25% had little effect on the critical current density. It is well known, however, that decreasing chromium content below about 12% would cause a large increase in the critical current.

The effect of interstitial content on critical current densities for these alloys was somewhat surprising in that the vacuum-melted alloys of the lower molybdenum content had higher critical currents than their commercial purity counterparts. At the highest molybdenum content, this behavior was reversed and the vacuum-melted alloys had lower critical currents than did the commercial purity steels. Of course, the adverse effects of higher carbon and nitrogen contents on mechanical properties and resistance to pitting corrosion outweigh the slight beneficial effect on critical current.[4-6]

A few corrosion tests in oxygen-saturated acids were conducted[7] to demonstrate the significance of the critical current values (Table 2). It was observed that alloys with a critical current density of about 1 mA/cm^2 or less in the given medium became passive. This is approximately the current density corresponding to the limiting diffusion

TABLE 2 — Results of 24-Hour Exposure Tests in Oxygen-Saturated Acids at 25 ± 1 C on Vacuum-Melted Alloys

Nominal Composition	Corrosion Rate ($g \cdot m^{-2} \cdot h^{-1}$) in Indicated Medium[(1)] 0.1N HCl	1.0N H_2SO_4
17% Cr	13.3	67
17% Cr-1% Mo	0.62	10
17% Cr-3% Mo	Nil	5.0
25% Cr	18.8	83
25% Cr-2% Mo	0.07	10.4
25% Cr-3.5% Mo	—	Nil

[(1)]Specimens were activated with magnesium at the start of the test.

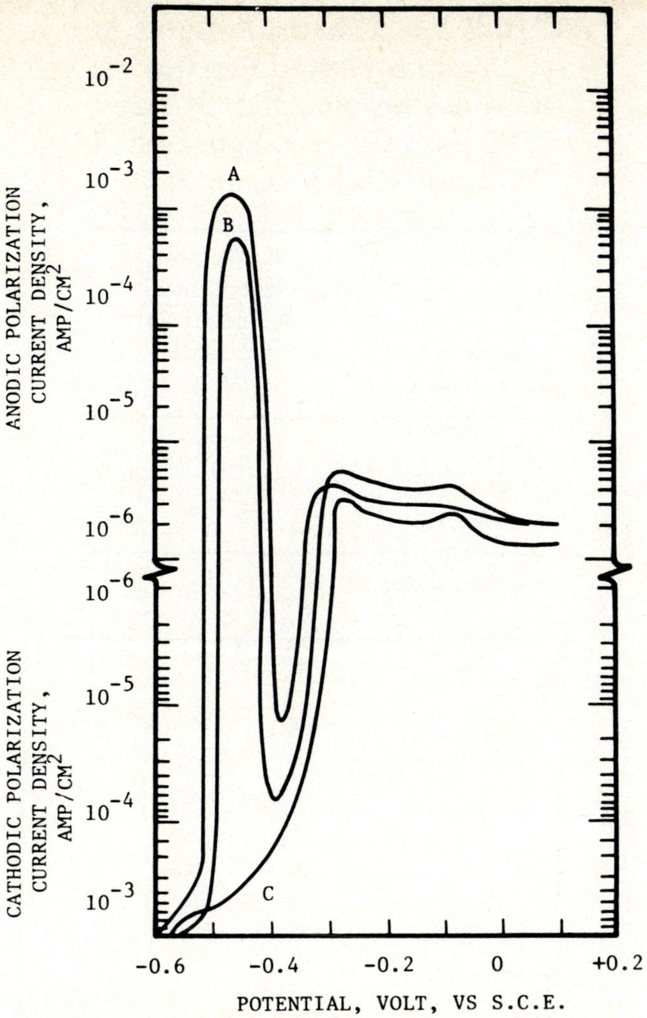

FIGURE 2 – Potentiodynamic polarization curve for 25% Cr-3.5% Mo stainless steels in 1N H_2SO_4 at 29.8 C. Curve A, 25% Cr-3.5% Mo; Curve B, 25% Cr-3.5% Mo-1% Ni; and Curve C, 25% Cr-3.5% Mo-2.5% Ni.

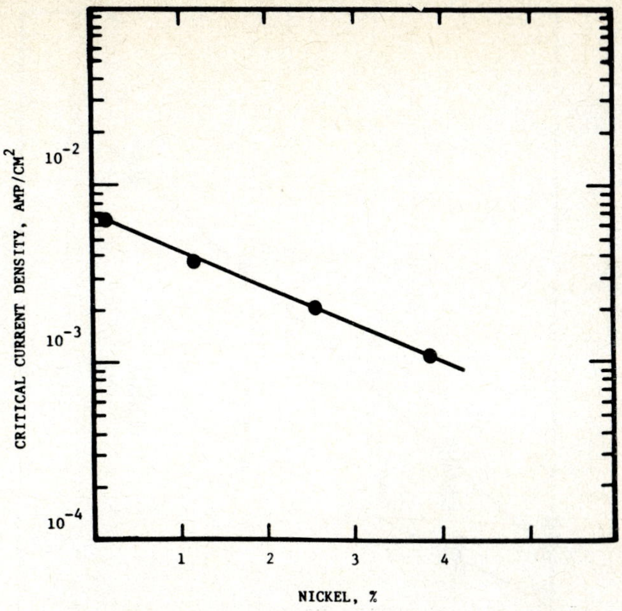

FIGURE 3 – Dependence of critical current density on nickel content for the vacuum-melted 25% Cr-3.5% Mo stainless steels in 1N HCl at 29.8 C.

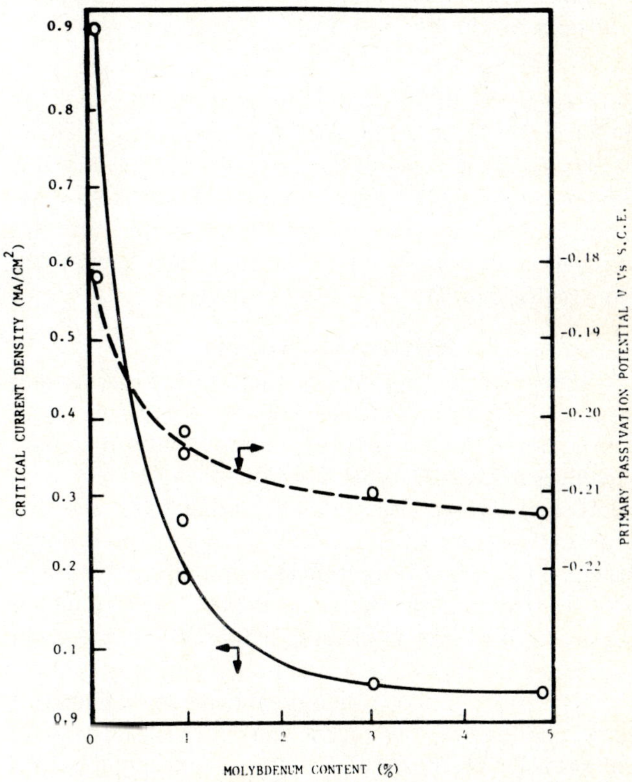

FIGURE 4 – Critical current densities and primary passivation potentials obtained from potentiodynamic scans at 600 mV/h in argon-saturated 0.1N hydrochloric acid on 20% Cr-39% Ni steels.

current for oxygen under the applied conditions. It should be noted that none of the 17% Cr alloys became passive in 1N H_2SO_4 although the corrosion rates decreased greatly as the molybdenum content of the alloys increased.

Nickel also has a pronounced tendency to lower the critical current density for passivity of ferritic stainless steels.[9] Tests on 25% Cr steels in 1N H_2SO_4 (Figure 2) showed that a nickel content of 2.5% completely suppressed the active peak. In 1N HCl, all steels tested displayed active peaks, but the critical current density decreased regularly with increasing nickel content (Figure 3).

The effect of molybdenum on critical current for austenitic alloys is very similar to the effect on ferritic stainless steels.[8] As an example, Figure 4 shows results on a series of alloys with the base composition 20Cr-39Ni in 0.1N HCl. It is clear that increasing molybdenum content decreases the critical current markedly. The effect of molybdenum on primary passivation potential is also definite although not large. The passivation potential shift in the active direction suggests that passivation would be easier as molybdenum increases.

In general, the primary passivation potentials of the ferritic steels studied did not vary in a systematic way with composition and this parameter was not considered especially useful in alloy development of the ferritic steels. Typical values for the primary passivation potential

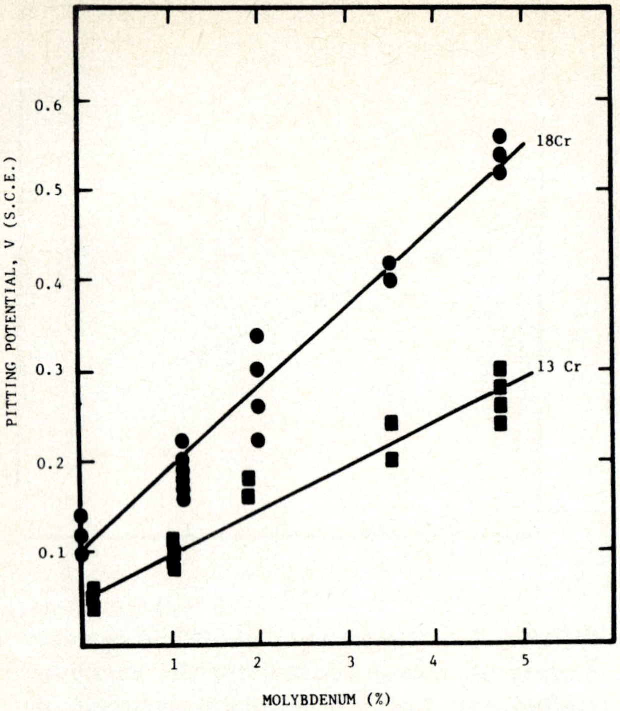

FIGURE 5 – Pitting potentials for vacuum-melted 13 and 18% Cr ferritic stainless steels in 1M NaCl at 25 C.

of 18% Cr vacuum-melted ferritic alloys were −0.45 V (SCE) in 1N H_2SO_4 and −0.34 V (SCE) in 1N HCl.[9] These values were essentially unaffected by molybdenum content over the 0 to 5 range, and 13% Cr alloys gave similar values. However, it is of interest to note that the passivation potentials of the ferritic steels are considerably more active than those for austenitic steels.

Pitting Corrosion

Immersion testing can be used to evaluate pitting resistance of a group of alloys, but it is normally very time consuming because a series of immersion tests of varying severity is required in order that a pitting threshold can be established for each alloy. On the other hand, electrochemical determination of the pitting or breakdown potential in a suitable electrolyte at the proper temperature can yield rather quickly a relative ranking of the resistance of a series of alloys to pitting. This is true even for small differences in corrosion resistance.

The usefulness of pitting potential is still a matter of some controversy, but recent work indicates that, with the proper experimental technique, meaningful results can be obtained.[10] In our work with ferritic alloys, the scratch technique developed by Pessall and Liu[11] has been found useful over the entire range of alloys studied. However, for alloys containing no more than 18% Cr or 3.5% Mo, scanning at 200 to 600 mV/h gave identical results with the scratch technique.[12]

An example of the pitting potential data obtained on some vacuum-melted ferritic stainless steels is shown in Figure 5. These clearly show the important role of both molybdenum and chromium in improving resistance to pitting. At both chromium contents, the addition of molybdenum raised the pitting potential, but at the higher

TABLE 3 – Corrosion Rates of Vacuum-Melted Ferritic Stainless Steels After 14 Days in 180 ppm Cl^- + 60 ppm Cu^{++} Saturated with O_2 at 90 C

Nominal Content Wt%	Corrosion Rate,(1) mdd — Nominal Cr Content, Wt%	
	14	18
0	160	0.5
2	2.5	0

(1)Only pitting occurred.

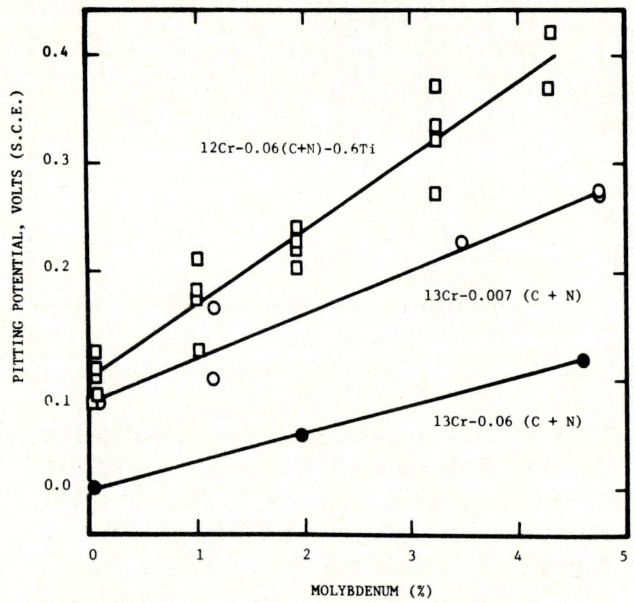

FIGURE 6 – Effect of titanium and interstitial content on pitting potentials of 12-13% Cr ferritic stainless steels. The pitting potentials were determined by scanning at 600 mV/h in 1M NaCl at 25 C.

chromium content, the increase caused by each increment of molybdenum was considerably greater. This increase in slope with chromium content indicates that chromium and molybdenum interact and the total effect of adding both chromium and molybdenum is greater than the sum of their individual contributions. Exposure tests on similar steels confirm the effects of chromium and molybdenum on pitting corrosion (Table 3). The results in Table 3 are reported as weight loss, but only pitting corrosion occurred.

Pitting potential determination can also detect fairly minor alloying effects as well as the major effects such as are associated with chromium and molybdenum. For example, the effects of carbon and nitrogen content or titanium content on pitting potential are shown in Figure 6. It is clear that lowering the carbon and nitrogen con-

TABLE 4 — Effect of Delta Ferrite Produced by Heat Treating at 1345 C on Pitting Potentials in 45 C 1M NaCl of 18Cr-10Ni-2.5Mo-0.159N Alloy

Heat Treatment	Volume % Ferrite	Pitting Potential(1)
1120 C (2050 F)	0	0.30
1345 C (2450 F) 1/2 h	0.55	0.19
1345 C (2450 F) 1/2 h + 1120 C (2050 F) 6 h	0	0.29

(1)Average of four tests.

TABLE 5 — Results of Controlled Potential Crevice Corrosion Tests in 1N NaCl on Ferritic Stainless Steels

Steel Type	Presence of Crevice Corrosion at Indicated Test Temperature 21-22 C	30 C	40 C	50 C	60 C
20Cr-2Mo	Yes	—	—	—	—
22.5Cr-2Mo	Yes	—	—	—	—
25Cr	Yes	—	—	—	—
26Cr-1Mo	No	Yes	Yes	—	—
25Cr-2Mo	No	Yes	Yes	—	—
28Cr-2Mo	No	No	No	Yes	—
25Cr-3.5Mo	No	No	No	Yes	—
25Cr-5Mo	No	No	No	No	Yes

tent had a consistently beneficial effect. The addition of titanium to the higher carbon and nitrogen content alloys increased the pitting potential presumably in part by reacting with the carbon and nitrogen. In addition to neutralizing the effect of carbon and nitrogen, the titanium itself apparently made a further small positive contribution to pitting resistance. Rather small effects, such as those illustrated in Figure 6, are readily detected by pitting potential tests but would require very careful selection of exposure conditions to be detected by simple immersion tests.

In actual alloy design, effects on mechanical and intergranular corrosion properties, as well as weldability, may be more important in selecting carbon, nitrogen, and titanium contents than are the relatively small effects on pitting resistance. However, control of the former properties often allows alternate approaches so that consideration can be given to fairly small effects on other corrosion properties.

Another example of detecting fairly small changes in pitting resistance involves the effect of the presence of delta ferrite on the behavior of austenitic stainless steels. An alloy that could be heat treated so as to contain delta ferrite or be free of it was examined, and delta ferrite definitely depressed the pitting potential (Table 4). After determining that delta ferrite did affect pitting resistance, it was possible to devise exposure tests that also showed this effect; but without the pitting potentials as guides, it would have been difficult to have run the correct exposure tests.[13]

Crevice Corrosion

Crevice corrosion can be studied electrochemically by polarizing the test specimens to a preselected potential in the passive region and then placing the rounded end of a glass rod in contact with the specimen. A rise in current after application of the glass to metal crevice indicates the onset of crevice corrosion.[14] However, it was found that there is no definite potential above which crevice corrosion occurs and below which it does not. Thus, in contrast to pitting potential determination, the electrochemical crevice test yields only go-no-go results.

Typical results of the electrochemical corrosion test on a series of ferritic steels are shown in Table 5. These results clearly demonstrate the large increase in crevice

TABLE 6 — Crevice Corrosion Tests in 10% $FeCl_3$-$6H_2O$ + 0.1N HCl

Steel Type	Weight Loss in mdd Computed Over Entire Coupon Area (Test Duration in Days Given in Parentheses) Test Temperature 25 C	30 C	40 C	50 C
20Cr-2Mo	64 (1)	197 (1)		
22.5Cr-2Mo	0 (7)	3 (1)	— —	— —
25Cr	670 (2.5)	— —	— —	— —
26Cr-1Mo	0 (23)	6.2 (4)	229 (1)	764 (1)
25Cr-2Mo	0 (7)	0 (1)	0 (1)	210 (1)
25Cr-3.5Mo	0 (24)	— —	— —	7 (5)
25Cr-5Mo	0 (63)	— —	— —	0 (11)

corrosion resistance with increases in both chromium and molybdenum content.[15] Very similar results were obtained on the same set of alloys by exposing them in 10% $FeCl_3 \cdot 6H_2O$ + 0.1N HCl with the crevice created by elastic bands (Table 6). Since the exposure tests give essentially the same data as the electrochemical tests, while requiring simpler equipment, no real advantage for the electrochemial test can be argued for routine testing.

Conclusions

Electrochemical testing, specifically measurement of critical current density for passivity and determination of pitting potential in chloride environments, yields data that are very valuable in developing stainless steels with improved corrosion resistance. The primary advantage of the electrochemical measurements cited above is that they yield scalar data as opposed to the go-no-go result usually obtained by exposure tests designed to evaluate ease of passivation or resistance to pitting corrosion.

References

1. E. A. Lizlovs, Corrosion, Vol. 22, No. 11, p. 297t, 1966.
2. E. A. Lizlovs, A. P. Bond, J. Electrochem. Soc., Vol. 116, p. 574, 1969.
3. E. A. Lizlovs, A. P. Bond, J. Electrochem. Soc., Vol. 118, p. 22, 1971.
4. W. O. Binder, H. R. Spendelow, Jr., Trans. ASM, Vol. 43, p. 729, 1951.
5. M. Semchyshen, A. P. Bond, H. J. Dundas, Effects of Composition on Ductility and Toughness of Ferritic Stainless Steels, Proc. Symp. Toward Improved Ductility and Toughness, Kyoto, Japan, October 25-26, 1971, Climax Molybdenum Development Company, Japan, Ltd., pp. 239-253.
6. R. N. Wright, Welding J., Vol. 50, p. 434, 1971.
7. A. P. Bond, E. A. Lizlovs, Werkst. Korros., Vol. 21, p. 336, 1970.
8. A. P. Bond, E. A. Lizlovs, J. Electrochem. Soc., Vol. 115, p. 1130, 1968.
9. E. A. Lizlovs, A. P. Bond, J. Electrochem. Soc., Vol. 122, p. 719, 1975.
10. A. P. Bond, Localized Corrosion—Cause of Metal Failure, ASTM STP 516, ASTM, Philadelphia, Pennsylvania, p. 250, 1972.
11. N. Pessall, C. Liu, Electrochim. Acta, Vol. 116, p. 1987, 1971.
12. E. A. Lizlovs, A. P. Bond, Corrosion, Vol. 31, No. 6, p. 219, 1975.
13. H. J. Dundas, A. P. Bond, "Effects of Delta Ferrite and Nitrogen Content on the Resistance of Austenitic Steels to Pitting Corrosion," CORROSION/75, Paper No. 159, National Association of Corrosion Engineers, Houston, Texas, 1975.
14. E. A. Lizlovs, J. Electrochem. Soc., Vol. 117, p. 1335, 1970.
15. E. A. Lizlovs, Localized Corrosion—Cause of Metal Failure, ASTM STP 516, ASTM, Philadelphia, Pennsylvania, p. 201, 1972.

A Potentiodynamic Study of Aluminum-Lithium Alloys in an Aqueous Sodium Chloride Environment

C. T. TSAO and P. P. PIZZO*

Abstract

The purpose of this study is to explain the electrochemical behavior of aluminum-lithium (Al-Li) alloys, to relate the potentiodynamic behavior to stress corrosion behavior, and to propose a possible means of accelerated stress corrosion testing. Potentiodynamic polarization tests in 3.5% NaCl aqueous solution of varying oxygen content were conducted for five Al-Li alloys having ingot or powder metallurgical alloyed product forms. Oxygen content of the sodium chloride electrolyte was a key factor in determining the electrochemical behavior, and the current-voltage characteristics are defined by this parameter. In general, the potentiodynamic behavior of Al-Li alloys was found to be quite insensitive to variation in composition and heat treatment, factors that profoundly affect the stress corrosion behavior. Large variations in the character of pitting and associated intergranular attack were observed. Surface redox reactions involving oxygen and hydrogen, however, dominated the current voltage characteristics, and the localized corrosion was not revealed therein. Alternate-immersion exposure prior to potentiodynamic polarization may provide a means of assessing susceptibility to localized corrosion.

Introduction

In the future, the development of advanced aluminum alloys for aerospace structural applications will continue to be of prime importance. Advanced alloys will be used in the design or major redesign of space, missile, and aircraft systems.[1-3] To reduce operational expense, alloys will be evaluated in terms of specific strength and specific stiffness to achieve major weight savings. Fatigue endurance and environmental degradation must also be determined to meet durability criteria. The stress corrosion cracking (SCC) of high-strength aluminum alloys has been a prevalent service problem in the aircraft industry, so localized corrosion and stress corrosion performance must be assessed prior to alloy implementation.

At present, the two most widely used structural aluminum alloys are the Al-Cu base alloy AA 2024[(1)] and the Al-Zn-Mg-Cu base alloy AA 7075. In the aged condition (-T3 and -T4 for AA 2024 aluminum, and -T6 for AA 7075 aluminum), the alloys are characterized by a low stress corrosion threshold stress value (less than 55 MPa/m^2) in the short-transverse direction.[4] Service problems are associated with this low stress corrosion resistance.

Of the many advanced aluminum structural alloys being developed, aluminum-lithium alloys offer the best potential for successful near term use in the aerospace industry. The specific benefits of Al-Li alloys include a potential 20% increase in both the stiffness-to-weight and strength-to-weight ratios. However, when they are exposed to chloride-containing aqueous solutions, active lithium bearing phases located at grain boundaries act as sites for localized corrosion.[5]

Polarization data have been used for many years to study electrochemical corrosion processes active in metallic systems. The potentiodynamic technique has been used for a variety of investigations such as basic corrosion research,[6] alloy development,[7] localized corro-

*Dept. of Materials Engineering, San Jose State University, San Jose, California.

(1)Aluminum Association (AA), Washington, DC.

sion,[8] and stress corrosion cracking (SCC) susceptibility of stainless steel[9] and brass.[10] Limited polarization data have been reported, however, for aluminum alloys in general and Al-Li alloys in particular.[5] Also, the current-voltage (I-V) curves are often simply reported without reference to the underlying mechanisms responsible for the observed electrochemical behavior.

Investigators studying aluminum polarization behavior have reported that the polarization data are influenced by oxygen content in the electrolyte. Tohma and Takeuchi[11,12] estimated the corrosion rate and pitting resistance of a variety of Al alloys (Al, Al-Mg, Al-Mn, Al-Fe, Al-Si, Al-Cu, Al-Mg-Si). Under nitrogen-deaerated conditions, the zero-current potential (E_O) fell below -1.0 V (E_{SCE}) and the pitting potential (E_p) varied between -0.7 and -0.8 V. However, under aerated conditions, both the zero-current potential and the pitting potential were found to increase from -0.6 to -0.8 V. Richer and Duquette[13] have compared Al-Mg-Li and AA 7075 (Al-Zn-Mg-Cu) alloys under aerated and nitrogen-deaerated conditions. They found that aeration of the solution reduced the passive-current density and increased the zero-current potential, while the pitting potentials remained essentially unchanged.

This behavior is not only a characteristic of Al alloys; AISI 304 stainless steel (SS) can exhibit similar tendencies. Indig and McIree[14] reported that dissolved oxygen is the major factor affecting the corrosion potential of AISI 304 SS in high-temperature, high-purity water. A significant increase in corrosion potential (approximately 0.5 V) is observed for dissolved oxygen concentration between 10 and 50 ppb.

The objectives of this paper are: (1) to explain the characteristics of the potentiodynamic curves for Al-Li alloys in aqueous NaCl solution, and (2) to correlate the electrochemical parameters of the potentiodynamic technique to observed pitting and intergranular cracking behavior. The overall goal of the research program is to propose a possible means of accelerated stress corrosion testing for Al-Li alloys. Powder metallurgy (PM) processed Al-Li-Cu alloys are the primary focus of this study.

Experimental

Materials

Five different Al-Li alloys having ingot or powder metallurgical product forms were investigated. The nominal composition and specific product form of each are shown in Table 1. Emphasis in the present study is on the two Al-Li-Cu alloys processed using the conventional PM technique. Their compositions are similar except for the magnesium content. The PM Al-Li-Cu alloy, herein

TABLE 1 — Nominal Composition and Product Form of the Aluminum Alloys Studied(1)

Alloy	Product Form	Composition (wt%) Li	Cu	Mg	Zr	Mn
PM, Base Al-Li-Cu	12.7 × 65 mm, round corner, rectangular section bar	2.6	1.4	0.006	0.09	—
PM, Mg-bearing Al-Li-Cu-Mg	12.7 × 65 mm, round corner, square section bar	2.6	1.4	1.60	0.09	—
MA(2) Al-Li-Cu	19 × 19 mm, round corner, square section rod	2.5	—	1.0	—	—
MA(2) Al-Li-Cu	19 × 19 mm, round corner, square section rod	3.0	2.1	—	—	—
Alloy X2020(3) Aluminum	Wing-skin from retired RA-5C US Navy aircraft	1.3	4.5	0.01	—	0.70

(1)PM designates the conventional powder metallurgical processes.
MA designates the mechanically alloyed powder metallurgy process.
(2)Carbon determined by combustion/thermal conductivity is 1.21%; oxygen determined by inert-gas fusion is 0.021%. These concentrations result from the mechanical alloying procedure and can be assumed to be similar in both the copper- and magnesium-bearing alloys.
(3)Additionally, the alloy X2020 contains 0.21 cadmium.

referred to as the base alloy, contains 2.6% Li and 1.4% Cu. The magnesium-bearing alloy, PM Al-Li-Cu-Mg, contains 1.6% Mg.

Powders were produced by cooling an atomized molten stream of the target composition in an inert gas (argon) atmosphere. High-purity (99.99 wt%) aluminum and a 20 wt% aluminum-lithium master alloy were combined to yield an approximate 2.6% Al-Li melt. High-purity element additions were made to adjust the melt to the desired alloy compositions. The cooling rate of the atomization process was estimated to be 10^3 K/s. The resulting powders were spherical, and about 150 μm in diameter (sized to 100 mesh). Powders were packed in a AA 6061 aluminum alloy can (13.7 cm OD × 0.32 thickness), vacuum degassed, and cold isostatic pressed to 415 MPa. The compacted powders were then hot-upset at 755 K (482 C) against a blind die and extruded at 672 K (399 C) through a 55.9 × 14.7 mm die. This resulted in an approximate 10:1 extrusion ratio.

The microstructure, aging response, and mechanical properties of the Al-Li alloys used in this study have been thoroughly documented elsewhere.[5]

Heat Treatment

All of the alloys except alloy X2020, were solution-treated at different temperatures for one hour, then cold-water quenched. The solution treatments were performed at 515 C for the base alloy, at 555 C for the Mg-bearing alloy, at 560 C for mechanically alloyed (MA) Al-Li-Cu alloy, and at 520 C for MA Al-Li-Mg. Solution treatment was performed in air prior to specimen fabrication. The alloy X2020 was tested in the as-received T6 condition.

After fabrication, aging was performed in a forced air circulation oven maintained at ±2 C of the desired temperature. All of the alloys were aged at 170 C, except the Mg-bearing alloy, which was aged at 190 C, for the indicated time.

Anodic Potentiodynamic Polarization

The standard polarization cell[15] was used as shown in Figure 1. Cylindrical specimens (working electrodes) with an electrode surface area of 1.4 cm^2 were machined from the solution-treated and quenched alloys. These specimens were aged, polished through a series of emery papers (Grade 2 through double aught), and turned on a bench lathe with kerosene as lubricant to a 1.0 μm diamond finish. They were degreased in acetone, and then rinsed in freon. After being attached to the working electrode holder of the cell, specimens were immersed in the polarization cell solution. Sodium chloride solution (3.5 wt% reagent grade NaCl) was prepared using distillated and deionized water. To achieve deaerated conditions, the solution was purged, prior to immersion of the test specimen, for a minimum of one hour with He gas. The oxygen content of the electrolyte was monitored by an oxygen meter. The oxygen concentration in the solution, after proper He-gas purging, was less than 1 ppm. Different oxygen concentrations ranging from 5 to less than 1 ppm were maintained in some experiments by the addition of purified oxygen gas through a flow-gas controller.

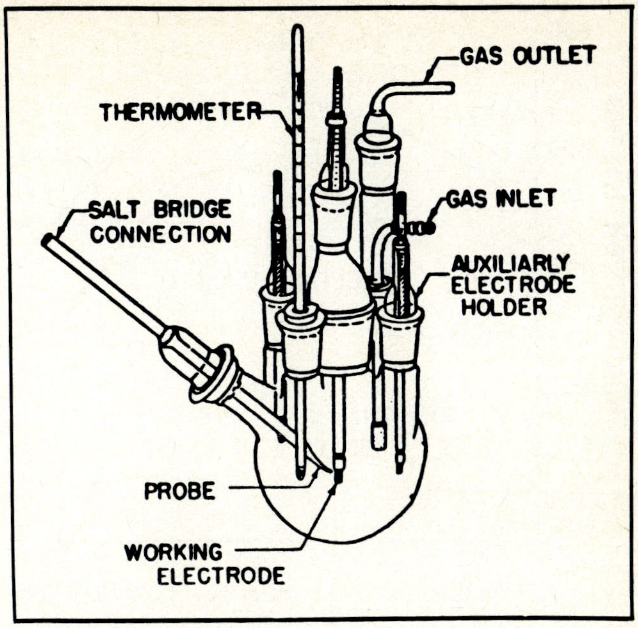

FIGURE 1 — Standard polarization cell.

A scanning Petrolite M410 Potentiodyne Analyzer[(2)] was used for all potentiodynamic measurements. A record of the potential vs the natural log of the current was plotted continuously using an X-Y recorder and a logarithmic converter. A scan generator controlled the scanning speed and polarity. A scanning rate of 100 mV/min was used in the study.

Cathodic Potentiostatic Charging in HBF_4

Coupon specimens with a surface of 1 cm^2 and polished with 1.0 μm diamond paste were used to study cathodic potentiostatic behavior of the PM Al-Li-Cu alloys in HBF_4 electrolyte. A solution of 5 mL HBF_4 (48%) and 200 mL of distilled water was used. Different active voltages (10 to 40 DC V) and immersion times (10 to 120 s) were investigated.

Alternate-Immersion Treatment

Alternate-immersion exposures (in accordance with ASTM[(3)] Standard G 44) were used to precondition some specimens prior to potentiodynamic polarization studies. Cylindrical specimens were alternately immersed in 3.5% NaCl solution for 10 min and then air dried for 50 min. The temperature of the salt solution was maintained at 30 ± 1 C. Pre-exposure conditions varied from 1 day to 2 weeks.

Chemically Induced Pitting

$AlCl_3$ and KOH additions to 3.5% NaCl aqueous solution were made to investigate the influence of high anion and cation concentration on localized pitting of the PM Al-Li-Cu alloys under constant immersion conditions. In addition, the susceptibility of these alloys to

(2)Registered trade name.

(3)ASTM, Philadelphia, Pennsylvania.

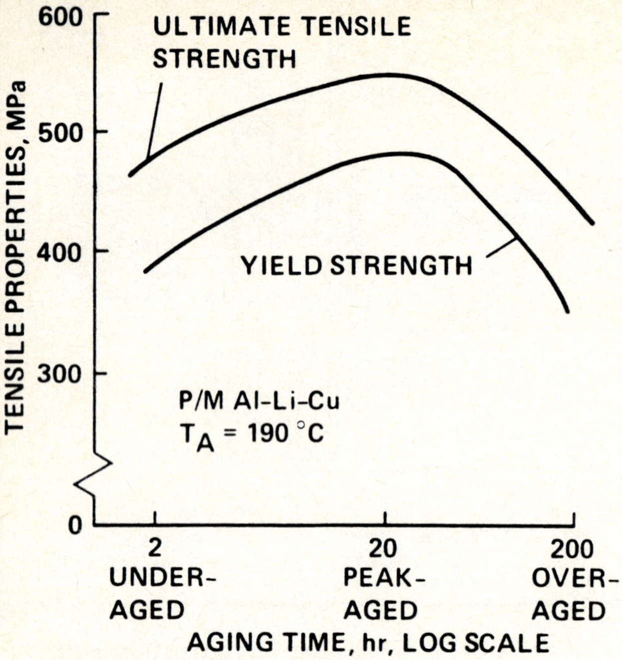

FIGURE 2 – Mechanical properties as a function of aging time.[5]

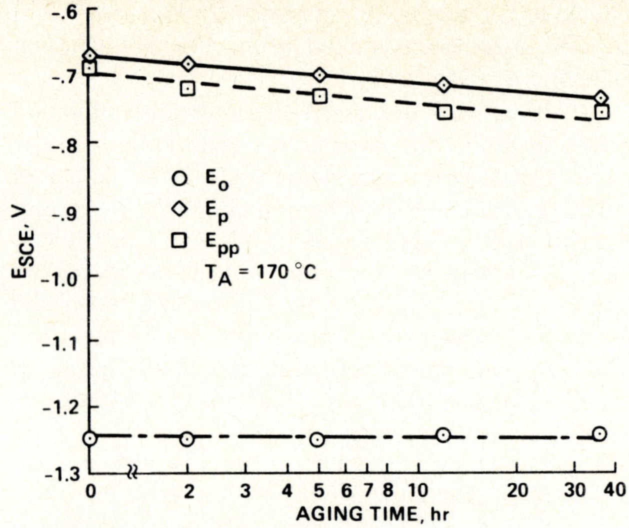

FIGURE 3 – The electrochemical behavior of PM base alloy as a function of aging time at 170 C.

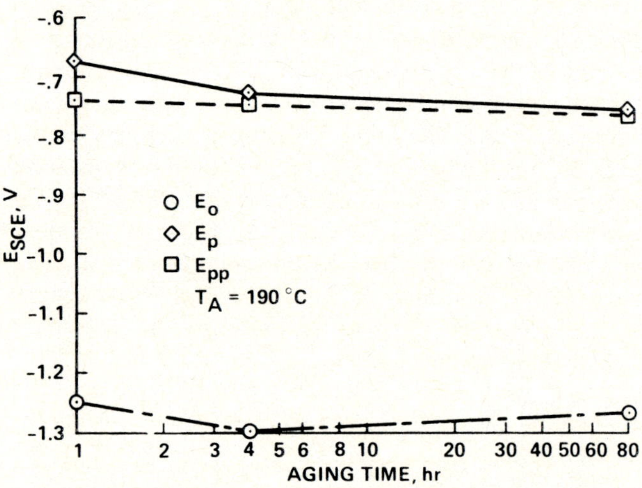

FIGURE 4 – The electrochemical behavior of PM Mg-bearing alloy as a function of aging time at 190 C.

intergranular corrosion per the accelerated test method of Military Specification MIL-H-6088F,[(4)] used to evaluate commercial high-strength aluminum alloys, was investigated. Per MIL-H-6088F, the alloys were immersed in an etching solution of HNO_3 (70%), 50 mL; HBF_4 (48%), 5 mL; and distilled water, 945 mL, for 1 min at 90 C to produce a uniform surface condition. The specimens were then rinsed, cleaned, and exposed in a solution containing NaCl, 57 g; H_2O_2 (30%), 10 mL; and distilled water, 1000 mL; for 6 h at 30 ± 5 C. After immersion, a cross section of the samples was metallographically examined.

Metallographic Evaluation

The objective of metallographic evaluation was post-exposure examination of general and localized corrosion (pitting and intergranular corrosion). The exposed surface and transverse sections were examined using conventional optical metallographic techniques.

Results and Discussion

Anodic Potentiodynamic Polarization

Effect of Heat Treatment. One of the objectives of this research was to investigate whether the polarization data provide a means of identifying aging conditions that render the alloys sensitive to SCC. The typical aging response in terms of mechanical properties for the PM Al-Li-Cu alloy is illustrated in Figure 2. In a concurrent study of the stress corrosion behavior of these PM alloys,[5] the underaged condition has been found to exhibit the highest SCC susceptibility.

Figures 3 and 4 show the electrochemical parameters of PM Al-Li-Cu alloys as a function of aging time. Generally, the zero-current potential (E_O) is relatively insensitive to the aging time; the pitting potential (E_p) and protection potential against pitting (E_{pp}) generally decrease (become more negative) with longer aging time, but the variation is minor (E_{SCE} is 0.06 V over the entire aging response). Therefore, the aging response in terms of polarization is relatively independent of aging time in contrast to the trends observed for the mechanical properties and stress corrosion susceptibility. In fact, a correlation with stress corrosion performance cannot be anticipated.

Effect of Composition. The main electrochemical parameters for the five Al-Li alloys of this study are summarized in Table 2 so that the compositional dependence

---

[(4)]Military Specification (MIL), G.P.O., Washington, DC.

TABLE 2 — Electrochemical Behavior of Al-Li Alloys Under He-Deaerated Conditions

Alloys	Nominal Compositions (wt%) Li	Cu	Mg	E_o, SCE (V)	E_p, SCE (V)	E_{pp}, SCE (V)	Slope of Passivation Region $\left(\frac{\mu A}{cm^2 \cdot V}\right)$	Shape of I-V Curve 
PM Base	2.6	1.4	—	−1.250	−0.660 to −0.680	−0.683 to −0.689	6 to 8	
PM Mg-Bearing	2.6	1.4	1.6	−1.250	−0.675	−0.740	30	
MA Al-Li-Mg	2.5	—	1.0	−1.020	−0.720	−0.735 to −0.740	29	
MA Al-Li-Cu	3.0	2.1	—	−1.040	−0.768 to −0.777	−0.778 to −0.782	20	
X2020-T6	1.3	4.5	—	−1.100 to −1.200	−0.690 to −0.730	−0.730 to −0.745	5 to 8	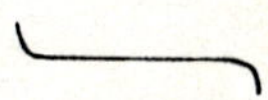

can be assessed. Systematic variations in key electrochemical parameters occur as a function of composition. Under He-deaerated conditions, for example, the PM Al-Li-Cu alloys are found to have a more negative zero-current potential (E_o) than alloy X2020-T6 aluminum, and the MA Al-Li-Cu and Al-Li-Mg alloys exhibit an even more noble zero-current potential. However, the pitting potential (E_p) and protection potential against pitting (E_{pp}) have reversed tendencies. That is, the PM Al-Li-Cu alloys have more noble (positive) values of E_p and E_{pp} than X2020-T6 aluminum, which in turn has more noble E_p and E_{pp} values than MA alloys. In effect, the Al-Li alloys exhibit a decreasing passivation range as follows: PM Al-Li-Cu, X2020-T6, MA alloys. This tendency may be more a function of process history than composition, as the PM Al-Li-Cu base alloy composition and that of the MA Al-Li-Cu alloy are quite similar.

Another important difference among these alloys is the shape of the current-voltage (I-V) curve. The PM and X2020 alloys have relatively flat (voltage-independent current density) passivation regions, while that of the MA alloys is comparatively sloped. In other words, the passivation films of PM and X2020 alloys are more stable than those of the MA alloys.

When the PM base alloy is compared with the PM Mg-bearing alloy, the Mg-bearing alloy systematically exhibits a more active (negative) protection potential against pitting (E_{pp}) than the base alloy. This suggests that the pit growth begins to occur at more negative potentials than is the case in the base alloy. For the MA alloys, the Al-Li-Cu alloy has more active potentials (E_o, E_p, E_{pp}) than is the case for the Al-Li-Mg composition, and the MA Al-Li-Mg alloy has the more voltage-independent passivation behavior.

Effect of Oxygen. The experimental results discussed above are for electrochemical tests performed in He-deaerated 3.5% NaCl aqueous solution. Experimental data for tests in aerated solution (summarized in Table 3) demonstrate that oxygen content of the electrolyte plays an important role in the electrochemical behavior of Al-Li alloys. Under the oxygenated condition, the zero-current potential (E_o) of PM alloys have more noble (positive) potentials than alloy X2020, and alloy X2020 exhibits a more noble value of E_o than the MA Al-Li-Cu and Al-Li-Mg alloys. This reverses the tendency under He-deaerated conditions reported earlier. It is important to note that the shape of the I-V curve undergoes a sharp transition from cathodic to anodic behavior, and that the passive portion of the curve is hidden (except for the MA Al-Li-Cu alloy with 1.5 h aging). The zero-current potential (E_o) shifts to a more noble (positive) potential by 0.4 to 0.5 V, which is very close to E_p. The effect of aging treatment on the electrochemical parameters of Al-Li alloys under oxygenated conditions shows the same general tendencies that were displayed for the deaerated conditions. That is, the zero-current potential (E_o), pitting potential (E_p), and protection potential against pitting (E_{pp}) become more negative with longer aging time (Figures 3 and 4).

Next, the PM Al-Li-Cu alloy was studied to deter-

TABLE 3 — Electrochemical Behavior of Al-Li Alloys Under O_2-Aerated Condition

Alloys	Nominal Compositions (wt%) Li	Cu	Mg	E_o, SCE (V)	E_p, SCE (V)	E_{pp}, SCE (V)	Slope of Passivation Region ($\frac{\mu A}{cm^2 \cdot V}$)	Shape of I-V Curve (I vs. V)
PM Base	2.6	1.4	—	−0.675	−0.665	−0.690	900	
PM Mg-Bearing	2.6	1.4	1.6	−0.730	−0.720	−0.747	200	
MA Al-Li-Mg	2.5	—	1.0	−0.740	−0.720	−0.740	1800	
MA Al-Li-Cu	3.0	2.1	—	−0.850	−0.790	−0.785	17	
X2020-T6	1.3	4.5	—	−0.760	−0.750	−0.760	600	

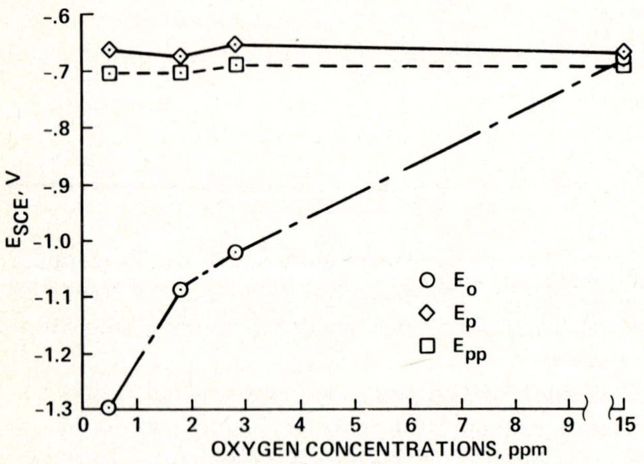

FIGURE 5 — The electrochemical behavior of solution treated PM Al-Li-Cu alloy as a function of oxygen concentration.

mine how the oxygen content influences the electrochemical behavior. The experimental results for the PM Al-Li-Cu alloys as a function of oxygen concentration are plotted in Figure 5. The zero-current potential (E_O) shifts to more noble (positive) potentials with higher oxygen concentration, while the pitting potential (E_P) is relatively unaffected. The passivation region shown by the scan thus becomes limited (narrow). When oxygen concentration reaches a critical value (approximately 5 ppm), the passivation region of the scan disappears, and the current density increases exponentially as the anodic region is entered.

The following hypothesis explains variations of the typical polarization curve under deaerated conditions and the adjustment of electrochemical behavior to an increased oxygen concentration in the solution.

The reduction reactions involving hydrogen gas production and oxygen gas reduction occur in parallel. That is, both contribute to the total cathodic current transfer below a sufficiently negative potential. When the oxygen concentration in solution is driven below some critical value, the contribution of the oxygen reduction reaction to the net cathodic reaction is greatly decreased and the hydrogen reduction reaction dominates. Because the exchange current density of hydrogen is lower than oxygen (at least 2 decades), the oxygen reduction is the major cathodic reaction until the limiting oxygen concentration in the solution is reached. With higher oxygen concentration, the oxygen reduction will shift to more noble (positive) potentials. Therefore, the zero-current potential (E_O) shifts to more noble potentials with higher oxygen.[8] With further increase of the oxygen concentration, the passivation region disappears, and the whole corrosion process is hidden by the reduction reaction of oxygen at the surface.

Localized Corrosion and Intergranular Attack

In this study, the correlation of localized corrosion to microstructural conditions and the dependence of the character of attack on environmental variations (i.e., electrolyte chemistry, oxygen content, electrochemical potential) were investigated. In this section, optical metallographic results are presented. If the character of intergranular attack could be related to the potentiodynamic current-voltage (I-V) behavior, intergranular attack and perhaps stress corrosion behavior could be predicted in a convenient, accelerated test.

Figure 6 is a typical working electrode surface that has been polished with 1 μm diamond paste. The scratches indicate the polishing direction. Figure 7 represents the surface of PM Al-Li-Cu alloy after potentiodynamic scanning in the salt solution. The large surface pits (15 to 30 μm, diameter) are distributed randomly after potentiodynamic scanning. These pits result when the voltage exceeds E_P and the passive film breaks down.

FIGURE 6 — As-polished surface of PM Al-Li-Cu alloy, cylindrical polarization specimen.

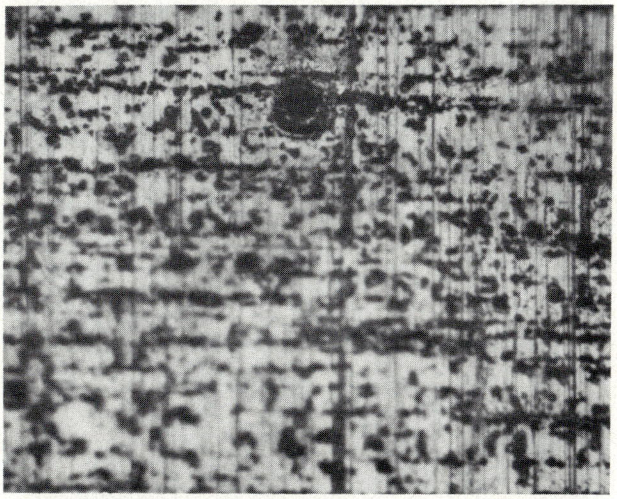

FIGURE 7 — Surface of PM Al-Li-Cu alloy after potentiodynamic scan in 3.5% NaCl solution.

If the voltage is maintained below E_p, these large pits are avoided.

However, localized corrosion still occurs below E_p. The pits along the extrusion direction are the results of stringer oxide particle attack. The stringer oxide particles lying along the extrusion direction are characteristic of PM Al-Li-Cu alloys,[16] and their nature is complex.[17] In powder metallurgical Al-Li-Cu alloys, the stringer particles contain Li_2O_2, Li_5AlO_4, $AlLiO_2$, LiCuO, $LiCO_3$, and δ (Al-Li) phase. Because of the presence of the active lithium and lithium-aluminum oxides, localized corrosion is promoted. Stringer particle attack was observed in all of the Al-Li alloys studied including the ingot product, alloy X2020.

In the constant-immersion polarization studies of this investigation, random pitting and stringer particle pitting were observed. However, in concurrent alternate-immersion tests, in which specimens are exposed to flowing sodium chloride solution for 10 min each hour (h), the stringer particle pitting is accompanied by intergranular attack. Pizzo, et al.,[5] showed that this intergranular corrosion was associated with stress corrosion crack initiation in the alternate-immersion tests.

These investigators[5] also ran comparative tests of stressed tuning-fork specimens under alternate-immersion vs constant-immersion conditions. For example, one of two tuning-fork specimens of PM Al-Li-Cu alloy (same heat treatment and stress level) was immersed in He-deaerated 3.5% NaCl solution, and a potentiostat was used to keep the voltage constant and in the passivation region. The other stressed specimen was alternate-immersion exposed. The specimen maintained under potentiostatic control did not fail during the duration of the 3000-h test, while the alternate-immersion specimen failed in just over 100 h. The ratio of the time-to-failure for alternate-immersion vs constant-immersion tests under potentiostatic control was typically 10:1.

Based on the above observation, it was decided to precondition cylindrical polarization samples with alternate-immersion treatment prior to the potentiodynamic scan to investigate the intergranular corrosion phenomenon. Figure 8 shows the results for the PM Mg-bearing alloy after preconditioning in alternate-immersion (48 cycle) conditions in 3.5% NaCl solution. The surface-normal and cross section of the cylindrical polarization specimen are illustrated in Figures 8(a) and (b), respectively. Intergranular attack, associated with localized pits, occurred. This tendency toward intergranular cracking is believed to be a result of chemical concentration mechanisms that are promoted during the wet-dry cycling of the alternate-immersion exposure.

After alternate-immersion (48 cycle) treatment to initiate intergranular attack, cylindrical specimens were transferred to the electrochemical cell, and potentiodynamic polarization was performed. Surface-normal and cross-section micrographs of the sample after potentiodynamic scanning are illustrated in Figures 9(a) and (b), respectively. Comparison of Figure 8(a) with 9(a) and Figure 8(b) with 9(b) shows that the potentiodynamic polarization accelerates the intergranular attack. The cracks become wider and deeper (100 to 350 μm). However, the same current-voltage (I-V) curve was obtained for the specimen with alternate-immersion pretreatment as compared to the one without alternate-immersion treatment. Therefore, the curve is insensitive to the intergranular attack and concomitant crack chemistry difference.

A comparison of Figures 8 and 9 also shows that much material is apparently lost during the potentiodynamic scan, and yet, the same I-V curve results. However, little material loss may accompany intergranular attack along the grain boundaries. The grains may simply pop out of the localized corrosion regions. In addition, the measured current is composed of contributions both from surface redox reactions as well as both localized corrosion processes. The surface redox reactions may dominate the net current. Additionally, the available equipment may be insensitive to the component of current associated with the intergranular corrosion attack at grain boundaries.

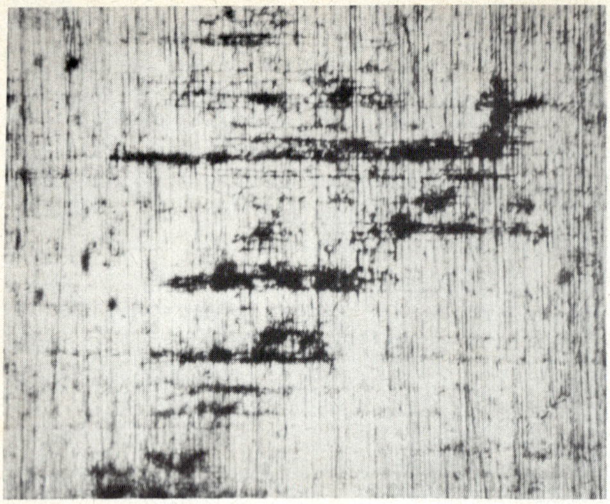

FIGURE 8(a) – Surface-normal view of underaged PM Mg-bearing alloy after alternate-immersion (48 cycle) in 3.5% NaCl solution.

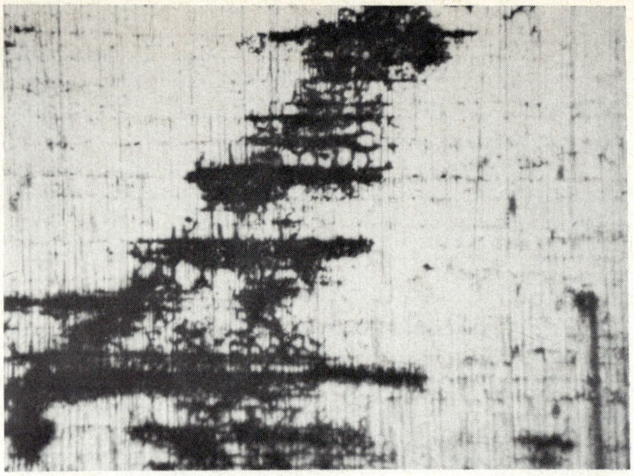

FIGURE 8(b) – Transverse section of underaged PM Mg-bearing alloy after alternate-immersion (48 cycle) in 3.5% NaCl solution.

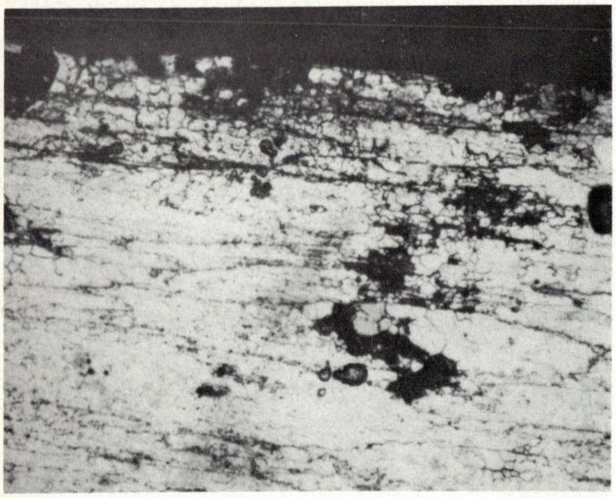

FIGURE 9(a) – Surface-normal view of underaged PM Mg-bearing alloy pre-alternate-immersion exposed (48 cycle), then potentiodynamically scanned in He-deaerated 3.5% NaCl solution.

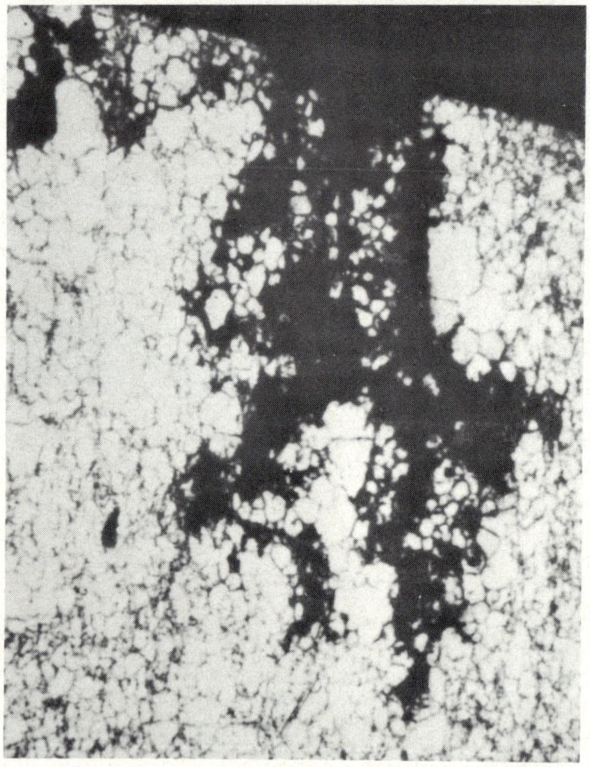

FIGURE 9(b) – Transverse section of underaged PM Mg-bearing alloy pre-alternate-immersion exposed (48 cycle), then potentiodynamically scanned in He-deaerated 3.5% NaCl solution.

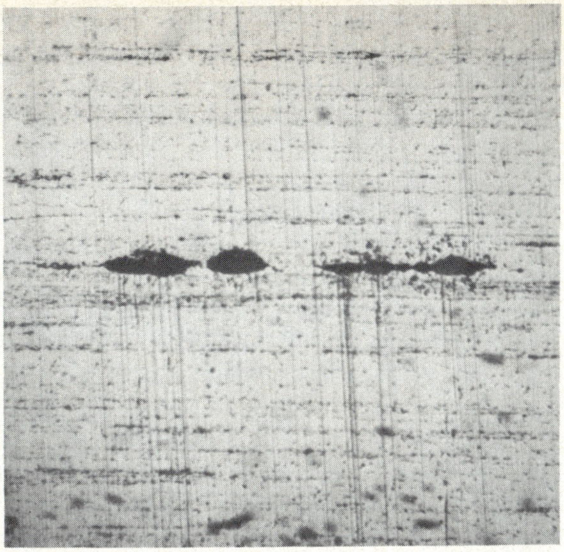

70 μm

FIGURE 10 — Surface of PM Al-Li-Cu alloy cathodically charged at 15 V in HBF_4 solution for 2 min.

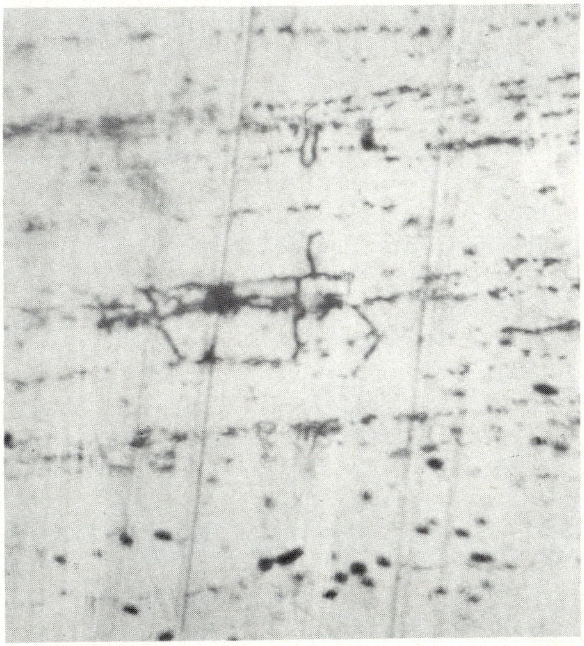

12 μm

FIGURE 11 — Intergranular cracking in PM Al-Li-Cu alloy, cathodically charged in HBF_4 solution for 40 s.

The chemical nature of aluminum is unusual compared to most metals. Aluminum is amphoteric; i.e., soluble in acid solutions as well as alkaline solutions. Therefore, the PM Al-Li-Cu alloys were also tested under cathodic charging in HBF_4 solution (5 mL HBF_4 + 200 mL H_2O). This results in the development of a high pH, alkaline, near-surface condition. Figure 10 shows the stringer particle attack that resulted, whose behavior was comparable to that observed under anodic charging (see Figure 7). Figure 11 illustrates the intergranular cracking

25 μm

FIGURE 12 — Surface of PM Al-Li-Cu alloy after potentiodynamic scanning in deaerated 3.5% NaCl plus 1N $AlCl_3$ solution.

25 μm

FIGURE 13 — Surface of PM Al-Li-Cu alloy after potentiodynamic scanning in deaerated 3.5% NaCl plus 1N KOH solution.

that sometimes occurred under cathodic charging. However, the phenomenon was limited and not reproducible.

It is apparent that under alternate-immersion conditions, some unique chemistry develops within and about stringer particle pits, perhaps a result of a chemical concentration mechanism, and intergranular attack results. Thus, in order to produce intergranular attack, different chemical environments were investigated. Figures 12 and 13 show the surface condition of PM Al-Li-Cu alloy after potentiodynamic scanning in He-deaerated 3.5% NaCl solution to which 15 mL of either 1N $AlCl_3$ or KOH were added, respectively. Local and severe pitting resulted, but no intergranular attack was observed.

Under constant-immersion conditions, reproducible intergranular attack was found to occur by following the accelerated intergranular corrosion tests of MIL-H-6088F. Figure 14 shows a transverse metallographic

25 μm

FIGURE 14 — Transverse section of PM Al-Li-Cu polarization specimen indicating localized corrosion and intergranular attack after treatment per MIL-H-6088F.

section of PM Al-Li-Cu after MIL-H-6088F treatment, demonstrating typical localized corrosion and intergranular attack. The uniqueness of this constant immersion exposure, which results in intergranular attack, is thought to be due to the required addition of H_2O_2 to the test solution. H_2O_2 is a powerful oxidizing agent, and this is known to have a pronounced influence on cracking tendencies.

Conclusions

From the results of this work it is possible to draw the following conclusions:

1. Oxygen content of the aqueous sodium chloride solution is a key factor in determining the electrochemical behavior of Al-Li alloys.

2. The zero-current potential (E_O) shifts to more noble (positive) potentials with higher oxygen concentration while pitting potential (E_p) is relatively unaffected.

3. Although mechanical properties are a strong function of aging time and temperature for age-hardened Al-Li alloys, polarization characteristics are found to be relatively unaffected, and a correlation with stress corrosion performance is thus not observed.

4. The electrochemical behavior of Al-Li alloys is relatively insensitive to compositional variations.

5. Stringer oxide particle attack and random pitting are the common phenomena for the five Al-Li alloys studied.

6. Intergranular corrosion is environmentally selective. Intergranular attack occurs after alternate-immersion treatment or for constant-immersion treatment per MIL-H-6088F. In these environments, the attack is pronounced and reproducible.

7. Intergranular attack initiated during alternate-immersion in aqueous NaCl solution is accelerated by subsequent anodic polarization. However, the electrochemical behavior is insensitive to the presence of the intergranular attack; that is, the current-voltage (I-V) curve is unchanged.

Acknowledgments

The authors acknowledge the National Aeronautics and Space Administration (NASA-Ames Research Center) for support of this research through Cooperative Agreement NCC 2-155 with San Jose State University. The authors extend their gratitude to H. G. Nelson (NASA, Ames Research Center) and M. E. Indig (General Electric Company, Vallecitos Nuclear Center) for their counsel. The authors express their appreciation to Sean Li for metallographic support and to Ching-Chi Sophia Chang for her assistance in manuscript preparation.

References

1. K. Webster, Metal Progress, Vol. 125, p. 33, 1984.
2. T. H. Sanders, E. S. Balmuth, Metal Progress, Vol. 113, p. 32, 1978.
3. R. E. Lewis, et al., Final Report AFML-TR-78-102, Lockheed Res. Lab., Palo Alto, California, Contract F 33615-77-C-5786, 1978.
4. M. O. Speidel, Met. Trans., Vol. 6A, p. 631, 1975.
5. P. P. Pizzo, R. P. Galvin, H. G. Nelson, Proc. 2nd Inter. Al-Li Conf., E. S. Starke, Jr., T. H. Sanders, Jr., Eds., TMS-AIME, AIME, New York, New York, 1984.
6. M. Stern, Corrosion, Vol. 14, No. 9, p. 440, 1958.
7. N. D. Greene, B. E. Wilde, Corrosion, Vol. 26, No. 12, p. 533, 1970.
8. M. Pourbaix, Corrosion, Vol. 27, No. 11, p. 449, 1971.
9. R. N. Parkins, M. Elices, V. Sanchez Galvez, L. Caballero, Corros. Sci., Vol. 22, p. 379, 1982.
10. R. N. Parkins, N. J. H. Holroyd, Corrosion, Vol. 38, No. 5, p. 245, 1982.
11. K. Tohma, Y. Takeuchi, J. Jap. Inst. Light Met., Vol. 29, p. 232, 1979.
12. K. Tohma, Y. Sugai, Y. Takeuchi, J. Jap. Inst. Light Met., Vol. 31, p. 157, 1981.
13. R. E. Richer, D. J. Duquette, Proc. 2nd Int. Al-Li Conf., E. S. Starke, Jr., T. H. Sanders, Jr., Eds., TMS-AIME, AIME, New York, New York, 1984.
14. M. E. Indig, R. McIlree, Corrosion, Vol. 35, No. 7, p. 288, 1979.
15. N. D. Greene, Experimental Electrode Kinetics, Rennsselaer Polytechnic Institute, Troy, New York, 1965.
16. M. Avolos-Borja, P. P. Pizzo, L. A. Larson, Proc. 2nd Int. Al-Li Conf., E. S. Starke, Jr., T. H. Sanders, Jr., Eds., TMS-AIME, AIME, New York, New York, 1984.
17. L. A. Larson, M. Avolos-Borja, P. P. Pizzo, Proc. 2nd Int. Al-Li Conf., E. S. Starke, Jr., T. H. Sanders, Jr., Eds., TMS-AIME, AIME, New York, New York, 1984.

Section 4

Coatings and Inhibitors

A Summary of Ellipsometric and Electrochemical Studies of the Delamination of Organic Coatings from Iron and Steel

*J. J. RITTER**

Abstract

Ellipsometric and electrochemical techniques for the *in situ* study of the corrosion of coated iron and steel are described. The results of these studies, an interpretation of the data, and the application of the technique to the study of inhibited coatings systems are summarized. It is concluded that interfacial oxide dissolution constitutes an important mechanism of coating delamination.

Introduction

The corrosion of metals protected by organic coatings (paints) is a major practical problem whose fundamentals are still being investigated and debated. This complex phenomenon involves combinations of processes such as ionic and molecular diffusion through polymers and polymer defects, chemical reactions on surfaces and in thin liquid layers, cohesive deterioration within polymers, the oxidation of the metal, the chemistry of polymer/substrate adhesion, and the chemistry of metals and metal oxides in high pH environments. In theory, if all of the pertinent phenomena could be monitored simultaneously for a given system, a complete picture of the corrosion process should emerge. Of course, because of the limitations of current technology, only a few of these processes can be observed concomitantly within a single experiment.

This publication focuses on the development and application of the combined techniques of ellipsometry and electrochemistry to the study of the corrosion of coated iron and steel. The material presented represents a summary of a four-year project conducted at the National Bureau of Standards.[(1)] For more detailed information concerning the approach and the interpretation of results, see References 1 through 5.

The objective of the project was to study both the interfacial phenomena and the chemistry responsible for the phenomena related to the corrosion of painted metals. To place the ellipsometric approach in perspective, it is important to recognize that paints are bonded to the native or air-formed oxide film on the metal and not directly to the metal itself. Thus, the native oxide film becomes an interface between the paint and the metal. Our feasibility studies showed that the ellipsometer would be most sensitive to changes in this interfacial region and thus be a valuable tool for the *in situ* monitoring of subcoating events as corrosion proceeds. The technique is readily extended to the study of corrosion inhibitor effects for those inhibitors that act near or within the interfacial region.

This paper deals with the highlights of the work and includes a description of the approach, the basis for the interpretation of the data, and an encapsulation of the significant results and conclusions concerning the cathodic delamination of organic coatings from iron and steel. Also included is a summary of chelating inhibitor effects on cathodic delamination.

Experimental

Only a brief summary of the experimental procedures is given here since more elaborate details have been published elsewhere.[1-5] Model computations for a multilayered system using the Drude equations indicated that substantial changes in the ellipsometric parameters Δ and ψ might be expected to arise from variations in inter-

*National Bureau of Standards, Inorganic Materials Division, Gaithersburg, Maryland.

(1)National Bureau of Standards, Gaithersburg, Maryland.

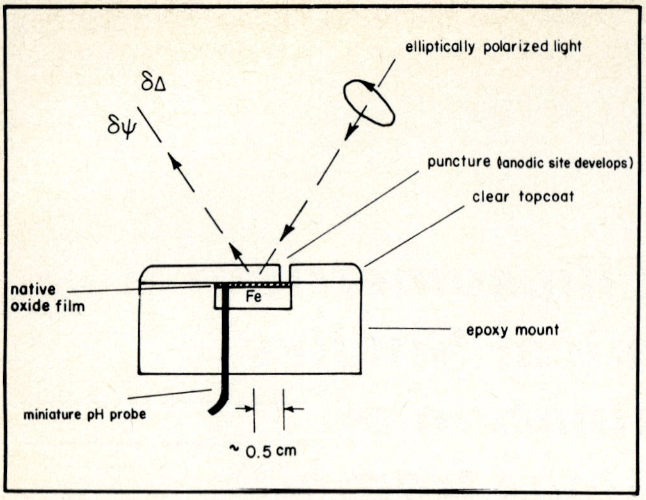

FIGURE 1 — Cutaway view of a typical specimen arrangement for simultaneous ellipsometric and pH measurements.

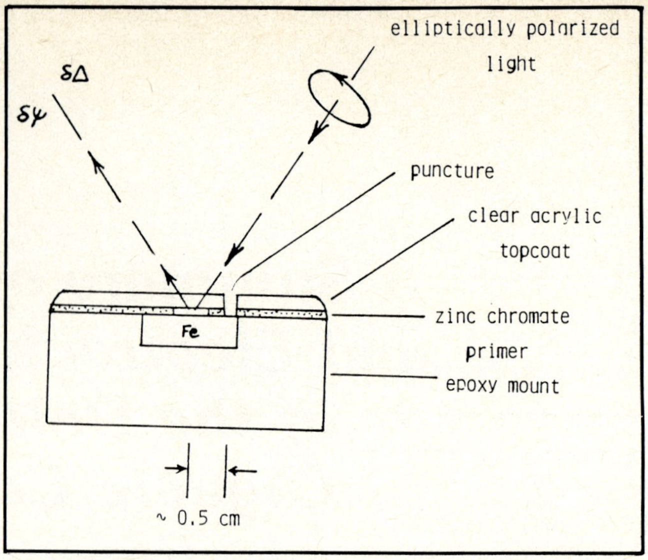

FIGURE 2 — Cutaway view of the specimen arrangement used for the zinc chromate primer/acrylic topcoat system.

facial oxide film thickness. It was determined experimentally that changes in surface texture, i.e., roughening or smoothing of the metal surface, also gave large changes in Δ and ψ. It is possible to distinguish between these two effects for the iron and steel systems by an analysis of the simultaneous changes in Δ and ψ.

The ellipsometric-electrochemical technique itself involves the passage of elliptically polarized light through the transparent coating and reflection from the substrate surface while a miniature pH probe simultaneously monitors the environment at the coating/substrate interface (Figure 1). Changes in the phase ($\delta\Delta$) and amplitude ($\delta\psi$) of the reflected light are indicative of alterations in surface texture (roughness) and changes in the thickness of the surface oxide film. The optical system, as configured for these studies, was relatively insensitive to changes in the coating and to the development of the micro-environment. Thus, the technique is valuable in studying the behavior of the substrate under the coating as the corrosion process proceeds.

Clear organic coatings such as cellulose nitrate, polystyrene, and acrylic lacquers were dip-coated on polished iron and steel surfaces to simulate painted metal. The coated specimens were supported in 0.05 M NaCl solutions and the coatings deliberately punctured to initiate the corrosion process.

Cellulose nitrate and polystyrene coatings placed on inert gold substrates were used to demonstrate that the ellipsometric responses arose only when active substrates underwent corrosion and not from changes within the organic coatings. Because the ellipsometric technique depends on the reflection of light from the substrate surface, anodic areas, where colored corrosion products accumulated, could not be examined for extended periods. Since the cathodic regions on the specimens generally remained highly reflective throughout the experiments, the technique became useful for studying events related to cathodic delamination.

A technique was also developed whereby the effects of adding opaque materials, e.g., pigments and inhibitors, to the coating system could be studied. For example, chromate salts[2] or chromate primers were applied directly to the specimen surface to allow for a small diameter (4 to 6 mm) "window" region for ellipsometric observation. The specimen face was then covered with a clear coating (Figure 2). This arrangement simulated a multilayered paint system.

The effects of organic chelating inhibitors on the subcoating processes were also examined. The chelating inhibitors were either distributed homogeneously in the coatings or applied in a two-step anodization process developed by Leidheiser, et al.,[6] and overcoated with the polymer.

Microsampling and microanalyses techniques were employed to obtain and analyze interfacial liquids.

Discussion

Because of the complex nature of the multilayered systems under study, the ellipsometric data were interpreted in a qualitative fashion.

Early experiments with concomitant pH measurements under cellulose nitrate and polystyrene coatings on Fe showed interpretable pH responses as the corrosion process occurred. When better coating materials, i.e., proprietary acrylic lacquers, were used in later work, pH changes were not observed or were erratic, presumably as a result of insufficient ionic transport through the coating to include the required reference electrode in the measurement circuit. The problem of locating a noncontaminating microreference electrode under the coating was not resolved and simultaneous pH measurements had to be discontinued on these coating systems.

Microanalyses on anodic and cathodic fluids showed a trend toward accumulation of Cl^- in anodic regions. This result suggested a mechanistic analogy between corrosion of painted surfaces and crevice or pitting corrosion, with the most significant difference being the presence of the organic layer.

Analysis of the data indicated that the regions being examined ellipsometrically became cathodic, and that the pH in these cathodic regions approached or may even exceed a value of 14. Accordingly, ellipsometric studies were performed on *uncoated* iron surfaces in saturated NaOH solution, a simulation of the subcoating environment in cathodic regions and in borate buffer solutions. As a result of these studies, ellipsometric responses could be qualitatively separated as a result of oxide film thickness changes from those resulting from surface texture changes, i.e., roughening or smoothing of the metal surface. For the iron system, it was found that when the ellipsometric parameters Δ and ψ changed simultaneously in *opposite* directions, the predominant process occurring was oxide film growth or dissolution. When Δ and ψ showed concomitant changes in the *same* direction, the predominent process was a surface texture change, i.e., the metal surface was becoming either rougher or smoother.

These relationships were used to interpret the ellipsometric data derived from coated samples undergoing corrosion. A typical response curve for a room temperature cured iron/acrylic system is shown in Figure 3. The coating was punctured at t = 0 to initiate the corrosion process. For a period of about 500 min, no changes in Δ and ψ were observed. At the end of this time, a rise in Δ and a concomitant decline in ψ was noted as corrosion set in (Region I). This type of behavior is interpreted as a change in oxide film thickness and the direction of change is indicative of a film-thinning process. The interpretation of the Δ and ψ changes is derived from the behavior of these parameters for uncoated iron in saturated NaOH as mentioned earlier. For coated iron, the initial changes in Δ and ψ arise from the interfacial oxide film dissolution in the highly alkaline cathodic subcoating environment. Since the coating is bonded to this oxide layer, delamination has occurred.

Within Region II, Δ and ψ decline sharply simultaneously. This simultaneous response is typical of a change in surface texture (roughening). The interpretation of the behavior of Δ and ψ is based on the observations of uncoated iron in borate buffer solution after potential cycling, which, in effect, gives a smoother surface. In the potential cycling experiments, Δ and ψ rise in concert, which is the opposite effect from that observed for coated metals undergoing corrosion. Since the phenomenon described in Region I was a thinning of the oxide layer, it is reasonable to postulate the presence of patches of oxide film-free metal. The high-pH environment promotes direct metal dissolution from these patches and the surface becomes roughened.

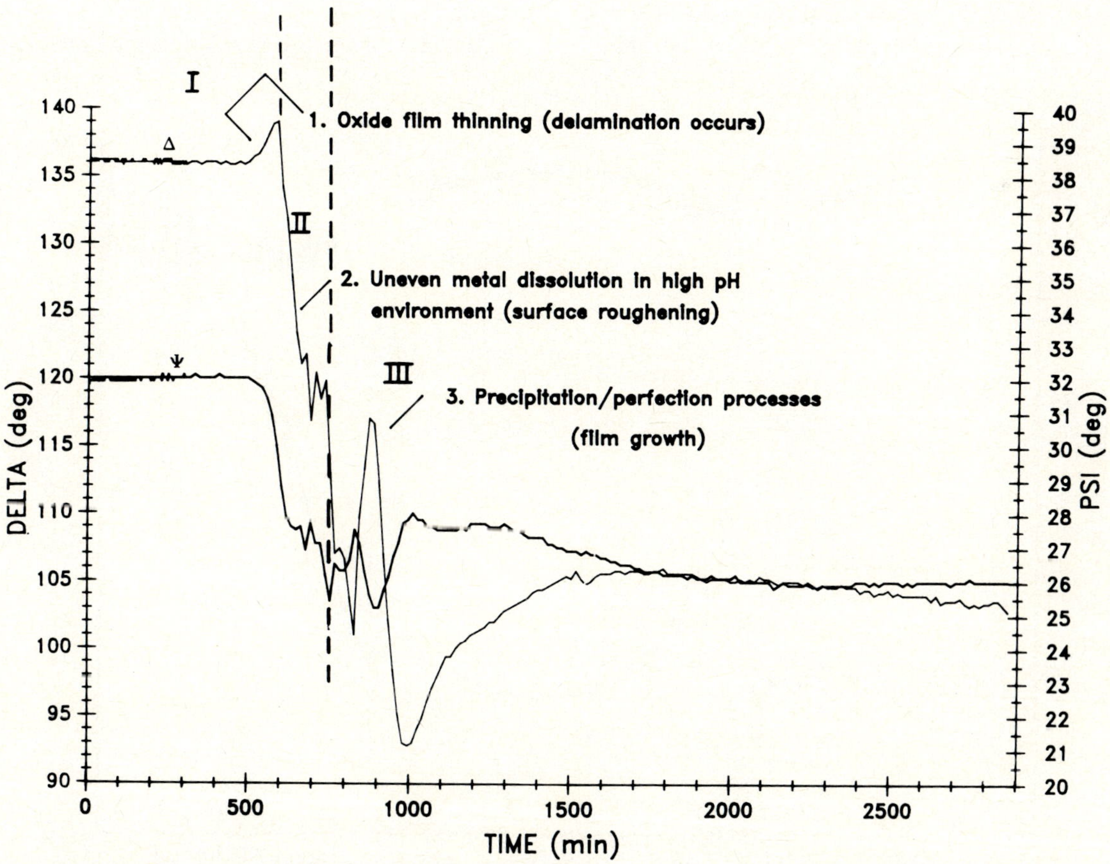

FIGURE 3 — Typical ellipsometric response curve for an acrylic-coated iron specimen undergoing corrosion in dilute NaCl solution.

Within Region III, △ and ψ are again in opposition. The overall result is indicative of an increase in film thickness. Since the phenomena in Regions I and II derive from the dissolution of iron either from the oxide or from the metal itself, it is postulated that that soluble iron species accumulate under the coating. Ultimately, solubility product constants are exceeded and precipitation occurs. The complexity of the ellipsometric response in Region III may arise from one or both of the following factors: the deposition of metastable films that partially redissolve and reprecipitate until a stable format is achieved; or the precipitation of a film that undergoes massive changes in its refractive index with respect to its immersion medium as it thickens.

It was determined that if an experiment of this type was stopped when surface roughening was first detected (Region II), the coating could be readily detached from the substrate surface. This evidence supports the assertion that delamination has occurred in Region I and indicates that the detection of surface roughening can be used to indicate that the delamination process has been completed in the region being monitored ellipsometrically.

If this iron/acrylic system is subjected to heat curing at 200 C for 40 min, a very different ellipsometric response curve is obtained when the system undergoes corrosion. The heat treatment process undoubtedly modifies the final polymer structure. Moreover, the oxide layer under the polymer thickens considerably during the curing process. Measurements by cathodic reduction of stripped specimens show typically a 14-nm oxide film, while an ordinary air-formed film will range between 2.5 and 3.5 nm. The utility of cathodic reduction to estimate oxide film thicknesses has been described previously.[2]

In examining the general features of the △ and ψ parameters for the heat-cured iron/acrylic system in Figure 4, it was noted that they move in opposition to each other throughout the experiment. The suggestion here is that the observed changes are due predominantly to alterations in the oxide layer thickness and that changes in surface texture (roughening) are insignificant. This interpretation is supported by evidence from potential cycling experiments for the absence of surface roughening at the conclusion of these measurements.[2]

A detailed interpretation of this type of curve must remain somewhat speculative until more evidence can be accumulated. In Region I, the initial increase in △ and decrease in ψ at ~1800 min indicate an oxide film thinning in the high-pH environment estimated at ~5 nm (total film ~14 nm). Coating delamination probably occurs at this point, but the metal itself is still protected by ~9 nm of oxide film. Thus, metal dissolution resulting in surface roughening does not occur.

The events beyond 2000 min are not well understood. The sharp rise of △ and decline in ψ maximizing at ~2500 min (Region II) is believed to arise from changes in the polymer near the interface. The substantial rise of △ and leveling of ψ along with the concomitant signal deteriorations in Region III (beyond 5300 min) may be related to the accumulation of a substantial liquid layer at the interface.

Since coatings are bonded to the native-oxide layer

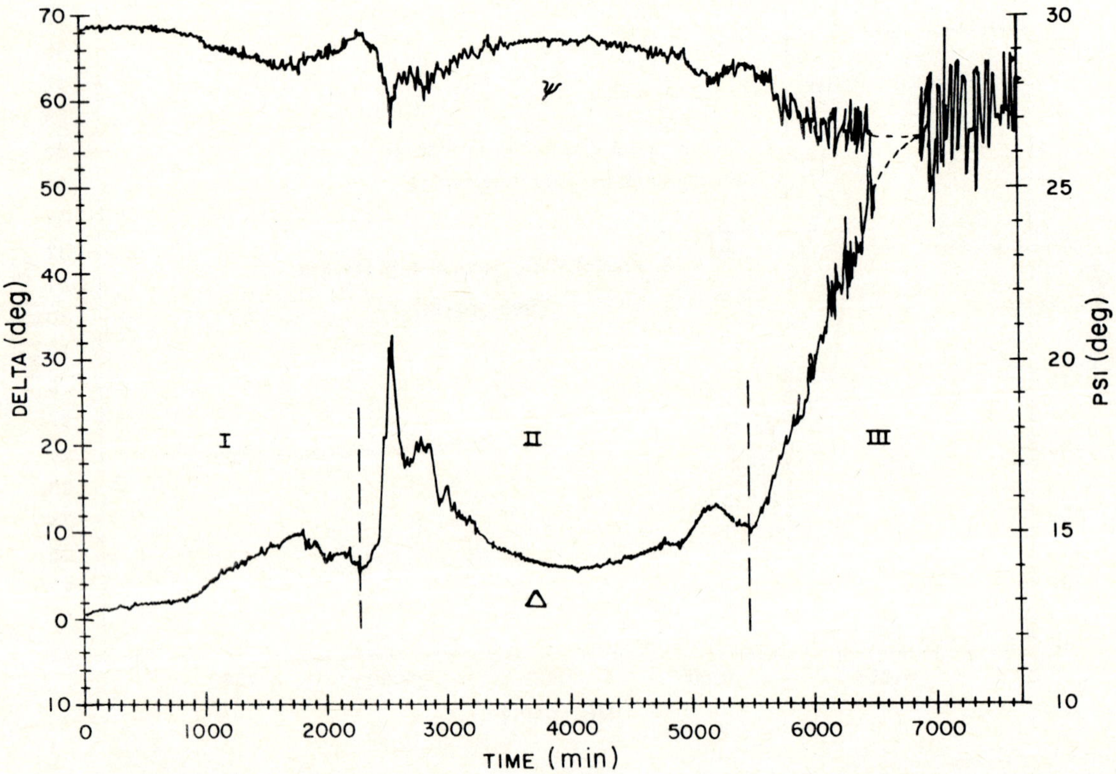

FIGURE 4 — Ellipsometric trace for a heat-cured iron/acrylic system.

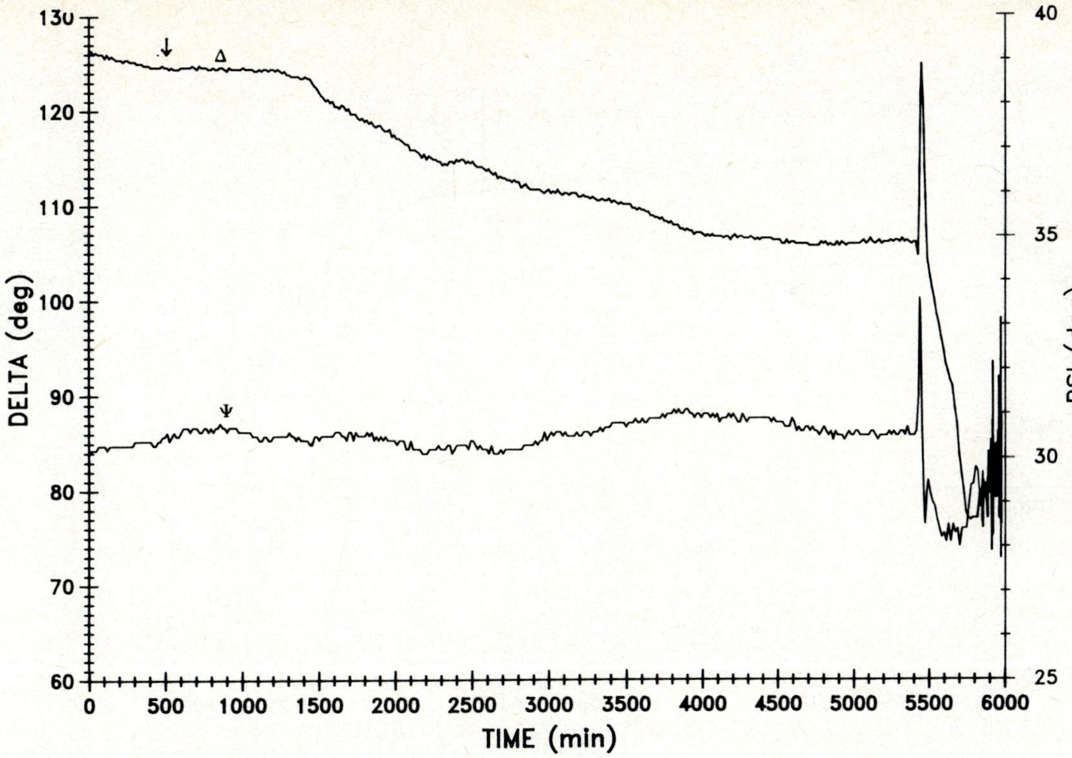

FIGURE 5 — Ellipsometric responses for corrosion in the iron/acrylic system with 1.7% 8-hydroxyquinoline dispersed in the acrylic coating. (Arrow ↓ indicates time of coating puncture.)

on iron and since the ellipsometric technique is useful for examining this interfacial region *in situ*, some interesting results were obtained when corrosion inhibitors were present. A brief description will be given of the results observed for two types of inhibitor systems, an acrylic coating containing 1 to 2% 8-hydroxyquinoline and a two-layer zinc chromate primer/acrylic topcoat system.

Figure 5 shows the case for inhibition by 8-hydroxyquinoline. Note that there is little prepuncture activity but that $\triangle$ and ψ converge very slowly after the coating is damaged. This convergence is interpreted as the development of an interfacial chelate layer. Finally, after ~5000 min, delamination sets in with a probable initial chemical destruction of the chelate/oxide interface, followed by the expected surface roughening phenomenon. This inhibitor delays the onset of delamination but probably does not alter the mechanism.

For comparison purposes, the zinc chromate primer/acrylic topcoat system was studied under comparable conditions. Figure 6 shows a decline in $\triangle$ and a very slight rise in ψ, suggesting a possible ingress of chromate into the interfacial region. Delamination is delayed by about 5000 min, a performance that is rather comparable to that of 8-hydroxyquinoline. Note that when the large ellipsometric changes occur, $\triangle$ and ψ move in concert, signifying the onset of surface roughening without preceding oxide dissolution; i.e., $\triangle$ and ψ do not show an initial divergence, as in Figure 3. Thus, in this system, delamination is delayed and probably occurs by some mechanism other than that of oxide dissolution.

Summary and Conclusions

The ellipsometric-electrochemical technique can provide valuable insights concerning the mechanism and kinetics of cathodic delamination processes. The interpretations of the data from this approach have identified the chemical dissolution of the interfacial oxide as an important mode of coating delamination. It has also been shown that the presence of inhibitor species in the interfacial region can delay the onset of delamination and in some cases subvert the normal mechanism of delamination for a given coating/substrate system. During the course of this work, it was recognized that there may not be a universal mode of coating delamination and that for a given coating/substrate system, one mechanism may predominate over others. This concept includes the possibility that two or more delamination mechanisms may operate simultaneously.

These studies on the interfacial microenvironment that develops as corrosion proceeds supports the findings of previous workers that cathodic areas under the coating become very alkaline. Moreover, the ellipsometric studies on *uncoated* iron in high-pH simulated micro-environments showed that native oxide film dissolution was favored. These results formed the basis for the proposed oxide dissolution mechanism of delamination. In addition, the microanalytical studies on Cl^- distribution in micro-environment samples indicated a mechanistic similarity between corrosion of coated iron and pitting or crevice corrosion insofar as the transport of ionic species was concerned.

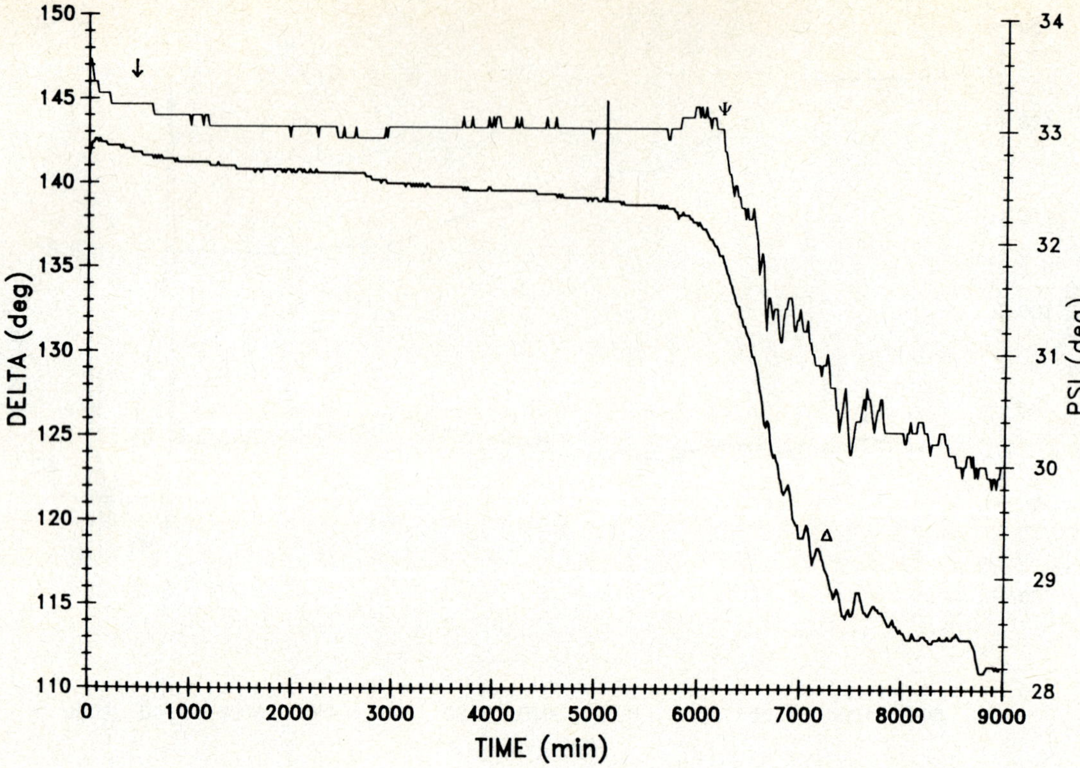

FIGURE 6 – Ellipsometric responses for corrosion in the iron/zinc chromate primer/acrylic topcoat system. (Arrow ↓ indicates time of coating puncture.)

Acknowledgments

The author thanks Jerome Kruger for suggesting the approach and for continued support with many helpful inputs during the course of the work. Appreciation is also extended to P. K. Schenck for automating the ellipsometer and electrochemical apparatus and to N. K. Adams for his valuable assistance in equipment maintenance and data acquisition. The project was partially funded by the Office of Naval Research under contract number NAONR 18-69 NRO 36-082.

References

1. J. J. Ritter, J. Kruger, Surf. Sci., Vol. 96, p. 364, 1980.
2. J. J. Ritter, J. Kruger, "Studies on the Subcoating Environment of Coated Iron Using Qualitative Ellipsometric and Electrochemical Techniques," Corrosion Control by Organic Coatings, H. Leidheiser, Jr., Ed., National Association of Corrosion Engineers, Houston, Texas, p. 28, 1981.
3. J. J. Ritter, M. J. Rodriguez, Corrosion, Vol. 38, No. 4, p. 223, 1982.
4. J. J. Ritter, J. Coatings Tech., Vol. 54, p. 51, 1982.
5. J. J. Ritter, J. Coatings Tech., Vol. 56, p. 55, 1984.
6. H. Leidheiser, Jr., H. J. Konnon, J. Electrochem. Soc., 1983.

The Application of Impedance Spectroscopy to the Evaluation of Corrosion Protection by Inhibitors and Polymer Coatings

M. W. KENDIG, A. T. ALLEN, S. L. JEANJAQUET, and F. MANSFELD*

Abstract

Electrochemical impedance spectra for metals in electrolyte solutions can be obtained rapidly over a wide range of frequencies and for impedance values ranging from several ohms to over $10^6\ \Omega$. This makes the technique applicable for evaluating complicated corrosion processes such as those occurring in neutral aerated media and the corrosion behavior of polymer-coated metals. In addition to the polarization resistance, information regarding surface film formation by inhibitors and water uptake by polymer coatings can be obtained from spectra analysis. Impedance spectroscopy has demonstrated the corrosion protection mechanism of a phosphonic acid/fatty amine mixture as an inhibitor for AISI 4340 steel in NaCl, Na_2SO_4, tap water, and deionized water. It has also shown how surface pretreatment can influence the corrosion resistance of polymer-coated Al AA[(1)] 2024 exposed to NaCl.

Introduction

Electrochemical impedance spectroscopy provides a valuable tool for evaluating corrosion kinetics and associated interfacial phenomena, since the frequency response of the electrochemical impedance spectrum allows for the determination of parameters such as the solution resistance, R_S, the interfacial capacitance, C, and the polarization resistance, R_P.[1,2] In addition, transport across interfacial barrier films such as inhibitor films or polymer coatings can be evaluated from the presence of multiple time constants in the impedance spectrum. This is particularly true for certain inhibitor films[7-9] and for polymer coatings, where a pore resistance, R_{PO}, as determined from the impedance spectra, provides a measure of the integrity of protective polymer coatings.[3-6]

In comparison to conventional DC techniques, additional information is provided by electrochemical impedance techniques in that a true determination of the steady-state corrosion resistance is possible, which is critical for evaluating very slowly corroding systems.[10] The use of a recently described optimized approach[11] has enabled more accurate and rapid evaluation of slowly corroding systems such as coated metals and metals in inhibited neutral environments.

The impedance spectra of AISI 4340 steel rotating-cylinder electrodes (RCE) in neutral solutions containing a phosphonic acid/fatty amine inhibitor (Inhibitor A), and the impedance spectra of polyurethane-coated Al AA 2024 panels having different pretreatments demonstrate important uses of electrochemical impedance spectra for evaluating corrosion protection.

Optimization of the Collection of the Impedance Spectrum

Despite extensive work on the development of analytical methods for extracting corrosion-related data from impedance spectra,[12-16] some problems remained. Specifically, the low and time-dependent corrosion rates and high capacitance of well-inhibited systems made it necessary to extend impedance measurements to very

*Rockwell International Science Center, Thousand Oaks, California.
(1)Aluminum Association (AA), Washington, DC.

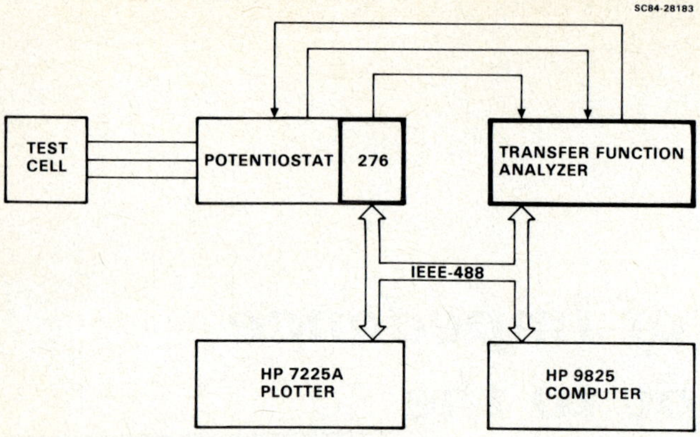

FIGURE 1 — Instrumentation used for the optimized collection of electrochemical impedance spectra.

low frequencies as rapidly as possible, while measuring impedances ranging over several orders of magnitude.

The automated data collection method, developed in this laboratory, optimizes the impedance measurement with respect to both speed and accuracy.[11] Figure 1 shows a schematic of the experimental arrangement. It includes a computer-controlled current amplifier (PAR Model 276 plug-in for the Model 173 potentiostat[(2)]) and a transfer function analyzer (TFA, the Solartron 1174[(3)]) connected to a HP9825[(4)] desktop computer via an IEEE-488[(5)] bus. This arrangement allows the computer to control the gain of the current amplifier and apply filtering. The computer can then set the gain of the current amplifier, depending on the magnitude of the observed impedance, so as to optimize the measurement. For frequencies below 1 kHz, the computer selects the current measuring resistor, R_m, of the Model 276 interface so that R_m relates to the impedance modulus $|Z|$ by:

$$\text{Log}(R_m) \geq \text{int}[\log(3|Z|)] \qquad (1)$$

where the integer function, int, determines the largest integer less than or equal to its argument. This procedure limits the output signal of the interface so as to avoid phase shift errors, but maintains a signal level high enough for accurate measurement of the current. At frequencies down to 1 Hz, ten cycles for current and potential, respectively, are sampled. Below 1 Hz, in addition to optimizing the gain of the interface, the computer signals the interface to introduce a filter for the current measurement and to reduce the data sampling to one cycle. This increases the speed of the low frequency data collection by a factor of 10 without sacrificing the signal-to-noise resolution. Figure 2 summarizes the methodology by presenting the schematic of the program logic. The method provides sufficiently rapid collection of impedance data and dynamic range for the accurate analyses described in this paper.

Experimental Approach

Steel in Inhibited Neutral Solutions

RCEs were prepared from AISI 4340 steel and ex-

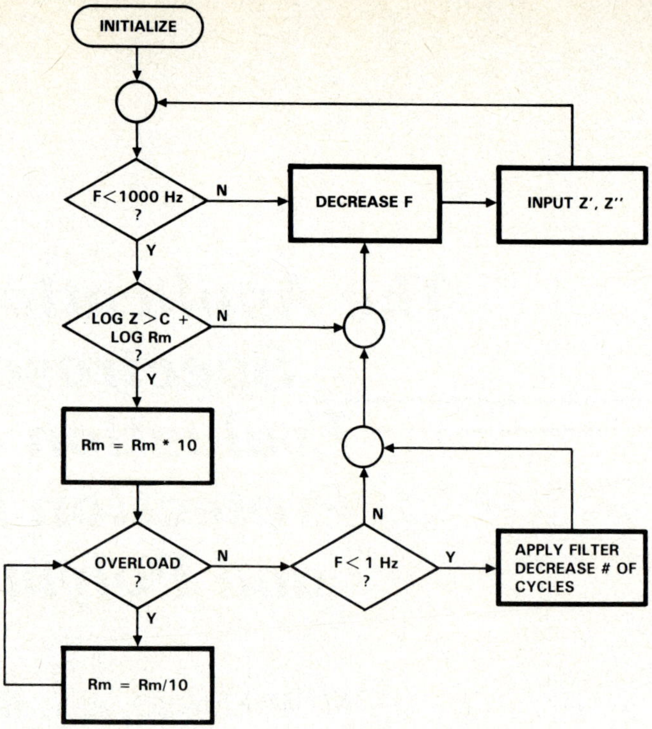

FIGURE 2 — Logic used for the optimized data collection.

periments were carried out with a modified IBM EC 219[(6)] rotating disk electrode system at 144 rad/s (1390 rpm) in solutions open to the air. A concentric Pt band or mesh served as the counter electrode. A Au wire pseudo-reference electrode was capacitively coupled to the saturated calomel reference electrode (SCE) placed in a Luggin probe. This arrangement minimizes high frequency phase shift resulting from possible slow response of the SCE. The impedance measurements were carried out as described previously[1,2] using the optimized approach.[11]

The solutions used for the tests described here included deionized (DI) water, tap water, 0.5 N Na_2SO_4, and 0.5 N NaCl. Each solution contained 0.1% of a mixture of phosphonic acid (trinitrilotrismethylene phosphonic acid) and fatty amine (N-oleyl-1,3-propane diamine), which will be referred to as Inhibitor A.

Coated Al AA 2024

One Al AA 2024 panel was pretreated with a commercial conversion coating and one panel was anodized in H_2SO_4 before coating with polyurethane. These panels were placed in a cell containing 0.5 N NaCl (open to air) for impedance measurements as described previously.[3] The exposed area of the coated Al was 20 cm^2. The impedance was measured periodically over a two-month period of exposure to 0.5 N NaCl.

Results and Analysis

AISI 4340 Steel in Inhibited Neutral Solutions

Figure 3 shows the impedance spectra for the AISI 4340 steel RCE in the neutral solutions containing In-

[(2)-(5)]Registered trade names.

[(6)]Registered trade name.

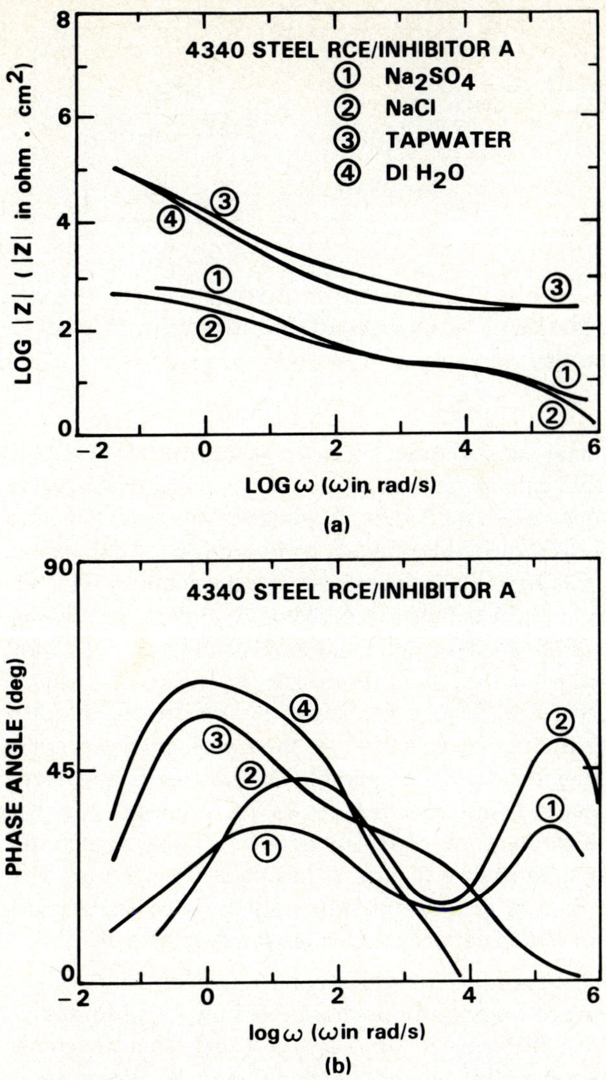

FIGURE 3 — Impedance spectra for AISI 4340 steel RCE after 24 h in neutral solutions containing Inhibitor A: (a) log|Z| vs log ω, (b) phase angle vs log ω.

hibitor A. As can be seen, the impedance for the steel in tap water and DI water is higher than in the chloride and sulfate electrolytes.

Above ω = 10,000 rad/s, the impedance modulus $|Z|$ for the tap water and DI water remains independent of frequency and therefore equals the solution resistance. As the frequency reaches very low values, $|Z|$ approaches the DC limit, R_{DC}, equal to the polarization resistance, R_p, plus the solution resistance, R_s. The DC limits for the tap and DI water are nearly equal and approach 2×10^5 $\Omega\bullet cm^2$, which is in considerable excess of the R_p value of 500 to 1000 $\Omega\bullet cm^2$ observed for the sulfate and chloride solutions [Figure 3(a)].

The results for chloride and sulfate show an additional feature at high frequencies. Above 10^5 rad/s, $|Z|$ drops [Figure 3(a)] and an additional maximum in the phase angle vs frequency curve occurs [Figure 3(b)]. This behavior, which is similar to that observed for polymer coatings,[3-6] is considered to result from the presence of porous inhibitor films.

Procedures have been developed for analyzing spectra containing either single or multiple time constants. The single-time constant data for tap and DI water can be analyzed by fitting the resulting complex impedance, Z, to the expression:

$$Z = R_s + R_p/(1 + (j\omega\tau)^{\alpha}) \tag{2}$$

where: $j^2 = -1$, R_s = solution resistance, R_p = polarization resistance, and $\tau = R_pC$, where C is the interfacial capacitance.

The α is a unitless parameter equal to 1 for ideal capacitive behavior, giving Nyquist plots of complete semicircles in the complex plane. However, for most real systems, $0 < \alpha < 1$. This causes Nyquist plots of circular arcs that have their centers below the real axis in the complex plane. The deviation of the impedance response of electrochemical systems from the ideal behavior has been attributed to a number of phenomena such as "frequency dispersion," etc. Juttner, et al., have recently proposed an explanation based on the concept of blocked surfaces under mass transport.[13] Theoretical impedance spectra derived using this concept have very similar features as those observed experimentally.[7-9]

Data analysis does not require measurement of Z down to the DC limit since it has been shown[14] that the following integral equation determines R_p from a spectrum extending from high frequencies down to ω_m, the frequency at which the out-of-phase impedance, Z'', reaches a maximum:

$$R_p = \frac{4}{\pi} \left| \int_{\infty}^{\log\omega_m} Z'' d\log_e\omega \right|$$

$$= 2.932 \left| \int_{\infty}^{\log_{10}\omega_m} Z'' d\log_{10}\omega \right| \tag{3}$$

The values of C and α are obtained from the following equations:

$$C = 1/\omega_m R_p \tag{4}$$

$$\alpha = \frac{4}{\pi} \tan^{-1}(2|Z''_m|R_p^{-1}) \tag{5}$$

where $|Z''_m|$ is the maximum in the absolute value of the out-of-phase impedance.

The analyses represented by Equations (3) through (5) have been performed for the tap water data and the results, in terms of the parameters R_p, C, α, and in addition, E_{corr}, appear in Figure 4. E_{corr} shows an increasing trend from about −385 mV to about −250 mV after 50 hours of exposure. The corrosion resistance, R_p, is plotted as a conductance, $1/R_p$, which is proportional to the corrosion rate. Initially, $1/R_p$ is near $50(M\Omega\bullet cm^2)^{-1}$. This corresponds to a corrosion current density, I_{corr}, equal to 1 $\mu A/cm^2$, assuming a B value of 20 mV. At 50 h, $1/R_p$ decreased to $2(M\Omega\bullet cm^2)^{-1}$, corresponding to a corrosion current density of 0.04 $\mu A/cm^2$ using the same B value. The capacitance C shows a general increasing trend rising

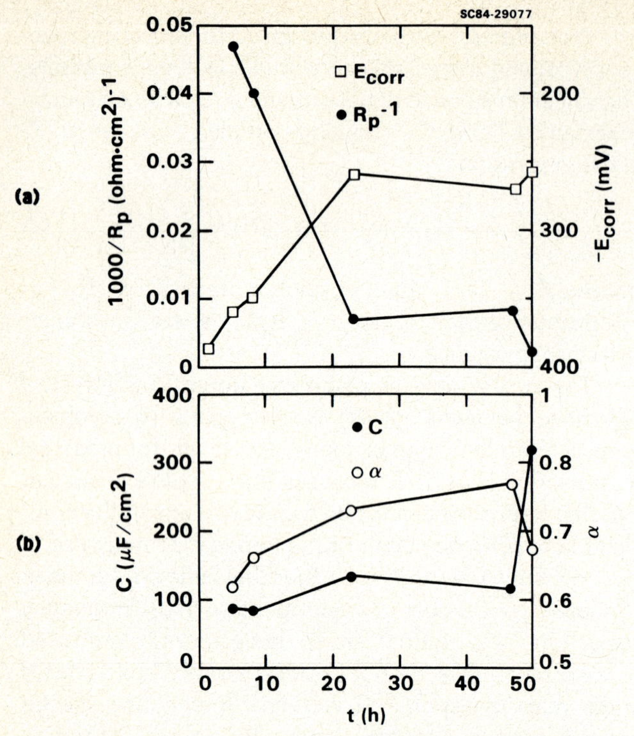

FIGURE 4 — The time dependence of: (a) $1/R_p$ and E_{corr}, (b) C and α for AISI 4340 steel RCE in tap water containing Inhibitor A.

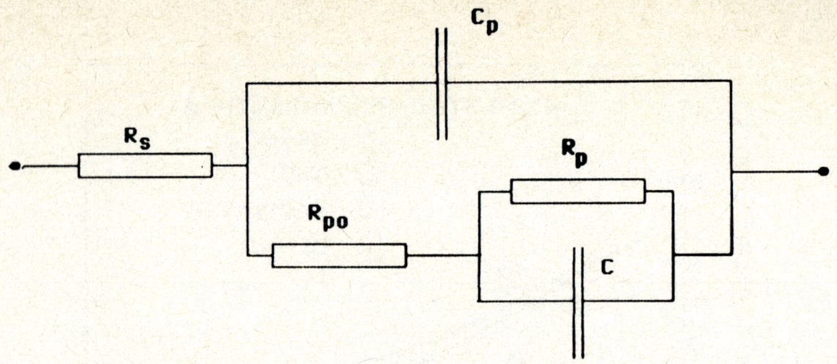

FIGURE 5 — Circuit analog for a filmed electrode.

from below 100 $\mu F/cm^2$ to values between 250 to 350 $\mu F/cm^2$ after 46 h of exposure. The estimated value for C may be influenced somewhat by the presence of an additional but unresolved time constant appearing as a shoulder in the phase angle vs log ω curve [Curve No. 3 in Figure 3(b)] which may be caused by formation of a porous inhibitor film.

The analysis for steel in inhibited chloride and sulfate solutions, where two well-resolved time-constants appear, is more complicated. The circuit analog shown in Figure 5 represents the behavior of the two time-constant system. The capacitance, C_p, represents the inhibitor film. Resistive paths, represented by the pore resistance R_{po}, are due to porous penetration of the film by electrolyte. C is a pseudo-capacitance resulting from the corrosion reaction in the pores of the inhibitor film. In general, C_p and C will both have values of α associated with them because of their nonideality. Analysis of the two time-constant spectra in terms of the elements shown in Figure 5 requires deconvolution of the spectra. At sufficiently high frequencies, and provided that the two processes are sufficiently resolved, Z can be approximated by an equation in the same form as Equation (2), except that R_p and C of Equation (2) are replaced by R_{po} and C_p, respectively. Since the integral analysis for R_{po}, C, and α, using Equation (3), needs to be applied only to the high frequency portion of the spectrum, the risk of error from the contribution to the integration from the lower frequency time constant is minimized. R_s can be obtained conventionally from the high frequency limit. The contribution of C_p and R_s can then be subtracted from the spectrum using an approach described previously.[12] The resulting deconvoluted spectrum represents the response for the remaining elements R_{po}, R_p, and C, which can in turn be determined with an additional application of the analysis provided by Equations (4) and (5). As an example, Figure 6 shows the frequency dependence of Z'' for steel in NaCl containing Inhibitor A. Before deconvolution of the spectrum, the two maxima in the $-Z''$ vs log ω plot result from the presence of the two time-constants (Curve 1 in Figure 6). After the deconvolution procedure, only the spectrum for the low frequency process remains (Curve 2 in Figure 6). The results of the analysis appear in Figure 7, which shows the Nyquist plot for the calculated and observed impedance. The small high frequency semicircle (Figure 7) having a $C_p = 1.5\ \mu F/cm^2$ and $R_{po} = 17\ \Omega\text{-}cm^2$ has the same form as typical spectra for filmed surfaces such as polymer-coated steel.[3-6]

The results for Na_2SO_4 containing Inhibitor A, which have been reported previously,[7,8] show $1/R_p$ to remain near 2 $(K\Omega \cdot cm^2)^{-1}$ (480 $\mu m/y^{(7)}$) with some decrease during the 50-h exposure period (Figure 8). The porosity of the film apparently increases with time since $1/R_{po}$ increases during the test. The capacitance for the film, C_p, decreases slightly from about 1.8 to 1.6 $\mu F/cm^2$. The calculated capacitance, C, in parallel with the polarization resistance, R_p, takes values that are initially at 500 $\mu F/cm^2$ and increase to over 2000 $\mu F/cm^2$. It is apparent, then, that a capacitance C does not provide a valid representation for the frequency dependence of the impedance representing the reaction in the pores. A more detailed "blocked surface" model[13] incorporating a Warburg element that describes mass transport seems more suitable for this analysis. For the present case, the model suggests that mass transport-controlled reduction of oxygen occurs in the pores of the inhibitor film.

The results for steel in 0.5 N NaCl containing the inhibitor appear in Figure 9. In contrast with the sulfate results, the film porosity, as measured by $1/R_{po}$, is initially high, but decreases to the values observed for the sulfate by the end of the 50-h test. Despite the fact that E_{corr} shows a decrease with time, the corrosion rate as measured by $1/R_p$ shows little change and remains at 1.5 $(K\Omega \cdot cm^2)^{-1}$ (360 $\mu m/y^{(8)}$), a value slightly below that observed for the sulfate [Figure 8(a)]. The film capacitance C_p shows a steady increase throughout the

(7)-(8) Assuming a B value of 20 mV.

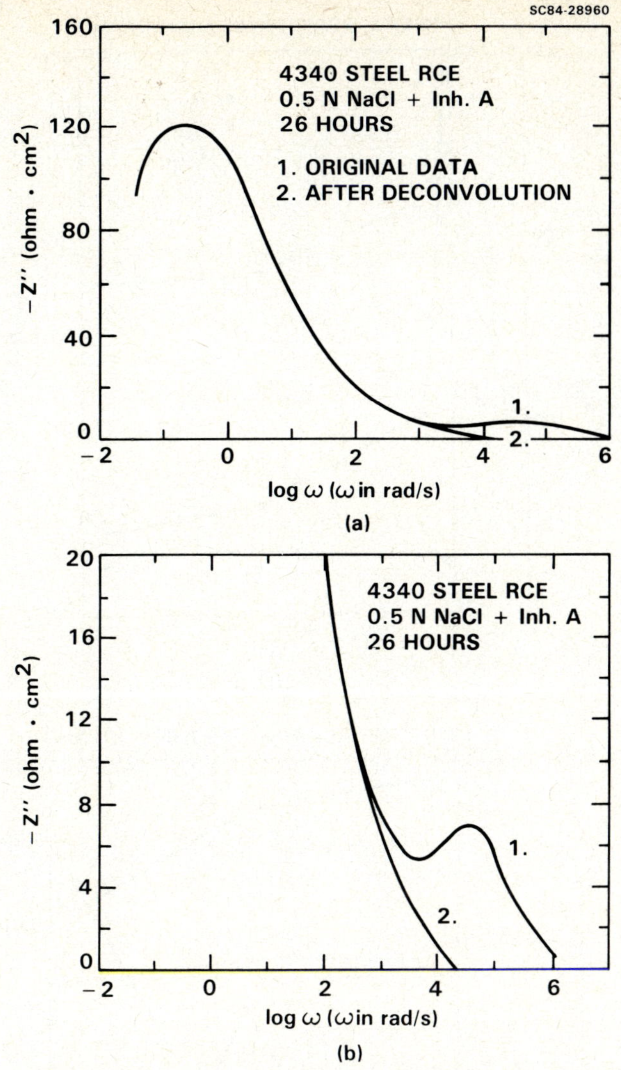

FIGURE 6 – Z'' vs log ω for AISI 4340 steel RCE in 0.5 N NaCl containing Inhibitor A. The curves shown are: (1) before, and (2) after deconvolution.

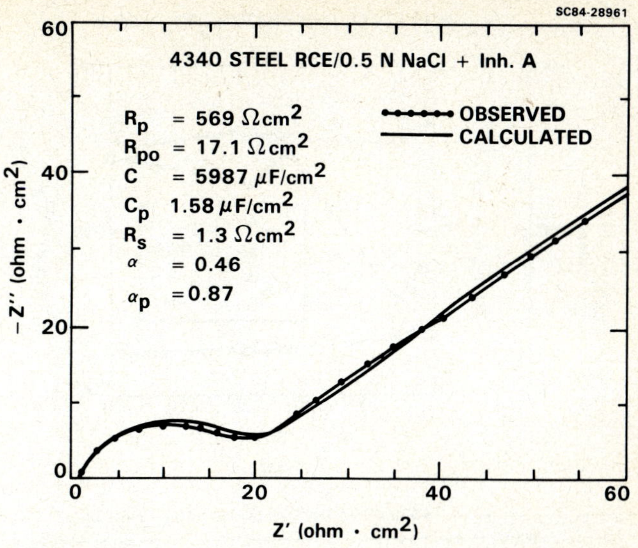

FIGURE 7 – Nyquist plot of the observed and calculated spectra for AISI 4340 steel RCE after 26 h in 0.5 N NaCl + Inhibitor A.

test [Figure 9(b)], but the interfacial capacitance, C, shows somewhat erratic behavior. As for the sulfate data, C_p for the chloride has values between 1 and 3 $\mu F/cm^2$, while C takes values in the 1000 to 6000 $\mu F/cm^2$ range. Also, the values for α associated with C remain near 0.5, as is characteristic for diffusion-controlled processes, while the values of α_p for C_p are initially near 0.9.

It is clear from the impedance data that inhibition of corrosion of AISI 4340 steel in aerated chloride and sulfate solutions by the phosphonic acid/fatty amine mixture involves the formation of a porous film having capacitance values between 1 and 3 $\mu F/cm^2$.

The similarity between the impedance spectra for the steel in chloride and sulfate solutions containing Inhibitor A and polymer coatings having a significant amount of porosity can be seen from the following section, which describes the results for polymer-coated Al AA 2024.

Polyurethane-Coated Al AA 2024

The Al AA 2024 panels were pretreated with a commercial conversion coating or a H_2SO_4 anodization and then coated with polyurethane before exposure to the 0.5 N NaCl electrolyte. The impedance spectra for the pretreated and coated Al samples appear in Figures 10 and 11. The polymer-coated panel with a conversion coating pretreatment develops a considerable amount of porosity after one week of exposure, resulting in the development of two time-constants, as seen from the appearance of two maxima in the phase angle vs frequency curves in Figure 10(b). The pore resistance R_{po} for the H_2SO_4 anodized and coated panel, however, does not become sufficiently low for the observation of the second time-constant (Figure 11).

Table 1 summarizes the R_{po} and R_p results for coated Al specimens. After one day, R_{po} for the conversion-coated and H_2SO_4 anodized and coated Al specimens have nearly equal values of 720 and 620 MΩ, respectively. These high, nearly identical values for R_{po}, observed after one day, show that there are no differences between the specimens with regard to the presence of initial defects. In fact, the conversion-coated specimen, which showed the greatest decrease in the R_{po}, actually had an initially lower capacitance, suggesting that it may have a somewhat thicker coating than the H_2SO_4 anodized specimen. Nevertheless, both specimens were considered to have a nominal 55-μm thickness. The specimen pretreated by conversion coating clearly shows the greatest rate of degradation, as evidenced by the greater drop in R_{po} and R_p throughout the two-month test (Table 1). After two months of exposure, the porosity of the conversion-coated specimen, as measured by $1/R_{po}$, is nearly 500 times greater than that for the H_2SO_4 anodized specimen, and the corrosion rate for the conversion-coated specimen, as estimated from R_p^{-1}, is more than 35 times that for the H_2SO_4 anodized specimen (Table 1).

Additional information can be obtained since C_p is proportional to the dielectric constant ϵ of the coating

$$C_p = \epsilon\epsilon_o A/d \quad (6)$$

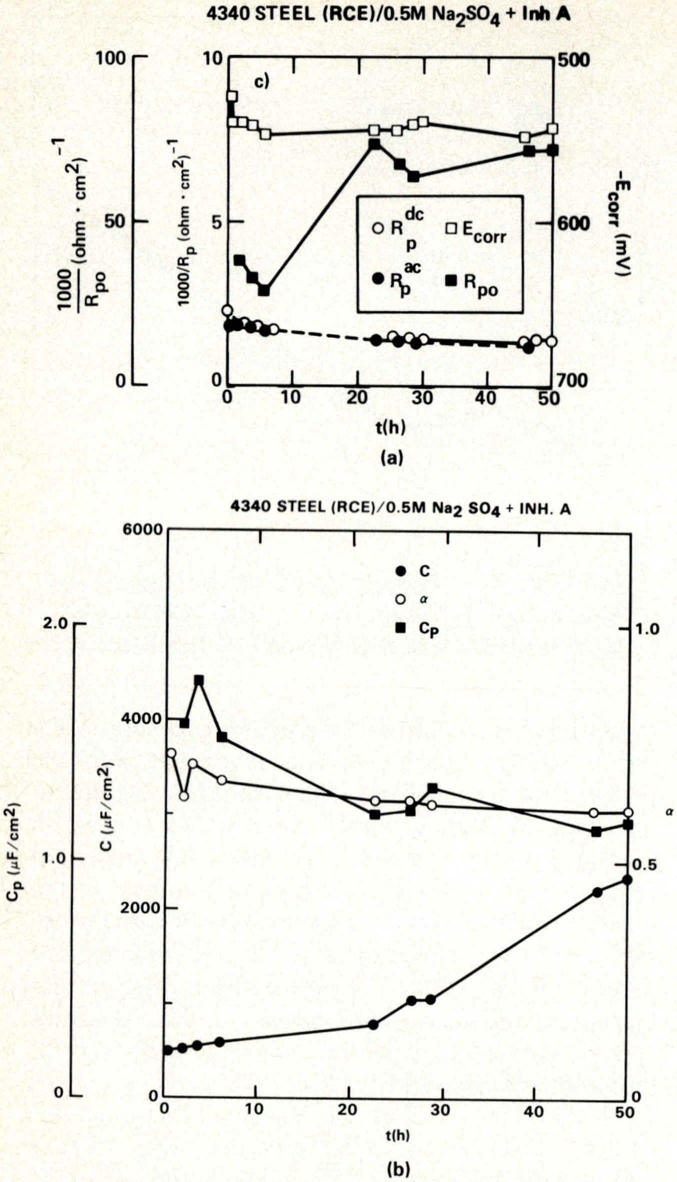

FIGURE 8 — The time dependence of: (a) $1/R_{po}$, $1/R_p$, and E_{corr}, (b) C_p, C, and α for AISI 4340 steel in 0.5 N Na_2SO_4 containing Inhibitor A.

where: ϵ_O = 8.85 10^{-14} F/cm, A = measured surface area, and d = coating thickness.

The capacitance used in Equation (6) must be that of the polymer dielectric and must not include faradaic or double layer components. This may be ensured by using sufficiently thick coatings or by evaluating C_p only from the ω^{-1} region of the impedance spectrum. Since ϵ is very sensitive to water uptake, a number of studies have used the capacitance to estimate the water uptake by polymer coatings.[17-19] Using the equation of Brasher and Kingsbury,[17]

$$C_p(t)/C_p(0) = 80^x \qquad (7)$$

where x = volume fraction of water, Leidheiser and Kendig[20] showed that for sufficiently thick coatings, such that

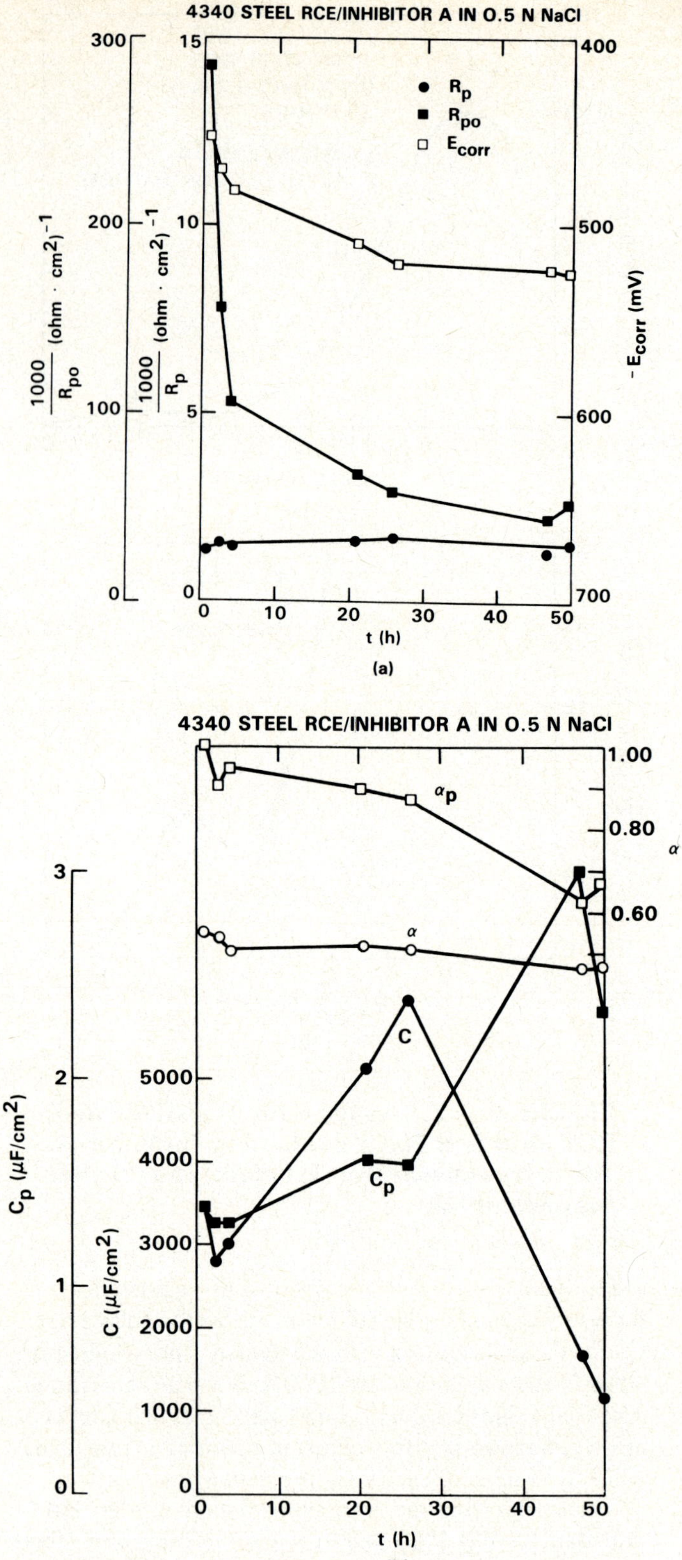

FIGURE 9 — The time dependence of: (a) $1/R_{po}$, $1/R_p$, and E_{corr}, (b) C_p, C, α, and α_p for AISI 4340 steel in 0.5 N NaCl containing Inhibitor A.

the measured capacitance at 1000 Hz could be ensured as being that of the dielectric coating, the calculated water content compared well with that observed from gravimetric data. By measuring the impedance for a coated

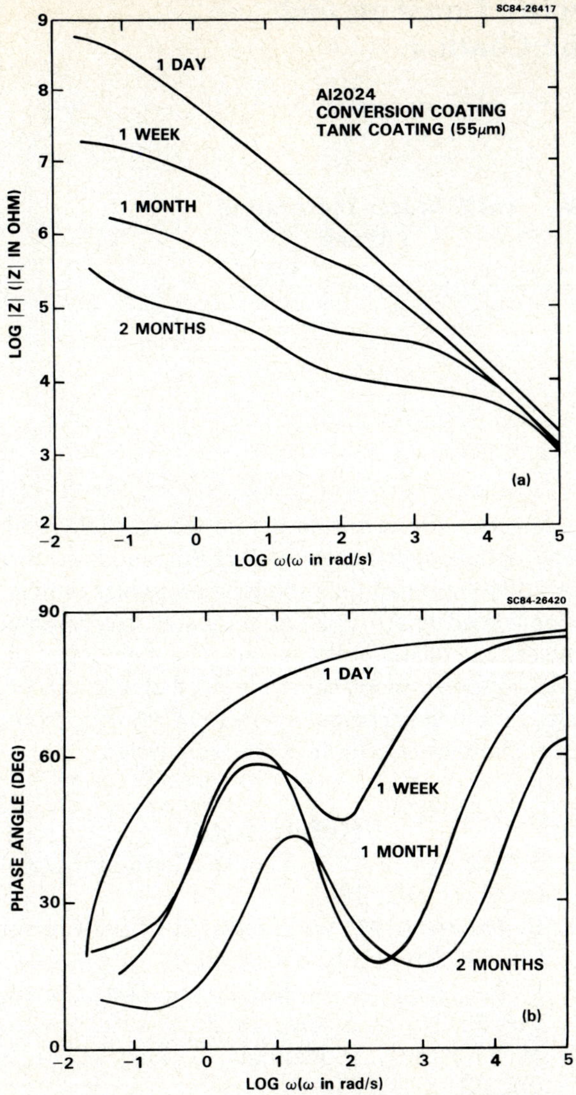

FIGURE 10 – The impedance spectra for conversion-coated and polyurethane-coated Al AA 2024: (a) log |Z| vs log ω, (b) phase angle vs log ω.

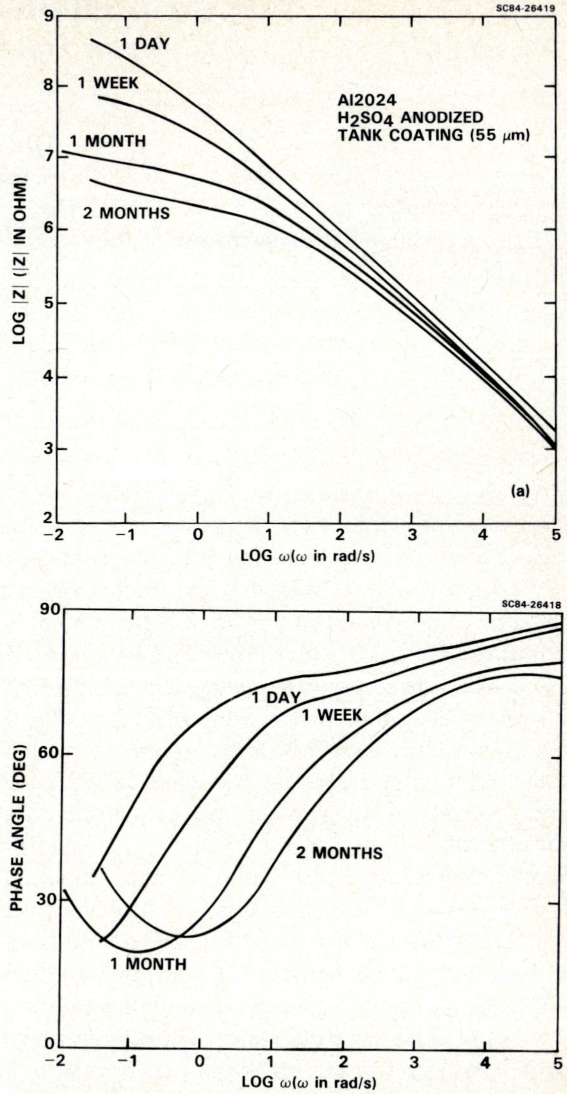

FIGURE 11 – The impedance spectra for H_2SO_4 anodized and polyurethane-coated Al AA 2024: (a) log |Z| vs log ω, (b) phase angle vs log ω.

TABLE 1 – Pore Resistance and Polarization Resistance Determined From the Impedance Spectra of Polyurethane-Coated Al AA 2024 (A = 20 cm²)

	Pore Resistance R_{po}		Polarization Resistance R_p	
Time	Conv. Coated (Ω)	H_2SO_4 Anodized (Ω)	Conv. Coated (Ω)	H_2SO_4 Anodized (Ω)
1 day	7.2×10^8	6.2×10^8	$>10^9$	$>10^9$
1 week	6.0×10^5	1.1×10^8	1.5×10^7	$>10^8$
1 month	4.2×10^4	9.7×10^6	1.6×10^6	$>10^7$
2 month	6.2×10^3	2.6×10^6	8.2×10^5	$>3 \times 10^6$

TABLE 2 – Effective Dielectric Constant and Water Uptake of the Coating

Pretreatment	Dielectric Constant 1 Day	Dielectric Constant 1 Week	vol% Water Increase in 1 Week
Conversion coating	8.7	19.3	18
Anodized H_2SO_4	10.3	16.5	11

metal over a sufficiently wide range, the ω^{-1} region can be determined and therefore the polymer capacitance can be extracted with great confidence. Sufficiently well-resolved ω^{-1} regions for the spectra in Figures 10(a) and 11(a) appear above 10^4 rad/s (1591 Hz) for the first day and first week data. C_p was readily evaluated using a least squares fit of the spectra from which the effective dielectric constants were obtained, assuming a 55-μm coating thickness and 20-cm^2 area (see Table 2). The volume fractions of water taken up by the coatings between one day and one week are shown in Table 2 as calculated from Equation (7). It must be noted that Equation (7) assumes a uniform distribution of water and is valid only for low volume fractions. Nevertheless, it is clear that the conversion-coated and polyurethane-coated material shows a greater uptake after one week despite its initially lower capacitance. The influence of the substrate chemistry on the uptake of water is consistent with earlier speculation that the water concentrates at the interface.[19]

Summary and Conclusions

The examples discussed here demonstrate the use of impedance spectroscopy for evaluation of corrosion protection by inhibitors and coatings. Impedance spectra are particularly well-suited for evaluating slowly corroding systems having high interfacial impedances. New optimized methods give the technique sufficient speed and dynamic range for evaluating polymer coatings and metals in the presence of corrosion inhibitors in neutral media. In addition to information concerning the corrosion rate, impedance spectra provide details of the protective mechanism and the nature of protective films.

Inhibition of AISI 4340 steel by phosphonic acid/fatty amine occurs by a film-forming mechanism in neutral chloride and sulfate. The corrosion rate of AISI 4340 steel in inhibited chloride is only slightly less than that in sulfate, but much higher than in tap or deionized water. No film formation could be detected for corrosion inhibition in deionized water, but some evidence of incipient film formation in tap water appears as a shoulder in the phase angle spectrum for the steel in tap water.

The results for the polyurethane-coated Al show that surface pretreatment can have a dramatic effect on coating integrity, the corrosion rate of the substrate, and the water uptake by the coating. A H_2SO_4 anodization of Al AA 2024 improved the protective properties of polyurethane-coated Al AA 2024 as compared to a commercial conversion coating pretreatment. The great value of impedance spectroscopy lies in the fact that reactions occurring at the metal/coating interface or resulting from different interfacial chemistries can be detected.

References

1. F. Mansfeld, M. W. Kendig, Corrosion, Vol. 37, No. 5, p. 301, 1981.
2. F. Mansfeld, M. W. Kendig, S. Tsai, Corrosion, Vol. 38, No. 11, p. 570, 1982.
3. F. Mansfeld, M. W. Kendig, S. Tsai, Corrosion, Vol. 38, No. 9, p. 478, 1982.
4. M. W. Kendig, F. Mansfeld, S. Tsai, Corros. Sci., Vol. 23, p. 317, 1983.
5. F. Mansfeld, M. W. Kendig, "Evaluation of Protective Coatings with Impedance Measurements," Proc. 9th. Int. Congr. on Metal. Corros., Toronto, Canada, Vol. 3, p. 74, 1984.
6. M. W. Kendig, F. Mansfeld, S. Tsai, Materials Performance, Vol. 23, No. 6, p. 37, 1984.
7. F. Mansfeld, M. W. Kendig, W. J. Lorenz, J. Electrochem. Soc., Vol. 132, p. 290, 1985.
8. W. J. Lorenz, F. Mansfeld, M. W. Kendig, "Corrosion Inhibition in Neutral Aerated Media. I. Theoretical Considerations," Proc. 9th. Int. Congr. on Metall. Corros., Toronto, Canada, Vol. 4, p. 110, 1984.
9. F. Mansfeld, M. W. Kendig, A. T. Allen, W. J. Lorenz, "Corrosion Inhibition in Neutral Aerated Media. II. Experimental Results," Proc. 9th. Int. Congr. on Metall. Corros., Toronto, Canada, Vol. 1-368, 1984.
10. F. Mansfeld, M. W. Kendig, Corrosion, Vol. 37, No. 5, p. 301, 1981.
11. M. W. Kendig, A. T. Allen, F. Mansfeld, J. Electrochem. Soc., Vol. 131, p. 935, 1984.
12. M. W. Kendig, E. M. Meyer, G. Lindberg, F. Mansfeld, Corros. Sci., Vol. 23, p. 1007, 1983.
13. K. Juttner, W. J. Lorenz, M. W. Kendig, F. Mansfeld, to be published.
14. M. W. Kendig, F. Mansfeld, Corrosion, Vol. 39,

No. 11, p. 466, 1983.
15. F. Mansfeld, M. W. Kendig, Werkst. u. Korros., Vol. 34, p. 397, 1983.
16. K. Hladky, L. M. Callow, J. L. Dawson, Br. Corr. J., Vol. 15, p. 20, 1980.
17. D. M. Brasher, A. H. Kingsbury, J. Appl. Chem., Vol. 4, p. 62, 1954.
18. R. E. Touhsaent, H. Leidheiser, Jr., Corrosion, Vol. 28, No. 12, p. 435, 1972.
19. M. W. Kendig, H. Leidheiser, Jr., J. Electrochem. Soc., Vol. 123, p. 982, 1976.
20. H. Leidheiser, Jr., M. W. Kendig, Corrosion, Vol. 32, No. 2, p. 69, 1976.

Computer Controlled AC Impedance Measurements for the Evaluation of Anti-Corrosion Pipeline Coatings

J. D. KELLNER*

Abstract

AC impedance measurements can be used for the determination of the expected performance of pipeline coatings for protection against corrosion. The techniques described in this paper for the evaluation of these coatings fall into two broad categories: (1) the nondestructive determination of the cathodic disbondment by measurement of the capacitance of the intentional holiday as a function of primer type, temperature, electrolyte composition, and holiday size; and (2) the determination of the dielectric properties of the coating as it is affected by water absorption.

Introduction

The importance of cathodic disbondment for the efficiency of anticorrosion protection by all types of organic coatings is well known. For coatings that are generally used under cathodic protection conditions, as are pipeline coatings, it is important to understand the process of cathodic disbondment and to develop rapid methods for its measurement.

Several workers, including Leidheiser,[1] have discussed the role of the hydroxide ion that accumulates at the boundary of a defect site. This accumulation results in an alkaline condition that may be responsible for the disbondment. Alexander[2] reported the effects of voltage, electrolyte composition, and other variables on the cathodic disbonding of pipeline coatings. These tests are time consuming and are not easily performed as a function of time. In order to study the phenomenon, a method for the nondestructive determination of the disbondment by a measurement of the capacitance at the defect site is reported.[3] The work cited used 2-in. (5.1-cm) diameter pipe samples and a potential step technique to determine the double layer capacitance, while the present work reports results over a 2-h period using coated steel plates that are easy to prepare and an AC impedance measurement of the capacitance. Although the relationship of the disbondment to the capacitance may not be linear or precise, the capacitance of pipeline coating samples on steel plates is shown in the present work to be an acceptable indication of the relative magnitude of disbondment to be expected using longer, more difficult tests. The capacitance measurement may be accomplished by obtaining the AC impedance spectrum of the sample. The AC technique may be used to monitor the capacitance and the effect of several variables on the extent of cathodic disbonding. The variables that are studied include experimental primer formulations, electrolyte composition, voltage, and defect size.

The second technique for the evaluation of pipeline coatings by AC impedance analysis uses the dielectric properties of the insulating coatings. Although much smaller in magnitude, the capacitance that forms across an intact coating can be used to monitor the coating's effectiveness. This is a result of the large difference in the dielectric constant of water and the insulating coating. A very small amount of water that penetrates the coating or the pores or defects in the coating can produce large enough changes in the capacitance and resistance to be detected by the use of AC techniques. The author has reported[4] the measurement of this capacitance increase of 2-in. (5.1-cm) pipe samples for several different pipeline coatings and has speculated on the mechanism of water penetration into these coatings. Again, in the present work, the samples are steel plates and the purpose

*The Kendall Company, Lexington, Massachusetts.

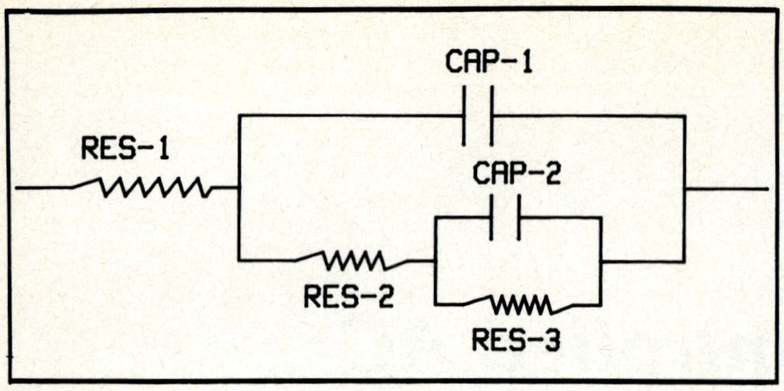

FIGURE 1 – Equivalent circuit of coated steel plate.

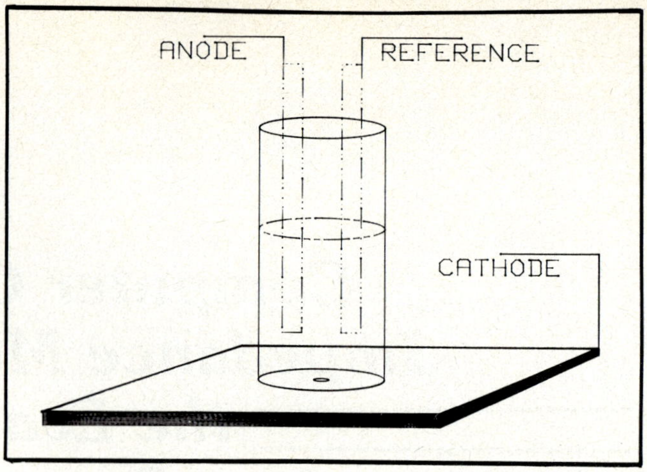

FIGURE 2 – Sample configuration of coated steel plate with holiday.

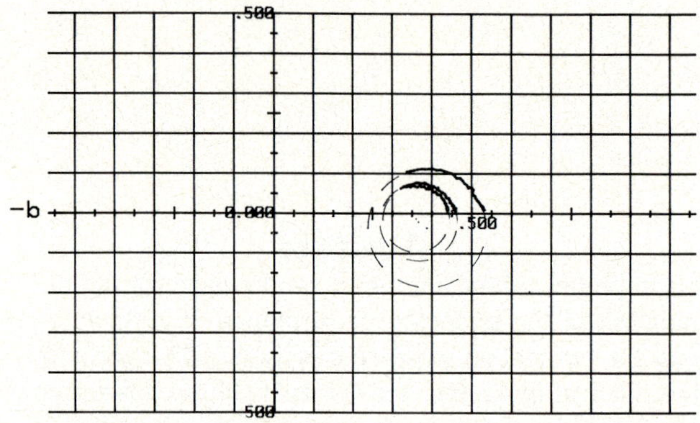

FIGURE 3 – Example of data for primer C, DIN electrolyte, – 1.50 V, 1 min, 60 min, and 90 min.

was to understand the mechanism of water penetration of pipeline coatings and to determine whether the water was available for a corrosion reaction under the coating.

General equivalent circuit models for coated steel systems have been discussed by Mansfeld, Kendig, and Tsai.[5] For a coating system with an intentional holiday, the equivalent circuit can be represented as shown in Figure 1. The resistor marked RES-1 represents the solution resistance, and CAP-1, the capacitance of an intact coating. Because pipeline coatings are generally quite thick, CAP-1 is small, in the range of about 10 pico (p) farads (F)/cm^2. The resistor marked RES-2 is the resistance through the coating, which is close to zero since there is an intentional holiday. Finally, the interface reaction at the holiday can be represented by a capacitance and a resistance in parallel. These are represented by CAP-2, the double layer capacitance, and RES-3, the transfer resistance. The double layer capacitance is much larger than CAP-1, about 20 to 40 μF/cm^2. The interface reaction in some cases may be replaced by a Warburg impedance if pore diffusion is present.

For a thick coating system without a holiday, as in the water penetration experiments, the double layer capacitance is close to zero and RES-3 is quite large.

Experimental Procedure

Sample Preparation

Stainless steel plates [3 × 5 in. (7.6 × 12.7 cm)] are used with a solvent cleaning only. The plates were primed and the tape applied after a 5-min drying time, and rolled with a 4-lb (1.8-kg) roller, applied once in two directions. For the cathodic disbonding measurements, a circular hole, or holiday, was cut into the coating with a hole punch, and all adhesive and primer was removed from the holiday area with solvent. For the water penetration measurements, two layers of tape were used, with varying degrees of overlap; no defects were applied. A 2-in. (5.1-cm) diameter polyvinylchloride (PVC) tube, 6 in. (15 cm) long, was cemented to the plate with silicon sealing compound, making a water tight seal. The PVC tube was filled with electrolyte and the counter electrode and the saturated calomel reference electrode were placed in it. During the test, a stream of air was used for stirring and to saturate the electrolyte with oxygen. A schematic representation of the experimental cell is shown in Figure 2.

AC Measurements

Cathodic Disbonding Experiments. AC impedance spectra of the coated steel samples were obtained as described earlier.[4] The data was fed into a computer for data manipulation and graphics display. Frequency scans from 1100 to 10 Hertz (Hz), 30 steps/dec, 10 mV AC potential, and integration times of three cycles were used. The sample panels were held at cathodic potentials of – 1.20 to – 2.00 V vs SCE during the entire test, with frequency scans taken every 2 min for 2 hours (h). Computer programs were written for real-time plotting and curve fitting of the Nyquist data to circle arcs and the calculation of the maximum frequency, diameter, and center locus for the circle. After the experimental data was plotted, the best fit circle was drawn to ensure that a proper calculation had been made. A typical plot is shown in Figure 3. The center of the circle did not appear on the real axis; nevertheless, the capacitance was calculated from the circle diameter and the maximum frequency as discussed in Mansfeld.[5]

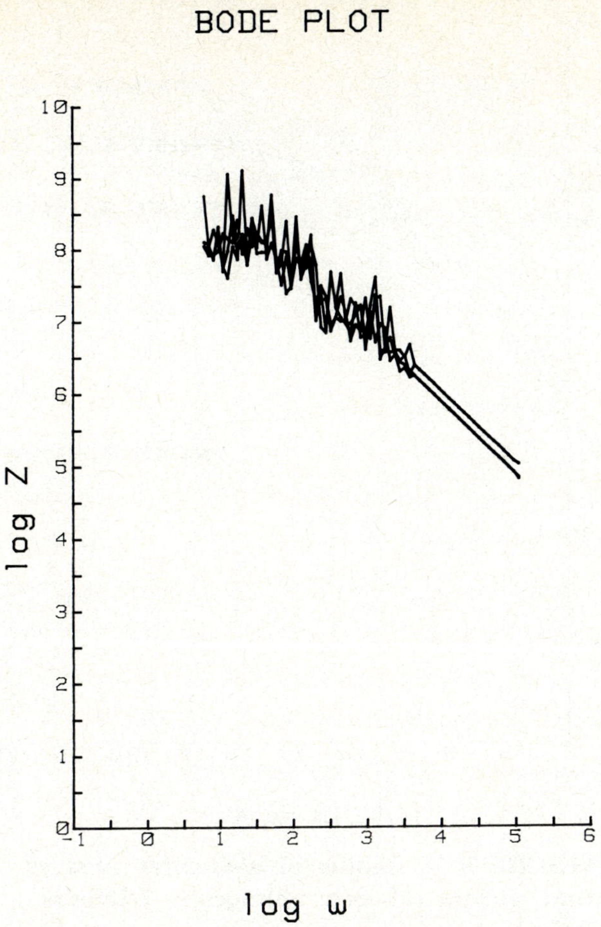

FIGURE 4 – Impedance spectrum of two samples: one with no overlap and one with a 1-in. (2.5 cm) overlap, intact coating.

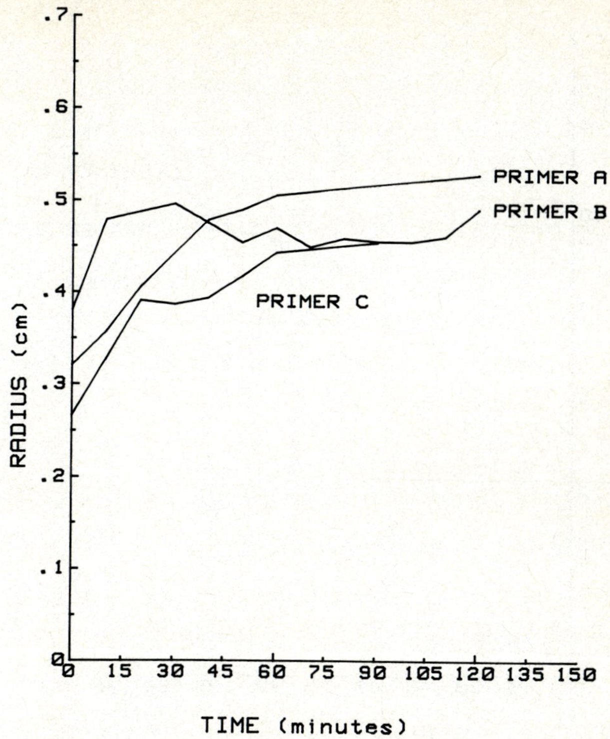

FIGURE 5 – Radius of disbonded area vs time for three different primer formulations in ASTM electrolyte.

Water Penetration Experiments. In this case, frequency scans were from 20 kHz to 1 Hz, with integration times at 3 cycles. The samples were electrically stressed at −6.0 V vs SCE as in ASTM G-9.[(1)] Because of the great resistance of the samples, AC voltages were 100 mV while the samples were held at −1.5 V vs SCE. Since the polarization resistance was many orders of magnitude larger than the cell resistance, a Bode plot was used to determine the sample capacitance. The slope of the linear portion[5] of this plot was approximately equal to −1. A linear regression was used to determine the slope and intercept precisely, in order to calculate the sample capacitance. Coating resistance was also obtained from the horizontal portion of these plots, where the slope is zero. A typical run for two samples is shown in Figure 4, with a capacitance of 150 pF, 115 pF, and a resistance of about 200 megohms.

Results and Discussion

Cathodic Disbonding Experiments

Effect of Primer. The capacitance measured is assumed to be proportional to the area of substrate surface exposed to the electrolyte; this area is assumed to be the disbonded area. The area is then calculated from the double layer capacitance and the radius of an equivalent circle is plotted vs time. Several primers exhibiting different results in full scale tests of cathodic disbondment using ASTM G-8 were used. Figure 5 shows the results for the ASTM electrolyte, which consists of 1% each of sodium chloride, sodium sulfate, and sodium carbonate. The results for the DIN[(2)] electrolyte, consisting of 3% sodium chloride, are shown in Figure 6. The results of the nondestructive 2-h test indicated the same relative order of cathodic disbondment as did the full 30-day test.

Effect of Electrolyte. The disbonded areas are slightly larger for the different primer types in the DIN electrolyte than in the ASTM electrolyte. Also, there was a rapid increase in the disbonded radius in all cases during the first 40 min and then a flattening of the curve in the ASTM electrolyte that was absent in the case of the DIN electrolyte.

Effect of Defect Size. The results for a smaller holiday size of 1/8 in. (0.32 cm) diameter can be seen in Figure 7. While the rate of the radius increase was about the same after 120 min, the radius was smaller for the smaller holidays.

Effect of Voltage. At a voltage of −1.2, no increase in the radius was observed, as can be seen in Figure 8. There was virtually no difference between the data for −1.4 and −1.7 V; however, there was some indication of a greater radius at −2.0 V, 60 min into the test. This behavior with voltage increase is also seen in

[(1)]ASTM, Philadelphia, Pennsylvania.

[(2)]Deutsches Institut fur Normung, Berlin, Federal Republic of Germany.

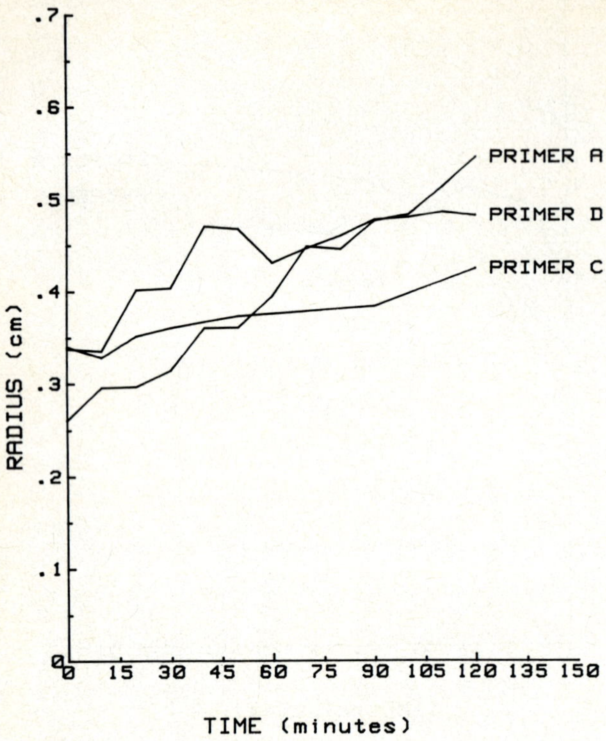

FIGURE 6 – Radius of disbonded area vs time for three different primer formulations in DIN electrolyte.

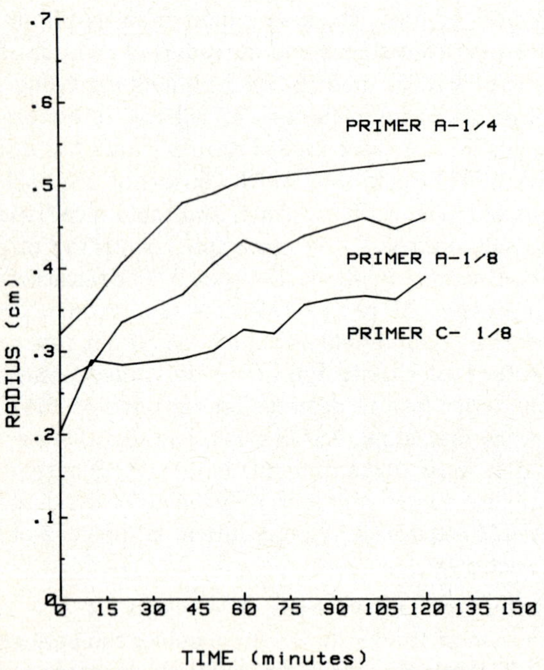

FIGURE 7 – Radius of disbonded area vs time in ASTM electrolyte, using 1/4 inch and 1/8 inch diameter holidays (0.64 cm and 0.32 cm).

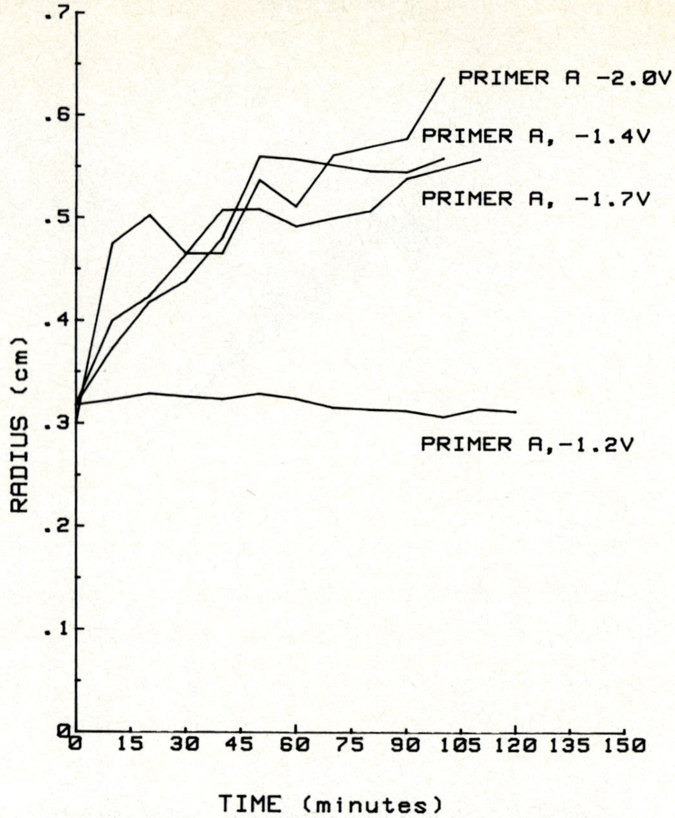

FIGURE 8 – Radius of disbonded area vs time at four different voltages in DIN electrolyte.

the large scale test, except that disbondment is observed at − 1.2 V.

Water Penetration Experiments

Two samples were used to test the effect of the overlap on the water penetration characteristics of a wrapped tape pipeline coating. One sample prepared on a steel plate had virtually no overlap, and the other had a 1-in. (2.5-cm) overlap. The expected capacitance can be calculated from[6]:

$$C = \frac{A \cdot k}{4 \cdot Pi \cdot d} \qquad (1)$$

where A is the area of coating in contact with the electrolyte, d is the average thickness of the coating, and K is the dielectric constant. Because only capacitance differences are significant in this test, it is unnecessary to determine the dielectric constant by using mercury. The expected capacitance can be calculated from Equation (1) since the capacitances resulting from the different thicknesses are in parallel. Also, the coating is not uniform, consisting of a polyethylene layer and a butyl adhesive layer; therefore, a dielectric constant of 2.3 for pure poly-

TABLE 1 – Areas and Calculated Capacitances

Sample	Total Area (cm^2)	Overlap Area (cm^2)	Calculated Capacitance (pF)
No overlap	23.4	1.42	120
1 in. (2.5 cm) overlap	23.4	14.2	83

TABLE 2 – Capacitance (pF) Observed

Sample/Day	1	6	10	24	32
No overlap	120	136	148	148	149
1 in. (2.5 cm) overlap	84	99	115	108	114

TABLE 3 – Percent Change of Capacitance

Sample/Day	6	10	24	32
No overlap	13	23	23	24
1 in. (2.5 cm) overlap	18	37	29	36

ethylene cannot be used. If an estimate of 3.7 is assumed, the capacitance expected can be calculated by adding the areas of different thicknesses:

$$C = \frac{A_1 \cdot K}{4 \cdot Pi \cdot d_1} + \frac{A_2 \cdot K}{4 \cdot Pi \cdot d_2} \qquad (2)$$

where the subscript "1" refers to those regions of no overlap and subscript "2" refers those areas that are overlapped. Table 1 lists the areas and capacitance calculated in this way and those actually observed are listed in Table 2. The percentage increase of the capacitance for each time interval is shown in Table 3.

The initial capacitances are very close to the calculated values, which simply indicates that 3.7 is a good average value for the dielectric constant. Both samples exhibit increases in capacitance with exposure to the ASTM electrolyte under electrical stress of −6.0 V vs SCE. The extent of the increase is not affected by the overlap area; in fact, the sample with the smallest overlap had the smallest increase in capacitance, indicating lower water absorption. Because of the extreme sensitivity of this test, a very small amount of water in the vapor phase absorbed by the coating could cause the observed capacitance increases. However, as shown in previous work,[4] for fusion bonded epoxy (FBE) and coal tar coatings, when the capacitance increase is accompanied by a resistance decrease, and the AC impedance spectrum shows evidence of a Warburg diffusion term, it can be shown that the water that penetrates is liquid water that can reach the metal surface and be available for a corrosion reaction. Therefore, of greater importance than the capacitance increase during this test is the fact that the resistance remains high and there is no evidence for diffusion in pores. Furthermore, the increase in capacitance and therefore, increase of water absorption stops after 30 days. Because the resistance of the coating remains extremely high (about 100 megohms), it is apparent that the water that has been incorporated into the coating is not available to support ionic conduction or a corrosion reaction.

Conclusions

1. The cathodic disbondment of pipeline coatings can be determined in a very short *in-situ* test. This nondestructive test shows the same relative order of cathodic disbondment as the large scale destructive test.

2. Primer formulations and defect size can significantly affect the rate of cathodic disbondment of protective coatings. Of less significance are the factors of voltage up to −2.0 V, and electrolyte composition.

3. Experiments with tape wrap overlaps indicate that water penetration of these coatings does not continue for more than 30 days, and that this water is not available for conduction or corrosion reactions.

References

1. H. Leidheiser, Jr., Indust. Eng. Chem. Prod. Res. Dev., Vol. 20, p. 547, 1981.
2. S. H. Alexander, "Variables That Influence Cathodic Disbonding Test Results," NACE 25th Annual Conference, Paper No. 13, 1969, National Association of Corrosion Engineers, Houston, Texas.
3. J. D. Kellner, 5th Inter. Conf. Intern. and Ext. Prot. Pipes, p. 365, 1983.
4. J. D. Kellner, ASTM Monograph, "Nondestructive Testing," to be published.
5. F. Mansfeld, Corrosion, Vol. 36, No. 5, p. 301, 1981.
6. Dielectric Materials and Applications, A. R. Von Hippel, Ed., Technology Press of MIT and John Wiley and Sons, New York, New York, 1954.

Porosity and Galvanic Corrosion in Precious Metal Electrodeposits

R. J. MORRISSEY*

Abstract

In a corrosive environment, precious metal electrodeposits often behave as barrier coatings. A galvanic cell, however, is formed at any coating imperfection in which the substrate is exposed. This paper describes the effects of deposit porosity on galvanic corrosion. Factors affecting the deposit porosity are also described, as well as conditions conducive to pore closure with minimum deposit thickness.

Introduction

A *barrier coating* is one that provides protection as long as its own physical integrity is maintained. Precious metal electrodeposits are ordinarily applied on less corrosion-resistant substrates. Thus, they function as barrier coatings, but with the added characteristic that a galvanic potential difference can be generated between the coating and the substrate to which it is applied. If a localized area of basis metal is exposed through some coating deficiency, its corrosion rate will be increased by the presence of the galvanic couple, and can be orders of magnitude higher than would have been the case had no coating been applied.

Ideally then, to provide complete corrosion protection, a precious metal electrodeposit should be completely defect-free. This is, of course, impossible in actual practice, but there are ways of minimizing the number and severity of defects that occur during electrodeposition.

The presence of defects, particularly very small defects, in electrodeposits is usually termed *porosity*. Numerous tests have been devised over the years for the determination of porosity, usually involving exposure of the plated piece to a corrosive atmosphere or solution to produce visible spots of corrosion product at the pore sites. In one variation of this test, the plated piece is made the anode in an electrolytic cell to accelerate the corrosion reaction.

Spot tests of the chemical and electrographic varieties described here can pinpoint the exact locations of pores or groups of pores in the coating. This is particularly useful information in examining selectively plated work or where the substrate surface finish contributes to the porous condition of the deposit. Corrosion products, however, tend to spread away from the actual pore sites; for this reason, then, it is difficult to use spot tests to obtain quantitative estimates of actual deposit coverage. This is particularly true when the deposits examined are thin. There is a class of porosity tests, called *electrolytic tests,* that are particularly useful in examining the coverage of relatively thin electrodeposits. These methods examine various aspects of the galvanic couple produced when the plated piece is immersed in a known electrolyte.

Theoretical

The electrochemistry of galvanic corrosion has been well researched.[1-3] For the galvanic current, that is, the current passing between both members of the couple at the corrosion potential, Mansfeld[3] writes:

$$\log i_G^A = \frac{\phi_{corr}^C - \phi_{corr}^A}{B_A + B_C} +$$

$$\frac{B_C}{B_A + B_C} \log i_{corr}^C \frac{A^C}{A^A} +$$

*Technic, Inc., Providence, Rhode Island.

$$\frac{B_A}{B_A + B_C} \log i^A_{corr} \qquad (1)$$

While for the corrosion potential itself,

$$\phi_G = \frac{B_A \phi^C_{corr} + B_C \phi^A_{corr}}{B_A + B_C} + \frac{B_A B_C}{B_A + B_C} \log \frac{i^C_{corr} A^C}{i^A_{corr} A^A} \qquad (2)$$

In the above expressions, ϕ^C_{corr} and ϕ^A_{corr} are the corrosion potentials of pure cathode and anode materials, and ϕ_G, the corrosion potential of the galvanic couple. The corrosion current densities of pure cathode and pure anode materials at their repsective corrosion potentials are represented by i^C_{corr} and i^A_{corr}, and i^A_G represents the corrosion current density of the galvanic couple. B_C and B_A are the Tafel constants for pure cathodic and anodic materials, and A^C and A^A are the cathode and anode areas.

Expressions (1) and (2) are considerably more complicated in appearance than they are in actual practice. Consider briefly the reactions occurring at an idealized galvanic couple in a mild aerated electrolyte. At the anode, oxidation of the basis metal occurs according to:

$$M \rightleftharpoons M^{z+} + ze^- \qquad (3)$$

Precious metals tend to act as oxygen electrodes in mild electrolytes, so that at the cathode:

$$1/2\, O_2 + H_2O + 2e^- \rightleftharpoons 2OH^- \qquad (4)$$

The total galvanic reaction is the sum of Expressions (3) and (4); it follows that either of these reactions can become the rate-limiting step in the entire galvanic process. The cathode process[4] requires that dissolved oxygen be available in the solution at the cathode surface. If such is the case, the reaction will be limited by the rate of electron transfer at the anode. This condition is referred to as *activation-controlled.* If, on the other hand, sufficient oxygen is not available at the cathode, the reaction becomes controlled by the rate at which oxygen can diffuse through the electrolyte. This is termed *diffusion-controlled.*

Expressions (1) and (2) can be simplified for either of these cases. For the corrosion potential, the simplified expressions are as follows:

For activation control:

$$\phi_G = K - \frac{B_A}{B_A + B_C} \log A_A \qquad (5)$$

For diffusion control:

$$\phi_G = \phi^A_{corr} - B_A \log A_A \qquad (6)$$

Expressions (5) and (6) show that the corrosion potential of the galvanic couple, that is, the potential that is read when the actual couple is compared to a reference electrode in the same solution, is a function of the logarithm of the anode area fraction, that is, the degree of porosity. This is true whether the couple is activation- or diffusion-controlled. In either case, a linear plot is obtained, but the slope will vary depending on the degree of diffusion control of the cathode reaction.

Experimental and Results

In any event, it is possible by fairly straightforward experimentation[4] to prepare calibration curves of corrosion potential vs log anode area fraction using known standards, and from these calibration curves to determine the porosities of unknown specimens by simply measuring their corrosion potentials in the same electrolyte. Figures 1 and 2 show calibration curves prepared for gold and palladium coatings on copper in 0.1 M NH_4Cl electrolyte. The slopes of these two plots are quite different, indicating that gold-copper couples are activation-controlled, while palladium-copper couples are diffusion-controlled *in the same electrolyte.* To explain this seeming anomaly, it is useful to examine the electrode behavior of gold, palladium, and copper in 0.1 M NH_4Cl in greater detail.

Figure 3 is a plot showing cathodic polarization

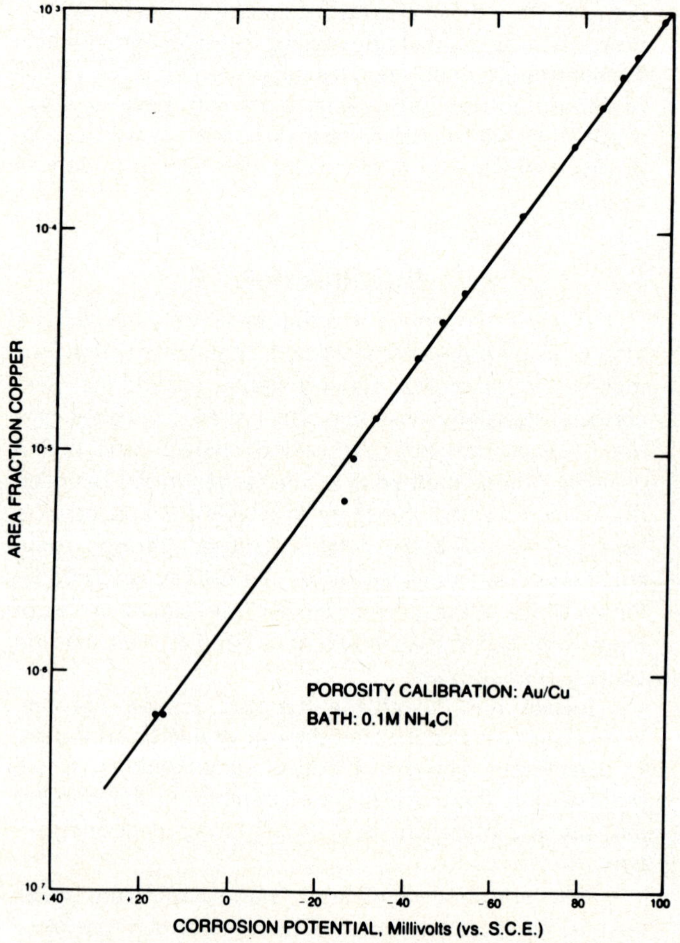

FIGURE 1 — Porosity calibration: gold on copper in 0.1 M NH_4Cl.

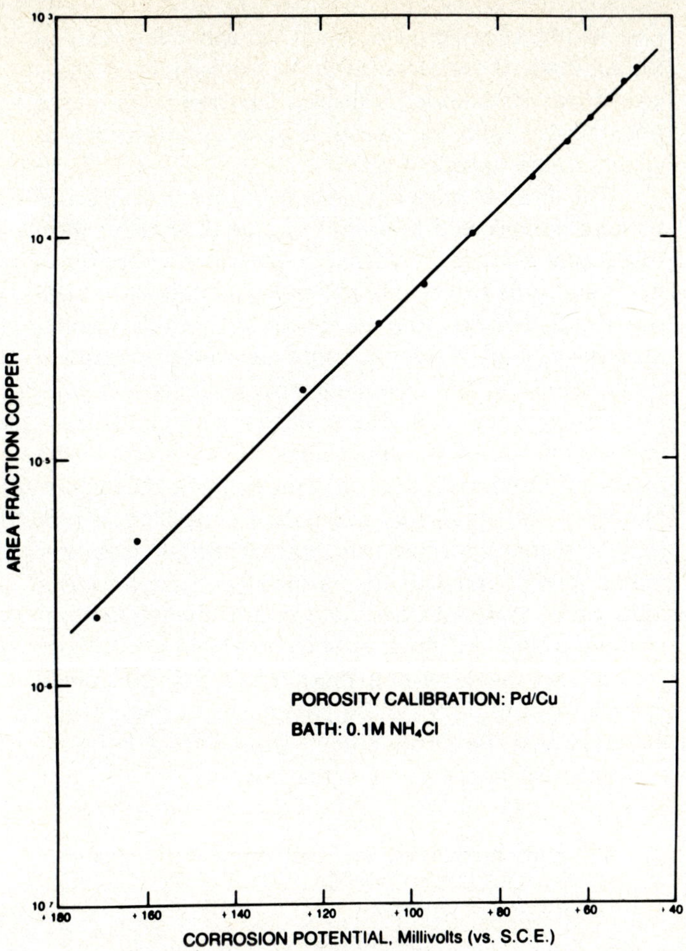

FIGURE 2 – Porosity calibration: palladium on copper in 0.1 M NH_4Cl.

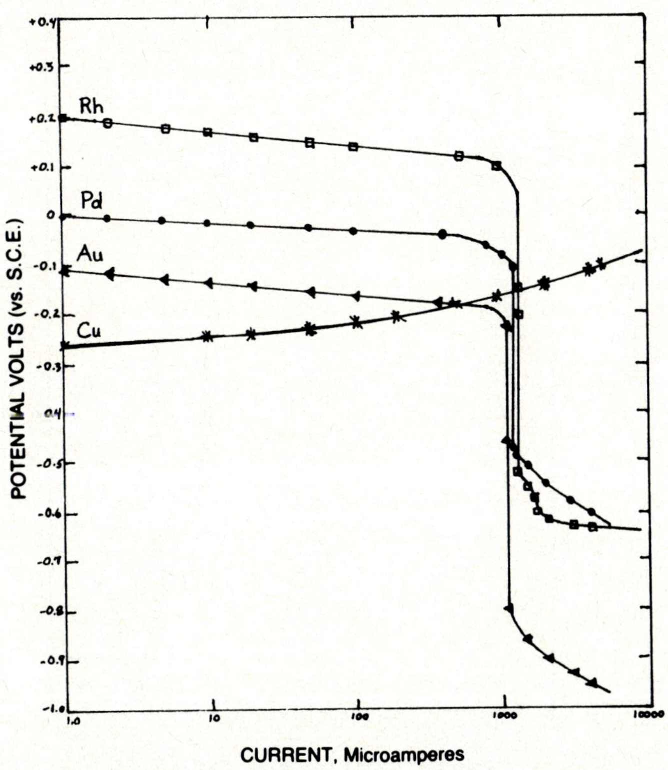

FIGURE 3 – Polarization curves in 0.1 M NH_4Cl.

curves of 50 cm^2 coupons of rhodium, palladium, and gold, together with an anodic polarization curve of a 2.5 cm^2 specimen of copper, in 0.1 M NH_4Cl. This sort of overlay plot of cathodic and anodic curves together is called an *Evans diagram*.[6]

The polarization curves shown for rhodium, palladium, and gold in Figure 3 are all characteristic of oxygen-electrode behavior. Note particularly the long near-vertical drops present in each of these. In these regions, the potential can be swept half a volt or more with no increase in current: this represents diffusion limitation of the cathode reaction. The point of intersection of the copper curve with palladium and rhodium occurs in this diffusion-limited region, while the point of intersection with gold occurs before diffusion-limitation. This accounts for the different slopes of the porosity calibration curves shown in Figures 1 and 2.

Ideally, a precious metal electrodeposit should be as free of defects as possible; it should also be able to seal over the defects such as do occur in minimum thickness. As might be expected, some deposits fulfill these functions markedly better than others, and this has led to a reexamination of the physical properties of precious metal deposits, and an attempt to correlate these properties with variables of the deposition process. Craig and coworkers[7] plated thick foils of gold from a number of electrolyte types, and observed variations in their crystallography, morphology, density, and bulk resistivity. Morrissey and Weisberg[8] surveyed the crystallography of gold deposits from a number of electrolyte and brightener systems, and found that several of these were highly oriented crystallographically with respect to the plane of the substrate. These authors then determined curves of porosity vs deposit thickness for several strongly oriented gold deposits, and found that both the covering power and the slopes of the porosity-thickness curves for these deposits were related to the crystal orientation. Specifically, deposits in which crystal planes of high packing density (low-index planes) were oriented parallel to the substrate surface showed high covering power (low porosity at low deposit thickness) and a rather slow decrease of porosity with increasing thickness. Deposits having less densely packed crystal planes (high-index planes) parallel to the substrate surface showed poorer covering power, but a more rapid rate of pore closure with increasing thickness. Some illustrations of this type of behavior in (111), (100), and (110) oriented gold deposits are shown in Figure 4. The center plot in Figure 4 is particularly interesting in that it shows a definite bend separating two distinct growth regimes. X-ray diffraction measurements of these specimens showed a transition from pure (100) to strong (111) orientation in these specimens with increasing deposit thickness, which was attributed to gradual exhaustion of the brightening agent during the plating run.

The effect of deposit crystallography is easily explained. A densely packed crystallographic plane can fit more atoms in a single unit area than one of lower packing density. Hence, it will grow more slowly per each atom added than its less densely packed counterpart.

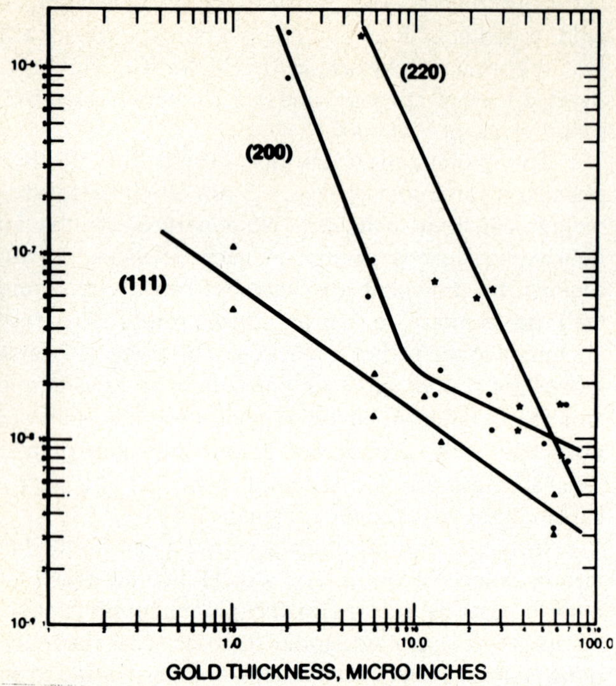

FIGURE 4 – Porosity vs thickness curves for various bright gold electrodeposits.

Considering lateral growth, that is, growth in the plane of the substrate, those planes with the highest rate of lateral growth per atom added will show the most rapid rate of pore closure with increasing deposit thickness; this is what is found experimentally.

The effect of grain size on porosity in electrodeposits is more complicated than that of crystallography. Morrissey and Weisberg[9] find that in general the porosity vs thickness curve for an unbrightened gold deposit on copper is of a z-shaped form as shown in Figure 5. Rehrig, Leidheiser, and Notis[10] found that pulsed deposition reduced the grain size of unbrightened gold deposits, with there being almost a linear correlation between the maximum grain size of the deposit and the reciprocal of the pulse "off" time, with the "on" time being held constant. Morrissey and Weisberg used pulsed deposition to prepare specimens for porosity and diffraction measurements, and found that the resulting porosity-thickness plots were displaced downward and to the left with increasing pulse "off" time, as shown in Figure 6. Diffraction measurements failed to detect any significant change in crystallography with pulse parameters in these specimens, so that the displacement observed had to be attributable to reduction of the grain size.

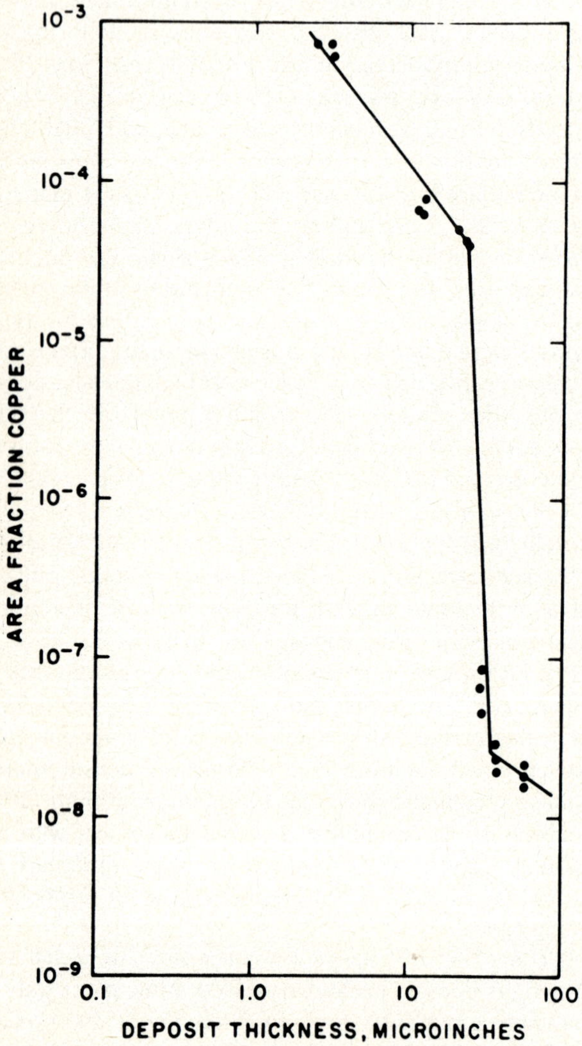

FIGURE 5 – Porosity vs thickness curve for unbrightened gold electrodeposit.

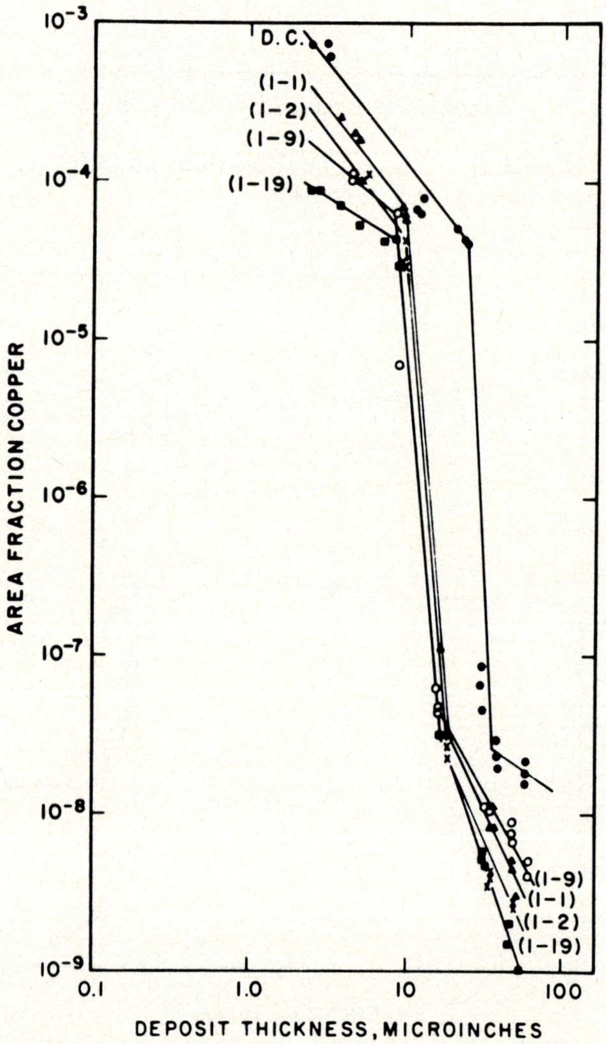

FIGURE 6 – Porosity vs thickness curves for pulse-plated gold electrodeposits.

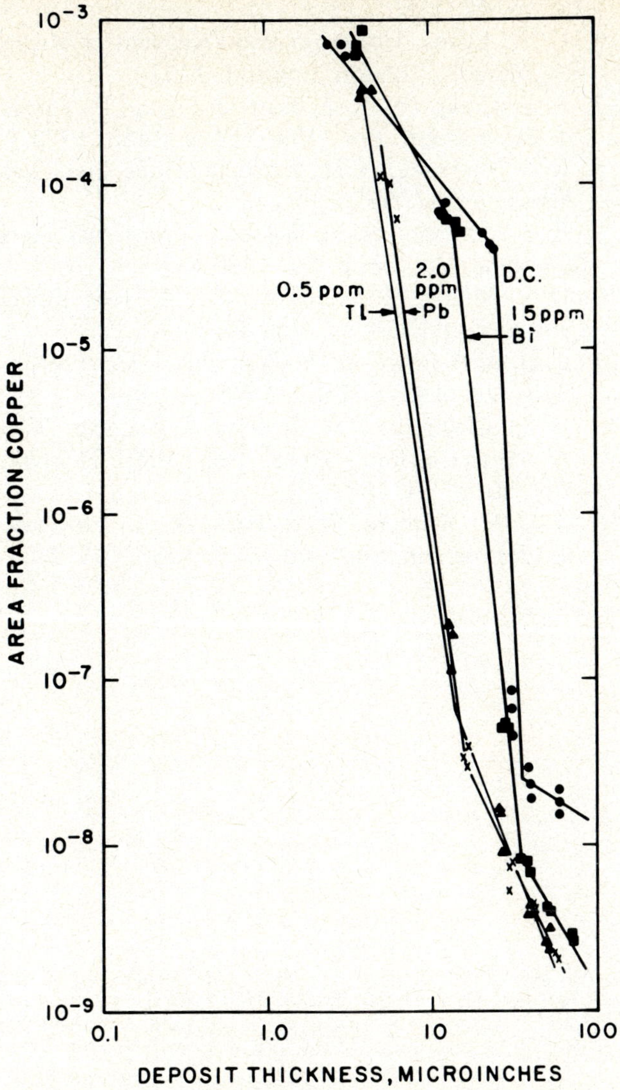

FIGURE 7 – Porosity vs thickness curves for semibrightened gold electrodeposits.

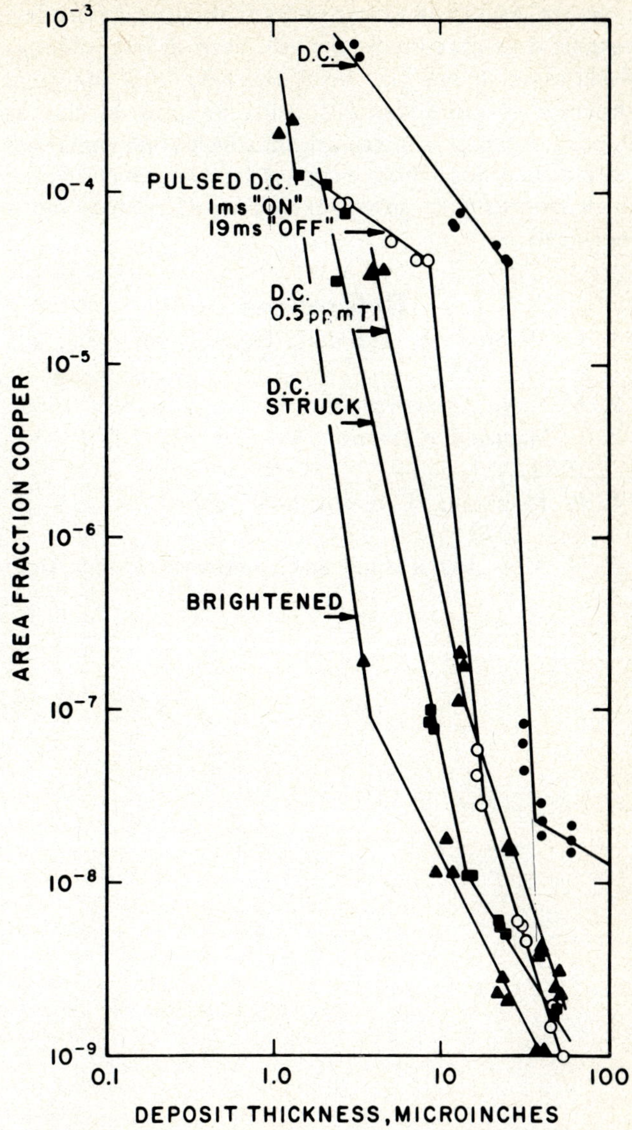

FIGURE 8 – Comparison of porosity vs thickness curves for various gold electrodeposits.

Similar experiments were performed using trace brightening agents of the thallium, lead, bismuth, and mercury group. These agents have been shown by McIntyre and Peck[11] to absorb strongly to gold surfaces, thus providing sites for preferred nucleation of crystallites, and thus yielding a smaller average grain size. Figure 7 shows the effects of these additives on the porosity thickness plots. Again, the porosity-thickness plots are displaced, with the order of effectiveness being Tl > Pb > Bi > Hg, which is also the order of their brightening (and hence, grain-size-reducing) effectiveness.

Additional experiments were conducted using a gold strike prior to plating, and also repeating the examination of cobalt-brightened gold at lower deposit thicknesses. The resulting porosity-thickness curves obtained over a fairly wide spectrum of deposit grain sizes are shown in Figure 8.

Interpretation and Conclusions

The data in Figure 8 allow some speculation to be made concerning the z-shaped form of the porosity-thickness plot. At low deposit thicknesses, crystallographic and porosity-thickness results suggest that the properties of the electrodeposited film are largely controlled by the surface condition and characteristics of the underlying substrate. This condition persists up to a limiting thickness, after which the properties of the film itself, primarily crystallographic properties, determine the rate of pore closure. Between these two regimes there occurs a very sharp, well-marked transition in which the porosity falls dramatically. The deposit thickness at which this rapid transition occurs varies with the deposit grain size.

The rapid transition in porosity appears at this point to represent the coalescence of individual clusters of clumped grains into a more or less continuous film. The fact that the thickness at which this coalescence occurs depends on grain size probably results from the observation[12,13] that the grain size *distribution* in fine-grained deposits is narrower than that of larger-grained deposits. Grains having a large distribution of sizes do not "fit together" as well as a more uniform array, and this effect persists as a measurable porosity even at large deposit thicknesses. It has also been observed[7] that the grain structure of electrodeposits tends to grow in size as deposit thickness is increased. Intuitively, one would expect lower porosity from a deposit in which a high ratio of nucleation to grain growth is maintained as thickness is increased.

References

1. C. Wagner, W. Traud, Z. Elektrochem., Vol. 44, p. 391, 1938.
2. M. Stern, Corrosion, Vol. 14, No. 7, p. 329, 1958.
3. F. Mansfeld, Corrosion, Vol. 27, No. 10, p. 436, 1971.
4. R. Morrissey, J. Electrochem. Soc., Vol. 117, p. 742, 1970.
5. R. Morrissey, Plating and Surface Finishing, Vol. 67, p. 44, 1980.
6. U. R. Evans, Corrosion and Oxidation of Metals, St. Martin's, London, England, 1953.
7. S. E. Craig, Jr., R. E. Harr, J. Henry, P. Turner, J. Electrochem. Soc., Vol. 117, p. 1450, 1970.
8. R. J. Morrissey, A. M. Weisberg, Trans. Inst. Metal Finishing, Vol. 53, p. 9, 1975.
9. R. J. Morrissey, A. M. Weisberg, Trans. Inst. Metal Finishing, Vol. 58, p. 97, 1980.
10. D. L. Rehrig, H. Leidheiser, Jr., M. Notis, Plating and Surface Finishing, Vol. 64, p. 40, 1977.
11. J. D. E. McIntyre, W. F. Peck, Jr., J. Electrochem. Soc., Vol. 123, p. 1800, 1976.
12. W. H. Cleghorn, J. A. Crossley, K. J. Lodge, K. S. A. Gnanasekaran, Trans. Inst. Metal Finishing, Vol. 50, p. 73, 1972.
13. J. D. E. McIntyre, W. F. Peck, Jr., S. Nakahara, J. Electrochem. Soc., Vol. 124, p. 302, 1977.

The Effect of Ion Implantation on the Corrosion Behavior of a High-Density Sintered Tungsten Alloy

F. C. CHANG, M. LEVY, S. S. LIN*

Abstract

The effect of Cr, Ni, Ta, and Ti ion implantation at a dose rate of 2×10^{17} ions cm^{-2} on the electrochemical corrosion behavior of a high-density sintered tungsten alloy has been investigated in Cl^--free and Cl^--containing aqueous solution buffered to pH values of 4, 9, and 12. A three-sweep potentiodynamic polarization technique was used to compare the polarization behavior of unimplanted and implanted surfaces. The surfaces of the ion-implanted tungsten alloy were characterized by Auger electron spectroscopic (AES) analysis.

Introduction

Sintered tungsten alloys are candidate materials for certain US Army applications because of their high density. Until recently, the corrosion of tungsten alloys has not been considered a problem. Andrew, et al.,[1] reported that a 90 W, 7.5 Ni, 2.5 Co alloy readily corrodes when exposed to air-saturated water vapor. The authors in a previous paper[2] showed that a tungsten alloy 97.1 W, 1.6 Ni, 0.7 Fe, 0.5 Cu, 0.1 Co also undergoes corrosion when exposed to a chloride-containing aqueous solution. Some dissolution of the tungsten grains and localized attack of the matrix alloy were observed.[2] This paper reports on the effects of Cr, Ni, Ti, and Ta implantation on the corrosion behavior of a sintered tungsten alloy in aqueous solutions of varying pH values by electrochemical techniques.

Experimental Procedure

The sintered tungsten alloy (W2) was fabricated using powder metallurgy techniques (isostatic compaction of mixed metal powders followed by sintering in hydrogen). The basic tungsten content was 97.1%. The matrix composition included 0.7% Fe, 1.6% Ni, 0.1% Co, and 0.5% Cu. Figure 1 is a micrograph of the alloy showing rounded tungsten particles surrounded by a layer of matrix solid solution. Note the porosity in both the tungsten grains and the matrix alloy.

Specimens for electrochemical tests were flat discs, 5/8 in. (16 mm) in diameter and 1/8 in. (3.2 mm) thick. The exposed surface area was 1 cm^2. The specimens were ground to a 64 root mean square (rms) finish, rinsed with distilled water, acetone-degreased, air dried, and ion implanted.

The aqueous solutions used for electrochemical tests were buffered to pH values of 4, 9, and 12 with and without 0.1 molar (M) NaCl. They were prepared from laboratory reagent chemicals and distilled H_2O. The buffers used to fix the pH included citric acid, disodium phosphate, sodium hydrogen carbonate, and sodium hydroxide.

Electrochemical polarization measurements were made utilizing a three-sweep method of potentiodynamic polarization. Prior to sweeping the potential, the specimen was allowed to remain in the solution for at least 30 minutes to permit the potential to come to a steady-state corrosion potential. The three-sweep method involved a positive direction sweep from the corrosion potential to +1.6 V vs SCE. This was followed by a negative direction sweep to a potential 250 mV more negative than the corrosion potential. Finally, there was a second positive direction sweep to a potential of +1.6 V vs SCE. The sweep rate used was 4 V/h throughout. Polarization runs

*Army Materials and Mechanics Research Center, Watertown, Massachusetts.

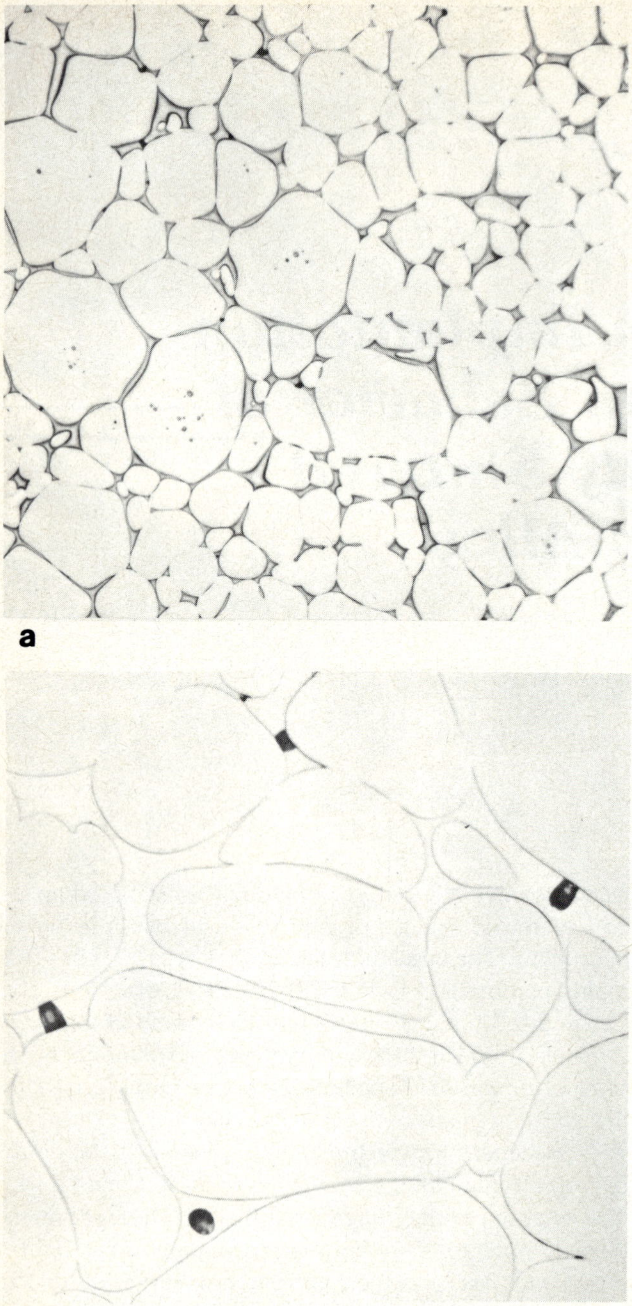

a

b

FIGURE 1 – Typical microstructure of the W2 tungsten alloy, unetched, showing porosity: (a) in the grain at 100X and (b) in the matrix at 500X.

were made with the tungsten alloy in both the virgin and ion-implanted conditions in argon purged solutions.

The ion implantation[(1)] was performed at a dose rate of 2×10^{17} ions cm^{-2}. The energy of singly charged positive ions used varied with the ion implanted: 156 KeV for Cr; 165 KeV for Ni; 160 KeV for Ta; and 142 KeV for Ti. The ion-implanted tungsten alloy specimens were analyzed with an ESCA/Auger instrument and scanning Auger microprobe (SAM) attachment before and after the polarization tests.

[(1)]Performed by the Naval Research Laboratory, Washington, DC.

The surfaces to be examined were carefully chosen so that the analysis was not made on cracks or depressions. Before the depth profile analysis was performed, the surface topology was examined and then an Auger electron spectrum was taken. From the spectrum, the surface elements and their concentrations could be obtained. The depth profile analyses were made at a minimum of two locations of the specimen surface. The variations of those implanted elements, plus tungsten, oxygen, and carbon and nitrogen signals along the depth were monitored throughout a sputtering period of from 20 to 100 minutes. The ion sputtering was continued until a steady-state signal of tungsten was obtained. The peak heights were then converted to the atomic concentrations using the available sensitivity factors.

Results and Discussion

Unimplanted W2 in Cl-Free Solutions—Effect of pH

The complete, three-sweep potentiodynamic polarization curves for the unimplanted tungsten alloy in aqueous solutions of pH values of 4, 9, and 12 are shown in Figure 2. In the pH 4 solution during the first positive direction sweep, the air-formed oxide film is repaired and replaced above 0.500 V by nonsteady-state film growth and a consequent increase in the anodic current density from ~50 $\mu a/cm^2$ to ~1000 $\mu a/cm^2$. A relatively well-marked hysteresis loop is observed upon a reversal of the sweep direction. Anodic current densities are substantially greater at corresponding potentials in the potential range between 0.500 and 1.400 V. It is believed that the hysteresis loop is attributable to crevice corrosion since crevice attack was observed near the masked surfaces and some porosity was present in the virgin W alloy (both in tungsten grains and in the matrix alloy) (Figure 3). Similar hysteresis loops have been reported for specimens containing synthetic crevices.[3,4] The major difference in the second positive direction sweep is the appearance of an anodic current density peak at ~0.500 V which must be due to an alloying element in the matrix material.

Three-sweep curves for the alloy in the pH 9 solution are similar to each other. They exhibit an active/passive transition at ~ −0.280 V and a secondary peak at ~ +0.200 V. A transpassive region is observed at ~1.400 V. There is no apparent hysteresis loop during the reverse sweep. Again, the three-sweep curves for the alloy in solution of pH 12 are essentially identical to one another. The anodic current density increases to ~560 $\mu a/cm^2$ where it remains fairly constant throughout the potential range −0.250 to 1.140 V, showing that the oxide film remains essentially unchanged. In general, increasing the pH from 4 to 9 to 12 serves to shift the corrosion potential in the more active or electronegative direction. The greatest shift is observed upon increasing the pH from 4 to 9, where the corrosion potential becomes ~330 mV more negative. According to the Pourbaix diagram for tungsten,[5] the passivation area ranges from pH 0 to pH 4 from the corrosion potential and above, to

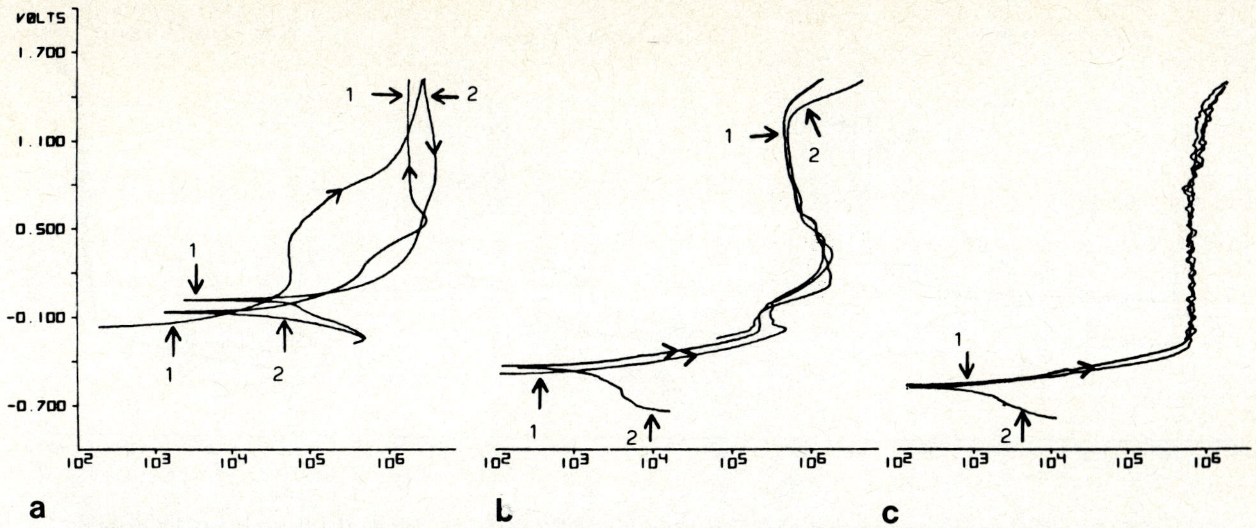

FIGURE 2 – Polarization curves of W2 in pH of (a) 4, (b) 9, and (c) 12 solutions without chloride.

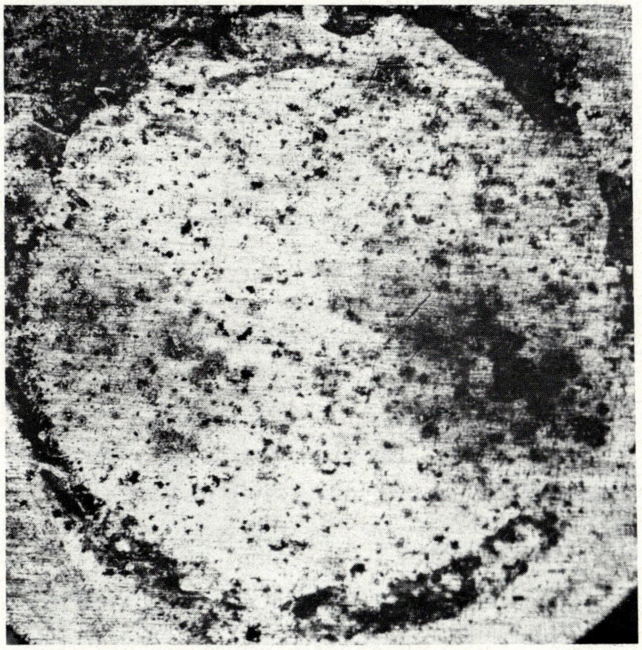

FIGURE 3 – Surface of W2 alloy after polarization scans showing crevice corrosion.

1.4 V where the tungsten is covered by WO_2 or higher oxides WO_3, W_2O_5. The corrosion area extends from the corrosion potential to ~ 1.4 V and above, at pH 4 to 16, and in the more alkaline solutions it has a slight tendency to decompose H_2O with evolution of H_2O dissolving as tungstic ions WO_4^{-2}. In addition, the curves at higher pH exhibit reversibility whereas the curve at pH 4 exhibits a marked hysteresis loop.

Chloride-Free Environment— Chromium-Implanted W2 Alloy

The potentiodynamic polarization curves for the W2 alloy implanted with 2×10^{17} Cr^+ ions cm^{-2} are shown in Figure 4. In this case, there are differences between the first positive sweep for Cr-implanted W2 and the corresponding sweep for unimplanted W2 at pH 4 and 9. At pH 4, the Cr-implanted W2 alloy is more difficult to passivate and the anodic current density is greater throughout the entire potential range. In solution of pH 9, however, the Cr-implanted alloy is more easily passivated while the anodic current density increases at potentials above 0.80 V. These data are consistent with the theoretical Pourbaix diagram for chromium,[5] considering $Cr(OH)_3$ or Cr_2O_3 in solutions not containing chloride. According to the Pourbaix diagram, chromium should be corroding in solution of pH 4 throughout the potential range of the first positive sweep. But at pH 9, passivation is indicated, except at corrosion potentials above 0.80 V where corrosion occurs. The films formed on Cr-implanted W in the pH 9 solution may be slightly more protective than those formed on the unimplanted W2. In solutions of pH 12, the shape of the curves for both the implanted and unimplanted alloy is essentially the same, indicating that the implanted specimen shows no change in the film by the process or the chromium.

Nickel-Implanted W2 Alloy

Figure 5 shows the potentiodynamic polarization curves for Ni-implanted W2 alloy. The curves for both the unimplanted and implanted material have similar shapes, but the anodic current density for the implanted alloy is greater throughout the entire potential range of the sweeps. The main area of corrosion from the Pourbaix diagram lies between pH −2 and 9. Also, a small area of passivation exists in the pH range of from 9 to 12. Therefore, the implanted Ni provides no beneficial modification of the oxide film throughout the pH range of 4 to 12.

Titanium-Implanted W2 Alloy

The potentiodynamic curves for the W alloy implanted with 2×10^{17} Ti^+ ions cm^{-2} (Figure 6) are similar to corresponding curves for unimplanted W. But in solu-

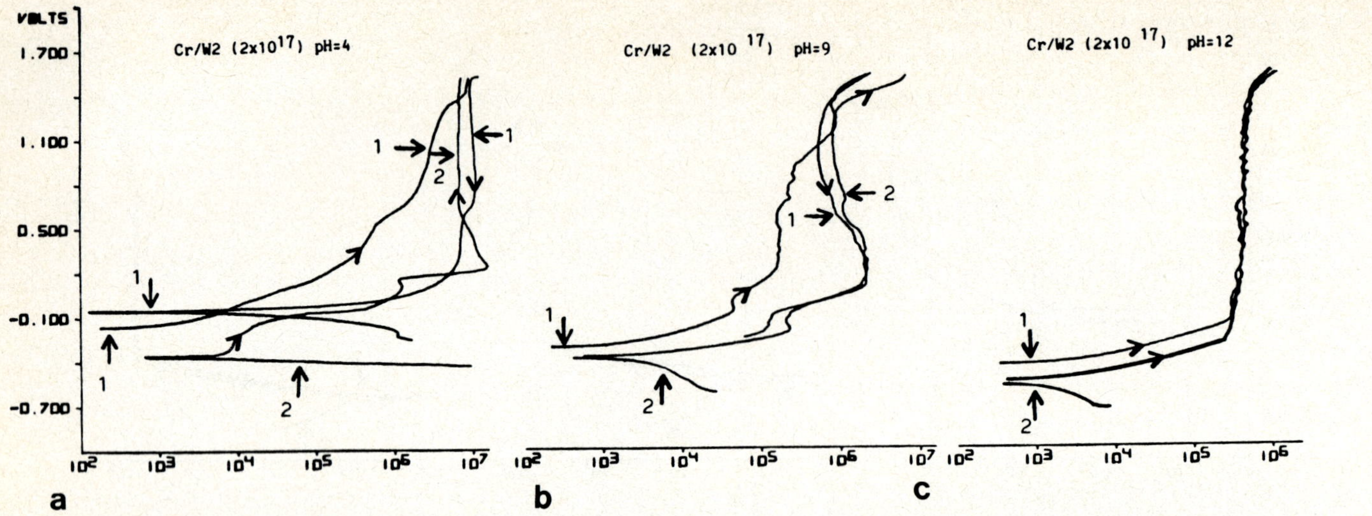

FIGURE 4 — Polarization curves of Cr-implanted W2 in pH of: (a) 4, (b) 9, and (c) 12 solutions without chloride.

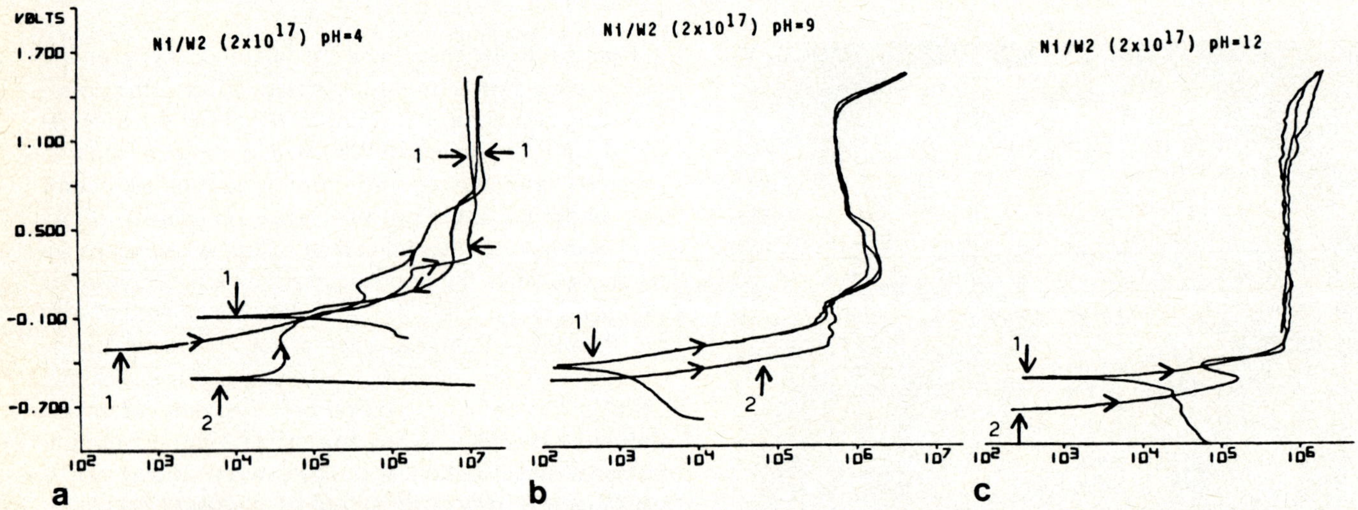

FIGURE 5 — Polarization curves of Ni-implanted W2 in pH of: (a) 4, (b) 9, and (c) 12 solutions without chloride.

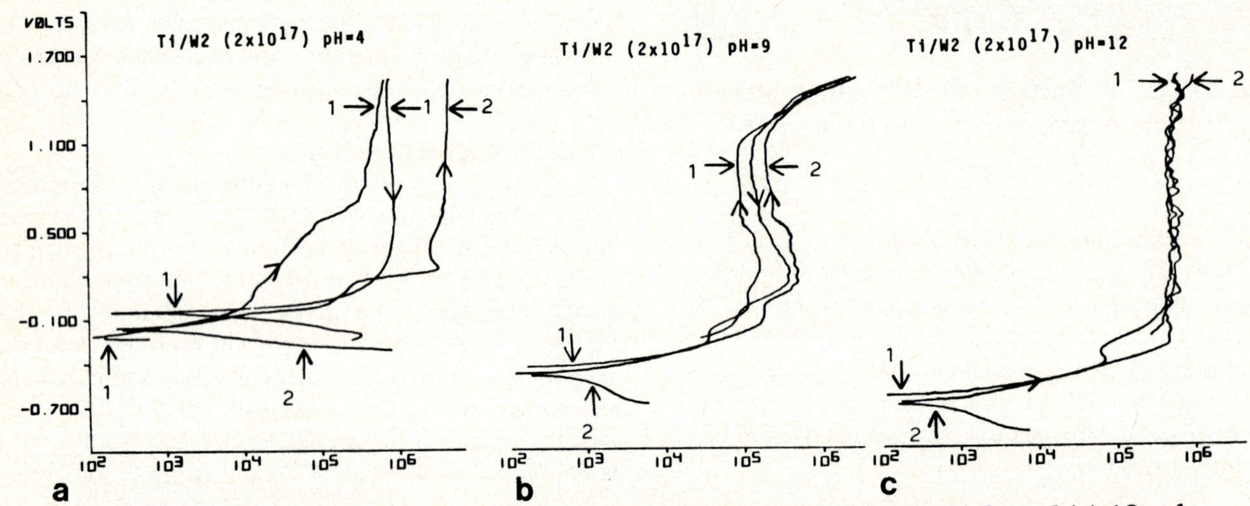

FIGURE 6 — Polarization curves of Ti-implanted W2 in pH of: (a) 4, (b) 9, and (c) 12 solutions without chloride.

FIGURE 7 – Surface of Ti-implanted W2 after exposure to potentiodynamic polarization scans in the pH 9 solution without chloride showing oxide film. (7X)

tions of pH 9 and 12, the magnitude of the anodic current density throughout the potential range of the sweeps is reduced by the implanted species, indicating that the films formed on Ti-implanted W are more protective than those formed on unimplanted W. (See Figure 7.)

Tantalum-Implanted W2 Alloy

Figure 8 shows polarization curves for the Ta-implanted tungsten alloy in solutions of pH values of 4, 9, and 12. The shapes of the curves are similar to those of the unimplanted alloy. The first positive direction sweep with the Ta-implanted alloy in the pH 4 solution is very similar to that observed with the unimplanted alloy. However, the second positive direction sweep in this figure differs from that observed with the unimplanted W alloy in that the current density of the implanted alloy is an order of magnitude greater. According to the Pourbaix diagram, Ta_2O_5, a stable protective oxide, is formed in aqueous solutions of the pH range of from 0 to 14. Nevertheless, larger currents were observed for the implanted alloy. This may be due to the high densities of the elements involved (W alloy ~19.3 g/cm^3, Ta-13.6 g/cm^3) and the magnitude of the ion energy (160 KeV) employed for implantation, which resulted in very low concentrations of Ta on the surface to a depth of 140 A. Tantalum implantation had little or no effect on the polarization curves for the alloy in the pH 9 and 12 solutions.

Unimplanted W2 in Chloride-Containing Solution—Effect of Cl and pH

The complete, three-sweep potentiodynamic polarization curves characteristic of unimplanted W2 alloy in 0.1 M NaCl solution buffered to pH values of 4, 9, and 12 are shown in Figure 9. These curves differ from the corresponding curves in a chloride-free environment as follows: the corrosion potentials are slightly more active, except at pH 9 where they are essentially identical; the passive current densities are markedly greater, and hysteresis loops are observed at all pH values. Increasing the pH increases the anodic current density, particularly from pH 4 to 9.

Cl-Containing Solution—Cr-Implanted W2 Alloy

Figure 10 shows the potentiodynamic polarization curves for Cr-implanted W2 alloy in the pH 4, 9, and 12 solutions. The general shapes of these curves are similar to the curves for the unimplanted alloy. The corrosion potentials remain essentially the same. In solutions of pH

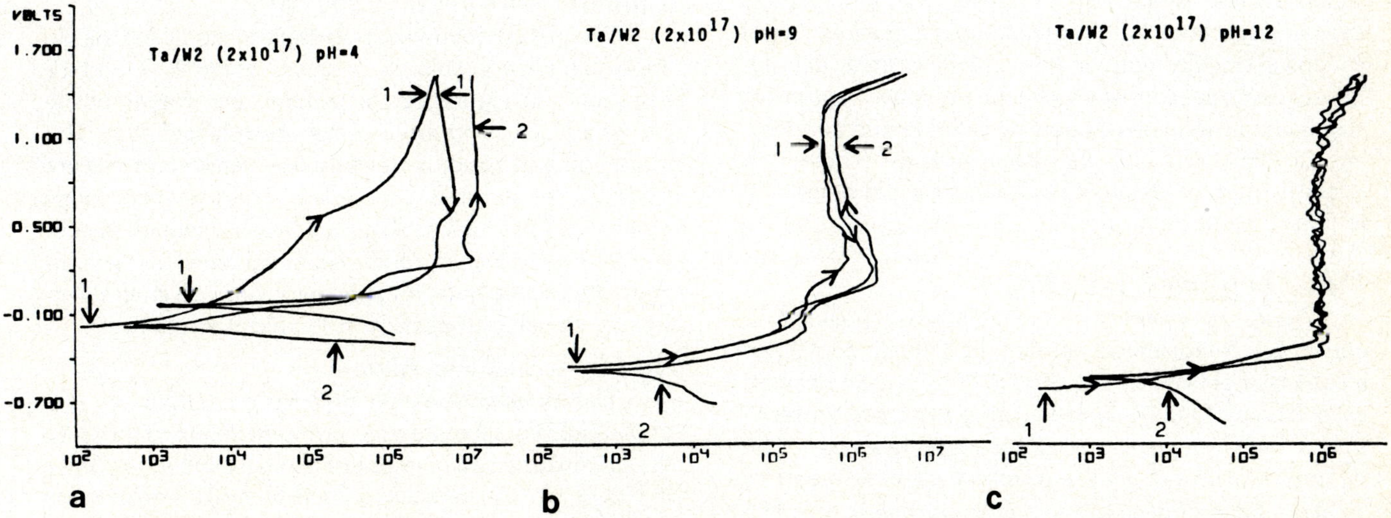

FIGURE 8 – Polarization curves of Ta-implanted W2 in pH of: (a) 4, (b) 9, and (c) 12 solutions without chloride.

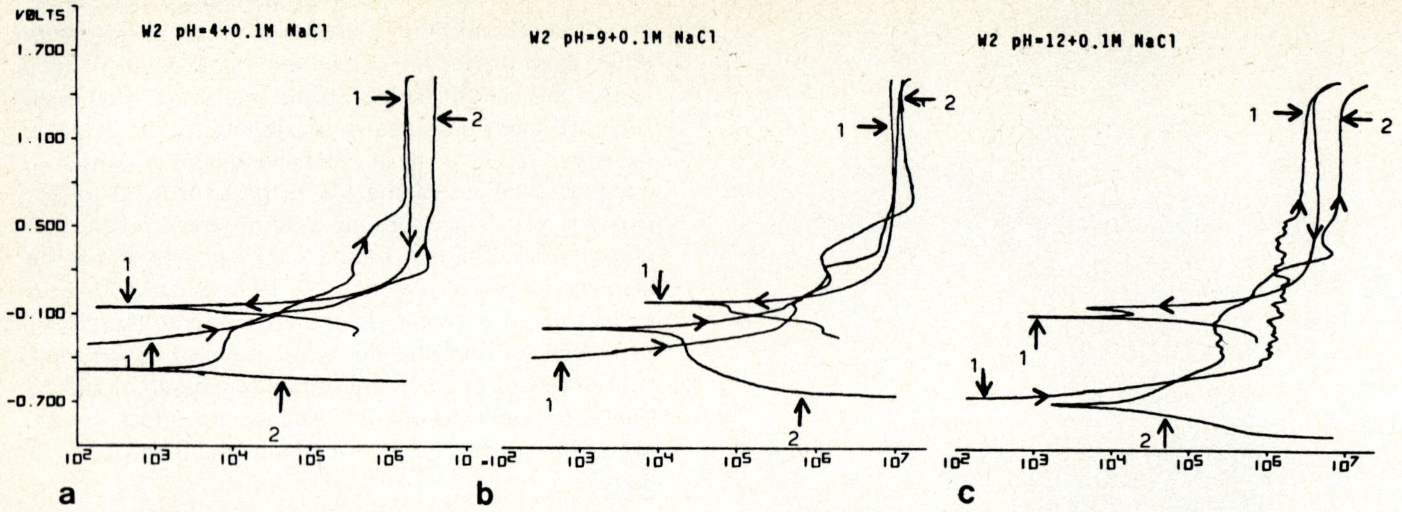

FIGURE 9 — Polarization curves of W2 in pH of: (a) 4, (b) 9, and (c) 12 solutions with chloride.

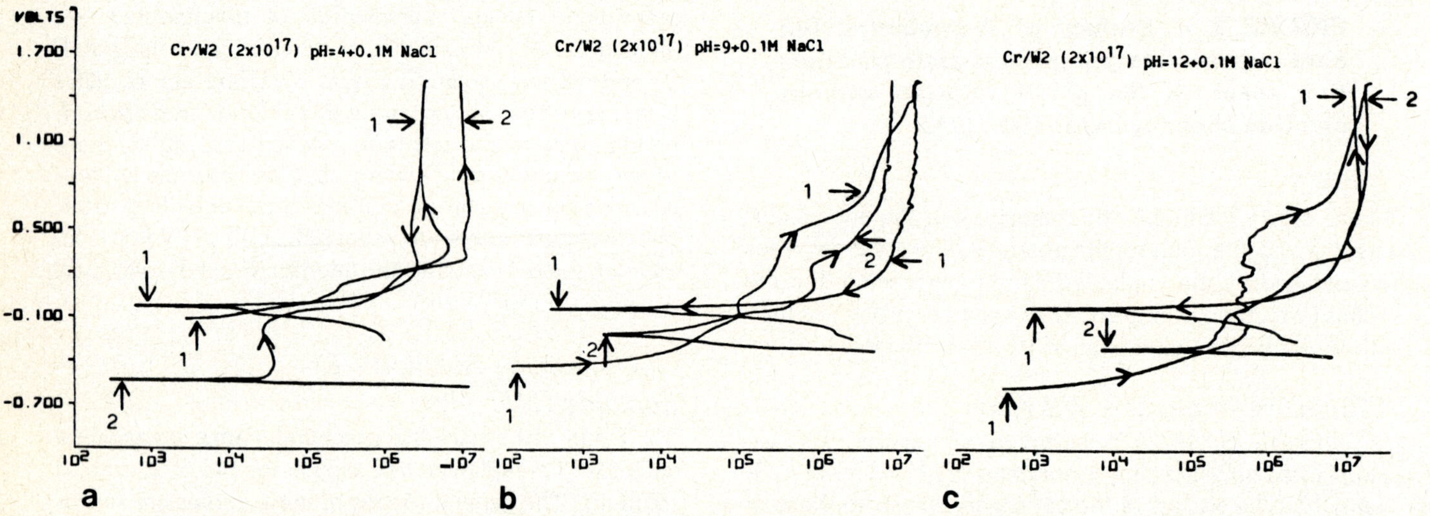

FIGURE 10 — Polarization curves of Cr-implanted W2 in pH of: (a) 4, (b) 9, and (c) 12 solutions with chloride.

4 and 12, the anodic current density for the implanted alloy is larger than that for the unimplanted alloy. At pH 9, however, Cr reduces the anodic current density. These results are consistent with the Pourbaix diagram for chromium in solutions containing chloride and considering $Cr(OH)_3 \cdot nH_2O$ as the stable species. There is a relatively narrow region of passivity in the pH range of 8 to 9 in Cl-containing solutions.

Nickel-Implanted W2 Alloy

Figure 11 shows the potentiodynamic polarization curves for Ni-implanted W alloy in chloride-containing solutions of pH values of 4, 9, and 12. The electrochemical behavior of the Ni-implanted alloy was quite similar to that of the Cr-implanted alloy previously described. The Ni implantation provided a beneficial effect in the pH 9 solution, but produced higher anodic currents in the pH 4 and 12 solutions. Again, results were consistent with the Pourbaix diagram.

Titanium-Implanted W2 Alloy

The potentiodynamic polarization curves for the Ti-implanted alloy in chloride solutions of pH 4, 9, and 12 are shown in Figure 12. In general, the shapes of the curves are similar to those of the unimplanted alloy. But more noble corrosion potentials and smaller currents are observed for the implanted alloy. The hysteresis loops observed in the reverse scans are broader than those observed in similar scans for the unimplanted alloy. In summary, the Ti implantation appears to have a beneficial effect on the alloy, especially in the pH 9 solution.

Ta-Implanted W2 Alloy

The potentiodynamic polarization curves for Ta-implanted W2 alloy are very similar in shape to those for the unimplanted alloy in Cl-containing solution as shown in Figure 13. The beneficial effects of the Ta-implantation are reflected in the smaller anodic currents at potentials up to +0.500 V in the pH 4 solution and to +1.400 V in the pH 9 solution.

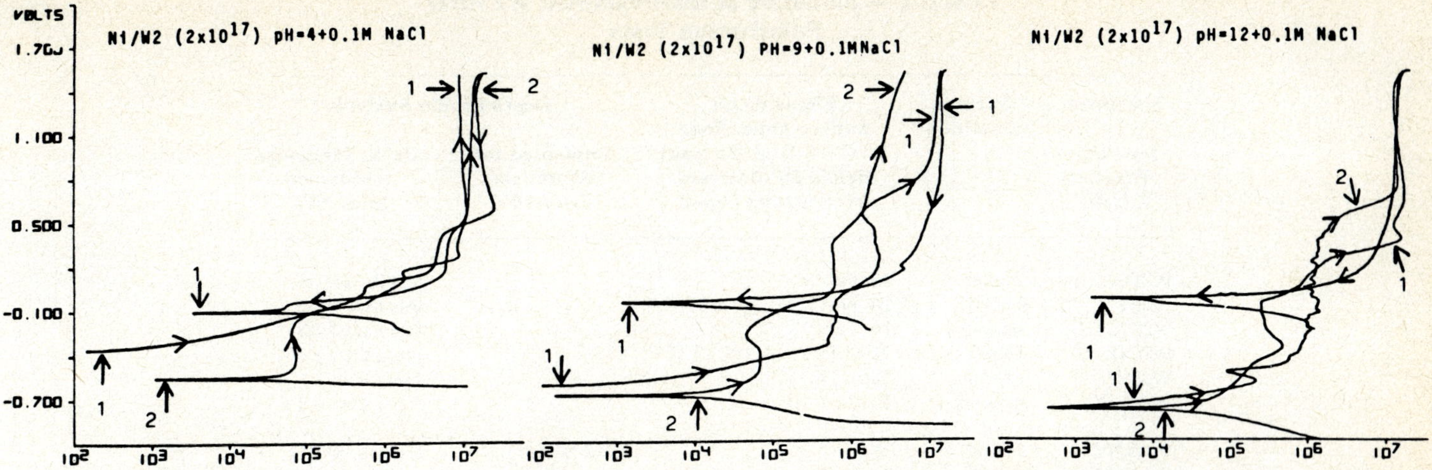

a b c

FIGURE 11 — Polarization curves of Ni-implanted W2 in pH of: (a) 4, (b) 9, and (c) 12 solutions with chloride.

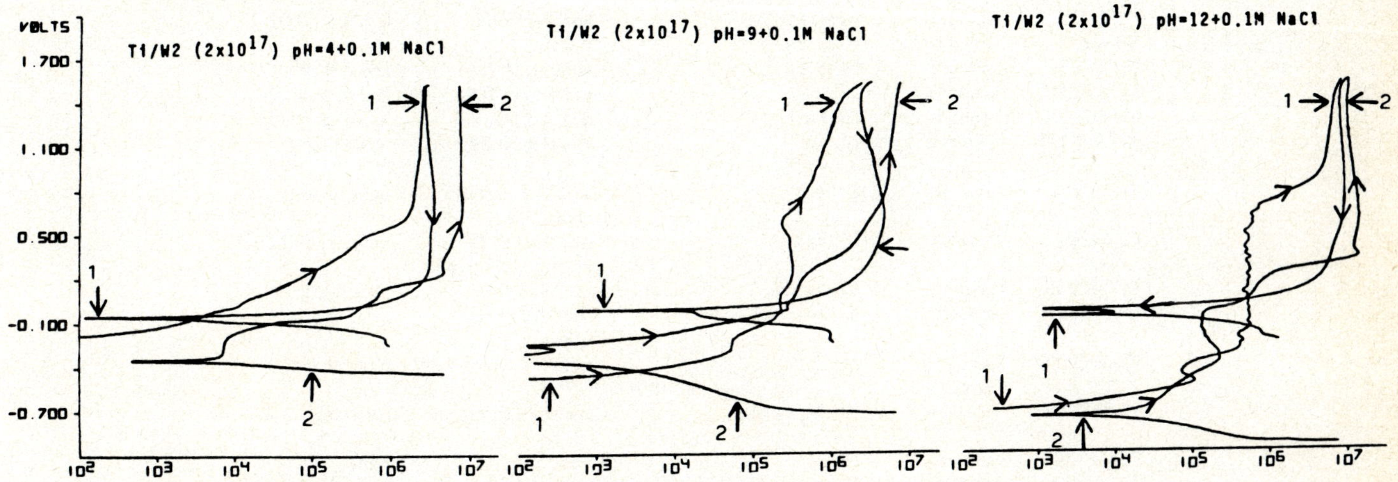

a b c

FIGURE 12 — Polarization curves of Ti-implanted W2 in pH of: (a) 4, (b) 9, and (c) 12 solutions with chloride.

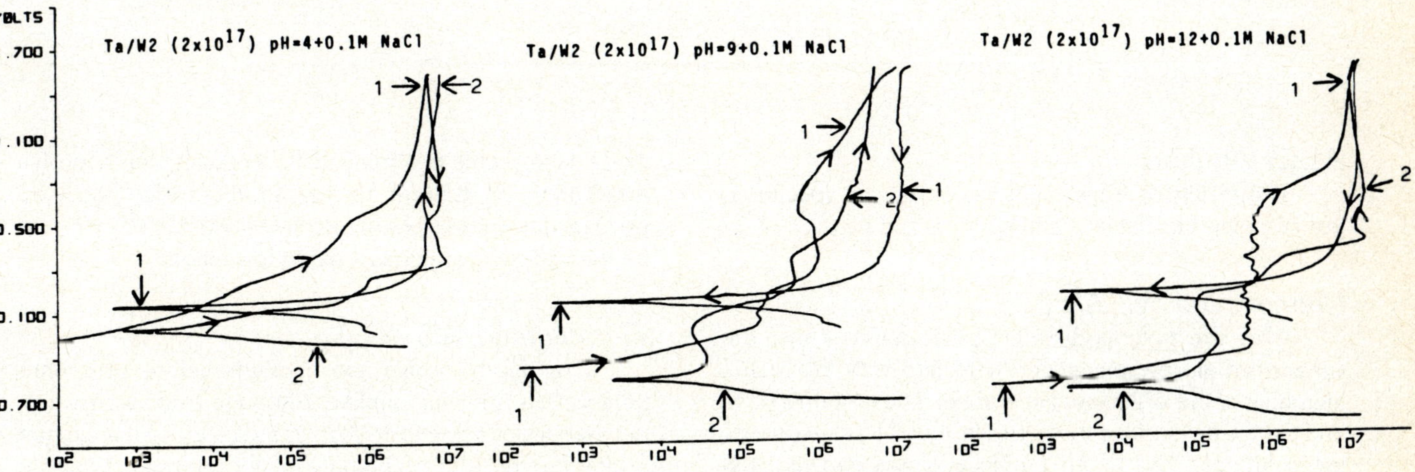

a b c

FIGURE 13 — Polarization curves of Ta-implanted W2 in pH of: (a) 4, (b) 9, and (c) 12 solutions with chloride.

TABLE 1 — Summary of Ion-Implanted W2 After Polarization Tests

Specimen Ion/W2-P (pH)-X (salt)	Surface Condition	Elements on Surface Other Than W, C, N, O, & Na (salt) Before DP Analyses (s) = Strong Signal	Depth Profile Analyses: Implanted Ion Width (μm) Error 10%	Depth Profile Analyses: Oxide Thickness Width (μm) Error 10%
W2P4	smooth	Ni, Fe		less than 0.05
W2P9	smooth	Ni, Fe		less than 0.01
W2P12	smooth	Ni, Fe		less than 0.05
W2P4X	deposits	Ni(s), Fe		more than 0.15
W2P9X	deposits	Ni(s), Fe		more than 1
W2P12X	deposits	Fe, Ni		more than 0.4
Ti/W2P4	smooth	Ni	0.12 & shift	less than 0.02
Ti/W2P9	smooth	Ti, Ni, Fe	0.1	less than 0.02
Ti/W2P12	smooth	Ti(s), Ni, Fe	0.12 & shift	more than 0.18
Ti/W2P4X	smooth	Ti(s), Ni, Fe	0.1 & shift	less than 0.02
Ti/W2P9X	deposits	Ti, Ni, Fe	0.12 & shift	about 0.1
Ti/W2P12X	deposits	Ti(s), Ni, Fe	0.13 & shift	about 0.15
Ni/W2P4	smooth	Ni(s), Fe	less than 0.03	less than 0.02
Ni/W2P9	smooth	Ni(s), Fe	0.08	about 0.03
Ni/W2P12	smooth	Ni(s), Fe	about 0.08	about 0.08
Ni/W2P4X	smooth	Ni(s), Fe	more than 0.08	less than 0.02
Ni/W2P9X	smooth	Ni, Fe	about 0.03	more than 0.15
Ni/W2P12X	deposits	Ni, Fe	less than 0.02	more than 0.1
Cr/W2P4	smooth	Cr, Ni	small Cr profile width 0.05	more than 0.1
Cr/W2P9	smooth	Ni, Fe	less than 0.01 extended Ni depth	more than 0.05
Cr/W2P12	coating	Cr, Fe, Ni	less than 0.01	about 0.05
Cr/W2P4X	smooth	Cr, Ni	small Cr profile width 0.05	more than 0.1
Cr/W2P9X	scales, layers	Ni, Fe	null	more than 1
Cr/W2P12X	coatings	Cr, Ni, Fe(s)	less than 0.05	more than 0.2
Ta/W2P4	smooth	Ni, Fe	trace	none
Ta/W2P9	smooth	Ni	trace	more than 0.04
Ta/W2P12	smooth	Ni	trace	about 0.03
Ta/W2P4X	spotty pits	Ni, Fe	trace	less than 0.01
Ta/W2P9X	scales coatings	Ni, Fe	trace extended Ni depth	more than 0.6
Ta/W2P12X	spotty layers	Fe(s), Ni	none	about 0.2

Auger Analysis

The results of Auger analyses of reacted specimens are contained in Table 1 and summarized below.

Unimplanted W2 Alloy

After the potentiodynamic polarization scans, the surfaces of all specimens were enriched with the matrix elements of the W2 alloy, Ni, and Fe. The amounts of Ni and Fe on the surface increased in the chloride-containing solutions. The depth profile analyses showed that anodic oxide films of varying thickness were present on all specimens. The oxide thickness varied widely from a minimum in the chloride-free pH 9 solution to a maximum of 1 μm in the pH 9 chloride solution. In the pH 9 chloride solution, two layers of Ni and W oxide films were found. The nickel film was present on top of the tungsten oxide layer, and it displayed a Gaussian depth profile distribution. In general, thicker anodic oxide films were produced in chloride solutions. It is likely that Cl ions contributed to the accelerated oxide formation.

Ti-Implanted W2 Alloy

The depth profile for titanium before and after polarization runs was similar. Although there were varying degrees of oxide formation as indicated by the oxygen profile curves, the magnitude of the titanium signal in these oxide layers was not changed appreciably and the widths of the profile remained the same (about 0.1 μm). (See Figure 14.) In some instances, a large concentration of titanium was observed on the surface. This was probably due to the displacement of the Ti toward the surface. The Ni and Fe matrix elements also concentrated on the

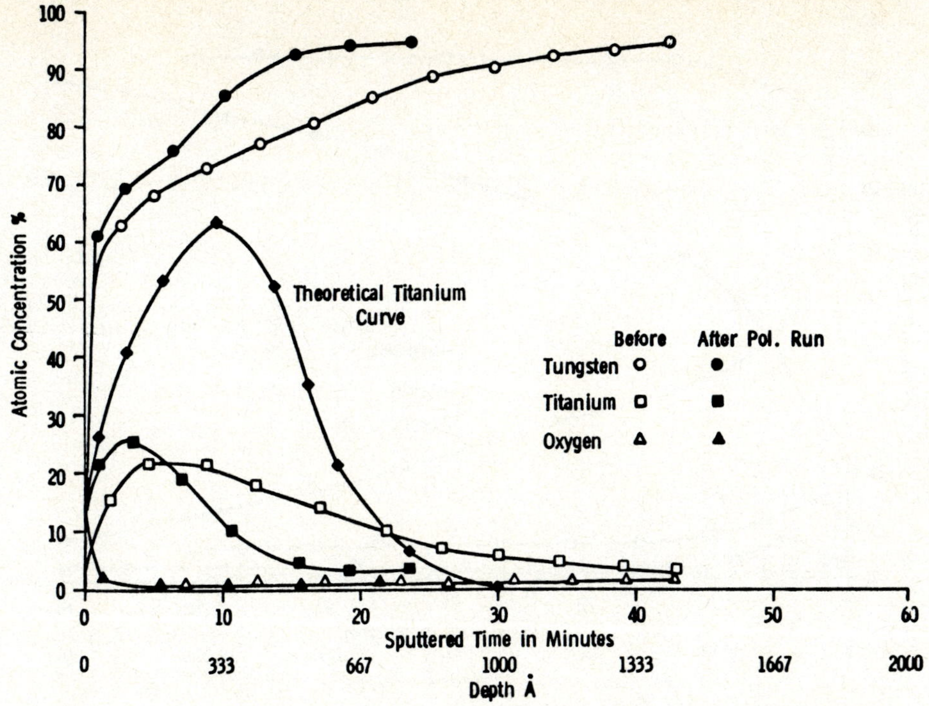

FIGURE 14 — Auger concentration depth profiles of Ti, W, and O before and after Ti-implanted W2 subjected to potentiodynamic polarization in pH 9 solution without chloride. Note the difference between the theoretical and actual profile of Ti.

top surfaces, and the enrichment of these metals on the surface was accelerated in chloride solutions.

Ni-Implanted W2 Alloy

The Ni-implanted W2 was unique in that the metal tended to accumulate on the surface. A high concentration of Ni on the surface was usually observed. The depth profile was found to alter considerably after polarization. In the Cl-free solutions, Ni was displaced significantly toward the surface so that only a portion remained. But in chloride solution, Ni disappeared. A significant oxide layer formation was observed as pH increased in Cl solutions. In the absence of Cl, the widths of the oxygen profile appeared to be associated with those of the nickel profile, that is, a nickel oxide had formed. But in the Cl solution (except for pH 9), the oxygen profile far exceeded that of nickel; that is, other oxides had formed.

Cr-Implanted W2 Alloy

The behavior of the ion-implanted Cr varied widely from solution to solution. The majority of Cr was found to concentrate near or at the surface together with Ni and Fe. The surface topology changed markedly in the high pH solutions with the formation of various oxide layers. In the Cl-free pH 9 solution, the formation of a Ni oxide layer was observed. Although small amounts of Cr, represented by a narrow profile width and a low concentration, were detected in the low pH solutions, no Cr was detected in the high pH Cl solutions.

Ta-Implanted W2 Alloy

Most of the Ta in the implanted surface was lost after the polarization runs (Figure 15). Even in the less corrosive, Cl-free solutions, the Ta profile had nearly disappeared, and the formation of Fe and Ni oxide layers predominated in Cl solutions.

An assessment of corrosion behavior based solely on Auger analyses (depth profile, thickness of oxide layer, distribution of W, and the matrix elements of the alloy) indicates that: (1) generally, corrosion increases with increasing pH and with the addition of Cl and (2) among the implanted species, Ti provides the best protection.

Conclusions

In brief summary, the unimplanted and implanted tungsten alloy exhibits an active/passive transition in the solutions studied, but the magnitude of the passive current density exceeded 100 $\mu a/cm^2$ in most cases. Generally, corrosion rates increase with increasing pH and in the presence of chloride ions. The hysteresis loops observed during reverse sweeps are attributed to crevice corrosion and porosity in both the tungsten grains and the matrix alloy. Among the Cr, Ni, Ta, and Ti implantations, Ti improved the corrosion resistance of the tungsten alloy. The magnitude of the current density throughout the potential range of the sweeps was reduced. The Auger depth profile for titanium before and after potentiodynamic polarization sweeps was similar; i.e., the magnitude of the titanium signal in the oxide layers and the width of the profile remained essentially the same. Generally, Pourbaix diagram data and Auger analysis results were in agreement with the electrochemical data.

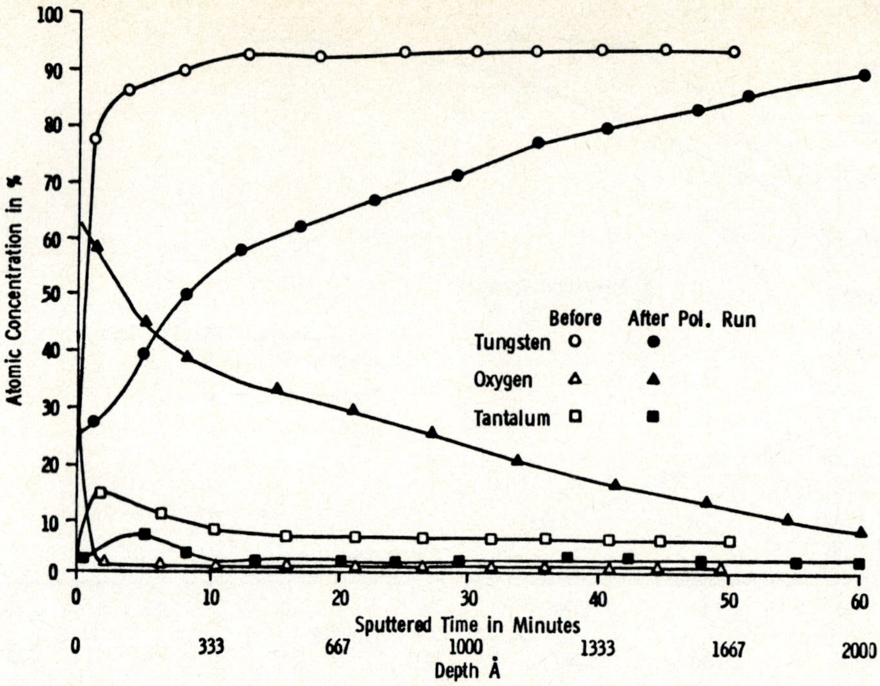

FIGURE 15 — Auger concentration depth profiles of Ta, W, and O before and after Ta-implanted W2 subjected to potentiodynamic polarization in pH 9, 0.1 M NaCl solution.

Acknowledgments

The authors acknowledge the assistance of S. Bell in performing the electrochemical experiments.

References

1. J. F. Andrew, M. T. Baker, H. T. Heron, Corrosion and Protection of Sintered Tungsten Alloy Ammunition Components, Proc. 2nd Charlottesville Conf. High Density Materials, October 1980.
2. M. Levy, F. Chang, Corrosion Behavior of High Density Tungsten Alloys, Proc. of the Second International Conference on Environmental Degradation of Engineering Materials in Aggressive Environments, September 21-23, 1981, pp. 33-42.
3. B. E. Wilde, "On Pitting and Protection Potentials: Their Use and Possible Misuses for Predicting Localized Corrosion Resistance of Stainless Alloys in Halide Media," Localized Corrosion, R. W. Staehle, B. F. Brown, J. Kruger, A. Agrawal, Eds., National Association of Corrosion Engineers, Houston, Texas, p. 342, 1974.
4. W. D. France, Jr., Crevice Corrosion of Metals, ASTM STP 516, Localized Corrosion—Cause of Metal Failures, ASTM, Philadelphia, Pennsylvania, pp. 170-171, 1968.
5. M. Pourbaix, Atlas of Electrochemical Equilibria in Aqueous Solutions, National Association of Corrosion Engineers, Houston, Texas, 1974.

Electrochemical Techniques for Corrosion Inhibitor Studies

G. R. CAMERON and A. S. CHIU*

Abstract

Alternating current (AC) impedance and direct current (DC) electrochemical measurements can be used to examine protective inhibitor films on mild steel. AC impedance provides information on the resistive and capacitive behavior, while linear polarization measurements determine polarization resistance, modelled Tafel slopes, and instantaneous corrosion rates. AC impedance data can be used to calculate corrosion rates only by incorporating linear polarization Tafel constants.

Background

Esso Resources Canada, Limited (ERCL) (Calgary, Alberta, Canada) currently operates approximately 1000 conventional rod-pumped wells and 250 submersibly pumped wells in their Western Provinces Operations. The primary method of controlling corrosion in these wells is by batch-circulation of film-persistent corrosion inhibitors. Since production rates vary from 5 to 700 m^3/day and casing/tubing annular fluid levels vary from 1000 m to pumped-off conditions, it is important to know proper inhibitor dosages and the effective inhibitor film life to achieve optimum corrosion protection.

Currently, there are no well-defined inhibitor application guidelines for batch-circulation treatments. Recommendations have been based on contact time and concentration, volume of production between treatments, the surface area of facilities, or previous failure history. Large discrepancies can exist in the inhibitor dosages required as a function of varying well parameters or production rates. In addition, the effectiveness of the inhibitor must be considered. Traditionally, film persistency wheel tests have been used to evaluate inhibitor performance. The "film, rinse, and run" steps in this procedure are intended to simulate the downhole batch-treatment application. Although this method is useful for screening the performance of a large number of candidate inhibitors, it has serious shortcomings when information on inhibitor film life or velocity effects is required. For these reasons, this study used DC linear polarization and AC impedance to evaluate film-persistent inhibitors, keeping in mind the specific application of batch treatments on wells with different production parameters.

Introduction

Linear polarization, or polarization resistance has been used extensively for corrosion rate measurements, and dates back to Tafels' original work[1] in 1905. A detailed overview of the development of this technique is beyond the scope of this paper, and excellent reviews have been written previously.[2,3] More recently, electrochemical techniques have been used to measure inhibitor effectiveness,[4-26] even though there have been reports in the literature[27-33] evaluating possible sources of error and the range of validity of use for this technique. AC impedance techniques have also been used to determine corrosion rates on bare metals[34-43] and coated materials.[44-48] Work has also been done recently comparing AC and DC methods for corrosion rate measurements.[49,50] Although this brief summary is obviously unable to cover all of the references to electrochemical measurements, it does illustrate the wide use of these techniques in corrosion measurement.

The primary objective of ERCL's use of electrochemical corrosion measurements was to utilize a technique that provided information on inhibitor effectiveness under conditions that could simulate production condi-

*Esso Resources Canada, Ltd., Calgary, Alberta, Canada.

tions. As noted earlier, film persistency wheel tests are useful for screening candidate inhibitors, but do not yield sufficient information on film life, optimum concentrations, or the velocity effects on inhibitor performance. DC electrochemical measurements can be used to determine instantaneous corrosion rates as a function of time. Measurements made in different velocity regimes can indicate its effect on inhibitor performance. AC impedance measurements were performed as a supplementary technique to attempt to correlate resistance data to film breakdown.

To summarize the objectives of this study, ERCL is concerned with determining a cost-effective inhibition program for field application by answering the following questions:

- What is the optimum inhibitor concentration required to give corrosion protection in the downhole and flowline production facilities?
- What is the optimum treatment frequency to maintain adequate corrosion protection?
- Are there significant effects of production velocity on inhibitor film life such that batch-circulation treatments need to be tailored to individual well/flowline conditions?

Experimental

DC linear polarization measurements were performed using an ECO Model 553[(1)] potentiostat, Model 560/A/Log Interface[(2)] and Model 567 Function Generator.[(3)] The potentiostat was interfaced with an HP Model 86[(4)] minicomputer with two HP 9130 A[(5)] flexible disc drives for data storage and manipulation.

AC impedance measurements utilized a Solartron Model 1250 Frequency Response Analyzer[(6)] and Model 1186 Electrochemical Interface.[(7)] This system was interfaced to an HP Model 9816[(8)] minicomputer with a Model 8290/M[(9)] dual flexible disc drive. Electrochemical cells were modified after Greene's design[51] and the rotating cylinder electrode was controlled by a Cole-Parmer Master Servodyne[(10)] with dual speed ranges of 0 to 300 rpm and 0 to 3000 rpm.

The inhibitors used throughout were commercially available "as-sold" strengths and could be classified generally as "oxyalkylated esters with acid-amine salts in mixed hydrocarbon and oxygenated solvents."

The inhibitor filming and rinsing steps were conducted for 1 h each on a wheel apparatus rotating at approximately 25 rpm. Produced brine (saturated with CO_2 and made up to 500 ppm H_2S) was used as the corrosive medium. Produced crude oil was added in a 10:90 ratio to the filming and rinsing stages. After filming in the appropriate inhibitor concentration, and rinsing in uninhibited produced fluids, the electrode was transferred to the electrochemical cell for the polarization measurements. The cylindrical electrodes consisted of AISI 1020 mild steel (1.27 cm in diameter × 1.27 cm in length) with a 600 grit surface finish.

DC linear polarization measurements were performed on the filmed rotating cylinder electrode using a combination of software programs written for the HP Model 86 minicomputer. After initial data entry identifying test parameters, the program determined the corrosion potential and polarized the working electrodes 20 mV anodic and 20 mV cathodic to E_{corr} in a "potentiostaircase" fashion. Each potential increment was 2 mV (at 500 μV/s) with a 10-s waiting interval before the potential and current values were stored. Modelled Tafel slopes were calculated from these values by a subroutine using the approach described by Mansfeld,[52] since weight loss calculated using this technique correlated better to actual weight loss than by using the three-point method described by other authors.[53,54] From this data, corrosion currents and rates, polarization resistance, and cumulative weight loss were calculated.

[(1)-(10)]Registered trade names.

Results and Discussion

Initially, four corrosion inhibitors were evaluated using the film persistency wheel test procedure and DC linear polarization measurements. This phase of the testing was intended to select the best candidate inhibitor and to provide a comparison between the two techniques. To obtain an indication of film life, the film persistency wheel tests were run for 24 and 72 h, and continuous polarization measurements were made. Figures 1 and 2 show these wheel test results, illustrating the decrease in percent protection between the 24 and 72 h intervals. Figures 3 and 4 show the equivalent results for the same intervals from integrated instantaneous corrosion rate measurements. The linear polarization results paralleled wheel test results in comparative inhibitor ranking, but the percent protection values were generally higher for the linear polarization measurements. Inhibitor A gave the best overall protection, but it was not chosen for additional testing because it was not commercially available for field use. Inhibitor B, currently used in the majority of wells, was selected for testing in the rotating electrode electrochemical cell.

The subsequent polarization measurements were made using filming concentrations of 500 and 1000 ppm at rotational speeds of 100, 260, and 1200 rpm. These conditions were chosen to simulate field conditions in a

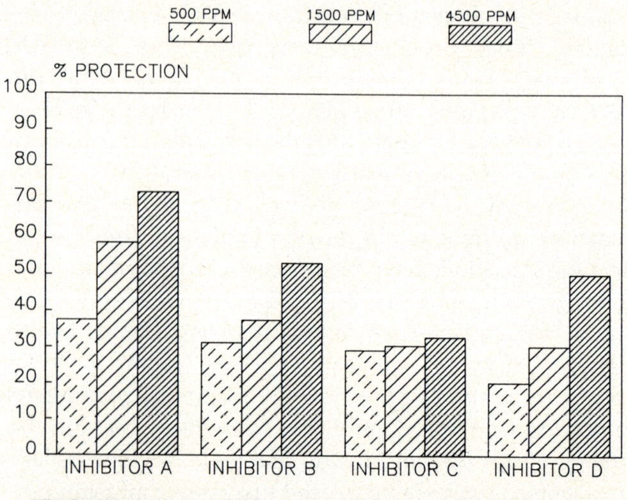

FIGURE 1 — Film persistency wheel tests (24-h exposure).

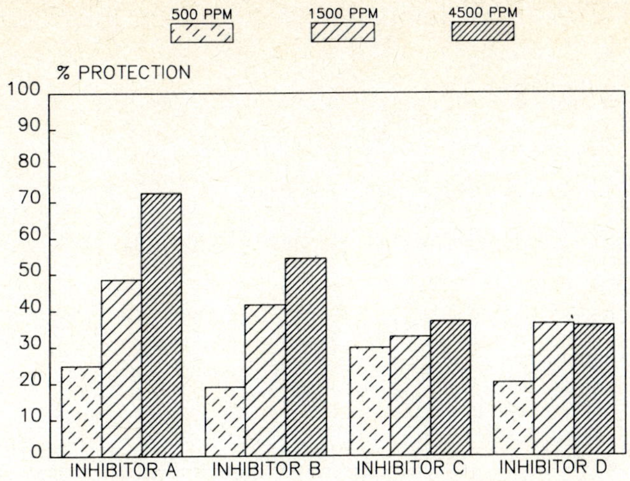

FIGURE 2 — Film persistency wheel tests (72-h exposure).

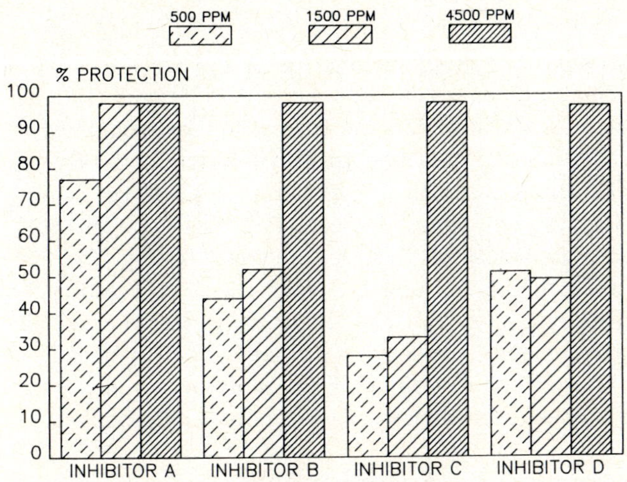

FIGURE 3 — Linear polarization measurements (24-h results).

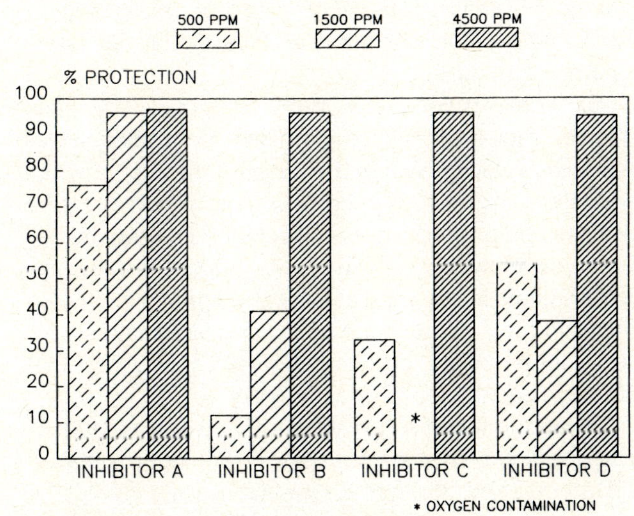

FIGURE 4 — Linear polarization measurements (72-h results).

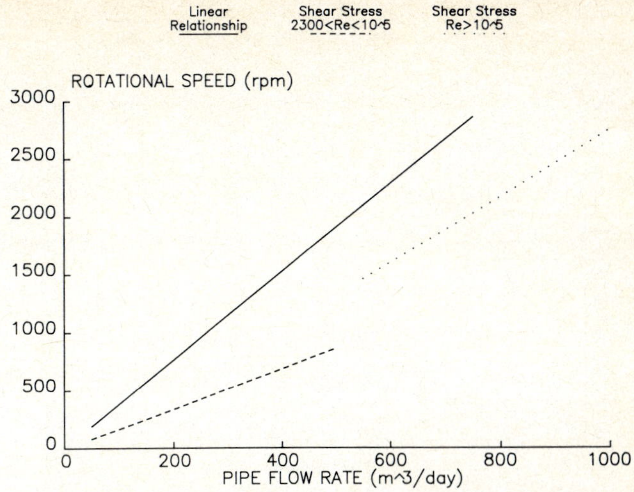

FIGURE 5 — Electrode rotational speed vs fluid flow rate (electrode r = 0.635 cm, pipe dia. = 7.62 cm).

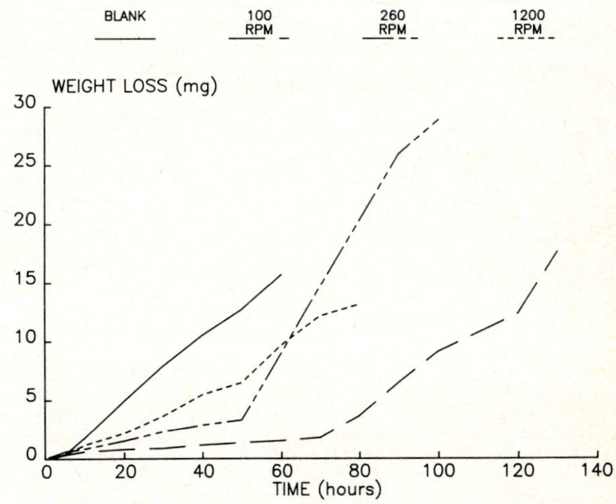

FIGURE 6 — Effect of velocity on inhibitor film life (500 ppm filming concentration).

test well where previous circulation tests and corrosion monitoring had been conducted. The three rotational speeds correlated to the calculated shear stress in the three line sizes of the field flow loop. (See Figure 5 and Appendix A.) Figures 6 and 7 show the cumulative weight loss data for the 500 and 1000 ppm velocity tests, respectively.

Polarization resistance data for the same two filming concentrations are shown in Figures 8 and 9. It can be seen that when the inhibitor film is protective, resistance values of approximately 1000 ohms are obtained. At higher velocities and/or lower filming concentrations, resistance values are in the range of 200 to 300 ohms. Unfilmed electrodes had resistances in the order of 50 to 70 ohms.

If the trend in polarization resistance readings is a valid measure of inhibitor protection, then AC impedance has the potential to monitor inhibitor film life in high resistance or low-aqueous media, such as is en-

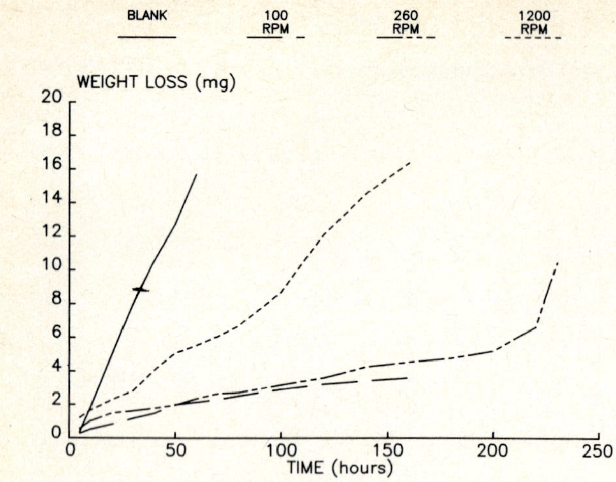

FIGURE 7 — Effect of velocity on inhibitor film life (1000 ppm filming concentration).

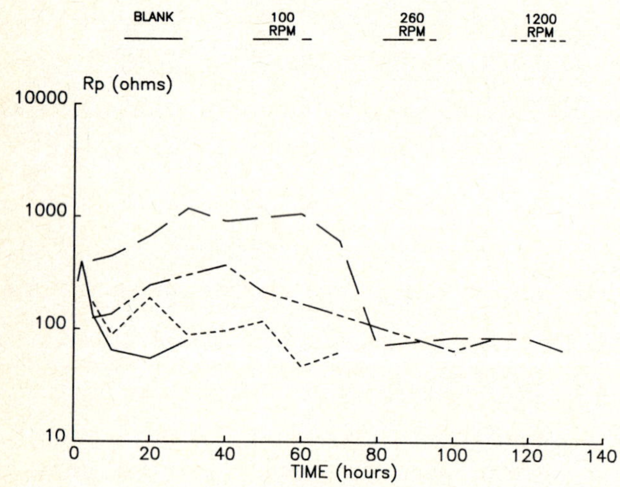

FIGURE 8 — Polarization resistance measurements (500 ppm filming concentration).

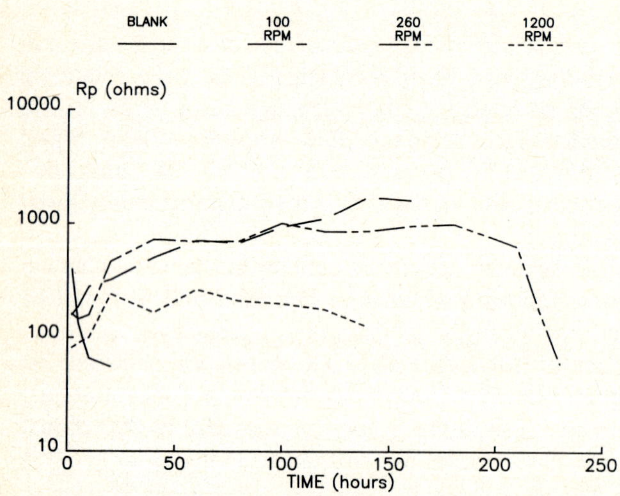

FIGURE 9 — Polarization resistance measurements (1000 ppm filming concentration).

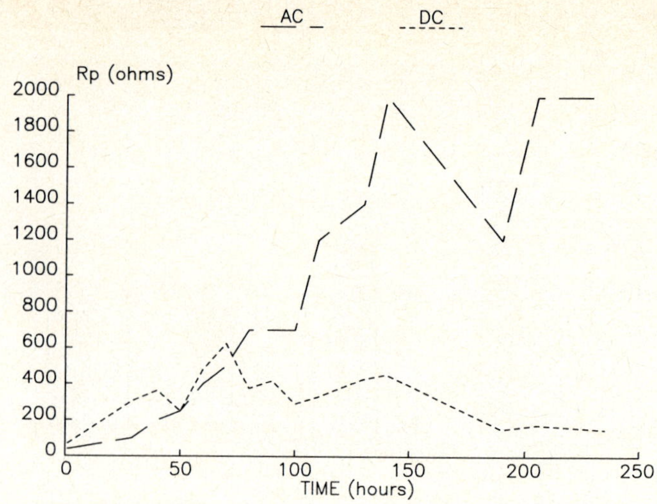

FIGURE 10 — AC impedance vs DC polarization (500 ppm filming concentration).

countered in a groupline after the free water knock-out. Even though grouplines contain over 90% crude oil, inhibition is still important because of the remaining water settling out and causing corrosion between the 5:00 and 7:00 o'clock positions on the interior pipe surface. Figure 10 shows a comparison between AC and DC polarization resistance values for 500 ppm inhibitor filming. For the first 75 h, the AC and DC values were in reasonable agreement; after 75 h, the DC readings decreased while the AC values increased. The high AC R_p values are usually difficult to estimate from Bode or complex plan plots since ideal semicircles were not always obtained.

Conclusions

Inhibitor performance, determined by film persistency wheel tests and DC linear polarization measurements, showed parallel trends of decreasing protection as a function of time. Although percent protection results obtained by linear polarization were generally higher, overall ranking of the inhibitors was reasonably consistent by both techniques.

Experiments conducted in the rotating cylinder electrode cell showed that filming inhibitor performance was affected by velocity. Within the range of the experimental parameters, the decreased corrosion protection at higher velocities could be somewhat offset by increasing the inhibitor concentration. These results indicate that the application guideline of using at least 1000 ppm filming concentration for a contact time of 1 h appears adequate for low to intermediate production rates (<100 m^3/day) with a treatment interval of two weeks.

Polarization resistance values (R_p) showed that effective inhibitor films had resistances of approximately 1000 ohms, and that film degradation was reflected by R_p values falling to unprotected (blank) values of approximately 50 to 75 ohms.

R_p values obtained from preliminary AC impedance measurements showed short-term correlation to DC values, but decrepancies developed between the techniques on longer term experiments. Semicircular complex plane plots were not always obtained, indicating that a simple circuit model of solution resistance, polarization resistance, and double layer capacitance may not be adequate to describe filming inhibitor behavior fully.

Work is continuing to compare alternating AC and DC measurements on the same rotating electrode cell, and to compare laboratory data on the effects of velocity to wellsite corrosion measurements after periodic inhibitor injections.

Acknowledgments

The authors thank Esso Resources Canada, Ltd., for granting publication rights and to V. S. Cass for typing this manuscript.

References

1. J. Tafel, Z. Fur Physiki Chemie, p. 641, 1905.
2. F. Mansfeld, Advances in Corrosion Science and Technology, Vol. VI, Plenum Press, New York, New York, p. 163, 1976.
3. Electrochemical Techniques for Corrosion, R. Baboian, Ed., National Association of Corrosion Engineers, Houston, Texas, 1977.
4. R. Annand, Corrosion, Vol. 22, No. 8, p. 215, 1966.
5. R. Klonowski, M. Jurozyk, "A Comparative Study Between Electrochemical and Classical Techniques for Evaluating Corrosion Inhibitors," CORROSION/80, Paper No. 193, National Association of Corrosion Engineers, Houston, Texas, 1980.
6. R. Leganlt, M. Walker, Corrosion, Vol. 19, p. 222t, 1963.
7. M. Walker, W. France, Mat. Prot., Vol. 8, p. 47, 1967.
8. M. Walker, Mat. Prot., Vol. 12, p. 29, 1973.
9. H. Lee Craig, Jr., "Electrochemical Methods of Determining Inhibitor Effectiveness for Aluminum Corrosion Problems," CORROSION/74, National Association of Corrosion Engineers, Houston, Texas, 1974.
10. K. Pillai, Corrosion Prevention and Control, Feb., 1980.
11. F. Donahue, K. Nobe, J. Electrochem. Soc., Vol. 114, p. 1012, 1967.
12. A. Akiyama, K. Nobe, J. Electrochem. Soc., Vol. 117, p. 999, 1970.
13. H. Kaesche, N. Hackerman, J. Electrochem. Soc., Vol. 105, p. 191, 1958.
14. H. Kaesche, Z. Electrochem, Vol. 63, p. 492, 1959.
15. H. Fisher, H. Yamaoka, Chem. Ber., Vol. 94, p. 1477, 1961.
16. E. Duwell, J. Electrochem. Soc., Vol. 109, p. 1013, 1962.
17. G. Okamoto, et al., Con. Sci., Vol. 2, p. 21, 1962.
18. H. Yamaoka, H. Fisher, Electrochem. Acta, Vol. 10, p. 679, 196.
19. F. Donahue, et al., J. Electrochem. Soc., Vol. 114, p. 1006, 1967.
20. N. Hackerman, et al., Corrosion, Vol. 31, p. 240, 1975.
21. K. Pillai, R. Narayan, J. Electrochem. Soc., Vol. 125, p. 1393, 1978.
22. K. Heusler, G. Cartledge, J. Electrochem. Soc., Vol. 108, p. 732, 1961.
23. M. Ahmed, et al., J. Electrochem. Soc., Vol. 124, p. 1667, 1977.
24. R. Chin, et al., Corrosion, Vol. 29, No. 5, p. 185, 1973.
25. D. Altura, K. Nobe, Corrosion, Vol. 32, No. 2, p. 41, 1976.
26. H. Kerst, "The Use of Electrochemical Technique in Studies of Corrosion Inhibitors and Their Evaluation," CORROSION/74, National Association of Corrosion Engineers, Paper No. 121, Houston, Texas, 1974.
27. F. Mansfeld, Corrosion, Vol. 30, No. 3, p. 92, 1974.
28. F. Mansfeld, Corrosion, Vol. 32, No. 4, p. 143, 1976.
29. F. Mansfeld, Werkst. Korros., Vol. 28, p. 6, 1977.
30. M. Danielson, Corrosion, Vol. 36, No. 4, p. 174, 1980.
31. F. Mansfeld, J. Electrochem. Soc., Electrochemical Science and Technology, Vol. 20, No. 4, p. 515, 1973.
32. R. Leroy, Corrosion, Vol. 29, No. 7, p. 272, 1973.
33. R. Leroy, Corrosion, Vol. 31, No. 5, p. 173, 1975.
34. W. Smyrl, Electrochemical Corrosion Testing, ASTM STP 727, F. Mansfeld, U. Bertocci, Eds., ASTM, Philadelphia, Pennsylvania, p. 198, 1981.
35. I. Epelboin, M. Keddam, H. Takenouti, J. Appl. Electrochem., Vol. 2, p. 71, 1972.
36. I. Epelboin, C. Gabrielli, M. Keddam, H. Takenouti, Electrochemical Corrosion Testing, ASTM STP 727, F. Mansfeld, U. Bertocci, Eds., ASTM, Philadelphia, Pennsylvania, p. 150, 1981.
37. S. Haruyama, T. Tsuru, Electrochemical Corrosion Testing, ASTM STP 727, F. Mansfeld, U. Bertocci, Eds., ASTM, Philadelphia, Pennsylvania, p. 167, 1981.
38. L. Williams, R. Taylor, Corrosion, Vol. 38, No. 8, p. 125, 1982.
39. D. D. MacDonald, M. C. H. McKubre, Electrochemical Corrosion Testing, ASTM STP 727, F. Mansfeld, U. Bertocci, Eds., ASTM, Philadelphia, Pennsylvania, p. 110, 1981.
40. F. Mansfeld, "Recording and Analysis of AC Impedance Data for Corrosion Studies," CORROSION/81, Paper No. 155, National Association of Corrosion Engineers, Houston, Texas, 1981.
41. F. Mansfeld, Corrosion, Vol. 36, No. 5, p. 301, 1981.
42. F. Mansfeld, M. Kendig, S. Tsai, Corrosion, Vol. 38, No. 11, p. 570, 1982.
43. M. Kendig, F. Mansfeld, Corrosion, Vol. 39, No. 11, p. 466, 1983.
44. F. Mansfeld, M. Kendig, S. Tsai, "Evaluation of Corrosion Behavior of Coated Metals with AC Impedance Measurements," CORROSION/82, Paper No. 182, National Association of Corrosion Engi-

neers, Houston, Texas, 1982.
45. F. Mansfeld, M. Kendig, S. Tsai, Corrosion, Vol. 38, No. 9, p. 478, 1982.
46. J. Payer, "Electrochemical Methods for Coatings Study and Evaluation," Electrochemical Techniques for Corrosion, R. Baboian, Ed., National Association of Corrosion Engineers, Houston, Texas, 1977.
47. J. Scantlebury, A. Guiseppielie, D. Eden, L. Callow, Corrosion, Vol. 39, No. 3, p. 108, 1983.
48. J. Scantlebury, K. Ho, D. Eden, Electrochemical Corrosion Testing, ASTM STP 727, F. Mansfeld, U. Bertocci, Eds., ASTM, Philadelphia, Pennsylvania, p. 187, 1981.
49. W. J. Lorenz, F. Mansfeld, Corros. Sci., Vol. 21, No. 9, p. 647, 1981.
50. A. A. Aksu, W. Lorenz, F. Mansfeld, Corros. Sci., Vol. 22, No. 7, p. 611, 1982.
51. N. D. Greene, Experimental Electrode Kinetics, Rensselaer Polytechnic Institute, Troy, New York, 1965.
52. F. Mansfeld, Corrosion, Vol. 29, No. 10, p. 397, 1973.
53. M. Danielson, Corrosion, Vol. 38, No. 11, p. 580, 1982.
54. S. Barnartt, M. Donaldson, Corrosion, Vol. 39, No. 1, p. 33, 1979.

Appendix A – Rotating Cylinder Electrode

Correlation of Shear Stress

To find the suitable rotation speed in the laboratory to simulate fluid flow in a pipe under production conditions, it is necessary to determine the shear stresses of the two cases.

For Pipe Flow

$$Re_{pipe} = 1473.7 \frac{Q\rho}{D\mu}$$

where: Re is the Reynold's number (dimensionless); Q is the flow rate (m^3/day); ρ is the density of the fluid (g/cm^3); D is the inner diameter of the pipe (cm); and μ is the viscosity of the fluid (CP).

For the turbulent regime of:

$$2300 < Re_{pipe} < 10^5$$

the friction factor (f) is:

$$f_{pipe} = 0.079 \left(Re_{pipe}\right)^{-0.25}$$

$$= 0.0128 \left(\frac{Q\rho}{D\mu}\right)^{-0.25}$$

For the turbulent regime $Re_{pipe} > 10^5$:

$$f_{pipe} = 0.046\, Re_{pipe}^{-0.2}$$

$$= 0.017 \left(\frac{Q\rho}{D\mu}\right)^{-0.2}$$

To determine the shear stress (τ) in both cases, the following equation must be used.

$$\tau_{pipe} = 100 \frac{f_{pipe}}{2} \rho \left(0.1474 \frac{Q}{D^2}\right)^2$$

$$= 10.8583 \frac{f_{pipe}\, \rho Q^2}{D^4}$$

For the Rotating Cylinder Electrode

$$Re_{cyl} = 10.4720 \frac{r^2\omega\rho}{\mu}$$

where: ω is the rotational speed (rpm); and r is the radius of the electrode (cm).

The friction factor for the cylinder is:

$$\frac{f_{cyl}}{2} = 0.0791\, Re_{cyl}^{-0.3}$$

$$= 0.0391 \left(\frac{r^2\omega\rho}{\mu}\right)^{-0.3}$$

The shear stress of the cylinder is:

$$\tau_{cyl} = 1096.6 \left(\frac{f_{cyl}}{2}\right) \rho\omega^2 r^2$$

Equating the Shear Stresses (i.e., $\tau_{pipe} = \tau_{cyl}$) and Solving for Rotational Speed

(A) For $2300 \le Re_{pipe} \le 10^5$:

$$10.8583 \left(f_{pipe}\right) \frac{\rho Q^2}{D^4} =$$

$$1096.6 \left(\frac{f_{cyl}}{2}\right) \rho\omega^2 r^2$$

$$10.8583 \left(0.0128 \left(\frac{Q\rho}{D\mu}\right)^{-0.25}\right) \frac{Q^2}{D^4}$$

$$= 1096.6 \left(0.0391 \left(\frac{r^2\omega\rho}{\mu}\right)^{-0.3}\right) \rho\omega^2 r^2$$

$$r^{1.4}\, \omega^{1.7}\, \rho^{0.7}\, \mu^{0.3} =$$

$$3241.5 \left(\frac{Q^{1.75}\, \rho^{0.75}\, \mu^{0.25}}{D^{3.75}}\right)$$

$$\omega = \left(3241.5 \frac{Q^{1.75}\, \rho^{0.05}}{r^{1.4}\, D^{3.75}\, \mu^{0.05}}\right)^{1/1.7}$$

(B) For $Re_{pipe} > 10^5$:

$$10.8583 \left(0.0107 \left(\frac{Q\rho}{D\mu}\right)^{-0.2}\right) \frac{\rho Q}{D^4}$$

$$= 1096.6 \left(0.0391 \left(\frac{r^2\omega\rho}{\mu}\right)^{-0.3}\right) \rho\omega^2 r^2$$

$$r^{1.4}\, \omega^{1.7}\, \rho^{0.7}\, \mu^{0.3} =$$

$$2709.7 \left(\frac{Q^{1.8}\, \rho^{0.8}\, \mu^{0.2}}{D^{3.8}}\right)$$

$$\omega = \left(2709.7\, \frac{Q^{1.8}\, \rho^{0.1}}{r^{1.4}\, D^{3.8}\, \mu^{0.1}}\right)^{1/1.7}$$

Section 5

Electrochemical Tests

Electrochemical Methods of Corrosion Testing

S. W. DEAN, JR.*

Abstract

Electrochemical methods have been used for a variety of purposes in corrosion testing. The corrosion rate can be measured by electrochemical methods because aqueous corrosion is an electrochemical phenomenon. A variety of polarization techniques, including Tafel line extrapolation, small amplitude polarization, and alternating current techniques have been used for this purpose. Polarization is also useful for characterizing the general corrosion resistance of metals. Specific tests, including the electrolyte corrosion (EC) test, the Ford anodized aluminum corrosion test (FACT), and the impedance test have been developed to measure the corrosion resistance of high performance coatings. Corrosion potential measurements are useful for a variety of purposes, including determining the susceptibility of bimetallic couples to galvanic corrosion, the solid solution composition of aluminum alloys, and the time of wetness in atmospheric exposures.

Introduction

The corrosion of metal in aqueous electrolytes occurs by an electrochemical mechanism. This implies that there are distinct cathodic and anodic reactions that constitute the corrosion process. There are often other chemical reactions that participate in the process, but they do not alter the fact that the process is essentially electrochemical in nature.

Electrochemical corrosion tests have been attractive to many investigators because they operate on the mechanism of the corrosion process. Electrochemical techniques offer a clean way to accelerate the corrosion process. For example, they can be used to increase or control the oxidizing potential of an environment and thereby reveal specific characteristics of a material, such as passivity, without the complication of adding chemical oxidizing agents. Similarly, electrochemical techniques can be used to measure corrosion rates without removing the specimen from the environment or even significantly disturbing it. This is particularly attractive because it allows an almost instantaneous measurement, even when the rate is low.

There are, however, some basic requirements for electrochemical testing. The corroding sample must be in an electrolyte, and it is necessary that this environment be continuous at least to the extent that all the electrodes show some electrical continuity through the electrolyte. In addition, the primary corrosion mechanism must be electrochemical, and if the corrosion rate is being measured, then the rate obtained will be the average over the entire specimen. The electrodes in question must be capable of being electrically isolated from their surroundings so that they not be in electrical contact with other conductors present in the electrolyte.

This paper provides a review of electrochemical methods of corrosion testing, including the more recent developments in the field. Earlier review papers on this subject[1,2] have provided much of the foundation for this paper, but developments have occurred in the interim. The emphasis of this paper is on practical testing methods for material selection, quality control and acceptance, and process control. No effort has been made to present the various laboratory corrosion tests that have been used to evaluate specific corrosion mechanisms.

Electrochemical Corrosion Rate

Corrosion Current Density

Corrosion rates are generally expressed as either a

*Air Products and Chemicals, Inc., Allentown, Pennsylvania.

TABLE 1 — Values of Constants for Equations 1 and 2

Equation	Penetration Rate—Unit	Mass Loss Rate—Unit	i_{corr} Unit	ρ Unit	K_1	K_2
1	mpy	—	$\mu A/cm^2$	g/cm^3	0.1288	—
1,2	mm/y(1)	g/m^2d(1)	A/m^2(1)	kg/m^3(1)	327.1	0.8954
1	mm/y(1)	—	$\mu A/cm^2$	g/cm^3	3.27×10^{-3}	—
2	—	mdd	$\mu A/cm^2$	—	—	0.0895
2	—	mdd	A/m^2(1)	—	—	8.954×10^{-3}

(1) = SI unit.

rate of metal penetration or a rate of mass loss per unit area, usually depending upon the measurement technique used. Mass loss per unit area can be equated to an average penetration if the density and area of the metal are known, and such measurements are often preferred by investigators because of the ease and accuracy of mass measurements. Penetration rates are usually expressed in mm/y (1 mm/y = 39.37 mpy), while mass loss rate per unit area is expressed in g/m^2d [1 g/m^2d = 10 mg weight loss/d^2/day (mdd)]. However, corrosion rates can also be expressed in terms of current density by means of Faraday's Law. The conversion between metal penetration rate, CR, and corrosion current density is given below:

$$CR = K_1 \, i_{corr} \frac{eq\ wt}{\rho} \tag{1}$$

Where i_{corr} = corrosion current density, ρ = alloy density, eq wt = alloy equivalent weight, and K_1 = constant.

The conversion between mass loss rate per unit area, MR, and corrosion current density is given by a similar expression:

$$MR = K_2 \, i_{corr} \, eq\ wt \tag{2}$$

Values for K_1 and K_2 in various commonly used unit system are shown in Table 1.

Although Equation (1) is valid at every point on the specimen surface, it is not generally possible to measure the current density as such. Instead, the total corrosion current is measured or estimated; this is then converted to current density by dividing by the exposed area.

In the case of alloys, the equivalent weight should be calculated using the atomic fractions of each component element in an expression such as:

$$\text{Alloy eq wt} = \sum_{i=1}^{m} \frac{f_i M_i}{n_i} \tag{3}$$

Where f_i = atomic fraction of the i_{th} component element in the alloy, M_i = atomic weight of the i_{th} component element, n_i = electron loss required to oxidize the i_{th} component element under the conditions of corrosion process (usually equal to the stable valence of the element), and m = number of component elements in the alloy.

This expression assumes that all of the component elements oxidize when the alloy corrodes and that they are all oxidizing at essentially a uniform rate. In some situations, these assumptions are not valid, and in these cases, the calculated corrosion rate will be in error. For example, if an alloy is composed of two or more phases and one preferentially oxidizes, then the calculation must be made for the phase undergoing oxidation.

Although the expression shown in Equation (3) above is rigorously valid, it requires a two-step operation to compute atomic fractions and then the equivalent weight. For alloys where the components have similar atomic weight, such as brass where copper is 63.5 and zinc is 65.4, the atomic fractions are very close to mass fractions. In cases such as this, the process is often simplified to use mass fractions in equations rather than atomic fractions, and the error introduced by this simplification is minor.

An illustration of the errors that can occur by using weight average equivalent weights rather than atomic fraction average is shown in Table 2. These results clearly indicate that the errors that occur from the use of weight fractions rather than atomic fractions are often minimal. However, it is good practice to verify that the use of weight average values will not introduce significant errors.

The use of incorrect valence values in Equation (3) can produce more significant errors. To illustrate this, Table 3 has been prepared to show the range of equivalent weights that can be calculated for AISI 304 stainless steel (SS). It is unlikely that some elements in an alloy will be highly oxidized in the corrosion process while others will be in a relatively reduced state. The underlined values (in Table 3) represent more or less compatible valence states. It is clear that the wrong valence assumption can result in an error of up to 20% for AISI 304 SS. In cases where the corrosion products are soluble, analysis of the corrodent solution can provide information on the state of oxidation of the corrosion products, although further oxidation or reduction of corrosion product ions is possible and should be considered. In other cases, it is possible to determine the oxidation state of the corrosion products by x-ray or electron diffraction studies of the surface films.

TABLE 2 — Comparison of Equivalent Weights of Alloys Calculated by Mass Fractions and Atomic Fractions

Alloy	UNS No.	Equivalent Weights		
		Mass Fraction	Atomic Fraction	% Difference
Brass (CDA 260)	C26000	32.03	32.02	0.05
Stainless Steel (AISI 304)	S30400	26.11	25.98	0.52
Hastelloy Alloy C	N10276	27.14	27.02	0.44
Aluminum (AA 7075)	A97075	10.93	10.83	0.87

The Pourbaix equilibrium diagrams[3] can also be useful in estimating the stable valence of the corrosion products at the conditions of potential and pH in question. This approach is probably the easiest and most reliable because the potential and pH conditions are usually accessible in electrochemical tests.

In general, the corrosion current density shown in Equations (1) and (2) is not directly measurable because it is exactly balanced by a cathodic current density. However, it is possible to estimate the corrosion current density by a variety of techniques as discussed below.

Polarization Techniques for Estimating Corrosion Current

The rates of both cathodic and anodic reaction usually follow Tafel behavior, i.e.,

$$E = a + b \log i \qquad (4)$$

Where E is the electrode potential of the specimen, i is the current density of the electrochemical reaction, and a and b are constants.

The "b" constant in this equation is often referred to as the Tafel slope and is usually expressed in mV/decade (dec). The corrosion current can be estimated if the electrochemical corrosion processes follow this Tafel behavior. The net current density on the specimen is related to the current densities of the electrochemical reactions occurring on the specimen surface by the following equation:

$$i\,\text{net} = i\,\text{anode} - i\,\text{cathode} \qquad (5)$$

The exponental behavior of current density with potential means that the anodic contribution to the net current density can generally be ignored if the specimen potential is 50 mV more negative than the open circuit value, while the cathodic contribution is usually negligible at potentials 50 mV more positive than the open circuit value. These relationships are shown diagrammatically in Figure 1.[4]

Based on these observations, several techniques can be used to estimate corrosion currents: the specimen can be polarized to potentials at least 50 mV more negative

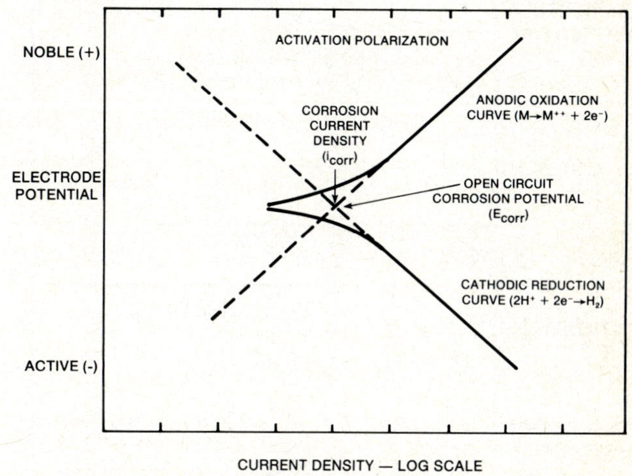

FIGURE 1 — Schematic cathodic and anodic polarization curve.

than the open circuit corrosion potential, and sufficient data points taken to obtain a reliable Tafel line. Then the current density data can be extrapolated back to corrosion potential using a semilogarithmic plot to obtain the corrosion current density value. The second technique is analogous to the first in that the specimen can be polarized anodically to potentials greater than 50 mV more positive than the corrosion potential, and the same process repeated. The third technique requires that both cathodic and anodic polarizations be carried out as described above, and the corrosion current is obtained at the intersection of the anodic and cathodic extrapolated lines. These three techniques are illustrated in Figure 2.

These techniques are used in laboratory tests to determine the corrosion rate of a metal or alloy in a specific environment.[5] However, they suffer from a variety of drawbacks that limit their usefulness. The resistivity of the solution can contribute significant errors to the estimation of Tafel slopes. The Luggin probe is intended to minimize the solution resistance error. The ideal Luggin probe is a fine capillary that is placed not closer than its outer diameter to the specimen. Closer placement causes errors because the probe shields the specimen. However, in dilute solutions or when high current densities are em-

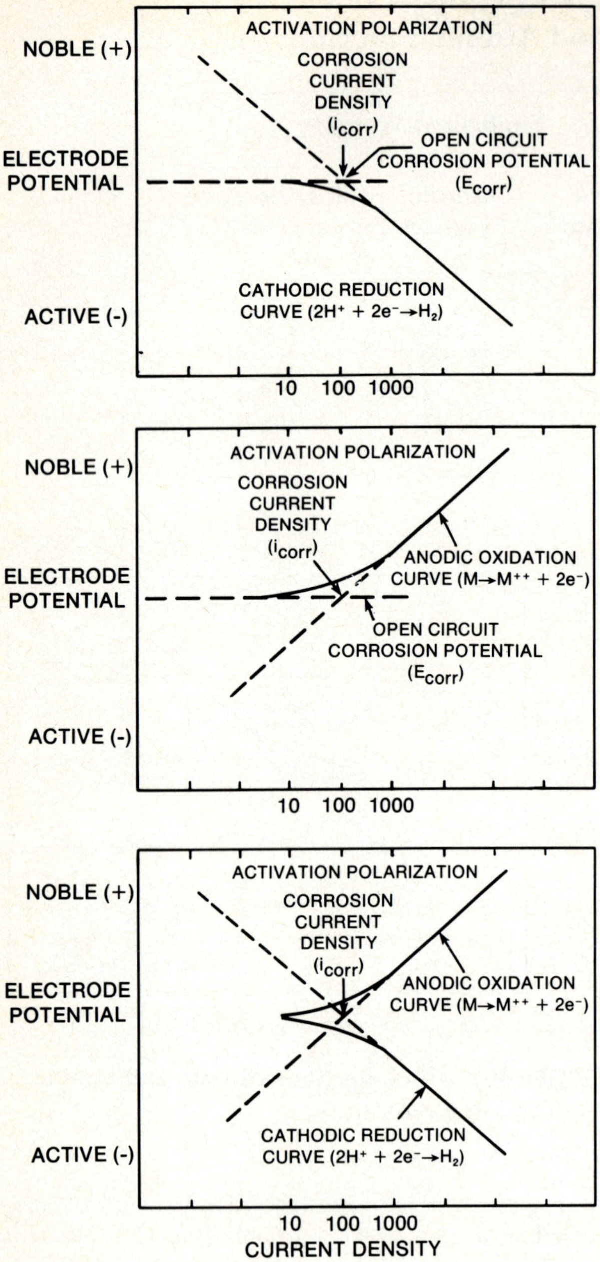

FIGURE 2 — Illustration of three Tafel extrapolation methods of estimating corrosion rate.

ployed, the solution resistance errors will be significant. In general, the effect of solution resistivity will be to introduce curvature in the Tafel line so that it is difficult or impossible to obtain an accurate linear portion of the curve that can be extrapolated. One approach has been to draw tangent lines to the measured curves at 50 mV away from the corrosion potential and consider these lines as the Tafel lines.

The effect of solution resistivity on the measured potential may be calculated from Ohm's Law:

$$\triangle E = \rho \, i \, l \qquad (6)$$

Where $\triangle E$ is the change in measured potential in V, ρ is the solution resistivity in ohm-cm, i is the applied current

TABLE 3 — Comparison of Calculated Equivalent Weights of AISI 304 SS with Various Valence Assumptions

Fe Valence	Cr Valence		
	2	3	6
2	27.7	26.1	24.6
3	26.4	24.8	23.2

Note: Underlined values represent compatible valence assumptions. Mass average used to calculate equivalent weights. Composition: 18% Cr, 10% Ni, balance Fe.

density in A/cm^2, and l is the distance from the specimen to the Luggin tip in cm.

This equation assumes that the potential gradient is linear with distance. The sign of $\triangle E$ will correspond to the Stockholm sign invariant convention for potential when the sign of i is based on positive for anodic currents and negative for cathodic currents as per ASTM Standard Practice G3.[4] Equation (6) will overestimate resistivity polarization errors in most laboratory cells because the auxiliary electrode area is usually substantially larger than the specimen electrode. However, this equation provides a useful approach for determining when resistivity errors will be significant.

Modern potentiostats often include resistance compensation based on an assumed solution resistance. This approach can help minimize resistance errors if the solution resistance is known and the current employed is relatively low. When high currents are used or tests are conducted in a highly resistive electrolyte such as relatively pure water, a better approach is to use a current interrupter. This is an electronic device that switches off the current briefly while the potential is being measured. The electrode potential decays relatively slowly while the solution resistivity contribution disappears instantly, thus allowing the specimen potential to be measured accurately.

Another source of error in using polarization curves to estimate corrosion rates is the scan rate effect. In order to obtain a true polarization curve, it is necessary to hold the potential constant until a stable current is achieved. In practice, it is normal to use a potentiodynamic scan, which gradually increases the potential and records the current continuously. When coupled with a logarithmic converter and an XY recorder, it is possible to obtain Tafel plots without the need for calculations or replotting of data. Scan rate effects may cause significant errors in Tafel curve determination when using a potentiodynamic technique. The magnitude of this effect can be estimated by rerunning the curve at a lower potential scan rate and comparing the current values obtained with the currents obtained at the higher rate. If the low scan rate curve shows substantially lower currents at comparable potentials, then it is likely that the scan rate errors are present in the system.

The use of polarization curves to determine corrosion rates has more errors than other techniques available, such as the polarization resistance approach. The method is useful because it does provide information on the electrochemical behavior of the system, but it is not unusual to encounter errors of a factor of 2 to 5 when comparing mass loss rates to corrosion current estimates from Tafel extrapolations.[6]

Polarization Resistance

Another approach to the problem of electrochemical corrosion rate measurement is to apply only a small potential difference to the specimen and measure the current this type of change produces. The potential-current density plot is approximately linear in the region within 10 mV of the corrosion potential. The slope of this plot in terms of potential divided by current density has the units of resistance-area and is often called the *polarization resistance*. This polarization resistance is related to the corrosion current density by the relationship:

$$i_{corr} = B \frac{i_{net}}{\Delta E} = \frac{B}{R_p} \qquad (7)$$

Where:

$$\frac{i_{net}}{\Delta E}$$

is the change in specimen current density per unit change in potential, R_p is the polarization resistance, B is the Stern-Geary constant of the system, and i_{corr} is the corrosion current density.

The Stern-Geary constant is related to the Tafel constants by the relationship:

$$B = \frac{b_a b_c}{2.303\,(b_a + b_c)} \qquad (8)$$

Where b_a and b_c are the anodic and cathodic Tafel constants respectively, measured on a decade scale.

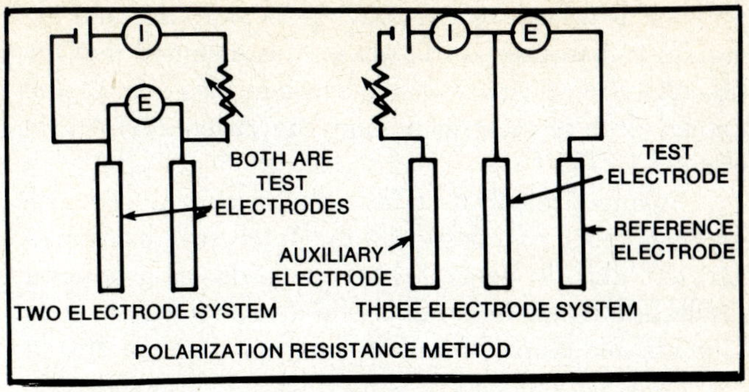

FIGURE 3 — Two types of polarization resistance probes.

The value

$$\frac{1}{R_p}$$

is sometimes referred to as *polarization admittance*. Typical units for these values are shown in Table 4.

In order to use this technique, it is, of course, necessary to know the Tafel slopes independently. However, in many cases, reasonable estimates can be made as to what these Tafel slopes should be, and in other cases, the method can be calibrated using mass loss data.

There are two different commercial systems in use for this purpose[5] and they are illustrated in Figure 3. One system employs three electrodes, usually of the same alloy.[7] One specimen is the working electrode, while the second electrode is the auxiliary electrode and the third is the reference electrode. The potential between the working and reference electrodes is monitored; in order to measure the corrosion rate, sufficient current is passed from or to the auxiliary electrode to displace the specimen potential either plus or minus 10 mV. The current flowing is then proportional to the corrosion rate. The size of the electrode is adjusted so that the same instrumentation can be used for each alloy. Instrumentation is available that can measure the current with either positive, negative, or alternating polarization.

The second approach is to use a two-electrode probe. In this case, both electrodes are of the same material. The

TABLE 4 — Typical Units for Polarization Resistance Measurements

Variable	Symbol	Conventional Unit	SI Unit
Corrosion Current	i_{corr}	$\mu A/cm^2$	A/m^2
Stern-Geary Constant	B	mV	mV
Polarization Resistance	R_p	ohm-cm^2	ohm-m^2
Tafel Constant	b_a, b_c	mV/dec	mV/dec
Polarization Admittance	$1/R_p$	mho/cm^2	siemens/m^2

potential between the electrodes is set at 20 mV and the current is measured. In this case, it is assumed that the potential drop divides equally between the cathode and anode. Both of these techniques are covered by ASTM Method D2776.[8]

Solution resistivity affects the accuracy of these methods much as it does with the Tafel curve extrapolation methods. In general, both methods will give good results as long as the product of the solution resistivity in ohm-cm and corrosion rate in mm/y is less than 250. If the three-element system is used with a close electrode spacing, the usable range is increased to a solution resistivity-corrosion rate product of 2500. The errors that occur from high resistivity electrolytes will decrease the measured corrosion rate below the rate actually experienced.

Both of these methods are very attractive in many applications, both in the laboratory as well as in plant situations. The corrosion rate can be measured directly and rapidly by commercially available instruments. Because the extent of polarization of the specimen is small, the measurement technique does not damage the specimen significantly, and so the measurement approach can be used for a long period of time as long as the general corrosion rate is not too high. Furthermore, it is possible to obtain mass loss data on the specimens themselves as a check on the corrosion rates measured. Generally, the correlation between mass loss rates and polarization resistance rates is between 10 and 50%, and much better than Tafel extrapolation techniques.

There are several problems with the polarization resistance approach. One problem is that the Stern-Geary constant is seldom known with accuracy because it is a function of the environment and alloy under study. Some values are available from literature sources.[9] It is also possible to run anodic and cathodic polarization curves as well as obtaining polarization resistance values on a specimen if the three electrode system is used. Another approach is to use the three point method of Barnartt[10] to obtain both Tafel constants and corrosion rate data. If alternating positive and negative polarization data are obtained, it is possible to use the method of LeRoy[11] to determine the value of D as given by the following equation:

$$D = \frac{1}{b_a} - \frac{1}{b_c} \qquad (9)$$

$$D = \frac{1}{\delta E} \log \left(\frac{i+}{i-} \right) \qquad (10)$$

Where δE is the polarization used (usually 10 mV), i+ and i− are the polarization currents measured in the positive and negative direction, and b_a and b_c are the anodic and cathodic Tafel constants.

When the D values become large, it is an indication that linear beahvior of the potential current relationship is not occurring. In this case, it is possible to use an averaging technique to obtain a better approximation of the corrosion rate.

$$R = \frac{2C \times A}{(C + A)} \qquad (11)$$

Where R is the corrected corrosion rate, C is the rate estimated by negative polarization, and A is the rate estimated by positive polarization.[8]

It should be noted that these methods apply only to the three-electrode configuration. In the two electrode system,[13] a difference between the current readings when the polarity is reversed indicates that localized corrosion is occurring and the highest value should be used for the corrosion rate.[8]

Both of these methods are useful to monitor the corrosivity of process streams. These devices can be used to signal upset conditions and to initiate the addition of corrosion inhibitors to a system. One large cooling system saved the cost of the monitoring system within one year by savings in inhibitor costs.

Other Small Amplitude Polarization Techniques

The polarization resistance technique uses a single finite polarization potential, usually 10 mV, to estimate the polarization resistance. If a potentiodynamic approach to polarizing the specimen is available, then more sophisticated analyses can be performed. The simplest approach is to use a constant scan rate polarization and obtain a plot of current versus potential. If the scan direction is reversed at some point, e.g., 10 mV, then a hysteresis curve will be generated. This approach is sometimes referred to as small amplitude cyclic voltammetry (SACV). If the hysteresis is significant, then the potential scan rate is probably too high and the run should be rerun at lower rates until hysteresis is no longer observed.[14] The slope of the potential-current curve at zero polarization is then the polarization resistance.[8]

A standard practice has been developed for the potentiodynamic polarization resistance technique by ASTM Committee G1. This standard, G59,[15] provides a reference test together with results for a reference alloy and results from an interlaboratory round robin program showing error bands of typical performance. The reference alloy is a lot of AISI 430 SS (UNS S43000) and the test environment is deaerated in sulfuric acid while the scan rate is 0.6 V/h. This method provides a technique that can be very helpful in allowing investigators to verify their techniques. The reference curve is shown in Figure 5. Potentiodynamic techniques are subject to significant errors from double-layer capacitance effects that cause underestimations of R_p and overestimations of the corrosion rate. On the other hand, uncompensated solution resistance causes overestimations of R_p. Mansfeld has discussed detailed procedures for estimating these errors.[16,17]

Another approach to small amplitude potentiodynamic polarization is to use a sinusoidal potential perturbation. In this case, the current will show a sinusoidal response somewhat out of phase with the potential signal. The resulting data output can be treated by conventional techniques using AC instrumentation to yield

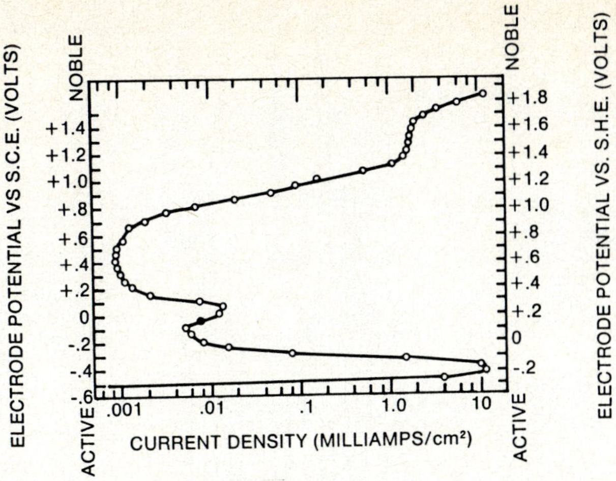

FIGURE 4 – Polarization curve for AISI 430 SS in 1.0 N sulfuric acid at 30 C.

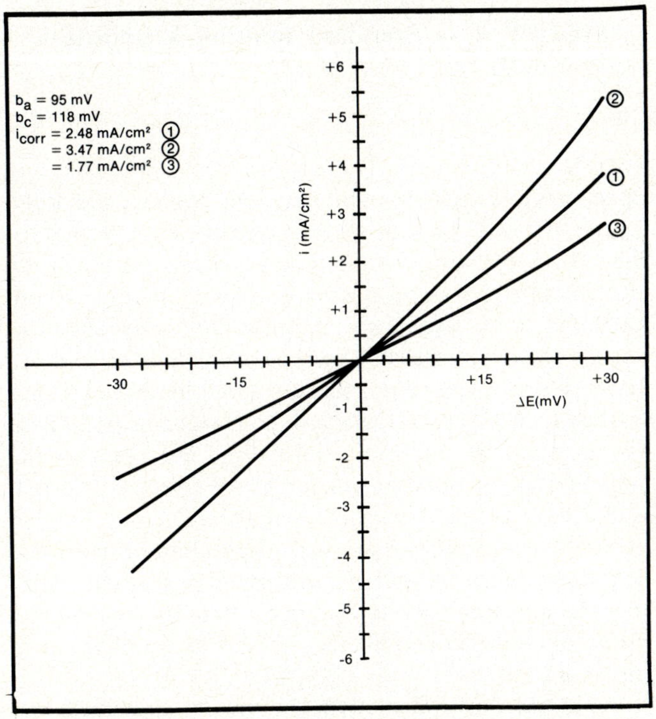

FIGURE 5 – Standard polarization curves for potentiodynamics polarization resistance measurements based on results from eight independent laboratories. Curve 1 is the mean result, while curves 2 and 3 represent +2 standard deviation.

Nyquist diagrams of the real and imaginary components of the polarization resistance. Figure 6 shows the equivalent circuit applicable for most electrode systems. See Figure 7 for an example of this type of data. A purely capacitive-resistive network will yield a true semicircle. Problems sometimes occur at the low frequency end of the data set.

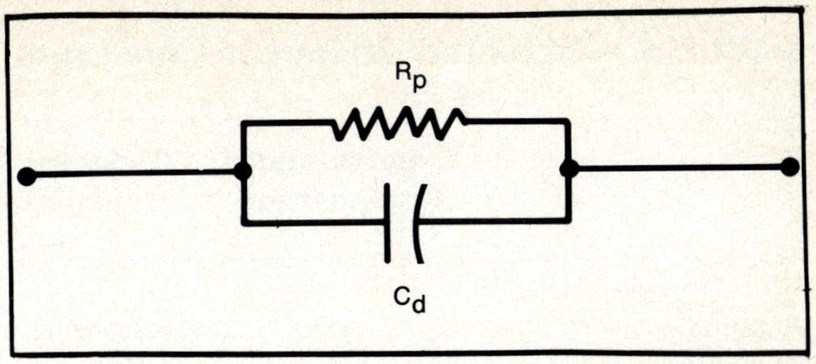

FIGURE 6 – Parallel component electrical model of electrode surface.

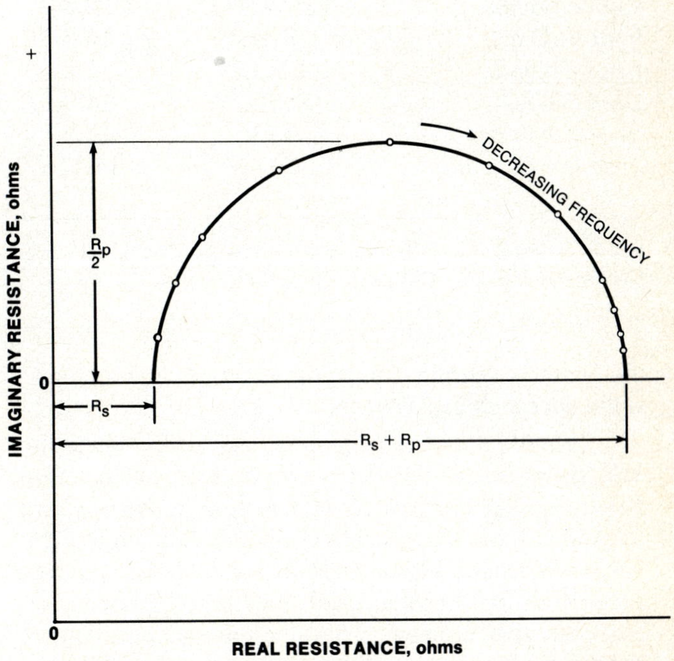

FIGURE 7 – Real versus imaginary resistance values measured as a function of frequency (Nyquist diagram).

This approach allows for the separation of solution resistance from the polarization resistance. It also provides a measure of the double-layer capacitance, which is in turn related to the thickness of the corrosion product or oxide layer on the specimen surface. The following equations can be used in those calculations.

$$C = \frac{1}{2\pi R_p f} \quad (12)$$

$$t - \frac{\epsilon A}{C} \quad (13)$$

Where R_p is the polarization resistance, f is the frequency at maximum value of the imaginary impedance, ϵ is the electric permitivity of the layer, A is the true specimen

TABLE 5 — Selected Dielectric Constants

	Approximate Composition	Dielectric Constant
Alumina	Al_2O_3	4.5-8.4
Aluminum Bromide	$AlBr_3$	7.1
Aluminum Chloride	$AlCl_3$	7.0
Antimony Trichloride	$SbCl_3$	5.34
Cupric Oxide	CuO	18.1
Cupric Sulfate	$CuSO_4$	10.3
Cupric Sulfate	$CuSO_4 \cdot 5H_2O$	7.8
Ferrous Oxide	FeO	14.2
Lead Chloride	$PbCl_2$	4.2
Lead Sulfate	$PbSO_4$	14.3
Silver Chloride	AgCl	11.2
Titanium Dioxide	TiO_2	14-110

SOURCE: Handbook of Chemistry and Physics, p. E63 (49th ed.), The Chemical Rubber Co., Cleveland, Ohio, 1968.

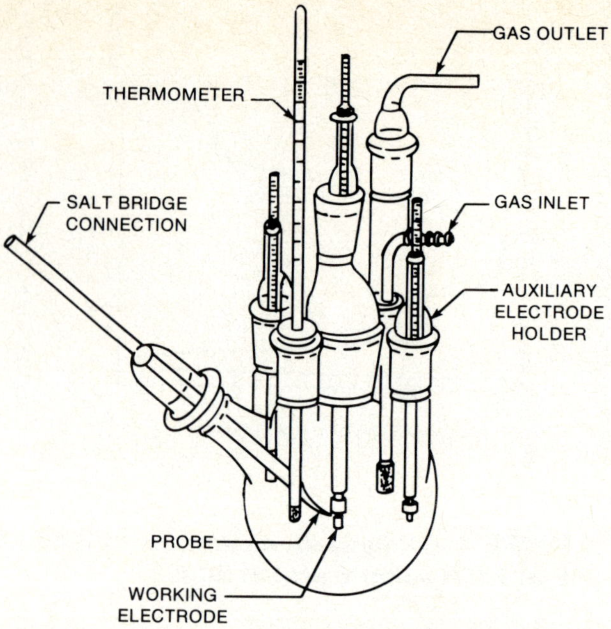

FIGURE 8 — Standard multineck polarization cell.

area of the electrode, t is the oxide film thickness, and C is the specimen capacitance.

The AC methods have been used to eliminate the solution resistance contribution to polarization, but they have not been exploited to estimate protective film thickness. One reason for this is that the true area of the specimen is often substantially greater than the apparent area as a result of pitting and roughening of the specimen. Another problem is that the dielectric constant of the film must be known. The ϵ term in Equation (13) is the product of the dielectric constant K and the electric permitivity of free space, ϵ_0.

$$\epsilon = K\epsilon_0 \tag{14}$$

Where $\epsilon_0 = 8.854 \times 10^{-12}$ farads/m.

Unfortunately, the data on dielectric constants are sparse and knowledge of the film composition is often not known. A few values are tabulated in Table 5.[18] Thus, film thickness calculations based on capacitance measurements are usually not reported.

The use of AC polarization techniques can allow the polarization resistive systems, either in specimens with protective coatings or in highly resistive electrolytes. In these cases, it is not necessary to run the polarization behavior over a full range of frequencies, but only a few points, for example, high and low frequencies. Haruyama, et al.,[19] have proposed using 10 KHz and 0.01 Hz for the frequencies and have obtained good results with this approach.

General Polarization Techniques

Laboratory Methods

Polarization techniques have been used for years to characterize the corrosion behavior of metal alloys in specific environments. The approach is to immerse the specimen in an environment that does not have oxidizing agents present, such as oxygen, ferric ions, chromates, etc. The metal specimen is then polarized anodically to determine its corrosion current as a function of oxidation potential. This approach is particularly desirable for metals showing passivity, such as stainless steel and related alloys. A typical polarization curve for AISI 430 SS is shown in Figure 4. The anodic current at potentials greater than 50 mV more noble than the open circuit potential should approximate the actual corrosion current at these potentials. However, it must be realized that the current values obtained are very dependent on both the solution conditions and the polarization method. This technique can be used to evaluate different alloys over a range of oxidizing conditions if the other conditions are held constant.

Modern instrumentation has greatly simplified the procedures required to obtain useful results. The potentiostat, electrometer, logarithmic converter, potential programmer, recorder and data acquisition system can be obtained as a single console. Similarly, the electrochemical cells used to conduct corrosion tests have been developed to allow both cylindrical and flat specimen configurations. The traditional polarization cell[20] is shown in Figure 8. In carrying out polarization diagram measurements, it is important to deaerate the system thoroughly and to guard against crevice corrosion. This is especially important in systems that show active-passive behavior.

ASTM Committee G1 has established a reference method, G5, using one lot of AISI 430 SS as a reference.[21] The results are shown in this standard. This technique has been particularly valuable to newcomers in the field because it gives them a way to verify their technique. The results in G5 are based on a composite of 15

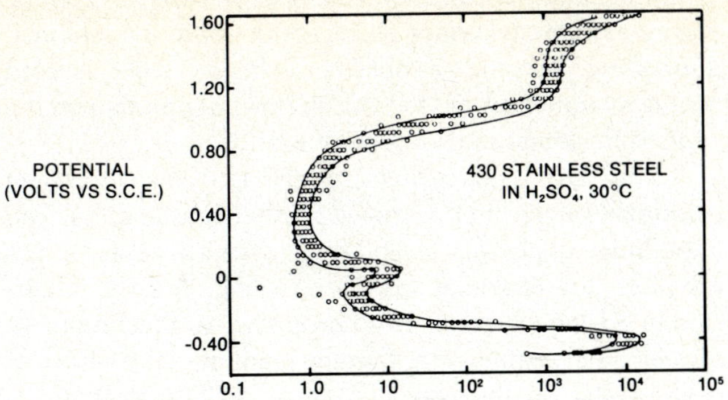

FIGURE 9 – Summary of potentiostatic anodic polarization curves from 15 laboratories for AISI 430 SS in 1.0 N H_2SO_4 at 30 C.

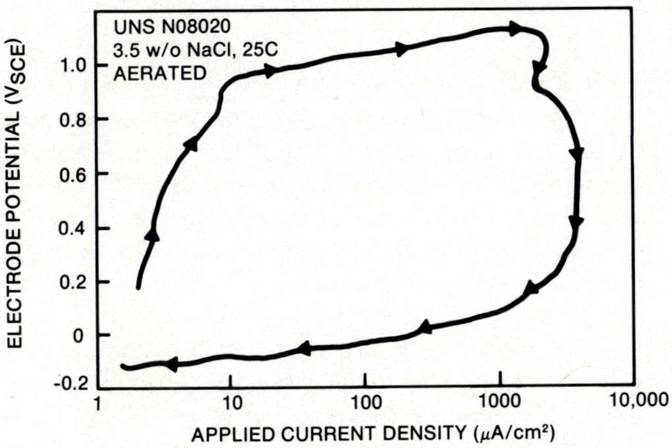

FIGURE 10 – Cyclic potentiodynamic anodic polarization curve for Carpenter 20-Cb3 containing a synthetic crevice. Note hysteresis loop.

independent laboratories, and it clearly demonstrates the degree of reproducibility obtainable with this method. A composite diagram showing the results of the interlaboratory program is presented in Figure 9. This method has been updated over the years to include modifications of the cell design and changes in the reference alloy as the old supply was exhausted.

Polarization techniques have become important in evaluating the tendency of alloys to show crevice or pitting corrosion. The concept here is to apply an increasing potential until the specimen undergoes accelerated attack as indicated by a rapid increase in current density. The potential scan is then reversed and the return of the current to the passive range is recorded. The characteristics of this potential-current hysteresis curve are related to the tendency of the material to pit or show crevice corrosion in the environment in question. A typical hysteresis curve[22] is shown in Figure 10. A standard method for conducting this type of cyclic potentiodynamic polarization measurement has been developed by ASTM Committee G1 as Standard Practice G61.[23] This practice includes reference alloys and reference curves showing error bands found in an interlaboratory round robin test.

This approach is based on the idea that when the potential reaches the point where the current begins to increase sharply, pitting or crevice corrosion is actually initiating. Consequently, this potential is often referred to as the *pitting* or *breakdown potential*. If the specimen has a crevice former present on it, this potential is related to the initiation of crevice corrosion. The solution within the pit or occluded area changes, usually becoming acidic and concentrated in chloride ions. When the potential scan is reversed, the ability of the alloy to repassivate and overcome the adverse chemistry within the occluded area is measured. This approach can also be used for marginally inhibited acid systems to show their tendency to cause pitting.[6]

EC Test

The electrolyte corrosion (EC) test[24-27] is an example of a polarization technique that is used to simulate extended atmospheric service on plated parts. This test was developed as a rapid evaluation technique for automotive parts with decorative multiple layer copper (optional)-nickel chromium plating systems. The specimen is masked to exclude insignificant areas from the test, and the test area is scrubbed with a MgO slurry to clean it. The specimen is then immersed in an acidic sodium nitrate, sodium chloride solution and polarized for a series of 1 min intervals. The polarization potential is +0.300 V relative to a saturated calomel electrode. After each interval, the corrosion is evaluated by means of an indicator solution that reveals penetration by the corrosion process to the base metal. Each period of two minutes is approximately equivalent to one year automotive service exposure in Detroit, Michigan. A commercial test instrument is available.[28]

The EC test has been standardized by ASTM Committee B8 as Standard Method B627 for evaluation the corrosion resistance of copper-nickel-chromium electrodeposits on steel and zinc base die castings designed for outdoor service.[29] The results of the test are reported as the number of corrosion sites that develop at the end of each exposure cycle.

FACT Test

The Ford anodized aluminum corrosion test (FACT) was developed as a rapid quality test for anodized aluminum in automotive applications.[30] This test also produces damage to the coating electrochemically by polarizing the sample in a specific manner. The extent of damage is measured by the current flow in the system over the test period. This has been standardized by ASTM Committee B8 as Standard Method B538.[31]

The specimen is made the cathode in an electrochemical cell placed on the specimen surface. The cell consists of a 1/4 in. (6 mm) inside diameter (ID) glass tube with a rubber gasket that seals the tube to the specimen and exposes a 1/8 in. (3.2 mm) diameter circle of the specimen to the test solution. The anode is a platinum wire, and the electrolyte is a 5% sodium chloride solution containing 1 g/gal (1 g/3.7 L) of cupric chloride. The solution is adjusted to pH 3.1 with acetic acid. A diagram

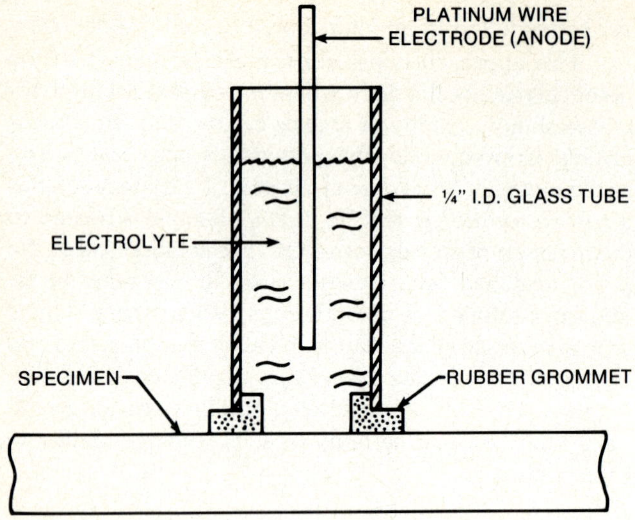

FIGURE 11 – Section diagram of FACT test cell.

of the cell is shown in Figure 11.

The cell is incorporated into an electrical circuit which impresses a high voltage on the specimen through a drop resistor. As the specimen coating is attacked, the cell current increases and the voltage decreases. The FACT number is obtained by integrating the voltage across the cell vs time for a three-minute period. FACT numbers of 600 volt-seconds (V-s) are obtained when no coating is present. 750 V-s indicate a coating of adequate thickness and seal.

The electrolytic action causes locally high pH regions at sites of high current density on the surface, and this shows the expected correlation with finishing variables, i.e., thicker coatings and longer sealing times give higher FACT numbers. The test is particularly sensitive to nickel acetate sealing, which gives substantially higher FACT numbers than hot water sealing for the same seal time. The test also gives significantly higher values on alloys with higher magnesium contents. This is probably a result of the low solubility of MgO in basic solutions. A commercial instrument is available.[32]

Impedance Test

The impedance test for anodized aluminum[33,34] is a commercial test used to evaluate the seal quality of anodized aluminum. This type of test is used both in the US and in Europe with minor variations. In the US, the test has been standardized by ASTM Committee B8 as ASTM B457[35] and is also known as the Aztac test. This test is similar to the FACT test in the sense that the test cell is placed on the specimen surface with a rubber gasket to seal in the liquid and define the specimen area at 0.20 in.^{2} (129 mm^2). The auxiliary electrode is either platinum or stainless steel. The electrolyte is a 3.5% sodium chloride solution.

Measurements are made with an impedance bridge employing a 1.0 V root mean square (rms) 1000 Hz source. A delay time of 2 min is specified between filling the cell and making the measurement. The results of the test are coating impedance values expressed in kilohms. A bare aluminum surface yields about 0.4 to 0.6 kilohms, depending on the anodizing voltage, while sealed coatings range from 50 to 500 kilohms, depending on the coating thickness and degree of seal.

This test yields results rapidly and they correlate reasonably well with the known finishing parameters. The impedance of partially sealed specimens increases with its age, i.e., the time elapsed since sealing. A 30-s immersion in boiling water has been proposed as a technique to activate the coating and thereby minimize the effect of coating age on the result. A commercial instrument is available.[36] The test is essentially nondestructive.

The area tested is small so that point-to-point variations in coating quality can be missed. Because leakage of the electrolyte will give erroneously low values, the cell should be checked carefully to see whether leakage has occurred and to ensure that bubbles have not blinded the test area and produced an erroneously high result.

EPR Test

The electrochemical potentiokinetic reactivation (EPR) test is an electrochemical test to quantitatively assess the degree of sensitization of austenitic stainless steel alloys. The test was designed to be rapid, nondestructive, and possibly to be applied in field situations. Failures in boiling water reactors in nuclear power plants had prompted the development of this test.[37-39]

The procedure is based on a reverse scan polarization technique similar to that discussed earlier. The specimen is exposed to a deaerated 0.5 molar (M) H_2SO_4 + 0.01 M KSCN solution at 30 C. The potential is first held in the active region, −0.300 V vs SCE for 2 min. The potential is then held in the passive range at +0.200 V vs SCE for 2 min. The potential is then scanned in the negative direction at 6 V/h for 6 min so that the potential returns to −0.400 V vs SCE. During this return scan, the current is measured and the total charge to reactivate the specimen Q is obtained. This charge is then normalized according to the grain boundary area exposed by the relationships:

$$GBA = A_S [5.095 \times 10^{-3} \exp (0.347X)] \qquad (15)$$

Where A_S is the specimen area and X is the ASTM grain size number determined at 100X.

$$P_a = \frac{Q}{GPA} \qquad (16)$$

The P_a values usually expressed in units of coulombs/cm^2. Values of P_a below 0.1 indicate a nonsensitized structure while values above 2 are likely to exhibit intergranular corrosion.

The use of this technique in the field is possible although difficult unless an experienced operator is available. Surface roughness can affect the results. Also, the method is not as sensitive as other techniques in heavily sensitized alloys.[40] This method is currently under development in ASTM Committee G1.

Corrosion Potential Measurements

Galvanic Series

Corrosion potential measurements in themselves have not been widely used because interpretation of the results of these measurements in the absence of other data is not possible. However, potential measurements are useful in determining galvanic relationships in an environment. The galvanic series in seawater[41] has been widely used because of its ability to reveal when galvanic corrosion is a potential problem. ASTM Committee G1 has prepared a Standard Guide, G82, for the development and use of a galvanic series.[42] This approach is useful for determining which metal in a couple will be subject to accelerated corrosion from galvanic effects. Different types of galvanic series are discussed and the techniques for developing them are presented.

Irreversible Corrosion Potential of Aluminum

The electrode potential of an alloy is determined by a number of factors, including the chemical composition and metallurgical condition of the alloy. With aluminum alloys in a salt peroxide electrolyte, the chemical and metallurgical factors contribute to the corrosion potential to the point where this potential can be used as a rapid method to determine the metallurgical condition of many alloys.[43] In this case, the solution used contains 58.5 g/L sodium chloride and 9 mL/L of a 30% hydrogen peroxide solution. The specimen is immersed in this solution with a reference electrode, i.e., saturated calomel electrode (SCE), and the potential is measured with an electrometer. The electrode potential stabilizes in 30 to 60 min so that an accurate value can be obtained in this period. For example, in the case of binary aluminum-copper alloys, the irreversible corrosion potential is determined by the solid solution copper content in the alloy. The results of potential measurements on a variety of binary aluminum-copper alloys are shown in Figure 12 after various thermal aging treatments. The results show clearly that, as the solid solution copper content increases, the specimen potential becomes more noble. Thermal treatments interrupt this behavior because of the precipitation of CuAl, which limits the concentration of copper in solid solution in the alloy. ASTM Committee G1 has standardized this method as Standard Practice G69.[44] A reference value of -748 ± 10 mV has been established for Alloy AA[(1)] 3003 aluminum sheet. This method is useful in verifying the heat treatment of aluminum-copper and aluminum-zinc alloys.

Atmospheric Corrosion Applications

The fact that dissimilar metals have different corrosion potentials has been employed by some investigators to monitor atmospheric conditions to determine when corrosion would occur.[45] The concept here is that atmospheric corrosion occurs only when the specimen surfaces are wet. A cell composed of gold and zinc or gold and copper elements on a thin insulator base is exposed.[46] After an initial period of exposure, zinc or copper corrosion products bridge the gap between the zinc or copper metal and the gold. Because these corrosion products are hygroscopic, they allow the passage of galvanic current between the copper and gold when the relative humidity of the atmosphere surrounding the panel approaches saturation. By proper calibration, it is possible to adjust the detector circuit so that a timer is activated when the local relative humidity exceeds about 85%. At this point, a timer is started which then records the length of time that wetness exists. The ratio between the time of wetness and the total exposure time then gives the time during which corrosion occurs in atmospheric environments. This time of wetness parameter has been found to correlate with atmospheric mass loss results in steels, and zinc and copper alloys in a number of atmospheric environments. The major factor causing irreproducibility in this method is freezing of the water to produce ice crystals on the cell surface.[47]

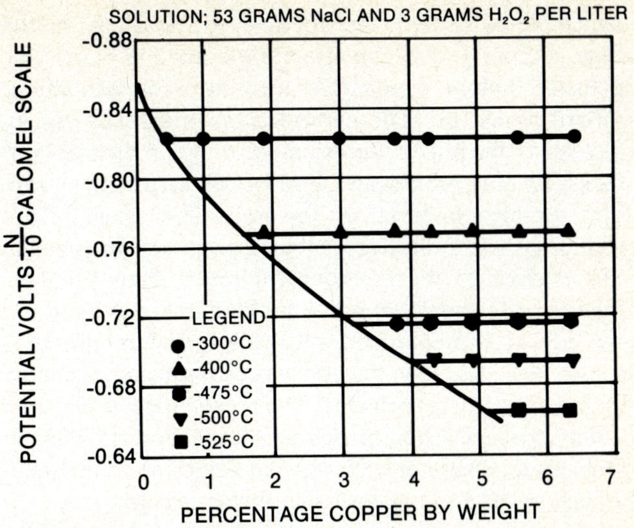

FIGURE 12 — Corrosion potentials of binary aluminum-copper alloys in a $NaCl$-H_2O_2 solution at various temperatures.

ASTM Committee G1 has standardized this approach in Standard Practice G84. This practice may be used both in calibrating environmental exposure sites as well as determining when unusual wetness occurs in specific exposure conditions. The G84 practice recommends using a potential of 0.1 V for gold-zinc and 0.01 V for gold-copper as the trigger potentials for signaling wetness. In actual tests, it is not important to adjust the trigger potential with great accuracy. For example, it was found that the time of wetness decreased only about 5% when the trigger potentials were increased by a factor of 2.[46]

A more sophisticated version of this concept employs a copper and iron cell in which the magnitude of the current is monitored as a function of time.[48] This approach is based on the fact that the corrosivity of moisture on the surface varies with time, depending on pollution levels, etc., and so the galvanic current is really a better

[(1)]Aluminum Association (AA), Washington, DC.

indicator than simply the presence of moisture. In this case, copper and steel plates are mounted in plastic and polished before exposure, much as a metallographic mount would be. The current levels measured are indicative of the rate of corrosion occurring on the steel surfaces. As time progresses, this rate will vary as the corrosion products build up on the steel. The results of this technique are indicative of the response of the steel or zinc surface to the environment rather than simply a measure of wetness time.[49] A problem that has been observed with this approach is that the current measured is substantially less than the corrosion current, as calculated by mass loss measurements. The reason for this disparity is that local cell action causes significant short circuiting in the system. Therefore, the current values are only qualitatively related to the corrosion that is occurring.

Conclusions

Electrochemical corrosion testing is continuing to evolve. Test methods are becoming more reliable as investigators gain more experience with the methods. The cost of electronic equipment has continued to decrease as solid-state technology has advanced, and this trend will probably continue although at a lower rate. However, automation has increased substantially and this development has tended to counteract cost reduction. The emphasis has been to develop tests that provide the best information. There has been a continuing effort to standardize test methods, both for characterizing material performance and environment corrosivity and this will certainly continue. Electrochemical methods will have a continuing impact on corrosion technology.

Acknowledgments

The author gratefully acknowledges helpful information from R. Baboian of Texas Instruments and Air Products and Chemicals, Inc. for permission to publish this paper.

References

1. S. W. Dean, Jr., W. D. France, Jr., S. J. Ketcham, Electrochemical Methods of Testing, Handbook on Corrosion Testing and Evaluation, W. H. Ailor, Ed., John Wiley and Sons, Inc., New York, New York, p. 171, 1971.
2. S. W. Dean, Jr., "Electrochemical Methods of Corrosion Testing," Electrochemical Techniques for Corrosion, R. Baboian, Ed., National Association of Corrosion Engineers, Houston, Texas, p. 52, 1977.
3. M. Pourbaix, Atlas of Electrochemical Equilibria in Aqueous Solutions, Pergamon Press, London, England, 1966.
4. ASTM G3-74, Standard Recommended Practice for Conventions Applicable to Electrochemical Measurements in Corrosion Testing, 1985 Annual Book of ASTM Standards, Vol. 03.02, ASTM, Philadelphia, Pennsylvania, p. 101, 1985.
5. NACE Publication 3D170, "Electrical and Electrochemical Methods for Determining Corrosion Rates (1984 Revision of 3D170)," National Association of Corrosion Engineers, Houston, Texas, 1984.
6. S. W. Dean, Jr., R. A. Woodroof, J. Nichols, "Electrochemical Methods for Evaluating Corrosion Inhibitors in Strong Acid Systems," Laboratory Corrosion Tests and Standards, ASTM STP 866, G. S. Haynes, R. Baboian, Eds., ASTM, Philadelphia, Pennsylvania, p. 228, 1985.
7. One example is the "Corrosion Rate Meter," Petrolite Corporation, Petreco Division, Houston, Texas.
8. ASTM D2776-79, Standard Test Methods for Corrosivity of Water in the Absence of Heat Transfer (Electrical Methods), 1984 Annual Book of ASTM Standards, Vol. 03.02, ASTM, Philadelphia, Pennsylvania, p. 72, 1984.
9. K. B. Oldham, F. Mansfeld, Corros. Sci., Vol. 13, No. 70, p. 813, 1973; also, F. Mansfeld, Corrosion, Vol. 29, No. 1, p. 10, 1972.
10. S. Barnartt, "Electrochemical Nature of Corrosion," Electrochemical Techniques for Corrosion, R. Baboian, Ed., National Association of Corrosion Engineers, Houston, Texas, p. 1, 1977.
11. R. L. LeRoy, J. Electrochemical Society, Vol. 124, No. 7, p. 1006, 1977.
12. One example is the "Corrator," Magna Corporation, Santa Fe Springs, California.
13. C. A. Arnold, "Instrumentation in Cooling Water Systems," presented at the Cooling Tower Institute Annual Meeting, Houston, Texas, 1975.
14. T. Y. Chen, A. Moccari, D. D. Macdonald, "Development of Controlled Hydrodynamic Techniques for Corrosion Testing," MTI Final Report, Ohio State University, Columbus, Ohio, 1984.
15. ASTM G59-78, Standard Practice for Conducting Potentiodynamic Polarization Resistance Measurements, Annual Book of ASTM Standards, Vol. 03.02, ASTM, Philadelphia, Pennsylvania, p. 331, 1984.
16. F. Mansfeld, M. Kendig, Corrosion, Vol. 37, No. 9, p. 545, 1981.
17. F. Mansfeld, Corrosion, Vol. 38, No. 10, p. 556, 1982.
18. Handbook of Chemistry and Physics, 49th ed., R. C. Weast, Ed., P. E. 63, Chemical Rubber Company, Cleveland, Ohio, 1968.
19. S. Haruyama, T. Tsuru, A Corrosion Monitor Based on Impedance Method in Electrochemical Corrosion Testing, ASTM STP 727, ASTM, Philadelphia, Pennsylvania, p. 167, 1981.
20. N. D. Greene, Experimental Electrode Kinetics, Rensselaer Polytechnic Institute, Troy, New York, 1965.
21. ASTM G5-72, Standard Recommended Practice for Standard Reference Method for Making Potentiostatic and Potentiodynamic Anodic Polarization Measurements, 1984 Annual Book of ASTM Standards, Vol. 03.02, ASTM, Philadelphia, Pennsylvania, 1984.
22. B. E. Wilde, Corrosion, Vol. 28, No. 8, p. 283, 1972.
23. ASTM G61-79, Standard Practice for Conducting

Cyclic Potentiodynamic Polarization Measurements for Localized Corrosion, 1984 Annual Book of ASTM Standards, ASTM, Philadelphia, Pennsylvania, p. 340, 1984.
24. R. L. Saur, R. P. Basco, Plating, Vol. 53, p. 33, 1966.
25. R. L. Saur, Plating, Vol. 54, p. 393, 1967.
26. R. L. Saur, R. P. Basco, Plating, Vol. 53, p. 981, 1966.
27. R. L. Saur, R. P. Basco, Plating, Vol. 53, p. 320, 1966.
28. This instrument, the EC Tester, is manufactured by Chuo Sei Sakosho Company, Ltd., Nagoya, Japan.
29. ASTM B627-79, Standard Method of Electrolytic Corrosion Testing (EC Test), 1983 Annual Book of ASTM Standards, Vol. 02.05, ASTM, Philadelphia, Pennsylvania, p. 489, 1983.
30. J. Stone, H. A. Tuttle, H. N. Bogart, Plating, Vol. 43, p. 877, 1965.
31. ASTM B538-70, Standard Method of FACT (Ford Anodized Aluminum Corrosion Test) Testing, 1983 Annual Book of ASTM Standards, Vol. 02.05, ASTM, Philadelphia, Pennsylvania, p. 371, 1983.
32. FACT Tester, S. Sterling Company, Southfield, Michigan.
33. E. T. Englehart, G. Sowinski, Jr., New Test Evaluates Anodized Aluminum, Society of Automotive Engineers (SAE) Journal, Vol. 72, p. 51, 1974.
34. E. T. Englehart, D. J. George, Materials Protection, Vol. 3, No. 11, p. 25, 1964.
35. ASTM B457-67, Standard Method of Test of Measuring Impedance of Anodic Coatings on Aluminum, 1983 Annual Book of ASTM Standards, Vol. 02.05, ASTM, Philadelphia, Pennsylvania, p. 272, 1974.
36. Z. Scope, Twin City Testing Corporation, Tonawanda, New York.
37. W. L. Clarke, V. M. Romero, J. C. Danko, "Detection of Sensitization in Stainless Steels Using Electrochemical Techniques," CORROSION/77, Paper No. 180, National Association of Corrosion Engineers, Houston, Texas, 1977.
38. W. L. Clarke, R. L. Cowan, W. L. Walker, Intergranular Corrosion of Stainless Alloys, ASTM STP 656, R. F. Steigerwald, Ed., ASTM, Philadelphia, Pennsylvania, p. 99, 1978.
39. W. L. Clarke, D. C. Carlson, Materials Performance, Vol. 19, No. 3, p. 16, 1980.
40. A. P. Majidi, M. A. Streicher, Corrosion, Vol. 40, No. 8, p. 393, 1984.
41. F. L. LaQue, Behavior of Metals and Alloys in Sea Water, The Corrosion Handbook, H. H. Uhlig, Ed., John Wiley and Sons, Inc., New York, New York, p. 383, 1948.
42. ASTM G82-83, Standard Guide for Development and Use of a Galvanic Series for Predicting Galvanic Corrosion Performance, 1984 Annual Book of ASTM Standards, Vol. 03.02, ASTM, Philadelphia, Pennsylvania, p. 485, 1984.
43. R. H. Brown, W. L. Fink, M. S. Hunter, Measurement of Irreversible Potentials as a Metallurgical Research Tool, Transactions, AIME, Vol. 143, p. 115, 1941.
44. ASTM G69-81, Standard Practice for Measurement of Corrosion Potentials of Aluminum Alloys, 1984 Annual Book of ASTM Standards, Vol. 03.02, ASTM, Philadelphia, Pennsylvania, p. 383, 1984.
45. P. J. Sereda, Weather Factors Affecting Corrosion of Metals, Corrosion in Natural Environments, ASTM STP 558, ASTM, Philadelphia, Pennsylvania, p. 7, 1974.
46. P. J. Sereda, S. G. Croll, H. F. Slade, Measurement of the Time of Wetness by Moisture Sensors and Their Calibration, Atmospheric Corrosion of Metals, ASTM STP 767, S. W. Dean, Jr., E. C. Rhea, Eds., ASTM, Philadelphia, Pennsylvania, p. 267, 1982.
47. ASTM G84-84, Standard Practice for Measurement of Time of Wetness on Surfaces Exposed to Wetting Conditions as in Atmospheric Corrosion Testing, 1984 Annual Book of ASTM Standards, Vol. 03.02, ASTM, Philadelphia, Pennsylvania, p. 503, 1984.
48. V. Kucera, E. Mattsson, Electrochemical Technique for Determination of the Instantaneous Rate of Atmospheric Corrosion, Corrosion in Natural Environments, ASTM STP 558, ASTM, Philadelphia, Pennsylvania, p. 234, 1974.
49. F. Mansfeld, T. V. Kenkel, "Electrochemical Studies of Atmospheric Corrosion," CORROSION/75, Paper No. 161, National Association of Corrosion Engineers, Houston, Texas, 1975.

Interaction of Microcomputers with Electrochemical Corrosion Testing Methods

D. G. TIPTON and D. G. MELTON*

Abstract

Computer data acquisition in the corrosion laboratory is described. These techniques can provide higher resolution, and be less labor intensive, faster, and more interactive than conventional data recording. Typical system design, component description and specification, interfacing, signal processing, and software considerations are presented. Several examples of computer data acquisition systems (DAS) in corrosion research applications are given.

Introduction

A major task in corrosion research involves the acquisition of electrical data, usually voltages, during the course of an experiment. Traditionally, strip chart and x-y voltage recorders have been used for acquiring data over relatively long time periods [e.g., longer than 1 second (s)]. Oscilloscopes have been used similarly for shorter time periods (e.g., 1 ms). With the advent of digital circuitry and the widespread use of microcomputers in the laboratory, new extremely powerful techniques are now available for data acquisition, numerical processing, data management, and data communication.

With their low cost, high performance, and ease of use, microcomputers are becoming widespread in the laboratory as test instruments. The modern microcomputer can be easily interfaced with experimental apparatus to provide automated monitoring of data signals and/or control of test input signals. A number of analog-to-digital (A/D) converters and digital-to-analog (D/A) converters are available to provide the communications hardware for interfacing. An additional advantage is the economic benefit of minimizing human labor costs involved in data collection, collation, computation, and storage.

Microcomputer data acquisition systems (DAS) have been used successfully in a wide variety of laboratory corrosion tests. These range in complexity from corrosion potential measurements to AC impedance frequency analysis. Simple acquisition and recording of signals such as corrosion potential, galvanic current, thermocouple voltage, or other transducer output with time utilizes the computer DAS as an automated electronic data storage strip chart. Used in potentiodynamic polarization tests, the computer can serve as an electronic storage x-y recorder with full graphic plotting of the auto-scaled data during the test. In AC impedance electrochemical testing, the computer can serve as a controller to operate the function generator, acquire the potential and current wave form data, carry out the complex mathematical analysis of the data, and graphically and numerically present the results.

All of these applications use low-cost, readily available, "user-friendly" microcomputers, A/D and D/A converters, and relatively simple computer programs. The present article describes computer applications for a wide variety of typical laboratory corrosion testing requirements. Hardware specifications, computer program design, and examples of computer data output are presented.

Equipment

A/D Converters

Real world data signals from most corrosion experiments are usually continuously variable voltages or currents. Analog-to-digital (A/D) converters are devices that

*LaQue Center for Corrosion Technology, Wrightsville Beach, North Carolina.

convert analog voltages or currents to numerical binary signals for computer input.[1]

Twelve-bit A/D converters (1 part in 4096 resolution) are generally considered of medium performance with respect to acquisition speed. A typical maximum speed is 20,000 samples/second (s). Eight bit A/D converters (1 part in 256 resolution) may be capable of very high speeds, e.g., up to 100,000,000 samples/s. A/D converters are available with up to 16 bits resolution (1 part in 65,536), but with correspondingly lower maximum speeds (1000 samples/s).

One other option in the selection of an A/D converter is a multiplexer. A multiplexer is a discrete device under digital control to allow one A/D converter to become a multichannel device. Multiplexed A/D converters are readily available for a 16-channel operation with a small penalty in maximum speed for multiplexer operation.

Real-Time Clocks

Most microcomputers are not configured with a clock for automatic input of time keeping or "real-time." A real-time clock is required for timing of data acquisition intervals and to allow management of acquired data as voltage-time data pairs. Real-time clocks are readily available with resolution of milliseconds that can time events for a year or more in duration. Although a relatively complex programming problem, many microcomputers can support interrupt signals that can be sent from a real-time clock to interrupt another running computer program. This scheme allows the computer to perform data acquisition routines at precise timing intervals while simultaneously performing other computation chores.

D/A Converters

For feedback control of processes, function generating, or other requirements for a variable output voltage under computer control, digital-to-analog (D/A) converters are available. D/A converters are the functional and operational inverse of an A/D converter. A 12-bit D/A converter with a 0 to 1 V range can ouput variable voltages in 0.0002 V increments.

Binary I/O

The most basic architecture of a digital computer can be utilized in simple binary input/output (I/O). Since information in a computer is stored at this level as binary digits of "on" or "off," these digits can be converted via an interface to off-on output control by relays. Binary input, or computer reading of the status of switches, can be obtained by a binary interface in an inverse mode to the output.

Microcomputer

The central system component is the microcomputer. Virtually any computer is a feasible controller for data acquisition. However, some microcomputers are more useful for this task than others. An ideal microcomputer should have a large number of peripheral equipment available, including magnetic storage devices, printers, and plotters. A number of microcomputers are available with 8-bit microprocessors, 64-K bytes of random access memory, magnetic data storage devices, and high level computer languages for ease in programming.

Mass Storage

An important peripheral device for a microcomputer DAS is a magnetic mass storage device for permanent, nonvolatile (not dependent on continuous electrical power) data storage. Mass storage devices usually employ magnetic media such as tape, flexible (floppy) disks, or hard disks. Magnetic tape is by far the least expensive and has the lowest performance; hard disks, the most expensive and the highest performance; and floppy disks, intermediate in both.

Printers and Plotters

Line printers are very useful for permanent hard copy output of acquired data and other information. Data are often used as the raw tabular printout of voltages measured. It is often possible, however, to obtain quality printout of engineering units, statistics, and other computations from the raw voltage values. Thus, a report-ready table of experimental results is available directly from the DAS.

Digital plotters are available for all microcomputers, which allow computers to translate numerical data to graphical plots in a report-ready format. Digital plotters allow final presentation of data acquired and analyzed by computer such that no manual manipulation of data is required.

Interfacing

The most important specification for any of the computer equipment and peripheral accessories described is the compatibility of the equipment with the microcomputer in question. All computers have unique logical and communications architecture that requires an interface to translate and/or facilitate communications between the device and the microcomputer. If left to the user unfamiliar with computer architecture and digital circuit design, fabrication and installation of these interface firmware are, at best, formidable tasks. Several interface conventions are in widespread use to solve this problem.

It should be noted that compatible interfacing ensures only that communication is possible between the computer and the peripheral device; software must be written and correctly employed for successful operation of an A/D converter or other device. Quality, "user-friendly" equipment should include software documentation and sample programs—preferably supplied as computer programs stored on floppy disk—for ease in software development.

Signal Processing

Like any data recorder, the maximum resolution of an A/D converter is possible only if the voltage to be measured reasonably matches the input range of the A/D converter. For example, a 12-bit A/D converter with a 0 to 10 V range has a resolution of ±0.0024 V. This is excellent precision for a high level input (> 1 V) but poor for the low level outputs (e.g., < 100 mV) typical of many transducers. Signal conditioners are available for many A/D converters to allow preamplification of low level signals with user-selectable input voltage ranges.

In addition to simple preamplification, modern digital and analog circuitry allows signal conditioning of specialized transducers such as thermocouples, resistance temperature detectors (RTD) for very sensitive temperature measurement, and bridge amplifier signal conditioners, based on the Wheatstone bridge, to allow input from strain gauges, load cells, pressure transducers, linear variable differential transformer (LVDT), etc.

Also, care should be taken when retrofitting computer DAS to analog systems, such as recorders. Many older signal transducers have outputs of current instead of voltage. Signal conditioners are available for conversion of 20 mA current loop outputs to voltage.

Signal Noise

Some of the qualities of computer data acquisition by A/D converters make the resulting measurements especially vulnerable to noise. Signal noise may not be evident in analog data acquisition devices, such as strip-chart recorders, since recorders act as mechanical filters. High frequency noise (> 100 Hz) typically produces no response on a strip-chart recorder. However, an A/D converter, with its typical sample-and-hold aperture of about 1 μs, can detect waveforms, including undesirable noise, in the kHz frequency range.

The origin of this noise offers some techniques for its reduction or elimination. Single-ended signal wiring is very susceptible to noise. In single-ended wiring, a single conductor is used for the signal, or a series of channels of signals, with a ground connection used as a reference or return. Figure 1 shows the physical connection between an instrument and the A/D converter using single-ended and differential wiring. If a voltage difference exists between the grounding point at the experiment and the grounding point at the A/D converter, then that difference will produce error in the measured signal. Grounding points can be different as a result of grounding wire discontinuities or high resistance connections. Also, the signal conductor of single-ended wiring can enter stray electromagnetic fields because of lights, motors, and other instruments. Differential wiring avoids many of the grounding problems by using a twisted pair of conductors for the signal and reference. Common node rejection techniques allow cancellation of the noise waveforms present in both conductors.

The input resistance of most A/D converters is very high (> 5 × 10^6 ohms). High input resistance, or impedance, can make signal cables act as antennae that can readily pick up radio frequency noise that pervades the environment. This effect is most pronounced if there is impedance mismatch between the instrument and the A/D converter, e.g., < 1 × 10^3 ohms output impedance for the instrument and > 5 × 10^6 ohms input impedance for the A/D converter. To match a low impedance output for an instrument, the inputs of the A/D converter may be connected to ground through appropriate value resistors R1 = R2(e.g., 1 × 10^4 ohms) as shown in Figure 2. Care should be exercised as damage may result to many instruments if the outputs are short-circuited. Equipment manufacturers should be consulted to determine whether impedance mismatch exists and to determine whether the above impedance changes are possible.

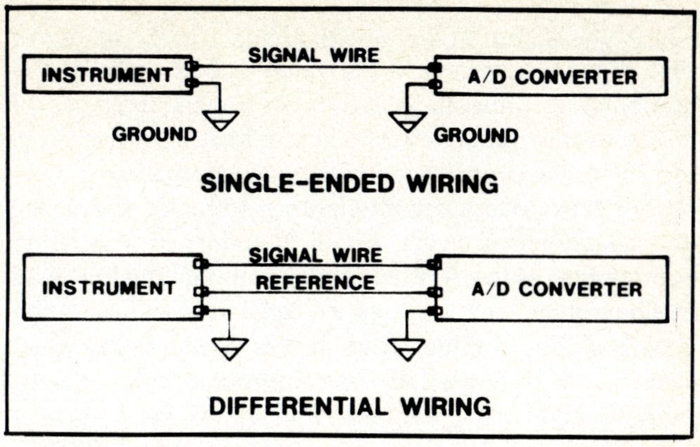

FIGURE 1 – Physical connection between instrument and DAS.

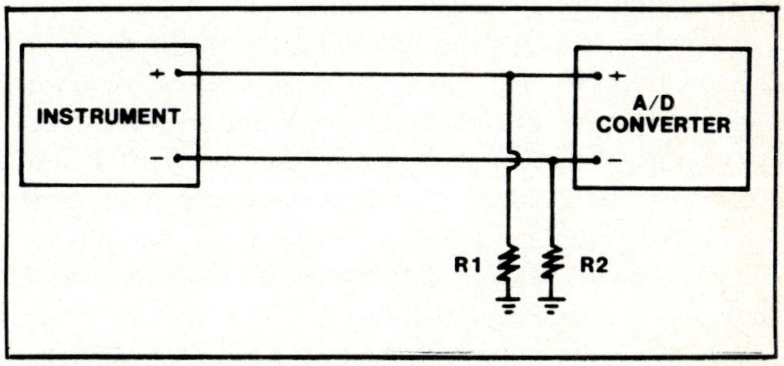

FIGURE 2 – Change resistance to ground to match impedance of instrument.

Shielded data cable offers further protection from induced noise. The shield should be grounded at one end only. The proper grounding point is not always obvious as it could be an "earth" ground to a driven grounding rod, a third wire utility electrical ground, or "analog" ground in the A/D converter. Often each grounding technique must be trial-connected, as an improperly connected shield can actually further degrade the signal-to-noise ratio.

All electrical conductors with a high ratio of length-to-diameter behave as low value capacitors. The longer the cable, the more pronounced the effect. If high frequency signals are being measured, this capacitance may attenuate the voltage and introduce an error difficult to detect as an unknown.

All the above problems suggest that, whenever possible, the signal conditioner, A/D converter, and computer should be located in as close proximity to the instrument as possible.

Another source of signal error not often cited is the computer itself. All data transfer and communication, as well as the output to the recording heads in the magnetic disc or tape drive and to the video monitor require the computer to generate a wealth of varying voltages of possible radio frequency from kHz to MHz frequencies. In fact, FCC[(1)] approval is required to ensure that computers do not cause damaging interference to radio, television, and telephone receivers. Some manufacturers separate components of the A/D converter such that the low level analog signals are converted to digital signals outside the chassis of the computer; then the converted digital signal is fed into the devices that allow communication to the information bus of the computer.

Correcting Signal Noise

When noise still results after the above avoidance techniques are employed, two filtering techniques may be used: analog and digital filtering.

Analog filtering is the simplest technique in the form of a low pass filter that attenuates a selected range of frequencies to near zero levels. These simple resistor-capacitor circuits, as shown in Figure 3, allow good noise suppression if the noise frequency is orders of magnitude away from signal frequencies. Some of the signal is inevitably attenuated along with the noise, and a systematic error source, large or small, results.

A more thorough treatment and removal of noise from the data signal is possible by digital filtering techniques. Since the data are converted to a number by the A/D converter, and since the microcomputer is particularly well suited for high-speed, repetitive computations, mathematical processing allows the incorporation of a numerical or digital filter. In its simplest form, several measurements may be averaged to eliminate cyclic noise. It should be noted that the sample interval must be sufficiently short to resolve the noise period with a sufficient number of samples to avoid aliasing.

The Nyquist sampling theorem states that:

$$\Delta t \leq 1/2f$$

where t is the sampling interval and f is the highest frequency to be observed in the signal. Thus, to avoid aliasing by 60 Hz AC line noise, a sample interval of $\leq 1/120$ s should be used for the group of measurements to be averaged. Still, if the AC noise does not have an average

[(1)]Federal Communications Commission (FCC), Washington, DC.

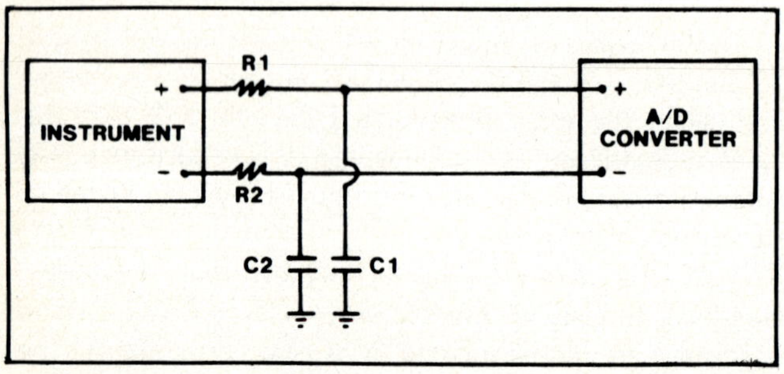

FIGURE 3 – Low pass filter for noise suppression.

value of zero, signal averaging will cause the fluctuations to disappear, but add an offset voltage.

Another digital filtering technique is known as the *moving window* or *boxcar averaging*. In this procedure, a new version of the data is created by taking a "window" of a number of successive data measurements and determining the average of that window. The sequence is repeated by moving the window down one point and repeating the procedure (add a new point and delete the oldest point). This technique is very effective at finding a DC signal obscured by AC noise.

The Fourier transform is a numerical technique mainly useful to determine an AC signal obscured by random noise. While not particularly effective for most corrosion data, the new applications of AC impedance electrochemical techniques require very sophisticated AC signal analysis. In Fournier analysis, the data are converted from the time domain to the frequency domain, spurious noise peaks are nulled out, and the inverse transform is made to produce a "clean" data set. The algorithms to convert from time to frequency domain involve very large numbers of complex multiplications and sine and cosine calculations. These computations are so large in number that even a high-speed mainframe computer requires a prohibitive time for completion. A number of so-called fast Fourier transform (FFT) algorithms are in use, with the Cooley-Tukey the most common.[2,3] To illustrate the time saving, a 4096-point conventional Fourier transform requires 16.7 million multiplications while the Cooley-Tukey method requires only 49,152 multiplications. FFT algorithms make it possible for microcomputers to perform these analyses. Several manufacturers have made available microcomputers interfaceable to high performance potentiostats, which perform FFT analysis of AC impedance data and display the results on CRT disply or plotter output.

The Data Acquisition System

A microcomputer based DAS is composed of the above equipment selected for the capabilities specific to the data acquisition task. Figure 4 shows a schematic diagram of a general purpose microcomputer DAS. A DAS may be selected and built "a la carte" as with accumulated A/D converters and other peripherals interfaceable to existing microcomputers. For example, TecMar[(2)] offers multichannel A/D converters with thermocouple preamps, D/A converters, and timers/counters directly interfaced to APPLE II, COMMODORE, AIM, KIM, TRS-80 and IBM-PC microcomputers.[(3)] Interactive Structures[(4)] has A/D converters with temperature, light, sound, and humidity sensors for APPLE II microcomputers. M. C. Miller Company[(5)] has a microprocessor controlled A/D converter with a general purpose RS-232C serial interface compatible with most computers.

Cyborg Systems,[(6)] Keithley Instruments, Inc.,[(7)] In-

[(2)]TecMar, Inc., Cleveland, Ohio.
[(3)]Registered trade name.
[(4)]Interactive Structures, Bala Cynwyd, Pennsylvania.
[(5)]M. C. Miller Co., Ringwood, New Jersey.
[(6)]Cyberg Systems, Chicago, Illinois.
[(7)]Keithley Instruments, Inc., Cleveland, Ohio.

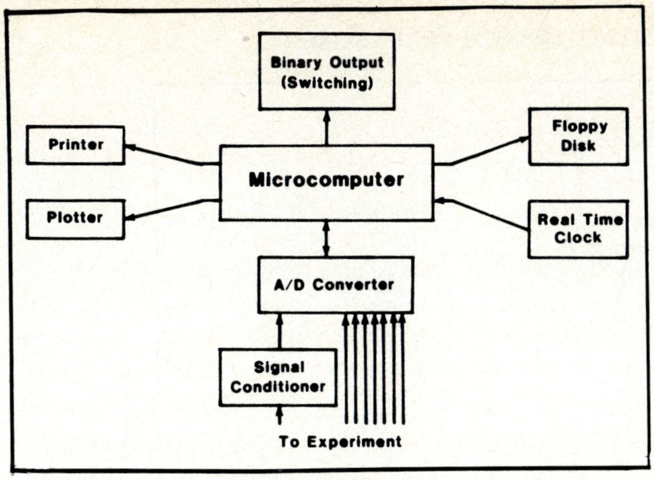

FIGURE 4 – Functional diagram of a microcomputer controlled data acquistion system showing key components.

teractive Microware,[8] and Optim Electronics[9] offer complete DAS packages compatible with APPLE II microcomputers with A/D conversion, D/A conversion, on-off binary control, on-off binary sensing, and variable amplification for many channels including software and presentation graphics to printers and plotters. Hewlett-Packard[10] and Data Translation[11] offer stand-alone DAS complete with computer. For the first time user, the complete system approach can be extremely flexible, can produce high performance results and is relatively easy to use.

Software Considerations

One fact not to be overlooked in microcomputer DAS is the requirement of computer programs for any useful tasks to be performed. Software support for A/D converters and other peripherals should be a primary consideration in equipment selection. Interfacing compatibility alone is useless until a computer program is available. Adequate software support can consist of written documentation and/or sample computer programs that illustrate all aspects of control, input, and output from the device. These illustrations of A/D converter use, for example, can then be used as functional subroutines in user-written software.

Some DAS hardware, such as the Issacs[12] DAS by Cyborg for the APPLE II microcomputer, employ a special modification of the computer language (BASIC). These modifications allow special commands to be implemented for ease in programming. The special commands establish software selectable options for DAS hardware operation, start and stop DAS operations, pass data into computer memory, and display data graphically on the monitor screen. DAS language implementations allow microcomputer programmers with average skill but no DAS experience, to immediately begin microcomputer data acquistion. From another viewpoint, the laboratory instrumentation technologist with no experience in microcomputer programming often finds computer DAS an effective and practical problematic approach to learn microcomputer programming.

Many A/D converters, such as those by TecMar and Interactive Structures, allow access and operation by the microcomputer using machine language instead of the typical high-level programming languages, such as BASIC or FORTRAN. Programming in machine language is extremely tedious, but programs run at speeds as much as 100 times faster. The A/D converters mentioned above are capable of sampling rates which are faster than the BASIC language can access them on the APPLE II. For very high-performance data acquistion, short subroutines can be written to perform some number of A/D measurements, write the results directly to a contiguous block of memory locations, and transfer control back to the calling program written in BASIC. Thus, the bulk of the programming chore is accommodated by the easily programmed BASIC language, while the very high-performance DAS task uses machine language at high speed.

Perhaps the most powerful attribute of computer data acquisition is the availability of mathematical computations to be performed on the raw voltage values. Such computations can range from simple engineering units conversion, to statistical analysis, to complex interactive computations involving several channels. Not only are mathematical computations available, but also mathematical logic operations. Computer DAS software, using IF...THEN logic, can control, effect changes, initiate alarm, or limit routines. Data can be integrated, differentiated, summed, etc., as it is being generated, or in "real-time," since the data already reside in the computer and, generally, the data acquisition procedures require very little devoted time from the computer.

Corrosion Laboratory Applications

Potential Measurement

Potential measurement is functionally a very simple use of microcomputer DAS. A real-time clock is used to time the acquisition interval and an A/D converter performs the measurements. An electrometer (a very high impedance voltmeter) is typically used for electrochemical potentiometry to avoid polarizing the reference electrode. A/D converters are typically fairly high in impedance, but more importantly they are connected to the specimen/reference electrode only for the duration of the A/C conversion, typically about 1 μs, avoiding the experimental problem of reference electrode polarization. For critical applications requiring higher impedance, simple buffer amplifier circuits can provide impedance as high as 1×10^{12} ohms.[4] The software establishes the time interval of data acquisition, starts an A/D conversion, retrieves the value, retrieves the time, and stores the voltage-time data pairs. Figure 5 shows an example of corrosion potential vs time for a freely corroding specimen of NiAl bronze in natural seawater. This example illustrates several advantages of computer data ac-

[8]Interactive Microware, State College, Pennsylvania.
[9]Option Electronics, Gaithersburg, Maryland.
[10]Hewlett-Packard, Rockaway, New Jersey.
[11]Data Translation, Inc., Marlboro, Massachusetts.
[12]Registered trade name.

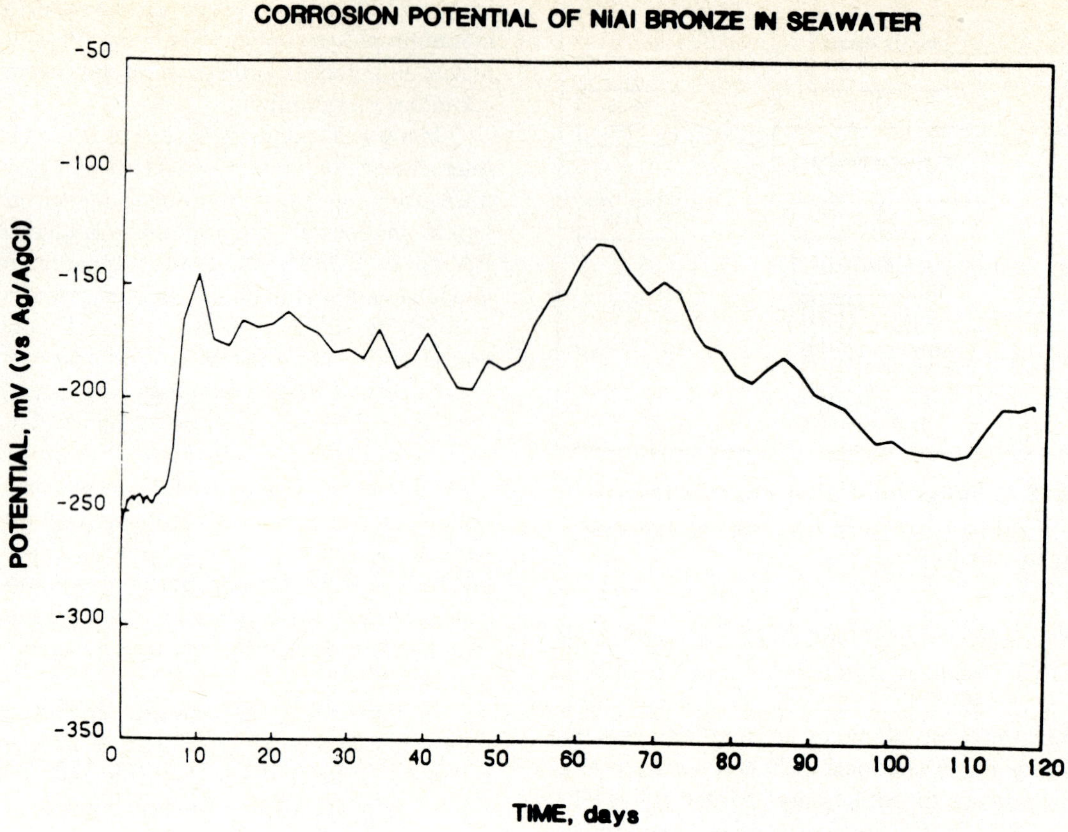

FIGURE 5 – Typical digital plotter presentation of potential versus time data from computer DAS.

quisition as compared to conventional data recording by strip chart voltage recorder. First, this specimen was but one of over one hundred whose potential was being monitored concurrently. Also, large quantities of data were collected, over 500 data pairs per specimen. Digital plotting the data then allowed compression of the time axis to show 120 days of data while still retaining the resolution of frequent measurements.

Current Measurement

Current measurement by computer DAS is analogous to potential measurement. For galvanic current measurement, a 1-ohm shunt resistor can be used for many engineering tests where the additional resistance (to the solution and interface resistance) represents little error. For more exacting demands, a zero resistance ammeter (ZRA) is required as a signal conditioner. ZRAs are available from electrochemical instrumentation suppliers, or may be fabricated by the user.[5] For potentiostatic experiments, a ZRA is not required and the current-to-voltage conversion of the potentiostat or a potential drop measurement across a shunt resistor in series with the counter electrode will provide an accurate measure of current.

Potentiodynamic Polarization

Potentiodynamic polarization curves generated by computer DAS require current and potential measurements by two channels of A/D conversion. Figure 6 shows apparatus for computer data acquisition from a research potentiostat. A multichannel A/D converter is utilized to monitor the potential and current outputs from the potentiostat. The computer graphics capabilities are utilized to draw in real-time the polarization curve, correctly scaled as shown on the monitor screen. After completion of the experiment, the curves may be replayed on the screen. For hardcopy output in a report-ready format, the data recorded on a floppy disk may be used as input to a separate graphics routine for a digital plotter interfaced to the computer. Figure 7 shows a typical polarization curve from the digital plotter. The current-potential

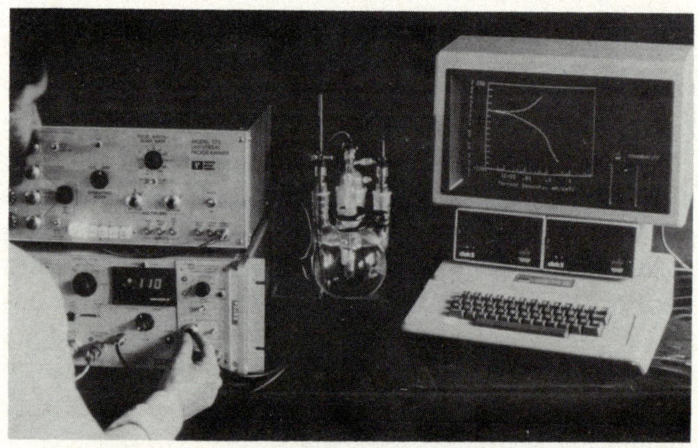

FIGURE 6 – Apparatus for computer data acquisition of potentiodynamic polarization data.

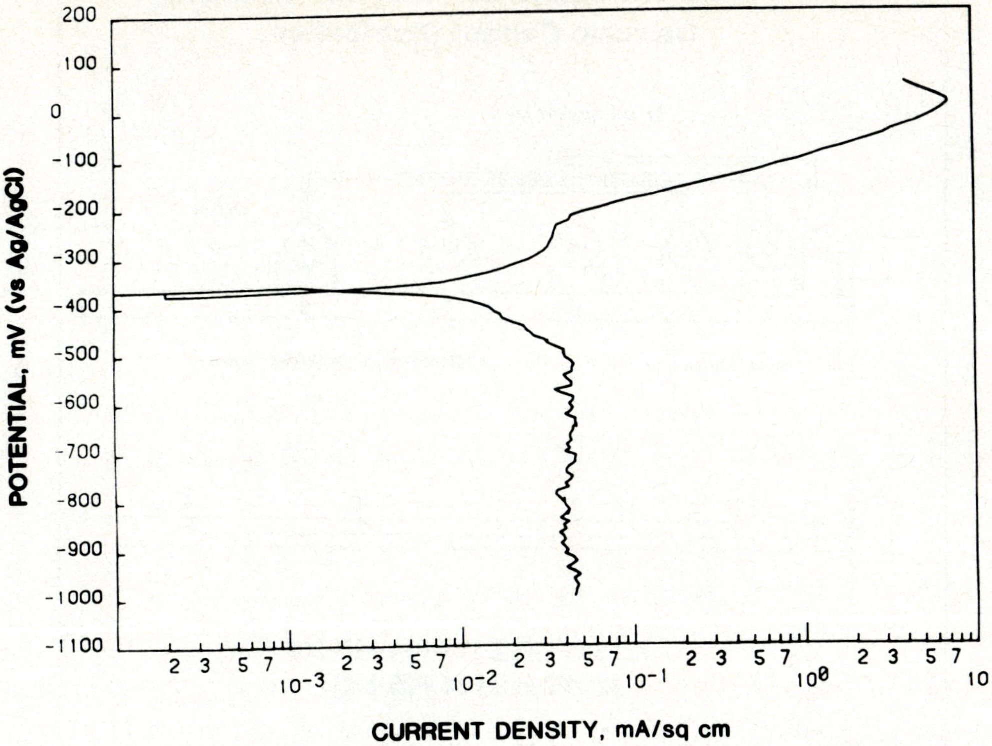

FIGURE 7 – Typical digital plotter presentation of potentiodynamic polarization curve from computer DAS.

data pairs are acquired in this example every 4 mV during the potential ramp. An anodic scan is taken by one experiment and a cathodic scan by another experiment on another electrode. The two data sets may be recalled and plotted together for the complete polarization curve as shown. The recall and replotting features allow simultaneous plotting of any number of curves and convenient rescaling of curves with no loss in the numerical accurary of the data.

Research potentiostats are now available with a built-in microprocessor dedicated to data acquisition, a D/A converter used to generate the reference voltage for the potentiostat (in lieu of a function generator or programmer), an A/D converter for data acquisition, a real-time clock for interval timing, and an interface for communication to and from a computer—all the ingredients for a DAS, but dedicated to the input and output instrumentation requirements for a research potentiostat. Here, computer DAS offers the advantage of high-resolution, graphics, data recording, and easy data recall.

Current Distribution

Figure 8 shows a schematic of a simulation of galvanic corrosion between a tube and tubesheet in a seawater cooled heat exchanger. ZRAs were used as signal conditioners for 10 channels of galvanic current measurements between the tubesheet and segments of the tube. The computations were made as indicated to determine the profile of current density down the tube from the tubesheet. Figure 9 shows a current distribution obtained after 120 days exposure to flowing seawater.

In this example, data recording would be difficult and tedious by conventional means. Ten channels of data were simultaneously recorded with incremental currents computed from cumulative raw data and current density computed from currents. Computer DAS allows the data to be accumulated as a three-dimensional array of current density, distance and time, of which these data in Figure 6 are but one data set at one time.

Fatigue Crack Growth

A microcomputer DAS has also been used to allow automated, unattended fatigue crack growth testing.[6] Here, an A/D converter reads 200 load-displacement data pairs during a single fatigue load cycle. Crack length is computed from load-displacement slope, geometry, and elastic compliance equations. The test method uses the high-performance characteristics of computer data acquisition in the data acquisition speed, resolution, and the ability to perform complex computations very quickly in real-time. The system software also contains validity criteria with respect to appropriate ASTM[(13)] Standard Methods (E647 and E399).

In addition to the crack growth data, computer DAS is also used to record environmental variables of the test such as corrosion potential and, in the case of potentiostated tests, the applied current and solution temperature.

[(13)]ASTM, Philadelphia, Pennsylvania.

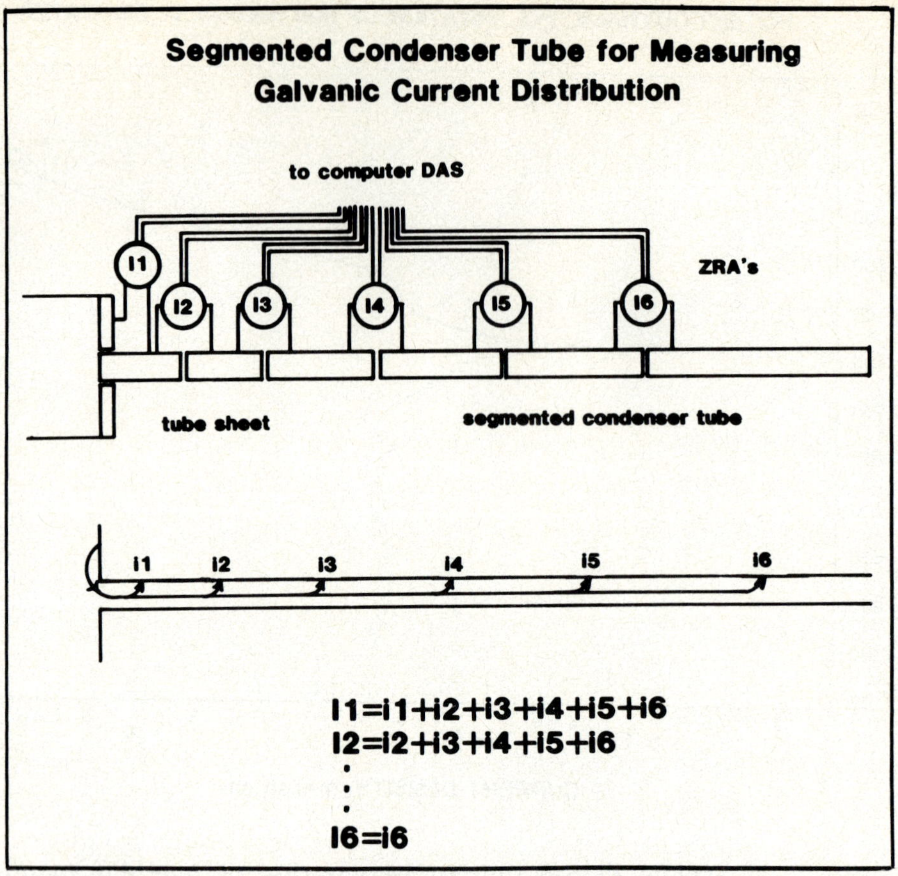

FIGURE 8 – Schematic diagram of segmented tube for determination of galvanic current distribution between tube and tubesheet in a heat exchanger.

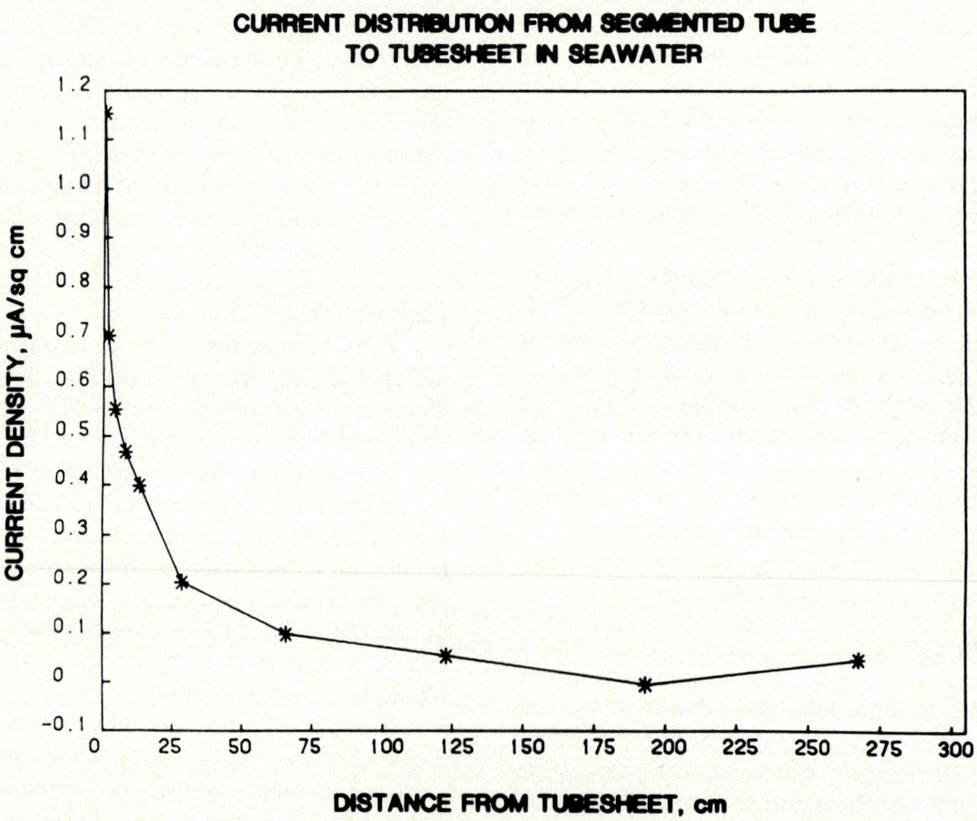

FIGURE 9 – Digital plotter presentation of galvanic current distribution in a simulated seawater cooled heat exchanger tube.

Summary

Microcomputers, already useful in the laboratory for data processing and management, are extremely valuable as controllers for corrosion data acquisition systems. The equipment is low cost, but effective software is required for any successful use. Computer DAS can provide higher resolution and be less labor intensive, faster, and more interactive than conventional data recording. Mathematical operations, computer graphics, and multichannel data management are features contributed by the computer that have no parallel in more traditional methods. The corrosion research laboratory can benefit from computer DAS techniques in all but the simplest, nonrepetitive, short-term test.

References

1. "Microcomputer Analog I/O Systems," Data Translation, Inc., Natick, Maine.
2. J. W. Cooper, The Minicomputer in the Laboratory, Wiley-Interscience, New York, New York, 1977.
3. J. W. Cooley, J. W. Tukey, Math. Comput., Vol. 19, p. 297, 1965.
4. J. I. Smith, Modern Operational Circuit Design, Wiley-Interscience, New York, New York, 1971.
5. J. A. Von Fraunhofer, C. H. Banks, Potentiostat and Its Applications, Butterworths, London, England, 1972.
6. D. G. Tipton, Environment Sensitive Fracture: Evaluation and Comparison, ASTM STP 821, ASTM, Philadelphia, Pennsylvania, 1983.

Nondestructive Electrochemical Tests for Detecting Sensitization in AISI 304 and 304L Stainless Steels

A. P. MAJIDI and M. A. STREICHER

Abstract

Three different electrochemical reactivation tests are compared with etch structures produced in the electrolytic oxalic acid etch test. There is a need for these nondestructive tests to evaluate welded stainless steel pipes and other equipment for susceptibility to intergranular attack. Sensitization associated with chromium carbide precipitates at grain boundaries can make these materials subject to intergranular attack in acids and, in particular, to intergranular stress corrosion cracking (IGSCC) in high temperature (289 C) water on boiling water (nuclear) reactor (BWR) power plants.

In the first of the two older reactivation tests, sensitization is detected by the electrical charge generated during reactivation. In the second test, it is measured by the ratio of maximum currents generated by a prior anodic loop and the reactivation loop. A third, simpler reactivation method is proposed in this paper. It is based on a measurement of the maximum current generated during reactivation.

If the objective of the field tests, which are to be conducted with portable equipment, is to distinguish between nonsensitized and sensitized material, this can be accomplished most simply, most rapidly, and at lowest cost by an evaluation of oxalic acid etch structures.

Introduction

The Fe 18Cr-8Ni stainless steels (SS) may become sensitized if they contain precipitates of chromium-rich carbides ($M_{23}C_6$) at the grain boundaries. These carbides are surrounded by metal whose chromium content has been depleted and is, therefore, much lower than that of the bulk analysis. In general, this precipitate forms in the range of 427 to 870 C. Temperatures in this range may be encountered during processing of mill forms, fabrication, and welding. Sensitization may make these alloys subject to rapid intergranular attack in a variety of acids and to intergranular stress corrosion cracking (IGSCC) in chloride-free, high-temperature water containing small amounts (0.2 to 8.0 ppm) of oxygen. An example of this is found in the cracking observed in weldments used to join AISI 304 SS pipes on boiling water (nuclear) reactor (BWR) power plants. These reactors operate at 289 C and are subject to a range of conditions during start-up, operation, and shutdown periods. IGSCC has been observed[1,2] in the heat affected zones (HAZ) of a large number of pipes ranging from 4 to 28 in. (102 to 711 mm) in diameter.

This problem has led to a need for a method of assessing the microstructures of not only weldments, but also the base plate or pipe to determine whether the pipes and weldments have been sensitized. If sensitized, they may be subject to IGSCC in service. Thus, a relatively simple and rapid nondestructive method is required that can be carried out on piping with portable equipment in the field. Such a method might, of course, also be applicable to a number of SS vessels, reactors, or various types of equipment on chemical or other process industry plants.

Of the evaluation tests specified for AISI 304 and 304L SS in ASTM[(1)] A 262, "Standard Recommended Practice for Detecting Susceptibility to Intergranular Attack in Austenitic Stainless Steels," only the electrolytic oxalic acid etch test[3] can be used as a nondestructive test. All other tests in this standard involve immersion of a small specimen in a hot acid solution and a determination of the degree of sensitization, either by weight loss deter-

*Dept. of Mechanical Engineering, University of Delaware, Newark, Delaware.

[(1)]ASTM, Philadelphia, Pennsylvania.

mination or by bending the tested specimen to examine it for the presence of fissures.

To provide a nondestructive, quantitative tool for assessing piping and weldments in piping, two versions of the electrochemical reactivation technique [also known as electrochemical potentiokinetic reactivation (EPR)] have been developed into routine procedures. These are the single loop (SL) and the double loop (DL) tests described by W. L. Clarke, V. M. Romero, and J. C. Danko[4] and by M. Akashi, et al.,[5] respectively. Recently concluded were detailed investigations of: (a) the parameters governing both of these tests, (b) a comparison of these two tests with the oxalic acid etch test and with the boiling ferric sulfate-50% sulfuric acid test, and (c) the mechanism of the two reactivation tests, including the role of residual (0.1 to 0.5%) molybdenum in AISI 304 and 304L SS.[6-8] The following is a brief comparison of the equipment, testing procedures, and properties of the two reactivation tests with the electrolytic oxalic acid etch test. In addition, a new and simpler reactivation test is described.

The purpose of this comparison is to provide a basis for the selection of the most appropriate method for a given field problem for assessing the metallurgical structure of piping or other equipment. Such equipment cannot be cut up for laboratory evaluation, but must be evaluated in the field, sometimes under difficult and even hazardous conditions. This data upon which this comparison is based were obtained primarily in laboratory tests.

Oxalic Acid Etch Test

For this test, a surface of about 1 cm^2 or greater is polished to a 600-grit finish with silicon carbide paper. The polished surface is made an anode in an electrical circuit consisting of an auxiliary electrode (the cathode, made of SS, graphite, or platinum) and a solution of 10% oxalic acid. An electromotive force (emf) source and a variable resistance are required to etch the surface for 1.5 min at 1.0 A/cm^2. Neither the length of time nor the current density is critical. Temperature control from ambient to boiling is also not a problem. No reference (calomel) electrode is required. The aim of the procedure is to produce a "heavy" etch, which clearly shows one of the three types of structures:

Step — Steps between grains. No ditches at grain boundaries and, therefore, no sensitization.

Dual — Some ditches or grooves at grain boundaries, but no grains completely encircled by ditches.

Ditch — One or more grains completely encircled by ditches or grooves.

These three structures are illustrated in Figure 1.

This simple, rapid test is specified in ASTM A 262-Practice A[9] for screening specimens of austenitic SS from testing in the four hot acid tests described in this standard. If a specimen has a step structure, it is not sensitized and will be immune to intergranular attack in the hot acid tests and, therefore, does not have to be tested in these solutions for 24 to 240 h. If there is a ditch structure, the specimen must be tested in the hot acid tests to determine the degree of sensitization. The procedures used on

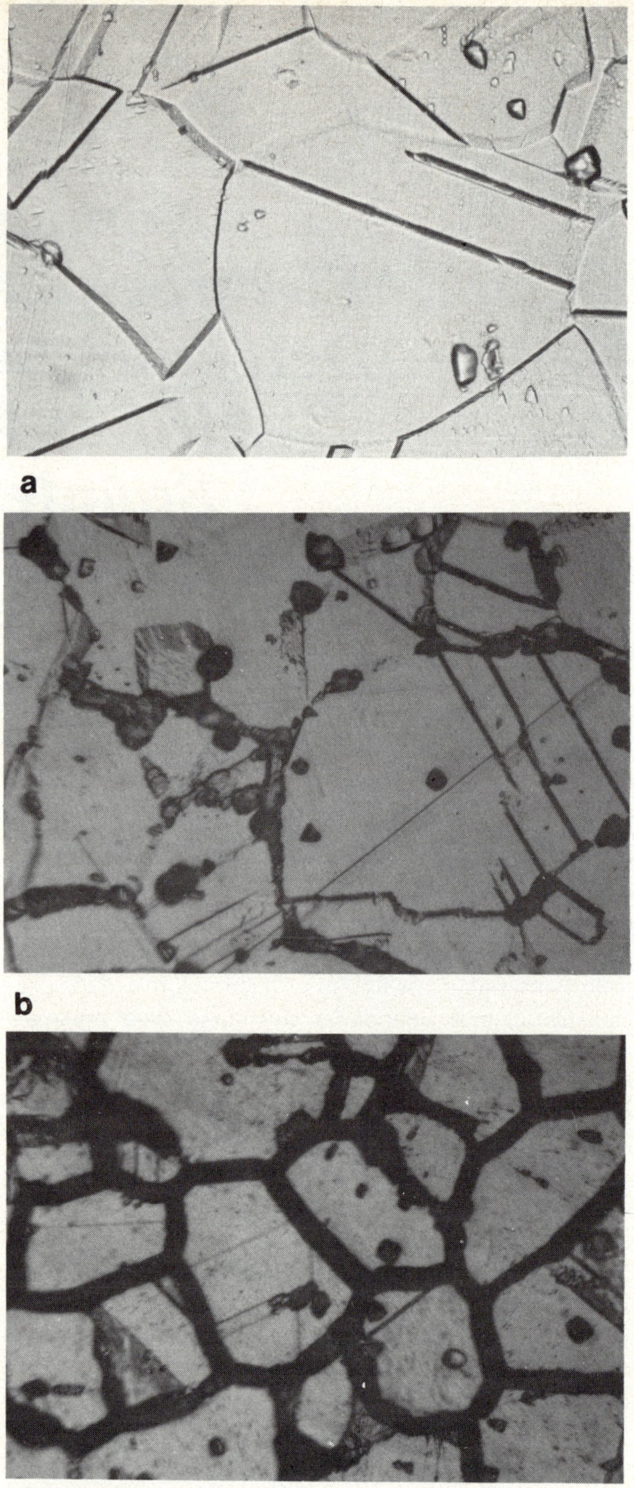

c

FIGURE 1 — Typical (a) step, (b) dual, and (c) ditch structures obtained in the oxalic acid etch test on AISI 304 SS. (400X)

specimens with dual structures vary with the hot acid tests.

The oxalic acid etch test has been used for 30 years as a rapid laboratory test.[3] It has also been used as a nondestructive test in the field on chemical plants for evaluating the microstructures of various types of equipment. The design of the cell, the electrical apparatus, and the requirements for surface preparation are all relatively simple. The acid solution can be prepared at the site because oxalic acid is an easily handled dry powder. A specially designed portable microscope has been developed, the DMR Rollscope,[(2)] for examination of microstructures on plant equipment. This can be used to classify oxalic acid etch test structures at 250 to 400 X. This microscope has accessories for taking pictures with a 35 mm or a Polaroid camera, which can be used as a permanent record.

The oxalic acid etch test is not a quantitative test. However, when the requirement is merely to differentiate between those materials completely free of all evidence of sensitization, i.e., pipe or weldments with a step structure, from those showing evidence of some degree of sensitization, dual or ditch structures, a clear distinction can be made. This classification can be learned readily by nonspecialists. Also, because a photographic record can be made, the basis of the decision can be reviewed by others and can be filed for future reference.

Electrochemical Reactivation Tests

When an AISI 304 SS specimen is immersed in 0.5 M H_2SO_4, there is hydrogen evolution corrosion. The corrosion potential is near −400 mV vs SCE. By connecting the specimen with a counter electrode, a calomel electrode, and a source of emf, it can be polarized anodically to more positive potentials. At potentials in the range of −50 to 700 mV vs SCE, the specimen becomes passive as a result of the formation of a protective film. When the direction of polarization is reversed, the surface is reactivated. It was found that the total electrical charge [coulomb = A × second (s)] passing through the surface during the scan can be related to the presence of sensitization in the specimen.[10,11] To facilitate the reactivation process, potassium thiocyanate (KSCN) is added to the sulfuric acid solution.[12]

Two reactivation procedures have been proposed[4,5] for routine, nondestructive, quantitative detection of sensitization. The solution for both reactivation tests is the same: 0.50 M H_2SO_4 + 0.01 M KSCN. The temperature control is specified at 30.0 ± 1 C for both tests. Because there is gradual decomposition of KSCN, this solution should not be older than about one week. Also, care must be used in handling solid KSCN because it readily absorbs moisture from the air.

Polishing equipment, electrochemical cells, a portable microscope, and specially designed and programmed, portable, electronic instruments for controlling electrode potentials and scan rates, and for measuring currents, current ratios, and total electrical charge have been developed and are available for field use to conduct the reactivation tests.

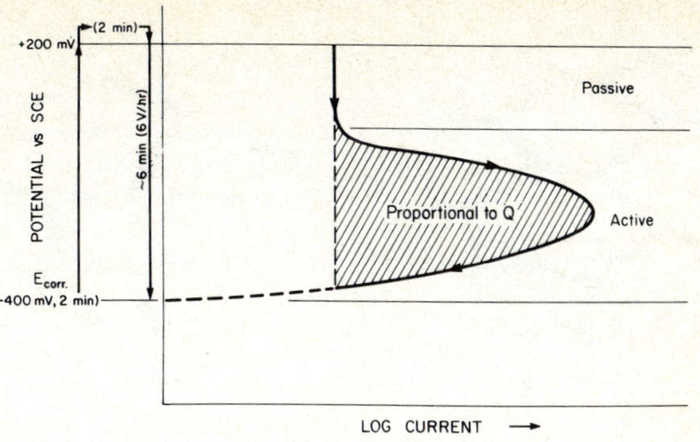

FIGURE 2 – Schematic diagram of single loop electrochemical potentiokinetic reactivation (SL-EPR) test for AISI 304 and 304L SS.[6]

Single Loop Test

The sequence of potential scans for this test is shown in Figure 2. First, the corrosion potential, E_{corr}, of the material is determined. For AISI 304 and 304L SS, E_{corr} is near −400 mV vs SCE. The specimen is then passivated by polarizing it to a potential of +200 mV and holding this potential for 2 min. After this passivation stage, the potential is decreased at a rate of 6 V/h (1.67 mV/s) until E_{corr} is reached. This reactivation results in the preferential breakdown of the passive film in the chromium-depleted grain boundaries of sensitized material and an increase in the current through the cell. A loop is formed in the E vs log I curve (Figure 2). Because the potential scan rate is constant, the area under this loop is directly proportional to the electric charge, Q, measured during the reactivation process. On nonsensitized materials, the passive film remains essentially intact at the scan rate used, and the size of the loop is much smaller than that for a sensitized material.

Q is a parameter that depends on the total area of the chromium-depleted zones. W. L. Clarke, et al.,[4,13] assumed an idealized model (Figure 3) for which: (a) all attack is confined to the grain boundaries, (b) the attack is uniformly distributed over all grain boundaries, and (c) the width of grain boundary attack is constant at 10^{-4} cm. On the basis of these assumptions, Clarke, et al.,[4,13] derived a formula for the grain boundary area (GBA) involving a measurement of the grain size of the material being tested. Q values are normalized into P_a values, coulomb/cm^2, by dividing them by the GBA.

$$GBA = A_s[5.09544 \times 10^{-3} \exp(0.34696\,X)] \quad (1)$$

where A_s is the surface area of the specimen and X is the ASTM micrograin size number measured in accordance with ASTM Practice E 112.[14]

Examination of the microstructures of sensitized SS after the SL test shows that there are variations in the width of the grain boundary attack and that there are large variations in the degree of grain encirclement by

(2)Registered trade name.

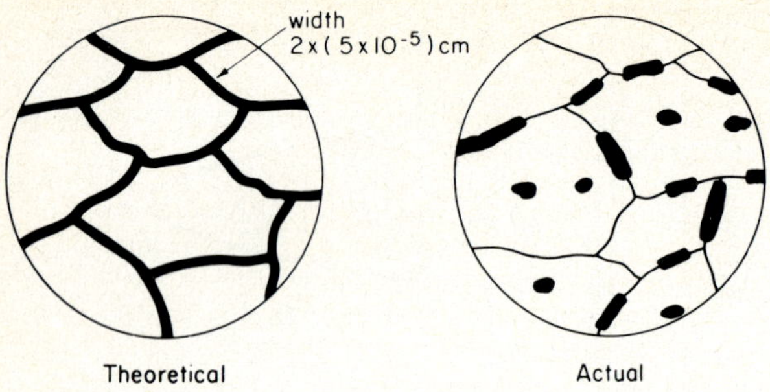

FIGURE 3 – Schematic SL-EPR test microstructures. Equation (1) for the calculation of P_a is based on assumptions illustrated on the left. An actual EPR microstructure for mild cases of sensitization, where P_a values are most useful, is described by the illustration on the right.[6]

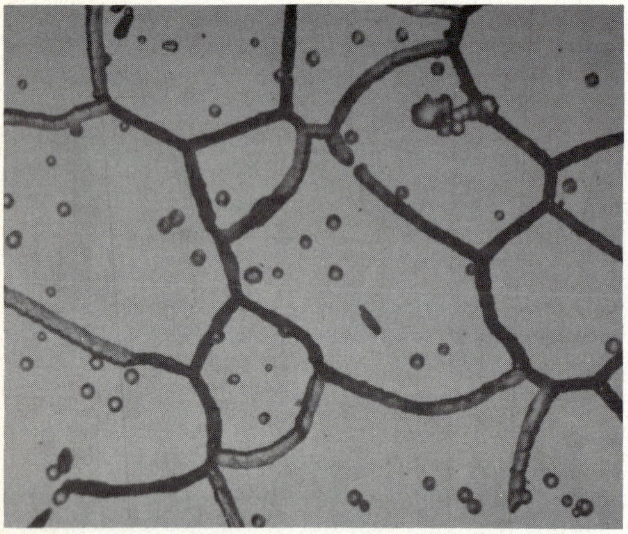

FIGURE 4 – SL-EPR etch structure of heavily sensitized material showing complete encirclement of all grains. Equation (1) for calculation of P_a assumes such an etch structure.[6] (400X)

intergranular attack, i.e., in cases of mild sensitization, only a fraction of the grain boundary is actually attacked preferentially (Figure 3). This schematic drawing also shows pits within the grains. These are caused by the nonmetallic inclusions in some heats. They also contribute to the charge, Q. Therefore, values of P_a (coulomb/cm^2) do not correspond to a literal, well defined corrosion process. Rather, they are more like a useful indicator, especially when there is pitting and when not all grain boundaries are surrounded by ditches. Complete encirclement of all grains happens only in relatively severe cases of sensitization (Figure 4).

The relationship of the SL test to oxalic acid etch test structures and to corrosion rates in the boiling ferric sulfate-50% sulfuric acid test (ASTM A 262-Practice B) is shown in Figure 5. These tests were made on a group of

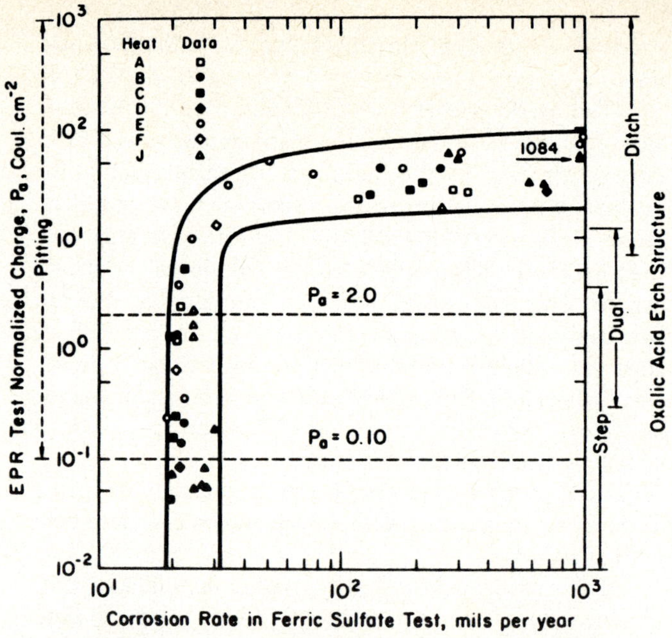

FIGURE 5 – Comparison of SL-EPR, ferric sulfate–50% sulfuric acid, and oxalic acid etch tests for AISI 304 and 304L SS. Solid triangles are data on additional heats given in Table 1.[6]

AISI 304 and 304L alloys whose compositions are given in Table 1. As mentioned above, corrosion is measured in this 120-h test by determining weight loss, which is primarily a measure of the number of metal grains that have been undermined and dislodged from the surface.[15] Thus, specimens with a step or dual structure in the oxalic acid etch test do not drop grains during the ferric sulfate test and the corrosion rates remain very low (Figure 5). However, in this range of low ferric sulfate rates and step and dual structures, there is a wide range of P_a values in the SL test.

In general, values of P_a less than 0.10 are characteristic of step structures, which are also free of pitting. Values of P_a greater than 0.10 are a result of pitting in step structure materials, or of some sensitization if greater than 0.4. Thus, in the range of 0.4 to 2.0, there may be pitting in material with a step structure; in material with a dual structure, there may be some sensitization or a combination of some sensitization with pitting. To establish whether or not the higher P_a values are caused by pitting, it is necessary to examine the etched microstructures after the SL-EPR test. Figure 6(a) shows a pitted microstructure after a SL test; Figure 6(b) shows the same area as Figure 6(a), but with an additional electrolytic etch in chromic acid. The latter shows that the distribution of the pits is random and that there is no ditching at the grain boundaries. The absence of sensitization is also shown by the oxalic acid etch structure on this specimen [Figure 6(c)]. Nevertheless, the P_a value of this specimen is 1.6 coulomb/cm^2, i.e., much greater than 0.10.

Figure 5 shows that the P_a value is a sensitive indicator of differences in mild degrees of sensitization, but that SL measurements do not readily distinguish between

TABLE 1 — Analyses of AISI 304 and 304L SS

Heat	Composition (wt%)								
	C	Cr	Ni	Mo	Mn	Cu	S	P	Si	N
B	0.061	18.22	8.97	0.06	0.56	0.09	0.011	0.025	0.58	0.034
E	0.060	18.68	10.16	0.08	1.69	0.08	0.012	0.022	0.50	0.040
AD	0.056	18.11	8.07	0.08	1.62	0.09	0.020	0.020	0.47	0.083
AN	0.055	18.17	8.76	0.11	1.37	0.20	0.015	0.019	0.51	0.039
AL	0.064	19.62	9.22	0.12	0.66	0.14	0.005	0.018	0.31	0.054
AF	0.050	18.46	8.46	0.13	1.73	0.15	0.005	0.034	0.55	0.039
AM	0.061	18.42	8.71	0.14	1.42	0.19	0.018	0.022	0.47	0.032
C	0.055	18.34	8.60	0.20	1.62	0.11	0.015	0.026	0.60	0.068
A	0.067	18.58	9.17	0.21	1.43	0.18	0.013	0.023	0.61	0.028
D	0.062	18.37	8.56	0.24	1.80	0.43	0.007	0.027	0.59	0.065
AK	0.066	18.22	8.42	0.26	1.66	0.24	0.008	0.020	0.59	0.039
AO	0.055	18.53	8.45	0.47	1.68	0.54	0.002	0.026	0.65	0.069
G	0.049	18.67	8.78	0.38	1.30	0.12	0.015	0.033	0.65	0.056
H(1)	0.021	18.83	10.27	0.58	1.38	0.11	0.009	0.027	0.59	0.020
BC(1)	0.028	19.16	10.86	0.02	0.94	—	0.033	—	0.35	0.033

(1)AISI 304L.

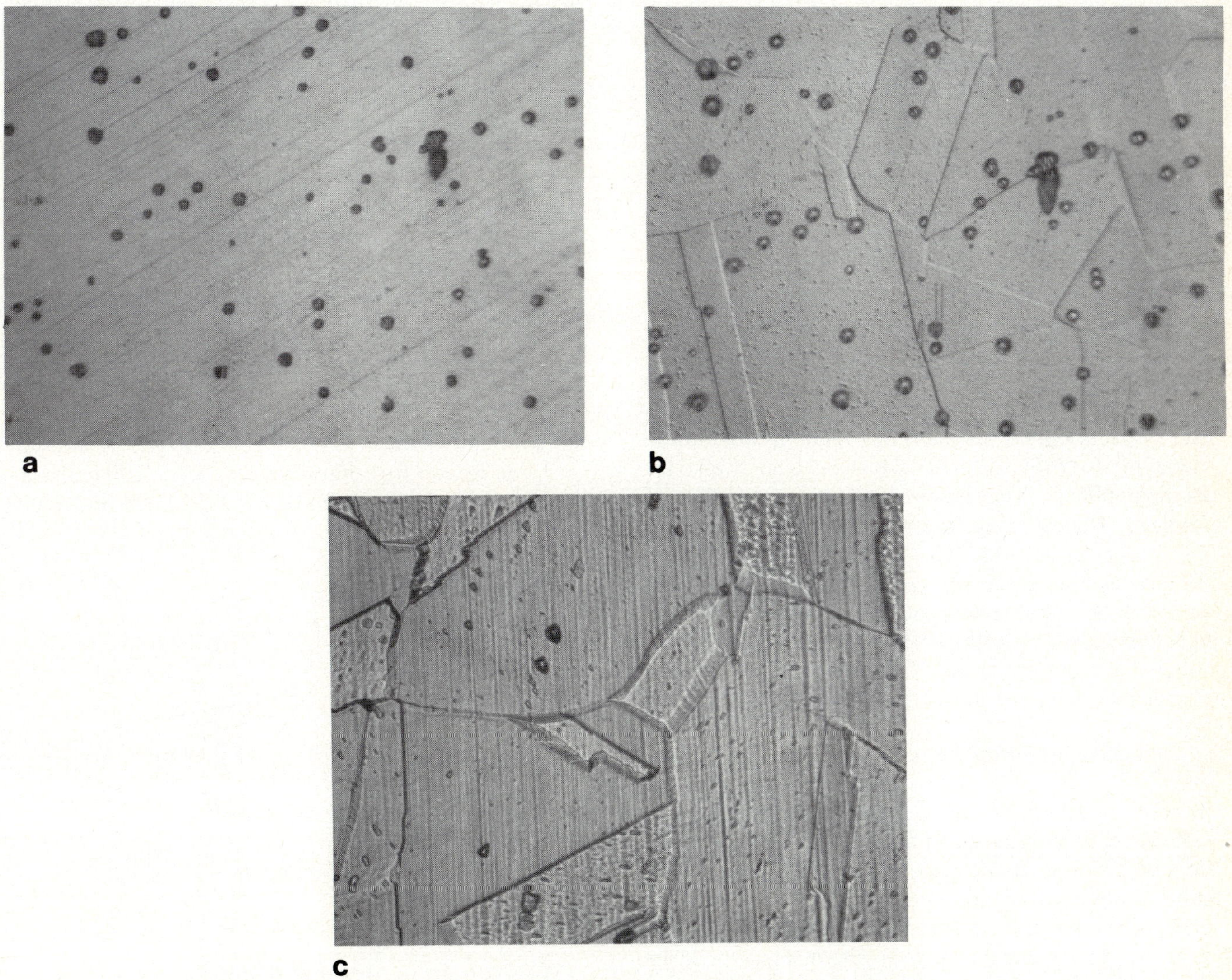

FIGURE 6 — Effect of random pitting on SL-EPR measurements: (a) structure after SL-EPR test, P_a = 1.60 C/cm², (b) chromic acid etch after SL-EPR test, same area as (a), and (c) oxalic acid etch structure (step structure). (Heat J, 400 X)

medium or severely sensitized materials. P_a values saturate in the range of severe ditch structures, i.e., when most of the grains are encircled by ditches. The overlapping in the P_a ranges of the oxalic acid etch structures in Figure 5 is a result of pitting. For example, the P_a value of 1.6 coulomb/cm^2 in the step structure specimen of Figure 6 also falls into the range of dual structures, which is from 0.4 to 10.0 coulomb/cm^2.

The line at 2.0 coulomb/cm^2 has been drawn in Figure 5 because Clarke[16] has selected it as the upper limit for acceptance of incoming pipes to ensure a "lack of discernible intergranular carbide precipitation." Values of P_a in the range of 0.1 to 2.0 coulomb/cm^2 on material having a step structure are a result of pitting. All etch structures produced in the SL test must be examined by microscope to establish whether the P_a value, when greater than 0.4 coulomb/cm^2, is indicative of pitting in a step structure, or whether there was some sensitization in a dual structure. Thus, it is necessary to make an oxalic acid etch to establish whether or not the specimen has a step or dual structure.

Clarke[16] has also concluded that 2.0 coulomb/cm^2 might be a "conservative number for evaluation purposes" on weldments on the *outside* of pipes. The implied assumption here is that if the outside of the pipe has a step structure, i.e., is completely free of sensitization, then the degree, if any, of sensitization on the *inside* of the pipe where the high temperature water is in contact with the AISI 304 SS weldment, is small enough to prevent IGSCC.

In a previous investigation,[6] P_a values from the SL-EPR test were found to be reproducible and comparable as long as the temperature is carefully controlled, e.g., at 30.0 ± 1 C, a surface finish of 1 μm is used and the scan rate is controlled at 6 V/h.

An apparent discrepancy between P_a measurements and oxalic acid etch structures has been reported in the literature[13] for AISI 304 material that had been heated for 24 h at 500 C. This problem was investigated[17] and resolved. The cause was sensitization confined to a very thin surface layer. This layer gave a high value of P_a. When the same specimen was repolished for an oxalic acid etch, this thin surface layer was removed and the bulk material underneath, which had not been sensitized, showed a step structure. Thus, there was no real discrepancy between the SL and the oxalic acid etch tests.

It was found[17] that the residual tensile stresses produced by mechanical cutting and grinding operations promote and facilitate rapid precipitation of chromium carbides at temperatures and times (500 C for 24 h) that do not produce any carbide precipitation in the unstressed bulk material underneath the surface layer. This effect of cutting and grinding on the susceptibility to sensitization has also been observed[17] after heating for 10 to 19 days at 300 C, which is near the operating temperature (289 C) of BWRs. There is, then, a possibility that sensitization may be detected in the thin surface layers of ground weldments that have been exposed for years at the operating temperature. Such sensitization may be only seven grains deep (0.5 mm), but might initiate IGSCC.

Double Loop Test

This test was developed in 1982 by M. Akashi, T. Kawamoto, and F. Umemura[5] to simplify the SL test. The solution composition, cell components, and temperature are the same as in the SL test. However, the sequence of polarization procedures, the surface finish, and the parameter used to measure the degree of sensitization are all different. Considerable time can be saved in surface preparation because this process can be stopped at the 100-grit silicon carbide level. There is no need to produce a 1-μm finish, as in the case of the SL test. The sequence of polarization steps is illustrated in Figure 7. The surface to be tested is exposed to the acid solution for about 2 min to establish the corrosion potential, which is near −400 mV vs SCE for AISI 304 and 304L SS. The surface is then polarized anodically to a potential of +300 mV vs SCE at a rate of 6 V/h; this takes 7 min. As soon as this potential is reached, the scanning direction is reversed and the potential is decreased at the same rate to the corrosion potential. The result (Figure 7), is two loops on a plot of electrode potential vs log of the current flowing between the test surface and the auxiliary electrode. This electrode may be made of graphite, SS, or platinum. The maximum currents for each loop are measured: I_a for the large, anodic loop, which is generated first, and I_r for the smaller loop generated during reactivation. The ratio of I_r:I_a is used as a measure of the degree of sensitization. In the form specified by Akashi,[5] there is no requirement for a grain size measurement.

In our investigation[7] of this test method, it was found that the values of the current ratio are indeed independent of the surface finish in the entire range of finishes from 1 to 150 μm (100-grit) on materials having step, dual, and ditch structures. Also, it was found that the test is readily reproducible, as long as the temperature, solution composition, and scan rate are carefully controlled.

The reason that the relatively rough (100-grit) surface finish produces reliable data in the DL test and not in

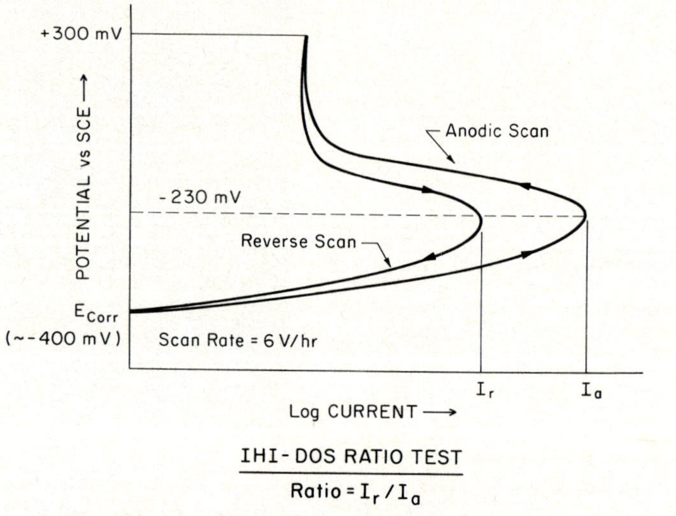

FIGURE 7 — Schematic diagram of the double loop (DL) - EPR test. Sensitization is evaluated from the current ratio, I_r/I_a.[7]

a

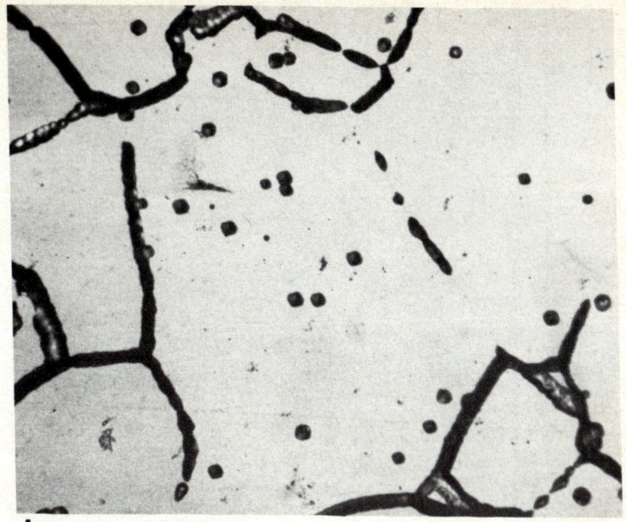

b

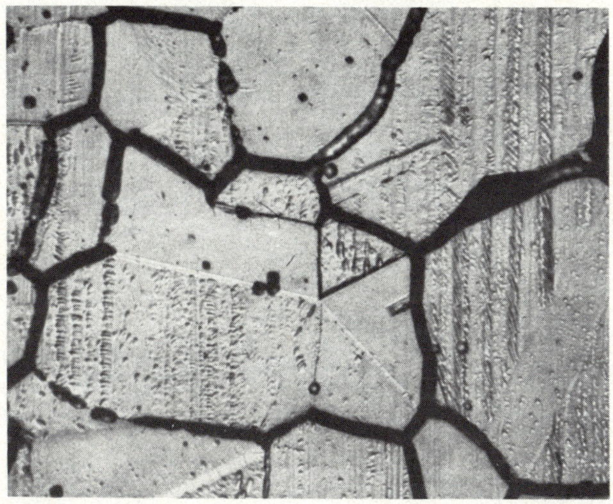

c

FIGURE 8 – Etching during DL-test. Finish = 1 μm. (400X). (a) Anodic loop only; (b) reactivation loop only; (c) double loop test.

the SL test is that the first (anodic) loop conditions or "cleans" the rough test surface before the reactivation scan by dissolving the flowed metal layer formed during grinding and polishing. This layer, when left in place, covers sensitized grain boundaries and thus retards reactivation of these boundaries during the reactivation scan in the SL test.[7,8] This is why a fine 1-μm finish is required for the SL test. The anodic conditioning scan of the DL test does not produce any ditching at sensitized grain boundaries (Figure 8).

The same alloys (Table 1) used for the correlation between the SL test with the oxalic acid etch test and the ferric sulfate test[15] were used for a comparison based on the DL test. The results have been plotted in Figure 9.[7] A comparison with Figure 5 shows that the DL results are also very sensitive in detecting small differences in relatively mild degrees of sensitization (dual structures), but that in the range of ditch structures, in which the ferric sulfate test readily distinguishes among various levels of sensitization, the ratios of I_r:I_a level off in a broad band. There is good agreement with the oxalic acid etch structures. Note that in Figure 9 the ranges for these structures do not overlap as they did for the SL test (Figure 5). The reason is that the DL ratios appear not to be affected by random pitting. The nonmetallic inclusions are dissolved or "cleaned-up" during the anodic scan. This is another advantage for the DL test.

Current ratios less than 0.001 correspond to step structures; ratios in the range of 0.001 to 0.05 to dual structures; and values greater than 0.05 fall in the broad range of ditch structures (Table 2).

The same alloys and specimens were used for SL and DL measurements. To compare the results of the two reactivation tests, the P_a values and the current ratios, I_r:I_a, have been plotted for each specimen in Figure 10.[7] It is apparent that there is good agreement between the two tests over the very wide range of values, even though the P_a values have been normalized by dividing the values of Q by the theoretical area of the grain boundary attack while the current ratios have not been normalized. Also, pitting in heats B and E increases only P_a measurements and does not affect the I_r:I_a ratios.

Nevertheless, to improve the correlation between

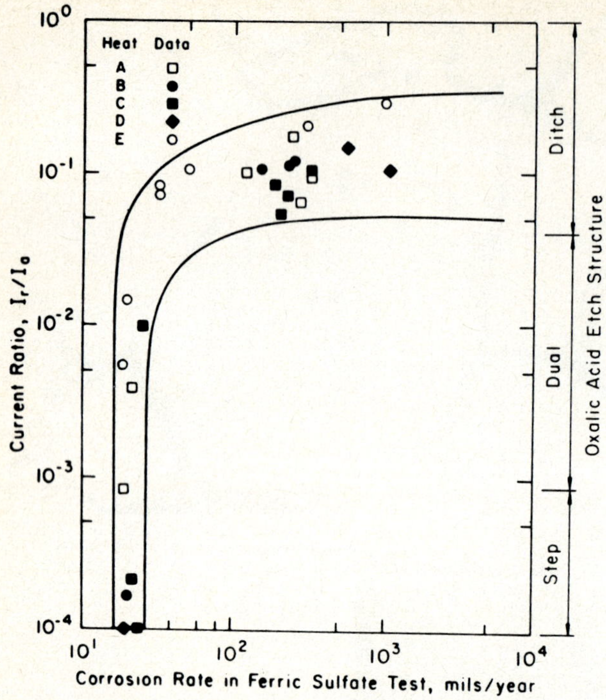

FIGURE 9 – Correlation of DL-EPR, ferric sulfate–50% sulfuric acid and oxalic acid etch tests for AISI 304 SS.[7]

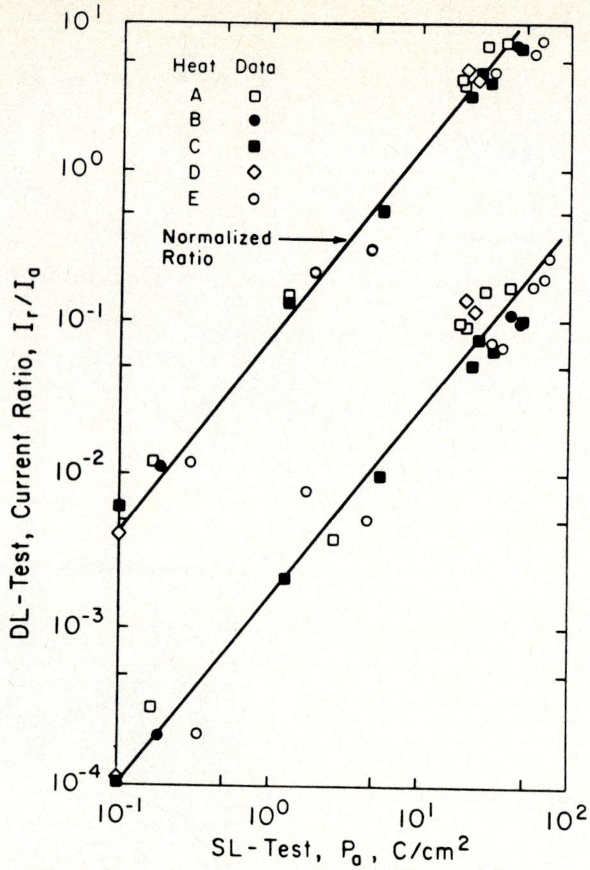

FIGURE 10 – Correlation of current ratios from DL-EPR tests with P_a values from SL tests for AISI 304 SS alloys.[7]

the tests, measurements of I_r have also been normalized by dividing I_r by the GBA [see Equation (1)] and I_a by the area of the specimen (A_s), because during the anodic loop the current comes from the entire surface of the specimen.

$$\text{Normalized Current Ratio} = \frac{I_r}{I_a[5.09544 \times 10^{-3}\exp(0.34696X)]} \qquad (2)$$

These normalized ratios have been plotted in Figure 10. It is seen that this improves the correlation between the SL and the DL tests.

A Simplified EPR Procedure

By using features from both the SL and DL tests, a new, simpler, and more rapid test has been developed. In a previous investigation,[8] the charge, Q, was shown to be directly proportional to the maximum reactivation current, I_r (Figure 11). For a scan rate of 6 V/h:

$$Q(\text{coulomb}) = 63(\text{sec}) \times I_r\,(\text{A}) \qquad (3)$$

TABLE 2 – Evaluation Criteria for AISI 304 and 304L SS in Reactivation Tests

Oxalic Acid Etch Structure	Equivalent Values in EPR Tests: Single Loop (P_a, coulomb/cm²)	Double Loop[(2)] (I_r:I_a ratio)	Simplified[(2)] (I_r:A_s A/cm²)
Step	≤0.10[(1)]	≤0.001	$\leq 10^{-5}$
Dual	0.4 to 10.0	0.001 to 0.05	10^{-5} to 0.001
Ditch	>10.0	>0.05	>0.001

Key: I_a—maximum current of anodic loop, I_r—maximum current of reactivation loop, and A_s—area of specimen (cm²).

(1)With pitting in material having a "step" structure, P_a values may reach 2.0 coulomb/cm².

(2)These values are for materials that have an ASTM grain size of about 3.5.

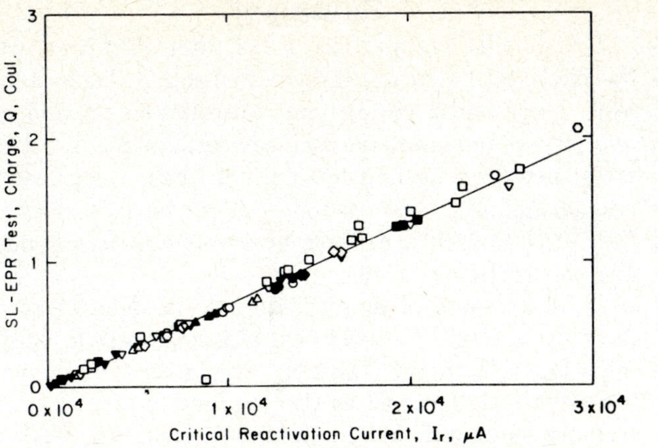

FIGURE 11 – The relationship between the charge, Q, and the maximum reactivation current, I_r, in the SL-EPR test for AISI 304 and 304L SS alloys with various degrees of sensitization.[11]

FIGURE 12 – Etch structure of a severely sensitized AISI 304 SS (1 h at 677 C, WQ) after the cleaning treatment of 2 min at −230 mV vs SCE in the S-EPR test. Note that there is only general corrosion of the grains (400X). In order to improve the quality of the photomicrograph, a surface finish of 1 μm was used. However, similar results were obtained with a finish of 100-grit.

This relationship was derived from a large number of measurements covering a wide range of degrees of sensitization. The equation is independent of degree of sensitization, the surface finish of the specimens, or the temperature of the solution; it is only a function of the scan rate. Thus, measurements of the maximum reactivation current, I_r, can be used in place of measurements of the charge, Q. This eliminates the need for a coulometer in the electronic instrument used to program this version of the reactivation method. This current, I_r, must be divided by the area of the test surface to give a value of A/cm^2.

The anodic scan in the DL test produces general dissolution which conditions the surface without intergranular attack on sensitized material (Figure 8). As a result, the maximum current, I_a, of the anodic loop is essentially the same for all AISI 304 and 304L materials, regardless of their susceptibility to intergranular attack. Therefore, the ratio of I_r:I_a is almost entirely a function of the reactivation scan, I_r, and there is no need for the anodic scan except for surface conditioning.

For greater reproducibility and comparison of measurements, the values of the maximum reactivation current/cm^2 can be normalized by a grain size measurement. This can be done either by dividing I_r by the GBA as given in Equation (1) or by dividing I_r/cm^2 by the ASTM grain size number (GSN). The larger the grain size, the greater the value of A/cm^2/GSN for a given value of I_r/cm^2. Such materials have smaller GSN and, therefore, a greater current flow per unit length of grain boundary for a given value of I_r.

The need for a 1-μm polish for the reactivation procedure can be eliminated by replacing the first or anodic scan of the DL test with a 2-min treatment of the 100-grit surface at a potential of −230 mV vs SCE. This is the potential at which the current is a maximum during the first anodic scan in the DL test (Figure 7). The conditioning or cleaning process is the most rapid at this potential. Figure 12 shows that this simplified procedure is as effective in conditioning the 100-grit surface as the regular DL procedure [Figure 8(a)].

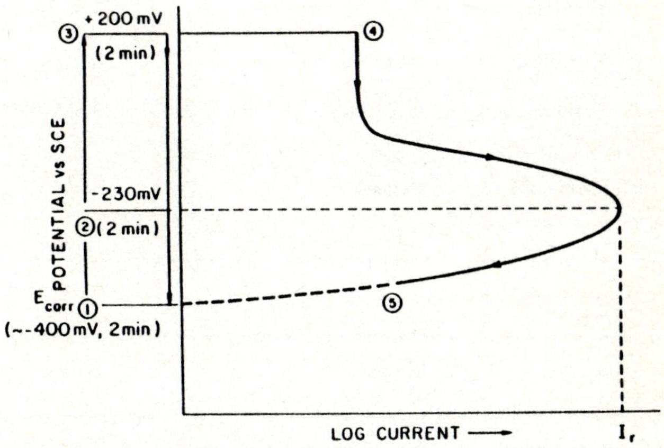

FIGURE 13 – Schematic diagram of the simplified, (S)-EPR, test for AISI 304 and 304L SS.

In preliminary tests, it has been found that this new procedure provides consistent and reproducible results on specimens having step, dual, and ditch structures. The procedures for the simplified EPR, or S-EPR, test are shown schematically in Figure 13.

A surface with 100-grit polish is allowed to establish its corrosion potential, about −400 mV vs SCE in 2 min. Its potential is then set at −230 mV and held for 2 min. After this, the potential is set at +200 mV and held for 2 min for passivation. After these treatments, the specimen is reactivated by scanning back to the corrosion potential at the rate of 6 V/h. During this reactivation scan, the maximum current, I_r, is measured. This value is divided by the surface area and can then be used as an indication of the degree of sensitization. If a more precise measure of the degree of sensitization is needed, the value of A/cm^2 can be normalized for grain size, either by dividing the maximum current itself (not A/cm^2) by the GBA as derived from Equation (1) or by merely dividing the actual ASTM GSN into the A/cm^2.

Values for I_r/cm^2 in the S-EPR test that correspond to the three types of oxalic acid etch structures are listed in Table 2, along with the corresponding values for the other two EPR tests.

Summary

A detailed comparison of the three EPR tests with the oxalic acid etch test is given in Table 3. Each of the tests is compared under nine categories to provide an overview of requirements for electronic equipment, solutions, polishing, testing time, procedures, calculations, and evaluation criteria. All four tests can be applied in the field with portable electronic instruments, microscopes, and electrochemical cells.

If the degree of sensitization is to be determined in the narrow HAZ of a weldment, it is necessary to determine first its location (Table 3). This is done by etching the weldment. This etch is also required to make a grain size measurement. When the position of the HAZ is

TABLE 3 – Comparison of the SL-, DL-, S-EPR Tests With Oxalic Acid Etch Structures for Detection of Sensitization in AISI 304 and 304L SS

Category	Test Method			
	SL-EPR	DL-EPR	S-EPR(1)	Oxalic Acid Etch(1)
Equipment				
A. Electrochemical	Potentiostat and scanner(2) with coulometer	Potentiostat and scanner(2) with current measurement	Potentiostat and scanner(2) with current measurement	EMF with ammeter and rheostat
B. Reference Electrode	Calomel electrode	Calomel electrode	Calomel electrode	Not required
C. Microscope	Required(3)	Required(3)	Required(3)	Required(4)
D. Solution	$0.5\overline{M}$ H_2SO_4 + $0.01\overline{M}$ KSCN	$0.5\overline{M}$ H_2SO_4 + $0.01\overline{M}$ KSCN	$0.5\overline{M}$ H_2SO_4 + $0.01\overline{M}$ KSCN	10% oxalic acid (source is a dry powder)
E. Source of Heat	Required	Required	Required	Not required
Temperature Control	30 ± 1 C	30 ± 1 C	30 ± 1 C	Not required
Actual Testing Time	10 min	17 min	12 min	1.5 min
Measurement of Surface Area	Required for calculation of P_a	Not required	Required	Estimate required for control of current density
Assessment Method	Calculation of P_a from Q-value and grain size	Calculation of I_r:I_a (normalized)(5)	Calculation of I_r/cm^2 (normalized)5	Classification of etch structures
Surface Finish	1-μm	100-grit	100-grit	600-grit
Grain Size Measurement	Required for calculation of P_a	In some cases	In some cases	Not required
Pitting	Must be monitored	Not a problem	Not a problem	Not a problem
Detection of Malfunction	Can be complex(6)	Can be complex(6)	Can be complex(6)	Readily detected by absence of etch

Key: SL—single loop, DL—double loop, and S—simplified, EPR tests.

(1) As described in ASTM A 262-Practice A.

(2) Portable, commercial electronic instruments and test cells are available to carry out these tests on plants.

(3) A microscope is required to locate the heat-affected zone (SL, DL, S), to measure the grain size (SL; in some cases DL and S) and to assess the EPR etch structure for evidence of pitting (SL).

(4) A microscope is required to classify etch structures and, when desired, to provide a photographic record.

(5) In some cases normalization for grain size is desirable.

(6) Malfunctions of calomel electrode, electronic instrument; masking of test surface by gas bubbles, absence of passivation.

located, the EPR test must then be made on an unetched portion of the polished surface to avoid testing a structure that has already been etched at grain boundaries. EPR etches are localized by placing a tape with a hole, 3 mm in diameter, over the HAZ. Because the width of the HAZ can be 2 to 3 mm, it may be difficult to place the hole in the tape so that either it exposes only the very narrow HAZ or, if necessary, it includes the same mix of HAZ and nonsensitized material to give a series of reproducible measurements. There may also be a range of degrees of sensitization across the HAZ.

This problem does not exist with the oxalic acid etch test. A relatively large area of the weldment can be etched. Evaluation is then made of the HAZ, which can be readily located, especially if there is any evidence of sensitization.

All three EPR tests provide quantitative information. However, it requires a costly electronic equipment and a high degree of surface finish (1 μm) in the case of the SL test. A proper understanding of the concepts involved in the EPR tests and the necessary skill required for operation of the procedures may be difficult for nonspecialists.

If the requirement is to evaluate pipe and the HAZ of weldments for "lack of discernible intergranular carbide precipitation,"[16] this can be met by an evaluation of the oxalic acid etch structures. A step structure is clear and readily identifiable evidence of the "absence of discernible intergranular carbide precipitation," and can, therefore, serve as a simple criterion for routine evaluation. Dual and ditch structures are evidence of carbide precipitation at grain boundaries and are, therefore, "not acceptable." Evaluations of oxalic acid etch structures are being made in laboratories and have been made on chemical plants by nonspecialists after a brief training period. Photographs of the etch structures can be used to provide a permanent record of the results.

The simple electrical apparatus required to produce oxalic acid etches, the requirement for a 600-grit finish, and the absence of the need for temperature control, for grain size measurements, for GBA calculations, or for concerns about pitting in the etch structures make the oxalic acid etch test the simplest and least costly test method.

Of the three quantitative EPR tests, the S-EPR test is the least complex. It requires only a 100-grit finish; it does not require a coulometer or difficult calculations. Pitting does not appear to be a problem.

The original EPR test, the SL test, is the most complex. It requires a 1-μm finish, a potential scanner with a coulometer, and a calculation for the GBA. Also, susceptibility to pitting, associated with randomly distributed nonmetallic inclusions, can have a significant effect on P_a values. Careful analysis of these values on the basis of a microscopic examination of the etch structure produced by the SL-EPR test and, perhaps, an additional oxalic acid etch, are required to arrive at a decision about acceptability or nonacceptability of the tested material in cases where there is pitting.

Thus, in the order of increasing complexity, time and cost, the four tests can be ranked as follows: oxalic acid etch, S-EPR, DL-EPR, and SL-EPR.

All of these tests provide a sensitive indication of relatively mild degrees of sensitization and can, therefore, readily distinguish nonsensitized material from material that is even slightly sensitized. None of the four tests is as useful in detecting differences in medium or severe degrees of sensitization as the various tests of ASTM A 262 in hot acids. However, these all require a small specimen for complete immersion in the test solutions. The theory, applications, and limitations of these tests have been described in detail in ASTM STP 656.[18]

References

1. A. Toboada, L. Frank, "Intergranular Corrosion in Nuclear Systems," Intergranular Corrosion in Stainless Alloys, ASTM STP 656, R. F. Steigerwald, Ed., ASTM, Philadelphia, Pennsylvania, 1978.
2. J. C. Danko, "Status of Boiling Water Reactor Owners Group Research Program," Paper No. 18. CORROSION/81, National Association of Corrosion Engineers, Houston, Texas, 1981.
3. M. A. Streicher, "Screening Stainless Steels from the 240 hr Nitric Acid Test by Electrolytic Etching in Oxalic Acid," ASTM Bulletin No. 188, February, 1953.
4. W. L. Clarke, V. M. Romero, J. C. Danko, "Detection of Sensitization in Stainless Steels Using Electrochemical Techniques," Paper No. 180, CORROSION/77, National Association of Corrosion Engineers, Houston, Texas, 1977.
5. M. Akashi, T. Kawamoto, F. Umemura, "Evaluation of IGSCC Susceptibility of Austenitic Stainless Steels Using Electrochemical Reactivation Method," Corros. Eng., Vol. 29, p. 163, 1980. (In Japanese—translated by BISITS.)
6. A. P. Majidi, M. A. Streicher, Corrosion, Vol. 40, No. 8, p. 393, 1984.
7. A. P. Majidi, M. A. Streicher, Corrosion, Vol. 40, No. 11, p. 584, 1984.
8. A. P. Majidi, M. A. Streicher, "The Mechanism of Potentiodynamic Reactivation Methods for Detection of Sensitization in Type 304 Stainless Steels," to be published.
9. ASTM Book of Standards, Vol. 10, A 262, ASTM, Philadelphia, Pennsylvania, 1978.
10. V. Cihal, A. Desestret, et al., "Tests d'Etude et d'Evaluation de la Sensibilite des Aciers Inoxydables a la Corrosion Intergranulaire," Communication au 5 ieme Congress, Europeanne de la Corrosion, Organize Per la Federation Europeanne de la Corrosion, Paris, France, September 24-28, 1973.
11. M. Prazak, Corrosion, Vol. 19, No. 3, p. 75+, 1963.
12. P. Novak, R. Stefec, F. Franz, Corrosion, Vol. 31, No. 2, p. 344, 1975.
13. W. L. Clarke, R. L. Cowan, W. L. Walker, "Comparative Methods for Measuring Degree of Sensitization in Stainless Steel," Intergranular Corrosion of Stainless Alloys; ASTM STP 656, R. F. Steigerwald, Ed., ASTM, Philadelphia, Pennsylvania, 1978.

14. ASTM Book of Standards, Vol. 11, E 112, ASTM, Philadelphia, Pennsylvania, 1979.
15. M. A. Streicher, "Intergranular Corrosion of Austenitic Stainless Steels, A Ferric Sulfate-Sulfuric Acid Test," ASTM Bulletin No. 229, ASTM, p. 77, April, 1958.
16. W. L. Clarke, "The EPR Method for the Detection of Sensitization in Stainless Steels," April (1981). Prepared for US Nuclear Regulatory Commission by General Electric Company. NUREG/CR-1095, GEAP-24888. Available from National Technical Information Service, Springfield, Virginia, USA.
17. A. P. Majidi, M. A. Streicher, Corrosion, Vol. 40, No. 9, p. 445, 1984.
18. M. A. Streicher, "Theory and Application of Evaluation Tests for Detecting Susceptibility to Intergranular Attack in Stainless Alloys—Problems and Opportunities," STP 656, p. 3, R. F. Steigerwald, Ed., ASTM, Philadelphia, Pennsylvania, 1978.

Modification of the Oxalic Etch Test for Field Nondestructive Testing of Welded AISI 304 Stainless Steel Pipes

*J. B. LEE and K. DIANATKHAH**

Abstract

This study describes a field nondestructive test method capable of detecting sensitized metallurgical structures in welded AISI 304 stainless steel (SS) pipes. The oxalic acid test (ASTM[(1)] A262-A) has been modified to improve its test temperature range to 100 C. The improved test method employs a 30-second etching in a 10% oxalic acid solution under an anodic current density of 1 A/cm^2 at the temperature above 60 C. Between 50 and 60 C the 30-second etching test should be used first. When the 30-second etching shows an under-etched grain boundary, the etching time should be increased to 90 seconds. At temperatures below 50 C, the 90-second etching, as described in ASTM A262-A, should be employed. A field test kit which includes a test cell is introduced along with the modified test method.

Introduction

Stainless steel (SS) has been a major engineering material primarily because of its excellent corrosion properties in many corrosive environments. However, sensitized metallurgical structures in stainless steels cause local corrosion along grain boundaries in environments where nonsensitized stainless steels do not suffer from this problem. The sensitized metallurgical structures in austenitic stainless steels, such as AISI 304 SS, are a result of chromium carbide (or carbonitride) precipitates along grain boundaries. When the steel is exposed to a temperature between 500 and 800 C, these precipitates result in depletion of chromium near the grain boundaries, causing the area to become more susceptible to corrosion. During welding, the base metal near the weld line is exposed to the above temperature range; therefore, this heat affected zone (HAZ) is susceptible to corrosion.

For the past 10 years, electrochemical test methods have been actively studied as a means of determining the degree of sensitization (DOS) in stainless steels.[1-7] The major advantage of the electrochemical methods over conventional immersion tests are: (1) The testing time is short and (2) the test method can be nondestructive.

However, most of the electrochemical methods that have been developed for the sensitization measurements are conducted under a controlled potential. The corresponding current representing the corrosion rate of a metal at a potential (or in a potential range) has been used as a measure of the DOS. The corrosion rate represented by the current is temperature-dependent. Therefore, the DOS measured by the current under a controlled potential is subject to the test temperature. This temperature dependency of the test results is one of the major limitations of the electrochemical methods for field use, since the temperature of the atmosphere varies from time to time. Also, the service temperatures generally differ from the temperature required for a specific electrochemical test. It has been reported that the DOS of a sensitized AISI 304 SS measured by the electrochemical potentiokinetic reactivation (EPR) test is doubled by every 10 C increase in temperature.[7]

Another laboratory test is the oxalic acid etch test (ASTM A262-A). This test consists of a 90-second etching in a 10% oxalic acid solution under an anodic current density of 1 A/cm^2. Figure 1 represents typical etched structures after the oxalic acid test. A ditch structure after the etch test [Figure 1(a)] indicates that the test sample is

*Standard Oil Production Co.

[(1)]ASTM, Philadelphia, Pennsylvania.

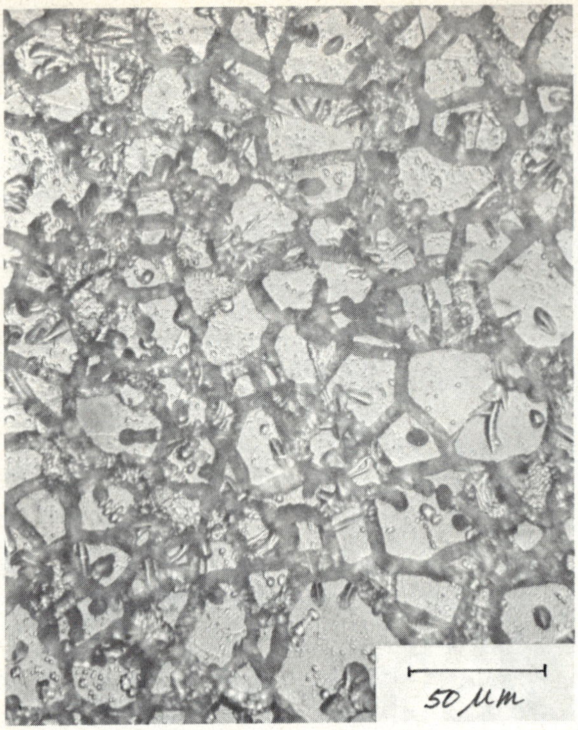

a

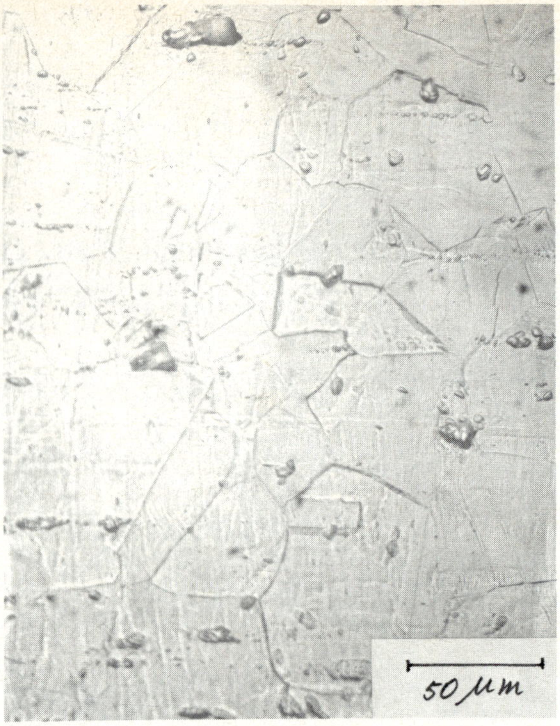

b

FIGURE 1 — Light micrographs of typical etched structures after the oxalic acid test: (1) Ditch structure showing sensitized metallurgical structure, and (2) step structure showing nonsensitized metallurgical structure.

sensitized and would suffer from intergranular corrosion in service. A step structure [Figure 1(b)] indicates that the sample is good. Regions near the grain boundary should be as corrosion-resistant as the interior of the grains. A dual structure, which is in-between the step and the ditch structures, would be marginally good. In a certain corrosive environment, the sample may suffer intergranular stress corrosion cracking (IGSCC) when the sample is under tensile stress. The sample may be more susceptible to pitting or crevice corrosion than the sample having the step structure. However, intergranular corrosion (IGC) in the sample is not expected. Definition of the dual structure, according to ASTM A262-A, is such that the etched structure has some ditches at grain boundaries, but no single grain is completely surrounded by the ditches.

Test Procedures

Welded AISI 304 SS was used for this study. The oxalic acid etch test was modified in order to improve its test temperature range. The temperature range investigated in this study was 0 to 100 C. The step by step testing procedures are as follows,

1. AISI 304 SS plate (30 × 15 × 1.2 cm) was gas tungsten arc welded (GTAW) to prepare the welded test sample. AISI 308 SS was used as a filler metal for the welding of AISI 304 sample.

2. The steel plate was cut into a 2.5 × 6.3 cm sample and was polished to a 600 grit finish with silicon carbide papers.

3. The specimen was degreased with acetone and rinsed with distilled water.

4. The specimen was located in a beaker that contained either ice or hot water for the testing at 0 C and high temperatures, respectively. The specimen was laid on a specimen support, which is located in the beaker, in order to immerse one-half of the thickness of the specimen in the hot water. Temperature of the hot water was controlled by a hot plate and an on-off temperature controller. Therefore, the actual skin temperature on the test site was slightly lower than the setting temperature.(2)

5. A SS tube (2 cm OD × 1.3 cm ID × 4 cm long) mounted on a polytetrafluorethylene (PTFE) base was used as a test cell. The PTFE base had an opening of 1 cm^2. Glass wool soaked in a 10% oxalic acid provided the electrolyte in the test cell (Figure 2).

6. An anodic current density of 1 A/cm^2 was applied to the SS specimen. Time for the testing was varied from 10 to 90 seconds in order to determine the optimum testing conditions at different testing temperatures. Fresh glass wool soaked in the electrolyte was used for each test.

Results and Discussion

Figure 3 shows the test results of the heat affected zone (HAZ) in the welded AISI 304 SS sample at room temperature using the standard oxalic acid testing procedure of ASTM A262-A. According to ASTM A262-A, this specimen shows the ditch structure in the HAZ, indicating that the specimen is sensitized. The same speci-

(2)This arrangement is a better simulation of the field service conditions. The service temperature of a SS pipe is generally defined by the temperature of the fluid flowing through the pipe.

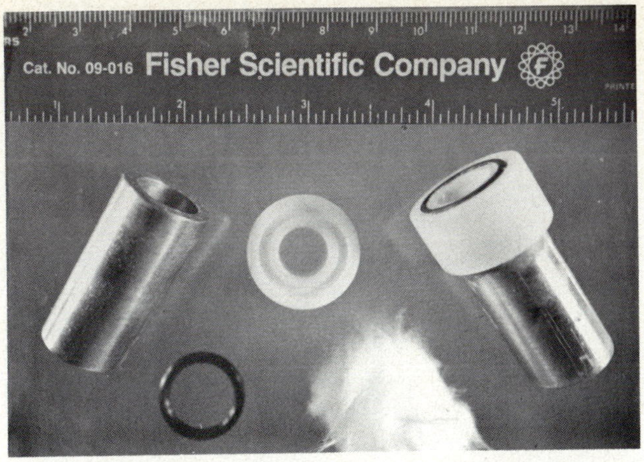

FIGURE 2 — Photograph of the test cell.

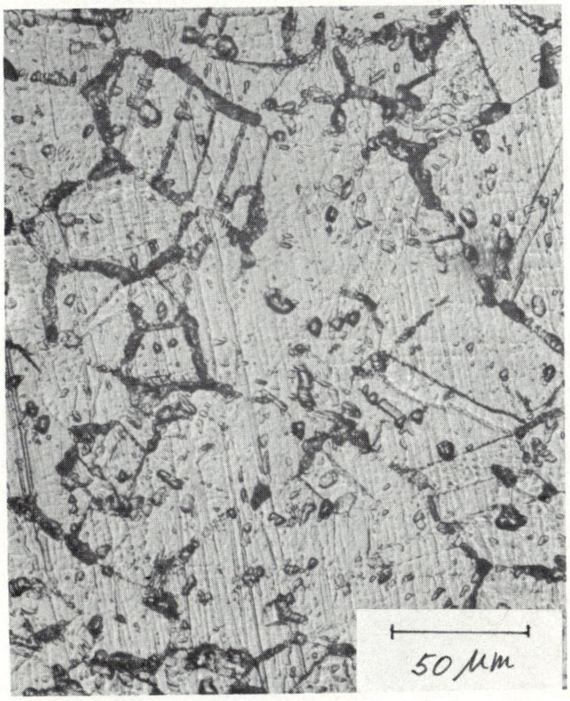

FIGURE 3 — The light micrograph of the oxalic acid etching test result of HAZ in the welded AISI 304 sample at room temperature (about 25 C).

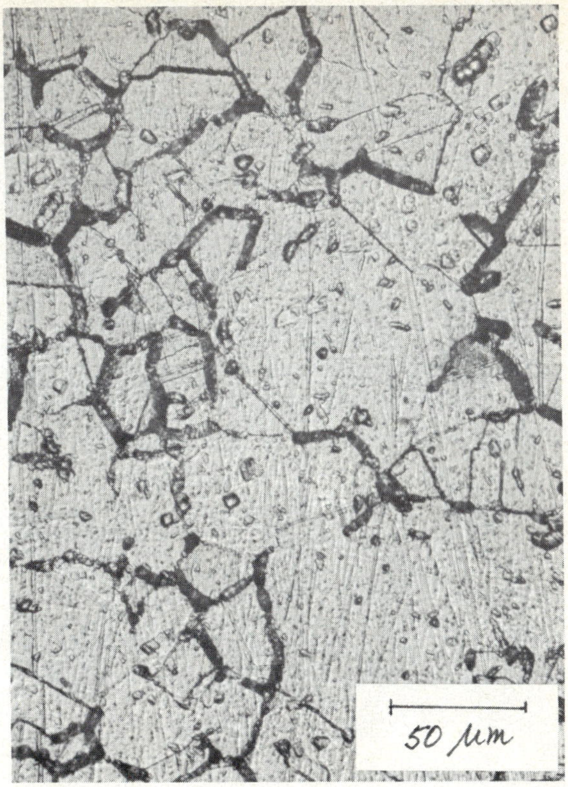

FIGURE 4 — The light micrograph of the oxalic acid etching test result of HAZ in the welded AISI 304 sample at 0 C (as tested, 400X).

men was tested at 0 C using the same test method, i.e., the 90-second etching in a 10% oxalic acid under an anodic current density of 1 A/cm^2. The result at 0 C (Figure 4) exhibits essentially the same etched structure as that of the room temperature (Figure 3). It is, therefore, concluded that the same procedure for the room temperature test can be equally applicable to the test at 0 C.

Figures 5 through 7 show the effect of etching time on the test result at high temperatures (50 to 70 C). The 90-second etching shows a clearer etched structure than the 30-second etching at 50 C (Figure 5). It should be noted, however, that both etching times show the same test results. Both test results in Figure 5 are the ditch structure, according to ASTM A262-A. This indicates that the etching time does not affect the final test result, although the width of grain boundary ditching is greater after the 90-second etching. Figure 6 shows clear grain boundary etching after either 30-second or 90-second etching at 60 C. On the other hand, the 30-second etching exhibits a clearer grain boundary etching at 70 C (Figure 7). At this temperature, the 90-second etching shows an over-etched test result due to the increased contribution of general dissolution within a grain.

Figure 8 shows the 30-second etching test results at 70 to 96 C. The 30-second etching shows clear grain boundary ditching in this temperature range. This investigation was conducted up to only 96 C to avoid boiling the water in the beaker. The test temperature was controlled by heating the water in the beaker as described in the preceding section of this paper. However, the same test method should be applicable to the boiling temperature of the test solution (about 100 C).

To determine the optimum etching condition in the temperature range of 50 and 60 C, one set of heat treated AISI 304 samples was prepared. The AISI 304 sample was severely sensitized after receiving the sensitization heat treatment at 730 C for 4 hours. The degree of sensitization in this heat treated sample was greater than that of the previous welded sample. On the other hand, a solution annealing heat treatment at 1100 C for 1 hour followed by water quenching produced a nonsensitized metallurgical structure in the steel. Figure 9 shows

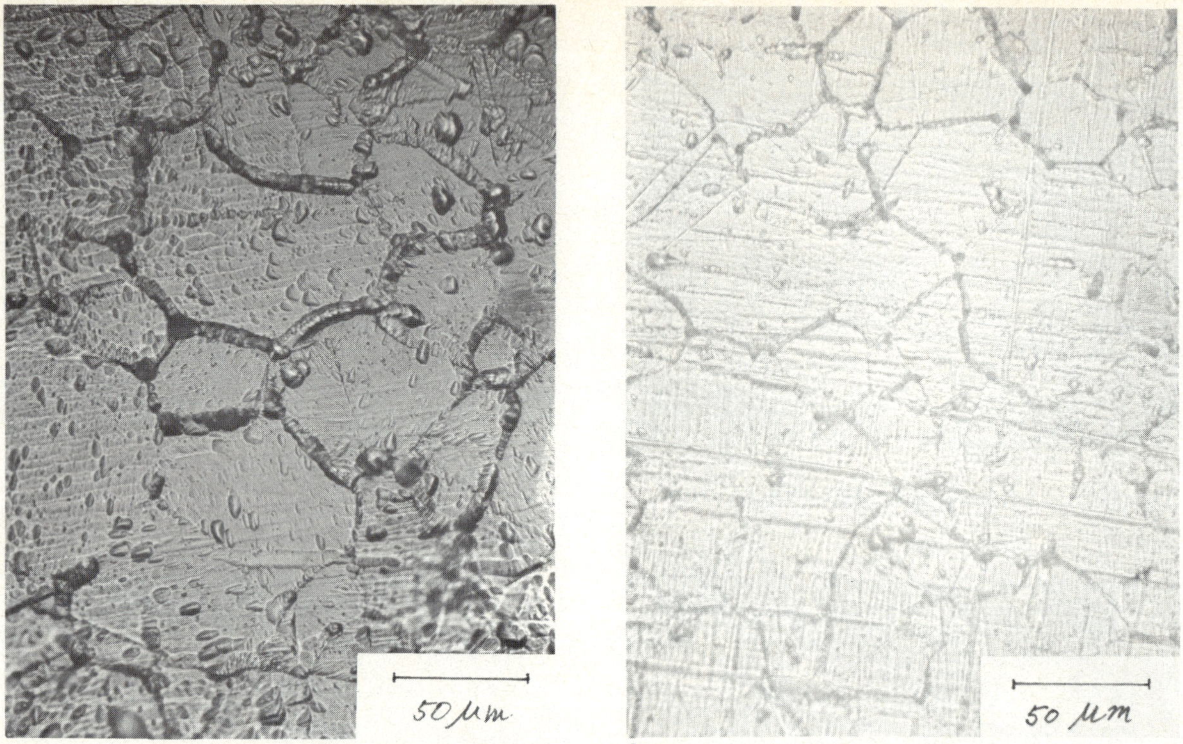

a b

FIGURE 5 – Light micrographs of the oxalic acid etching test results of HAZ in the welded AISI 304 sample at 50 C (as tested, 400X): (a) 90-second etching, and (b) 30-second etching.

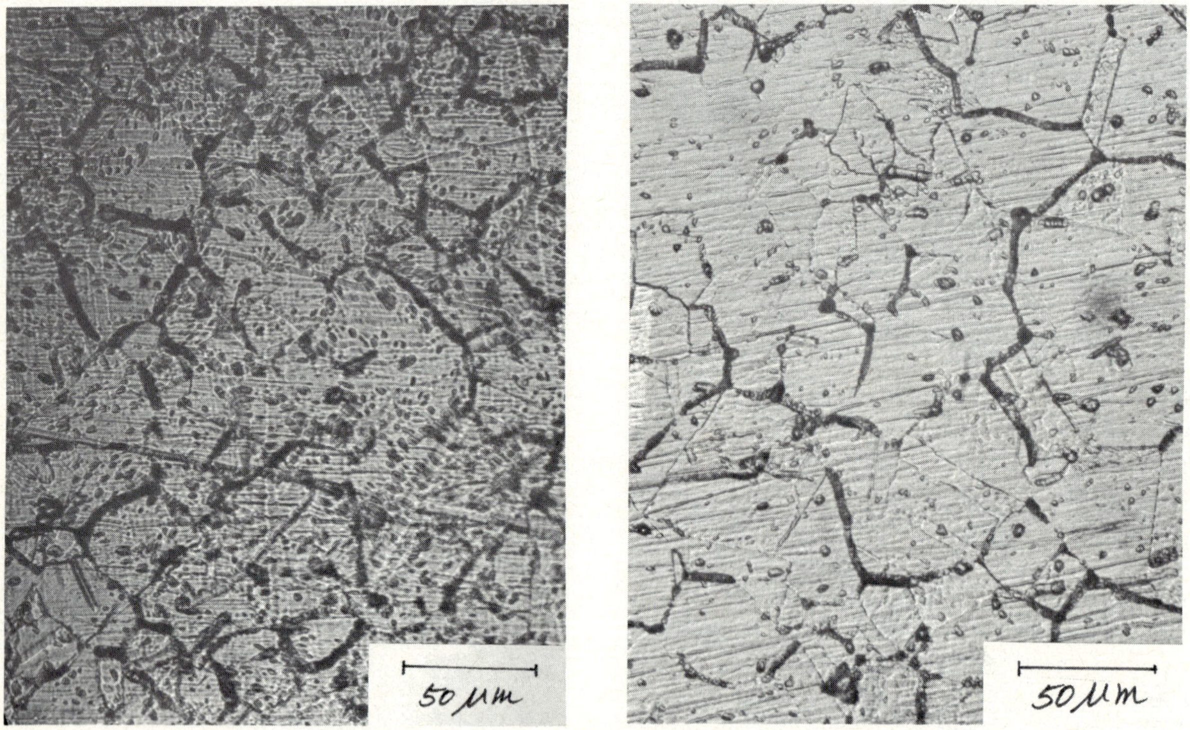

a b

FIGURE 6 – Light micrographs of the oxalic acid etching test results of HAZ in the welded AISI 304 sample at 60 C (as tested, 400X): (a) 90-second etching, and (b) 30-second etching.

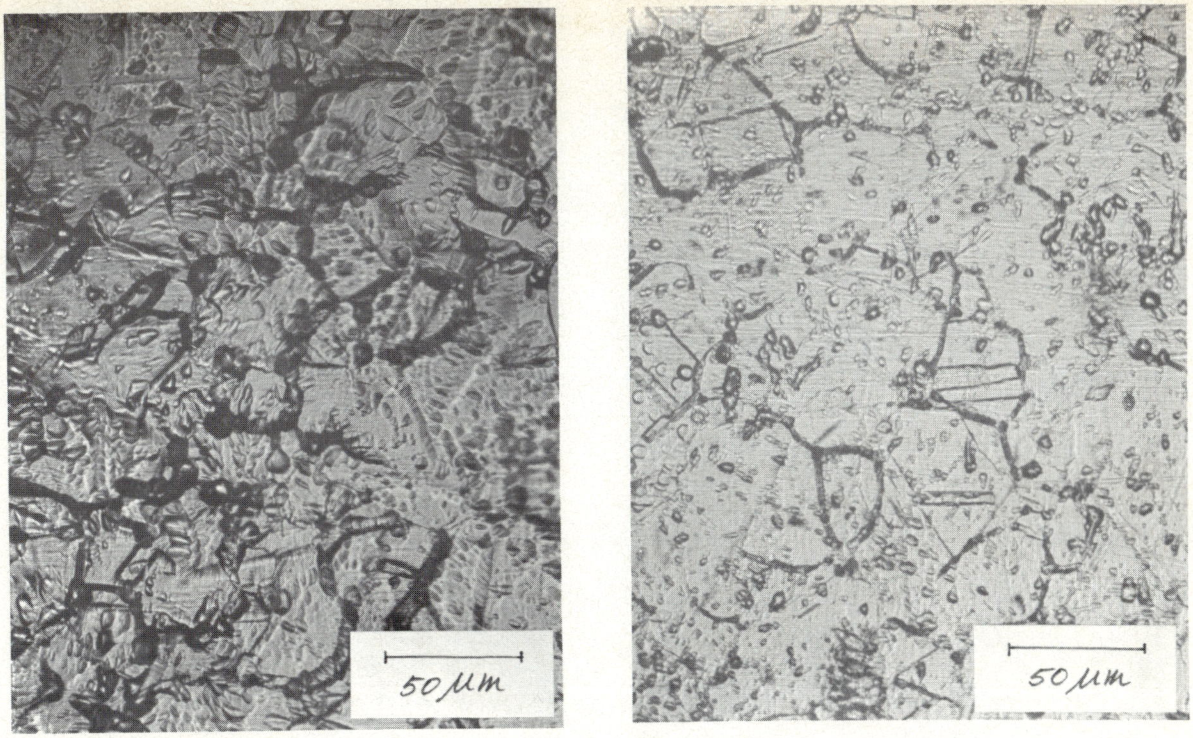

a b

FIGURE 7 — Light micrographs of the oxalic acid etching test results of HAZ in the welded AISI 304 sample at 70 C (as tested, 400X): (a) 90-second etching, and (b) 30-second etching.

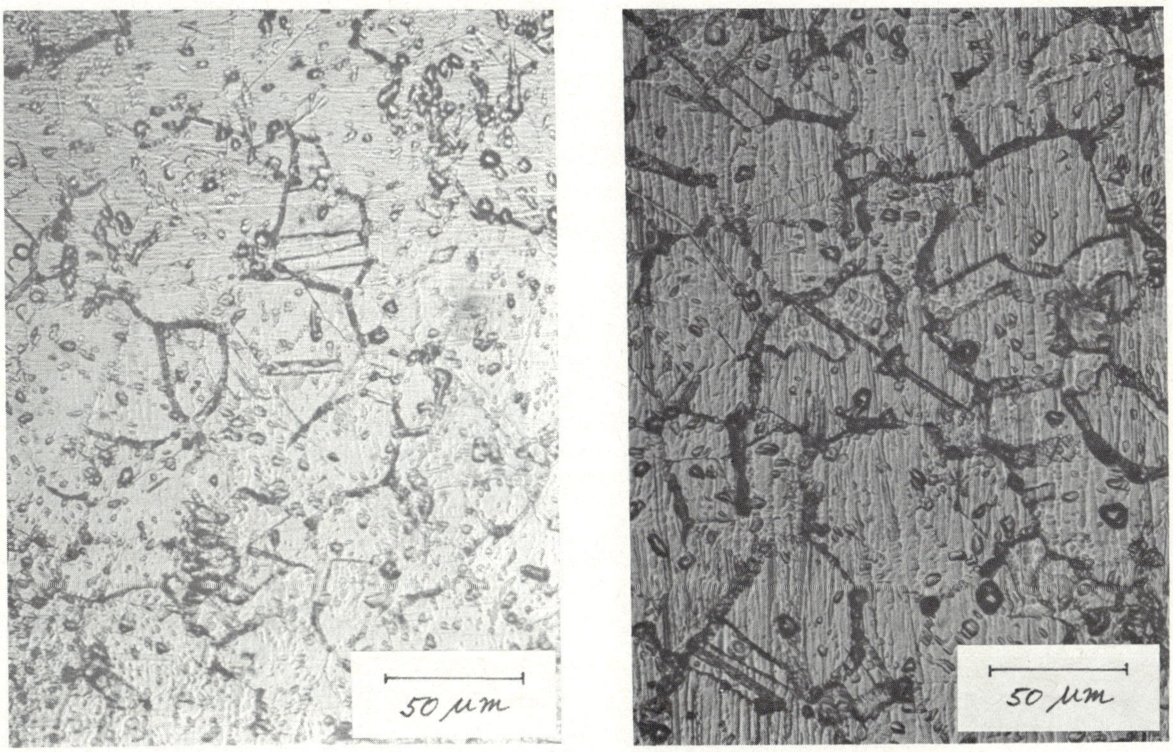

a b

(Figure 8 continued)

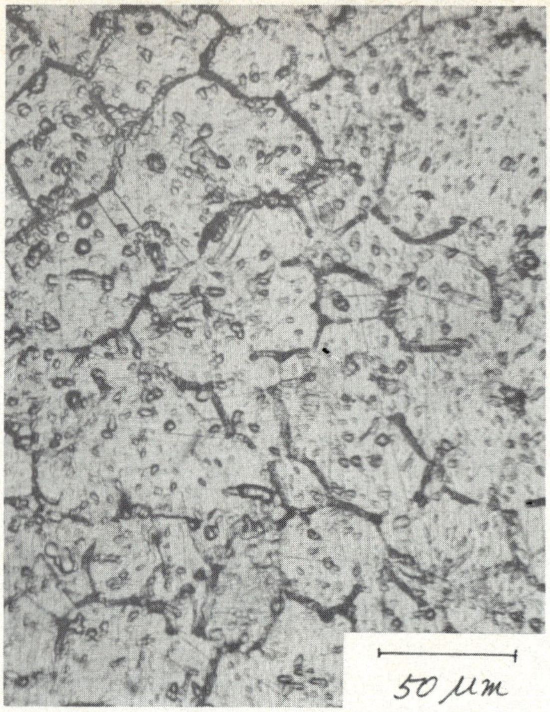

c

FIGURE 8 — Light micrographs of the oxalic acid etching test results of HAZ in the welded AISI 304 samples after the 30-second etching (as tested, 200X): (a) 70 C, (b) 80 C, and (c) 96 C.

a

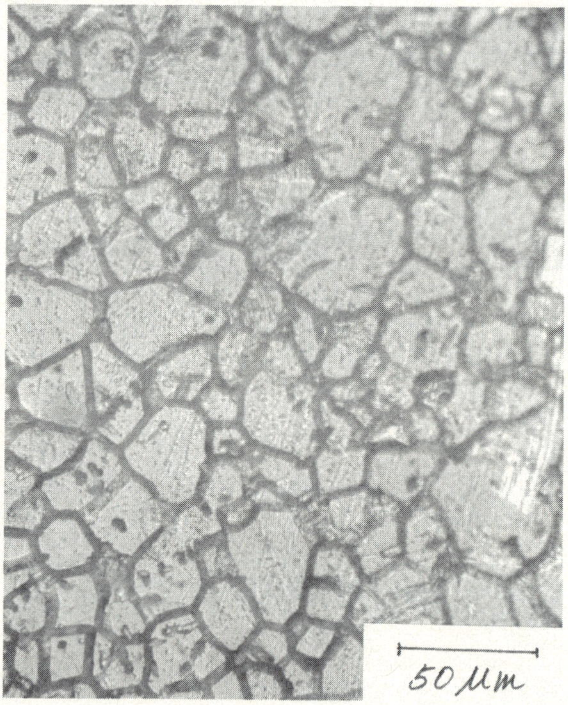

b

FIGURE 9 — Light micrographs of the oxalic acid etching test results of heat treated AISI 304 SS samples after the 30-second etching at 50 C (as tested, 400X): (a) nonsensitized, and (b) severely sensitized.

the test results of these two samples (nonsensitized and severely sensitized samples) after the 30-second etching. Unlike the welded specimen described in Figure 5, the severely sensitized specimen exhibits distinct grain boundary ditching after the 30-second etching at 50 C. The nonsensitized specimen does not show grain boundary ditching. The following etching times were, therefore, determined to be the best for the temperature range of 0 to 100 C:

1. At temperatures below 50 C, the 90-second etching time, as defined by ASTM A262-A, should be used.

2. Between 50 and 60 C, a 30-second etching should be used first. When the 30-second etching shows an under-etched grain boundary, the etching time should be increased to 90 seconds.

3. At the temperature above 60 C, a 30-second etching should be employed.

Application of the Test Method

The sensitized metallurgical structure in SS causes IGC and intergranular stress corrosion cracking (IGSCC) problems in a number of environments. For example, the sensitized SS suffers from IGC in oxidizing acids, such as nitric acid.[8] The SCC of austenitic SS in chloride solutions has been a well known corrosion problem for a long time. This corrosion is more severe when the steel is sensitized. The IGSCC in the sensitized AISI 304 SS pipe has also been a serious problem in the boiling water nuclear reactor operation. Here, the oxygen-contaminated, high-temperature, high-pressure pure water causes the IGSCC when the SS pipe is sensitized.[9] The polythionic acid cracking problem in petroleum refining industries is a result of the sensitized metallurgical structure in SS.[10] The sensitized AISI 304 SS also has cracking problems in fluoride solutions.[11] It should, however, be noted that the actual occurrence of these corrosion problems in service depends on the corrosivity of the individual environment where the material is used. In a certain mildly corrosive environment, only the heavily sensitized SS suffers from the corrosion problems. On the other hand, no sensitization in a SS component should be tolerated in a very corrosive application. The degree of sensitization tolerable in a specific application should be determined on the basis of the individual corrosive condition. The ASTM A262-A classifies the oxalic acid test results into three groups: the step, dual, and ditch structures as described in the Introduction section of this paper. This classification of the oxalic acid etched structure can be further refined as shown in Figure 10.[12] In this figure, C1 is the step structure according to ASTM A262-A. The C2, C3, and C4 in the figure are the dual structure. The ditch structure includes C5 and C6. This refined classification of the etched structure is more convenient when the allowable degree of sensitization in a SS component is determined for a specific corrosive application. The test method described in this paper is a useful tool for preventing an unexpected catastrophic corrosion failure of SS structures caused by sensitization. The reasons are as follows:

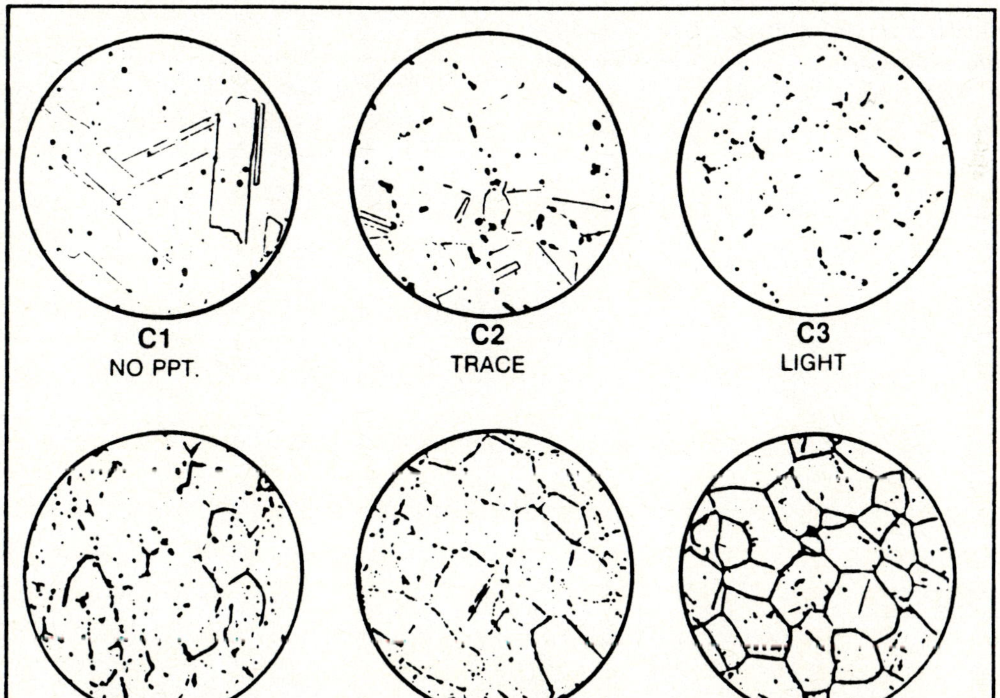

FIGURE 10 — The refined classification of the oxalic acid etching test results (Reference 12).

- The test is nondestructive; the measurement can be *in-situ* (or on-site).
- The testing time is short (30 or 90 seconds).
- Range of testing temperature is broad (0 to 100 C); the temperature change of atmosphere is not a limitation of the test method.
- The test method is simple and does not require a highly skilled person.

For the field nondestructive measurements, a DC power supply, field polishing equipment, a light microscope, and a camera in addition to the test cell shown in Figure 2 are needed. These devices (except for the test cell) are available commercially.

Conclusions

The oxalic acid test (ASTM A262-A) has been improved for use in field nondestructive detection of sensitized metallurgical structures in welded AISI 304 SS. The test methods are as follows:

- When the test temperature is between 60 and 100 C, a 30-second etching in a 10% oxalic acid under an anodic current density of 1 A/cm^2 should be used.
- Between 50 and 60 C, the 30-second etching should be used first. When the 30-second etching shows an under-etched grain boundary, the etching time should be increased to 90 seconds.
- At temperatures below 50 C the 90-second etching time as defined by ASTM A262-A should be used.

The field test kit which includes a test cell has been introduced. The test described in this paper is convenient for field measurements since it is a nondestructive, fast, and simple test. Also, the test temperature range is broad (0 to 100 C).

Acknowledgment

The authors thank M. A. Streicher of the University of Delaware for the helpful discussions on the test method. Standard Oil Production Company's permission to publish this paper is also acknowledged.

References

1. P. Novak, R. Stefec, F. Franz, Corrosion, Vol. 31, No. 10, p. 344, 1975.
2. W. L. Clarke, R. L. Cowan, W. L. Walker, "Comparative Methods for Measuring Degree of Sensitization in Stainless Steel," ASTM STP 656, p. 99, 1978.
3. V. Cihal, A. Deserstret, M. Froment, G. H. Wagner, Extend. Abstr. 5th Conf. Euro. Corros. Fed., p. 249, 1973.
4. L. A. Medvedeva, V. M. Knyazheva, Ya. M. Kolotyrkin, S. G. Babich, Zashch. Metal., Vol. 11, No. 6, p. 699, 1975.
5. J. B. Lee, "A New EPR Test for Determining Degree of Sensitization in Ferritic Stainless Steels," Proc. Inter. Symp. Honor. Professor H. H. Uhlig on his Seventy-Fifth Birthday, p. 201, R. P. Frankenthal, F. Mansfeld, Eds., Electrochem. Soc., Inc., Princeton, New Jersey, 1981.
6. A. Borrello, et al., Werkst. Korros., Vol. 32, p. 332, 1981.
7. A. P. Majidi, M. A. Streicher, Corrosion, Vol. 40, No. 8, p. 393, 1984.
8. A. J. Sedriks, "Corrosion of Stainless Steels," A. Wiley - Interscience Publication, p. 120, 1979.
9. H. R. Copson, G. Economy, Corrosion, Vol. 24, No. 3, p. 55, 1968.
10. C. Samans, Corrosion, Vol. 20, No. 8, p. 256+, 1964.
11. G. J. Theus, J. R. Cels, Corrosion Problems in Energy Conversion and Generation, Electrochem. Soc., Princeton, New Jersey, p. 384, 1974.
12. E. A. Loria, Corrosion, Vol. 40, No. 8, p. 441, 1984.

Recent Applications of Electrochemical Techniques for Corrosion Research in the Automotive Industry

R. L. CHANCE and M. S. WALKER*

Abstract

This paper describes how electrochemical techniques were used for a number of diverse applications in corrosion research over the past several years. Anodic polarization, for example, was useful in several studies: (1) to demonstrate how large carbides in mild steel caused accelerated corrosion when the metal was deformed; (2) to determine the corrosion characteristics of aluminum in ethylene glycol solutions; (3) to help characterize surface films on stainless steel at anodic potentials; (4) to show the improvement in breakdown potentials for steel implanted with phosphorus ions; and (5) to determine the corrosivity of exhaust gas condensates toward steels used in the automobile exhaust system. Current measurements taken at constant potential were effective for determining corrosion rates of aluminum in cavitating ethylene glycol solutions, and galvanic current measurements were effective for determining corrosion rates of dissimilar metal couples. The polarization resistance (linear polarization) method was used to measure corrosion rates of steel and zinc alloy in ethanol fuel as a function of inhibitor content.

Introduction

Electrochemical techniques have been used to evaluate not only the corrosion resistance of metals and alloys, but also the corrosivity of fluids associated with the automobile, its environment, and its manufacture. Occasionally, electrochemical tests are used for screening purposes, after which there may be a follow-up with approved industrial standard tests that are often much longer and more expensive. In many instances, particularly in research, there is no better way to determine quickly the effect of subtle differences in alloys and solutions than by electrochemical methods. In this article, highlights of some of the more recent and diverse applications of electrochemical techniques will be reviewed. The first application shows how electrochemistry was used to detect the effects of heat treatment and carbide formation on corrosion of deformed steel. Another application demonstrated how galvanic corrosion rates were determined for various dissimilar metal couples appropriate to the automotive industry. Still another section describes how an electrochemical method was used to compare the effects of inhibitors in preventing cavitation damage. A fourth method deals with compositional changes in anodic films and their effect on corrosion. A fifth application demonstrates the effect of phosphorus ion implantation on the corrosion characteristics of mild steel. The final two applications of electrochemical techniques show how differences in exhaust condensates and ethanol fuel can affect corrosion of metals.

*Physical Chemistry Dept., General Motors Research Laboratories, Warren, Michigan.

Effect of Heat Treatment and Carbide Formation[1]

This study was initiated because anodic polarization measurements indicated that deformed low carbon steel, in certain unexplained cases, exhibited significantly less resistance to corrosion than did undeformed steel. The subsequent investigation showed that the observed deterioration in corrosion resistance was related to heat treatments that produced large carbide particles. These particles, most of which were longer than 50 μm, suffered brittle fracture when torsionally deformed, and caused a parting of the carbide/ferrite interface. A scanning electron microscope (SEM) photomicrograph of one of these

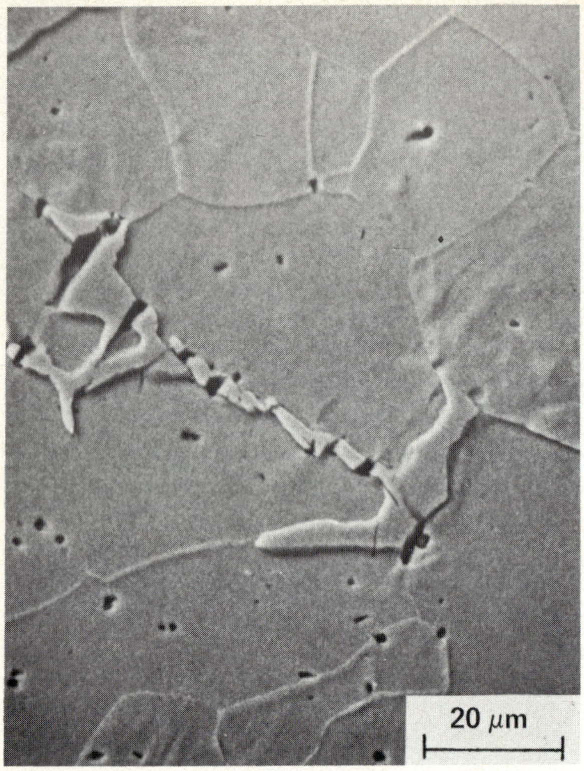

FIGURE 1 – Photomicrograph (SEM) of a carbide particle in AISI 1008 steel wire after torsional deformation. Heat treatment: (D) 800 C for 90 min, slow furnace cooled.

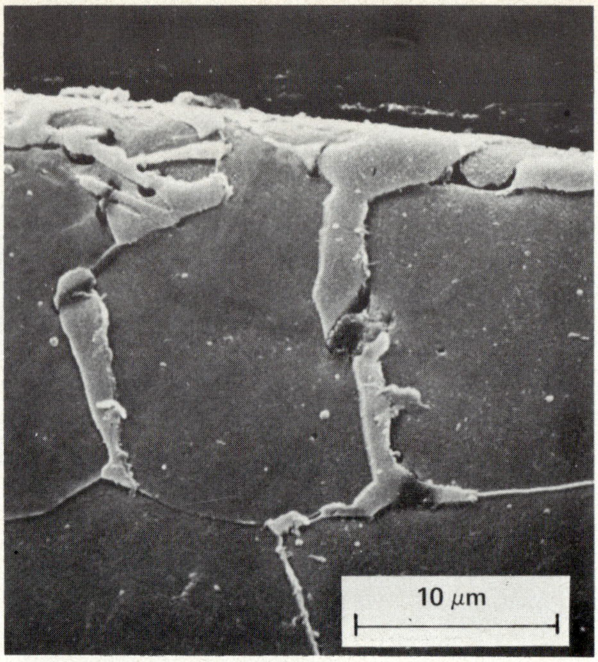

FIGURE 2 – Photomicrograph (SEM) of a cross section of AISI 1010 steel wire. Heat treatment: (E) sequential.

carbides in AISI 1008 stainless steel (SS) is shown in Figure 1. This anchor-shaped particle, about 65 μm long, was fragmented into a number of pieces and cracked in several places. Another photomicrograph (Figure 2), which shows carbides in a cross section of a piece of AISI 1010 SS, illustrates how capillary paths formed in the metal could act as crevice sites and cause accelerated corrosion.

The metal test pieces in this study were in wire form and were either vacuum-encapsulated in quartz tubes before heat treatment or simply heat treated in a protective, nonoxidizing atmosphere, such as 96% nitrogen and 4% hydrogen. Metal deformation was accomplished in each case by twisting the wire in a lathe. This procedure is mechanically reproducible and is a relatively simple and rapid method for obtaining severe triaxial deformation. Except for certain specific experiments, all test specimens were polished through 600-grit SiC paper to provide a uniform cross section and a reproducible surface. A standard polarization cell was used to contain the electrolyte for anodic polarization measurements. Ammonium nitrate (0.6 M) solution was used as the electrolyte because in that environment steels not only exhibit active-passive behavior but are also subject to such types of corrosion as crevice corrosion, pitting, intergranular attack, and stress corrosion cracking (SCC). The temperature of the electrolyte was 25 ± 0.2 C and the potential scan rate was 0.334 mV/s.

The steel wires were subjected to five different heat treatments, which resulted in five different sets of anodic polarization curves depending on the amount and size of carbides formed and the degree of deformation (number of twists imparted). The heat treatments are described in Table 1.

When heat treatment did not produce large carbides, as in cases A and B, the anodic polarization curves displayed only simple active-passive behavior with low current densities in the passive region ($< 10\ \mu A/cm^2$) and no significant excursions or peaks. However, with carbide-forming heat treatments, the passive region was reduced considerably and a large second current density maximum occurred between 0 and 0.4 V vs SCE. An example of this may be seen in Figure 3, where anodic polarization curves are shown for AISI 1010 SS wire, which had been twisted 0, 2, 4, 8, 16, and 32 times. These curves demonstrate the effect of progressive increases in plastic deformation. The active region did not change appreciably with deformation, but the length of the passive region did. In addition, there were generally large increases in current density in the passive region, particularly between 0 and 0.4 V vs SCE.

The effect of surface composition is shown in Figure 4. Although the three wires were heat treated by the sequential method (E), they were prepared differently in other respects. Ordinarily, the wire specimens were polished to provide a uniform diameter for area measurement and a reproducible surface for conducting electrochemical measurements. It was reasoned, however, that since torsional deformation produces strain that is greater at the surface of the wire than at the core, the effects of deformation would be more pronounced if the wires were not polished. Curves 1 and 3 in Figure 4 were obtained

TABLE 1 — Heat Treatments

A. Full Anneal (900 C for 15 min, furnace cooled at 50 C/h)

This resulted in very large ferritic grains and a few spheroidal iron carbide particles.

B. Normalizing Heat Treatment (980 C for 15 min, air cooled)

This resulted in small amounts of lamellar pearlite in ferrite matrix.

C. Modified Process Anneal (875 C for 30 min, furnace cooled to 690 C, soaked 16 h, furnace cooled at 50 C/h)

This resulted in the precipitation of iron carbides at grain boundaries that agglomerated into more massive shapes.

D. Heat Treatment at 800 C (800 C for 90 min, furnace cooled at 50 C/h)

This resulted in the precipitation and growth of numerous carbide particles, the largest of which were 60 to 80 μm and located mostly at grain boundaries.

E. Sequential Heat Treatment (870 C for 15 min, air cooled to 690 C, soaked 16 h, furnace cooled at 50 C/h)

This resulted in massive carbides dispersed in a ferrite matrix.

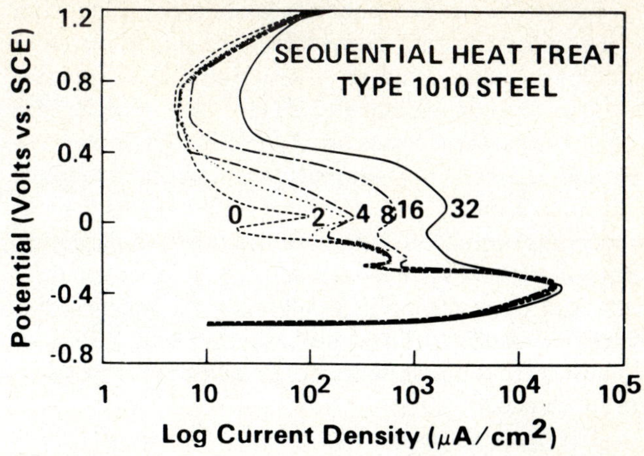

FIGURE 3 — Anodic polarization curves for AISI 1010 steel wire twisted 2, 4, 8, 16, and 32 times in 10 cm. Heat treatment: (E) sequential.

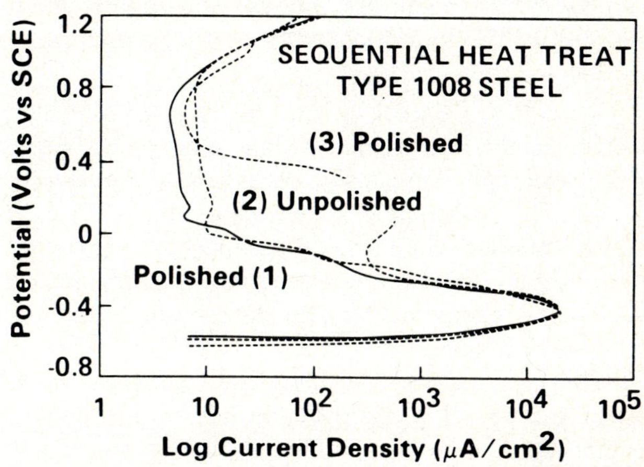

FIGURE 4 — Anodic polarization curves for AISI 1008 steel wire: (1) undeformed, polished; (2) deformed, unpolished; and (3) deformed, polished. Heat treatment: (E) sequential.

from polished wires, undeformed and deformed, respectively. Curve 2 was obtained from unpolished and deformed wire. If polishing decreases the effect of deformation, then Curve 2 would be expected to show even larger current densities in the passive region than Curve 3, but this was not the case. The reason became evident upon close examination of the microstructure of the unpolished wire. Even though precipitation and growth of large, irregularly shaped carbide particles had indeed occurred as a result of the heat treatment, none of the carbide growth was visible either at the surface or to a depth of approximately 0.05 mm from the surface. Whenever wires were polished prior to anodic polarization, the amount of metal removed was always at least 0.07 mm; thus, any existing carbides were always at the surface of the metal and exposed to the electrolyte.

This experiment showed that unless carbides that are fractured by deformation are exposed to the electrolyte, little or no change will be seen in corrosion characteristics, particularly in the passive region. In this study, plastic deformation by itself was not sufficient to produce any significant change in corrosion resistance.

In summary, the anodic polarization technique was used to study changes in corrosion resistance of plastically deformed, low carbon steel. The effects of deformation were shown to be potential-dependent and were related to heat treatment and the formation of carbides. When fractured, the large carbides created separations at the carbide/ferrite interface where accelerated corrosion occurred by a crevice mechanism.

Galvanic Corrosion Measurements Relevant to Automotive Applications[2]

When two dissimilar metals in metallic contact are exposed to a corrosive environment, there is a tendency for one of the metals to undergo accelerated corrosion and for the other metal to be protected. This type of corrosion is called *galvanic corrosion*, and it is one of the more severe forms of corrosion. The need to reduce vehicle weight introduced the use of many new galvanic couples on the automobile and led to this study of galvanic corrosion rates.

Galvanic corrosion behavior is often predicted from the measured corrosion potentials of uncoupled metals in a particular environment. It is assumed that the greater the difference in potential between two metals, the greater the galvanic corrosion. A table of these potentials (galvanic series) can be a useful indication of the tendency to corrode galvanically; however, other factors may influence the corrosion rate. For example, polarization behavior, cathode-to-anode area ratios, solution conductivity, and other factors can determine the kinetics of a corrosion reaction. To determine the actual rate of galvanic corrosion, it is necessary to measure the galvanic current directly.

The galvanic current was measured with a potentiostat that was set to maintain a zero potential difference between the two members of a galvanic couple. At zero potential difference, the current that flows is the true short-circuit galvanic corrosion current. The total amount of galvanic attack occurring during a test period is equal to the amount of electric charge that passes between the members of the couple. The amount of charge is either measured with a current integrator or calculated from a recording of current vs time.

Figure 5 is a photograph of the disassembled test specimens and holder used in the laboratory tests. Figure 6 is a photograph of assembled specimens and holder in a beaker. The test specimens, which measure 7.5 × 2.5 cm, were masked with electrical tape so that an area of 20 cm^3 was exposed to the electrolyte. A lucite spacer maintained a separation of 0.3 cm between the members of a galvanic couple.

After cleaning and weighing, the specimens were masked and placed in the specimen holder. The holder was placed in the beaker, and the electrical connections were made. The test solution was then added and the potentiostat immediately turned on. After the test, the specimens were cleaned of corrosion products, using standard ASTM[(1)] cleaning procedures, and reweighed.[3]

[(1)]ASTM, Philadelphia, Pennsylvania.

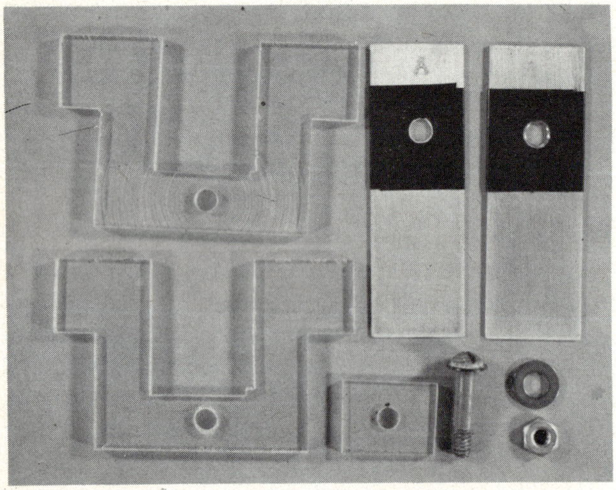

FIGURE 5 — Disassembled galvanic test specimens and holder.

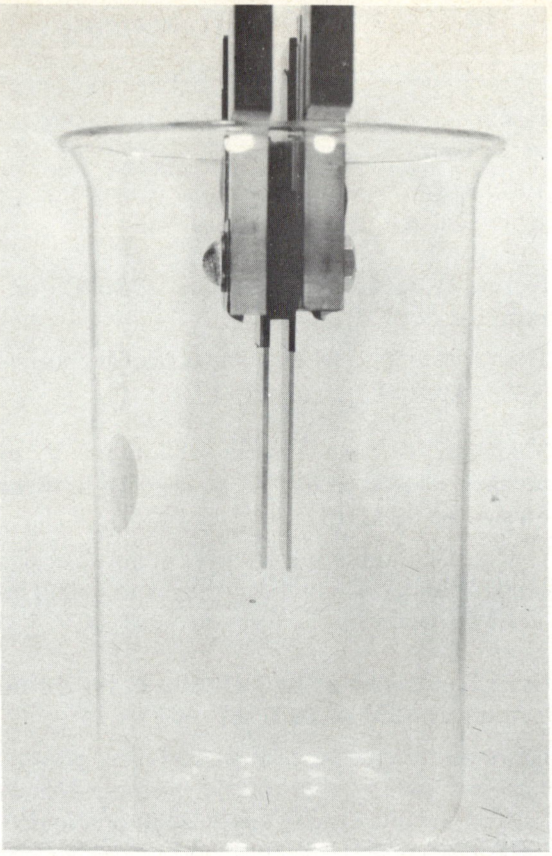

FIGURE 6 — Side view of test specimens and holder.

Figure 7 is a photograph of the galvanic corrosion test specimens that were installed on the underbodies of seven cars. To compare the results of the car tests with the laboratory tests, the galvanic corrosion rate (i_g) for the car test specimen was obtained by converting the difference in weight loss between a coupled and uncoupled specimen to a current density.

The laboratory tests were used to rank the metallic couples on the basis of galvanic current and to study the effects of pH, solution composition, length of exposure, area ratio, and surface conditions on galvanic corrosion rates. Results from the laboratory tests were compared with results from the car tests. The following few examples were taken from this study.

Figure 8 is a recording of the galvanic corrsion current for the steel-AISI 430 stainless steel (SS) couple in 3.5 wt% NaCl at pH 2. The inserted table lists the galvanic corrosion rate (i_g) and the total corrosion current (i_w). The galvanic corrosion rate was determined from the recording and the total corrosion rate was determined from the weight loss measured at the end of the test. The total corrosion current is equal to the sum of the galvanic current that flows between the two members of the couple and the corrosion current that flows between local anodic and cathodic sites on the surface of a single member of the couple.

Steel was the anode for about the first hour; then the polarity reversed and the AISI 430 SS was the anode for the next 6 hours; the polarity then reversed again and the

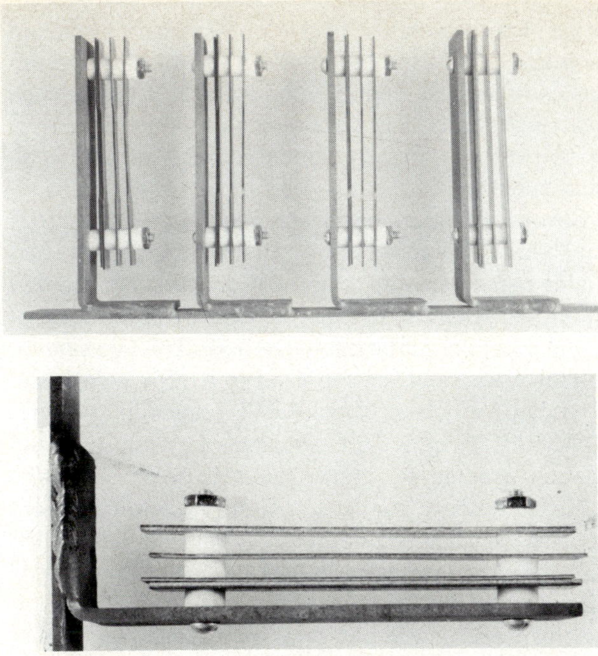

FIGURE 7 — (a) Test specimens bolted to a car mounting rack. (b) Closeup of test specimens.

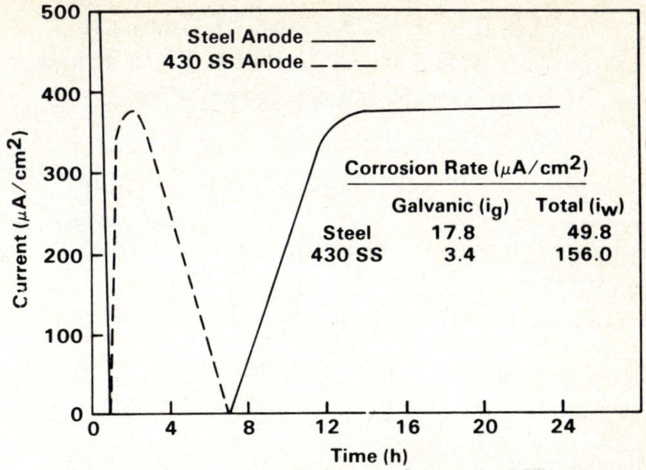

FIGURE 8 — Galvanic current flow for steel—AISI 430 stainless steel couple in 3.5 wt% NaCl solution at pH 2.

steel was the anode for the remaining 17 hours. Surprisingly, the total corrosion rate of AISI 430 SS (i_w = 156 μA/cm^2) was about three times greater than steel. The galvanic current contributed less than 4 μA/cm^2 to the total corrosion current of the AISI 430 SS, while local action currents added 152 μA/cm^2. Also, local action current was responsible for 32 μA/cm^2 of the 49.8 μA/cm^2 total corrosion rate of steel.

The effect of cathode-to-anode area ratio on galvanic corrosion was evaluated for a solder-brass couple at room temperature in 3.5 wt% NaCl. Solder specimens with areas of 2, 10, and 20 cm^2 were coupled to 20 cm^2 of brass. Table 2 lists the galvanic current (I_g), the total current (I_w), the galvanic current density for the solder anode (i_g), and the total current density for the solder anode (i_w). The galvanic corrosion current (I_g) decreased from 321 to 270 μA, only about 16%, when the area of the solder anode was reduced from 20 to 2 cm^2. In contrast, the penetration rate of the solder anode, which is proportional to the current density (i_g), increased by a factor of 135/16, or 8.4 times. This is an example of how catastrophic failure can occur when galvanic couples with a large cathode-to-anode area are exposed to a corrosive environment.

The galvanic corrosion rates in the 3.5 wt% NaCl in laboratory tests were about five times greater than the car exposure rates, and the ranking of the galvanic couples showed a reasonable agreement (Table 3). The only exception was the galvanized steel-steel couple, which ranked first in the laboratory test (i_g = 76.7 μA/cm^2) and sixth in the car test (i_g = 0.69 μm/cm^2). Disagreement between laboratory immersion test and car test can occur because metals such as zinc form protective coatings in the cyclic wet and dry environments on a car that do not form in continuous immersion in salt water.

TABLE 2 — Effect of Cathode to Anode Area Ratio on the Galvanic Corrosion Behavior of a Solder-Brass Couple in 3.5 wt% NaCl

Area (cm^2)		Cathode to Anode Area Ratio	Corrosion Rate of Solder			
			Current (μA)		Current Density (μA/cm^2)	
Cathode Brass	Anode Solder		Galvanic I_g	Total I_w	Galvanic i_g	Total i_w
20	20	1:1	321	377	16.0	18.9
20	10	2:1	320	372	32.0	37.2
20	2	10:1	270	273	135.0	137.0

TABLE 3 — Comparison of Galvanic Corrosion Rates from Car Test With Rates from the 24-h Laboratory Test

Couple	Corrosion rate ($\mu A/cm^2$) Car Test Rank	Car Test I_g	Laboratory Test Rank	Laboratory Test I_g
Galvanized Steel Steel	6	0.69	1	76.7
Aluminum Brass	2	5.18	2	33.8
Steel Brass	1	6.31	3	24.2
Aluminized Steel Steel	3	3.88	4	18.8
Aluminum Steel	4	2.73	5	14.6
Zincrometal Steel	8	−0.45	6	12.0
Steel AISI 430 SS	5	2.16	7	10.9
Zincrometal Aluminum	10	−0.98	8	3.28
Galvanized Steel Aluminum	7	−0.32	9	3.08
Aluminized Steel Aluminum	9	−0.55	10	2.49

Electrochemical Measurements in Cavitating Solutions[4]

In this investigation, a series of electrochemical tests were conducted in which the effects of cavitating ethylene glycol solutions on the corrosion of cast aluminum were studied.

Cavitation may be defined as the formation and collapse of cavities or vapor/gas bubbles in a liquid. It is a condition that occurs when liquid pressure falls below vapor pressure at localized sites in the system. Cavitation damage occurs when the bubbles collapse or implode at high impact velocities at or near the metal surface. There are two components that contribute to cavitation damage—the mechanical component and corrosion. The relationship between the mechanical and corrosion components is complex because the processes involved are frequently interrelated and synergistic. Nevertheless, investigators have made attempts to compare that relationship for various systems. In this study, where cavitation was induced ultrasonically, electrochemical measurements were used to determine how much of the damage to cast aluminum could be attributed to electrochemical corrosion and how much to the mechanical impact of collapsing bubbles.

Several electrochemical techniques were used in this study; namely, potential-time, anodic polarization, and rate-time measurements. However, only the data derived from anodic polarization and rate-time measurements at constant potential will be discussed here. The two electrolytes that were involved consisted of a 15% solution of well-inhibited ethylene glycol (GM[2] Engineering Standard 6038-M) or of a 15% solution of poorly inhibited ethylene glycol (containing 25 g/L sodium tetraborate).

Measurements involving anodic polarization were made using conventional equipment for controlling potential, such as indicated in ASTM G5, and the test cell consisted of a standard 1-L, multinecked, polarization cell, which was immersed in a transducer tank where ultrasonic waves were generated. Test specimens were cubical pieces of UNS A03190 cast aluminum alloy that were held in the polarization cell with a conventional compression gasket holder.

Results are shown in Figure 9 where anodic polarization measurements are compared for cavitating and noncavitating conditions using well-inhibited and poorly inhibited ethylene glycol/distilled water mixtures. The curves for the well-inhibited system represent very low corrosion rates and were nearly identical for cavitating and noncavitating conditions. Calculations based on electrochemical data showed that the average corrosion rate represented by each curve was 0.2 mm/y penetration. Actual weight-loss measurements showed a corrosion rate of 0.3 mm/y for each specimen. The curves for the poorly inhibited system represent much higher corrosion rates according to the current density values. Cavitation increased the corrosion rate by an order of magnitude. For example, the calculations based on electrochemical data showed that the average corrosion rate for aluminum in the poorly inhibited system under noncavitating conditions was 4 mm/y compared to 38 mm/y for cavitating conditions. Actual weight-loss measurements showed corrosion rates of 8 and 77 mm/y for noncavitating and cavitating conditions, respectively. This means that electrochemical corrosion (38 mm/y) contributed about one-half of the total damage to the specimen as a result of cavitation (38/77 mm/y).

Corrosion rates at constant potential represent those rates that occur in a specific oxidizing environment. As suggested by the anodic polarization curves, a potential of −0.6 V vs SCE is a moderately oxidizing environment in which corrosion rates would not be excessive. The results of some of these measurements are summarized in Figure 10, which shows corrosion rates, expressed as current density, for UNS A03190 cast aluminum alloy in well-inhibited and poorly inhibited ethylene glycol solutions under cavitating and noncavitating conditions. An initial spike of high current density was often observed,

[2]General Motors, Warren, Michigan.

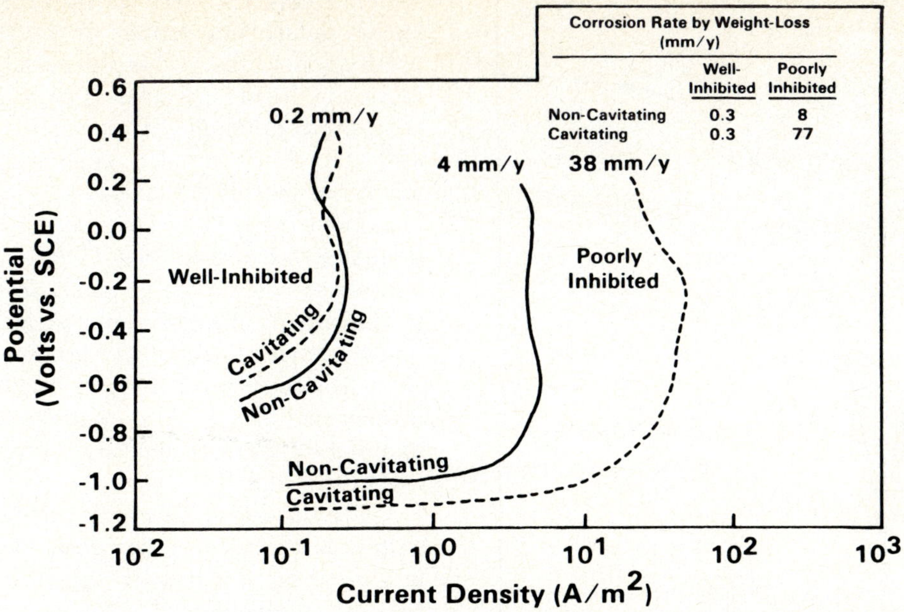

FIGURE 9 — Anodic polarization curves for aluminum in well-inhibited and poorly inhibited ethylene glycol/distilled water mixtures under cavitating and noncavitating conditions at 82 C.

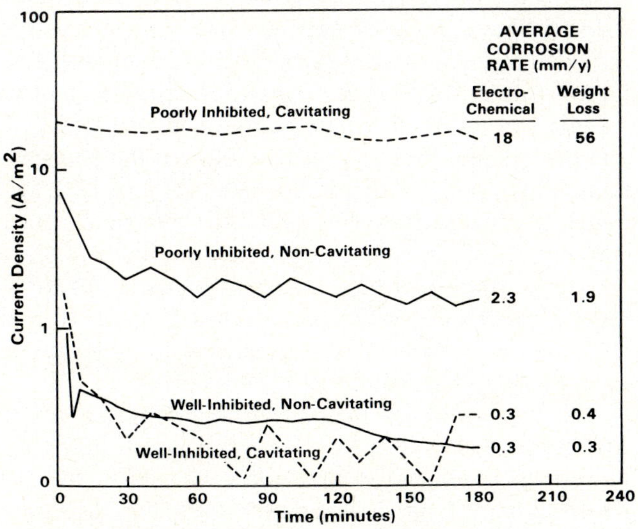

FIGURE 10 — Corrosion rates for aluminum at −0.6 V vs SCE as a function of time in well-inhibited and poorly inhibited ethylene glycol/distilled water mixtures under cavitating and noncavitating conditions at 82 C.

which apparently is characteristic of the system whether well-inhibited or not. After about 10 min, the high current densities declined significantly, and then after 30 min began to level off to a point where the longer term corrosion rates (proportional to the cavitation damage) could be estimated. Visual inspection and weight losses determined at the end of the test bore this out. The data in Figure 10 reinforce the anodic polarization data in Figure 9 by showing that cavitation not only increases the total damage to the metal in poorly inhibited systems, but also significantly increases electrochemical corrosion (2.3 vs 18 mm/y). In this test, electrochemical corrosion accounted for about one-third of the total metal loss (18/56 mm/y). It may be concluded that electrochemical measurements of corrosion rate can be related to the total damage of a metal in cavitating engine coolants, and these measurements provide a reasonably reliable means for estimating the degree of cavitation damage that will occur under specific conditions. Furthermore, because field experience has shown that poorly inhibited ethylene glycol engine coolants of the type used in this study contribute to serious damage of aluminum water pumps, this controlled potential test can be related to field service.

Characterization of Surface Films[5]

Anodic polarization measurements were made for SAE[(3)] J409 stainless steel (SS) (12% Cr) using deaerated sodium sulfate solution in which the pH, temperature, and molarity were varied. Electron spectroscopy for chemical analysis (ESCA) and Auger electron spectroscopy (AES) surface data were obtained to help explain unusual polarization phenomena, such as the extreme sensitivity to pH of the critical current (i_C).

Figure 11 is a graph of the critical current density of SAE J409 SS in 0.5 M sodium sulfate at 82 C as a function of pH. The critical current increased by more than two orders of magnitude (28 to 3600 $\mu A/cm^2$) as the pH was reduced from 3 to 2.8.

To find a reason for this large difference in critical current, surface films were examined by AES, and composition profiles were constructed. The results can be

---

[(3)]Society of Automotive Engineers (SAE), Warrendale, Pennsylvania.

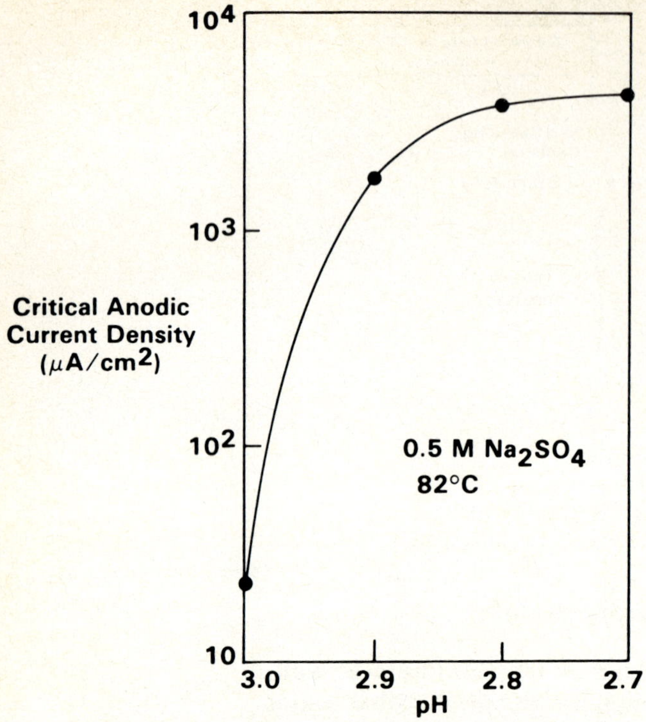

FIGURE 11 — Critical anodic current density of SAE J409 stainless steel in 0.5 M sodium sulfate solution at 82 C as a function of pH.

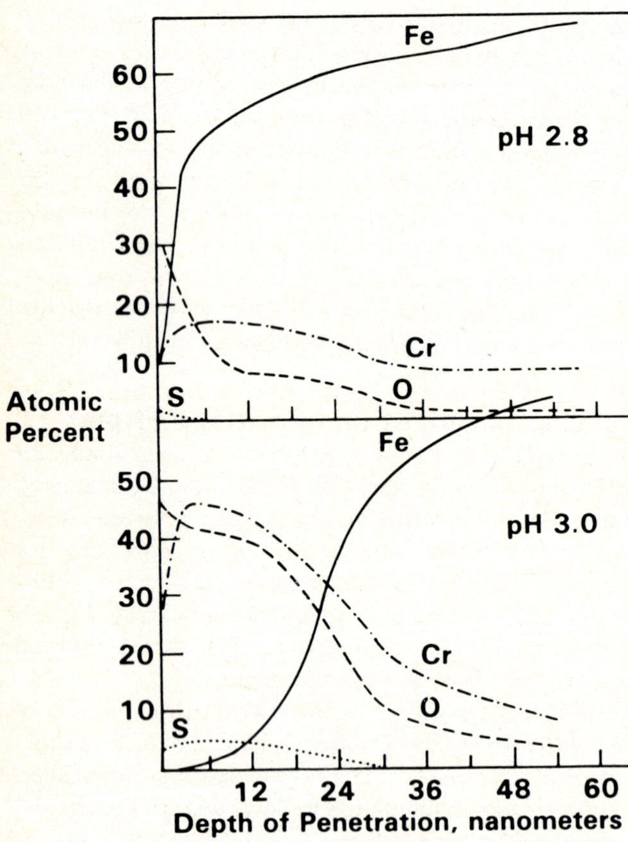

FIGURE 12 — Auger composition profiles of the surface film on SAE J409 stainless steel anodically polarized to the primary passive potential in pH 2.8 and 3.0, 0.5 M sodium sulfate solutions at 82 C.

seen in Figure 12, which shows the composition profiles for the surface films formed on the metal during anodic polarization to the primary passive potential. The film formed in the pH 2.8 solution (i_C = 3600 μA/cm^2) was composed primarily of iron and oxygen and was estimated to be about 5 nm thick; whereas the film formed in the pH 3.0 solution (i_C = 28 μA/cm^2) was composed primarily of chromium and oxygen and was estimated to be about 25 nm thick. The surface analysis appears to confirm the generally accepted concept that the passivating effect of chromium in SS depends largely on the formation of a protective chromium-bearing oxide or hydroxide. Without that film, high corrosion rates can be expected. In this metal-electrolyte system, the formation of the chromium film is very sharply defined around a pH value of 3.

The traverse of the anodic polarization curve starting at the open circuit corrosion potential (E_{corr}) and ending at a potential of minimum current in the passive region (e_{pass}) was investigated with respect to film composition and thickness by Auger depth profiling. The two solutions chosen as variables for this study were those in which the critical currents for SAE J409 SS differed by more than two orders of magnitude; namely, the 0.025 M (i_C = 30 μA/cm^2) and the 0.5 M (i_C = 3600 μA/cm^2) sodium sulfate solutions with pH at 2.8 and temperature at 82 C. Table 4 shows the potentials at which specimens were removed for surface analysis, and also the film thickness for those potentials. Figure 13 shows composition profiles for films on the metal at the corrosion potential (E_{corr}) and the primary passive potential (E_{pp}) in the two solutions. These profiles support the data in Figure 12 in that they show that surface films associated with small critical currents (0.025 M solution) are relatively thick and are composed largely of chromium and oxygen; whereas, those films associated with large critical currents (0.5 M solution) are relatively thin and are composed largely of iron and oxygen.

As the metal was polarized to a potential about 75 mV more positive than E_{pp} (i.e., to E_{corr} + 200 mV), two things happened to the surface film. First, the film began to show large fluctuations in thickness. As shown in Table 4, the thickness ranged from 18 to 340 nm in the 0.5 M solution. The second and most important occurrence was a compositional change that occurred to the film in the 0.5 M solution. This is shown in Figure 14 by the fact that at this prepassive potential, the film was composed largely of chromium and oxygen, whereas at the primary passive potential (E_{pp}), it was largely iron and oxygen. The composition of the film in the 0.025 M solution did not change, remaining primarily chromium and oxygen. The formation of a chromium-oxygen film appears to be a preprequisite to passivity for this system.

As the metal was polarized further into the passive region, the films became thinner as shown in Table 4 for E_{corr} + 300 and E_{corr} + 400. Film composition did not change significantly.

Finally, at 0.3 V in the passive region (E_{pass}), the films were estimated to be 8 and 3 nm thick in the 0.025 and 0.5 M solutions, respectively. After 30 minutes at 0.3 V in the 0.025 M solution, the film thickened to 13 nm

TABLE 4 — Film Thickness on SAE J409 Steel Anodically Polarized to Selected Potentials in pH 2.8 Sodium Sulfate Solutions at 82 C

	0.025 M Sodium Sulfate		0.5 M Sodium Sulfate	
	Potential (mV vs SCE)	Film Thickness[1] (nm)	Potential (mV vs SCE)	Film Thickness[1] (nm)
E_{corr}	−470	41	−625	15
E_{pp}	−390	49[2]	−555	19[3]
E_{corr} + 200 mV	−270	18-340	−425	15-40
E_{corr} + 300 mV	−170	32	−325	16
E_{corr} + 400 mV	−70	29	−225	5
E_{pass}	300	8	300	3
E_{pass} (30 min)	300	13	300	3

(1)Film thickness is calculated as the average of the sputter depth required to decrease the oxygen Auger signal to half maximum intensity and to increase the iron Auger signal to half maximum intensity. Sputter rate is calibrated as Ta_2O_5 because the sputter yields of these films are unknown. Therefore, film thicknesses are relative and should not be considered absolute.
(2)i_c = 27 μA/cm^2.
(3)i_c = 3600 μA/cm^2.

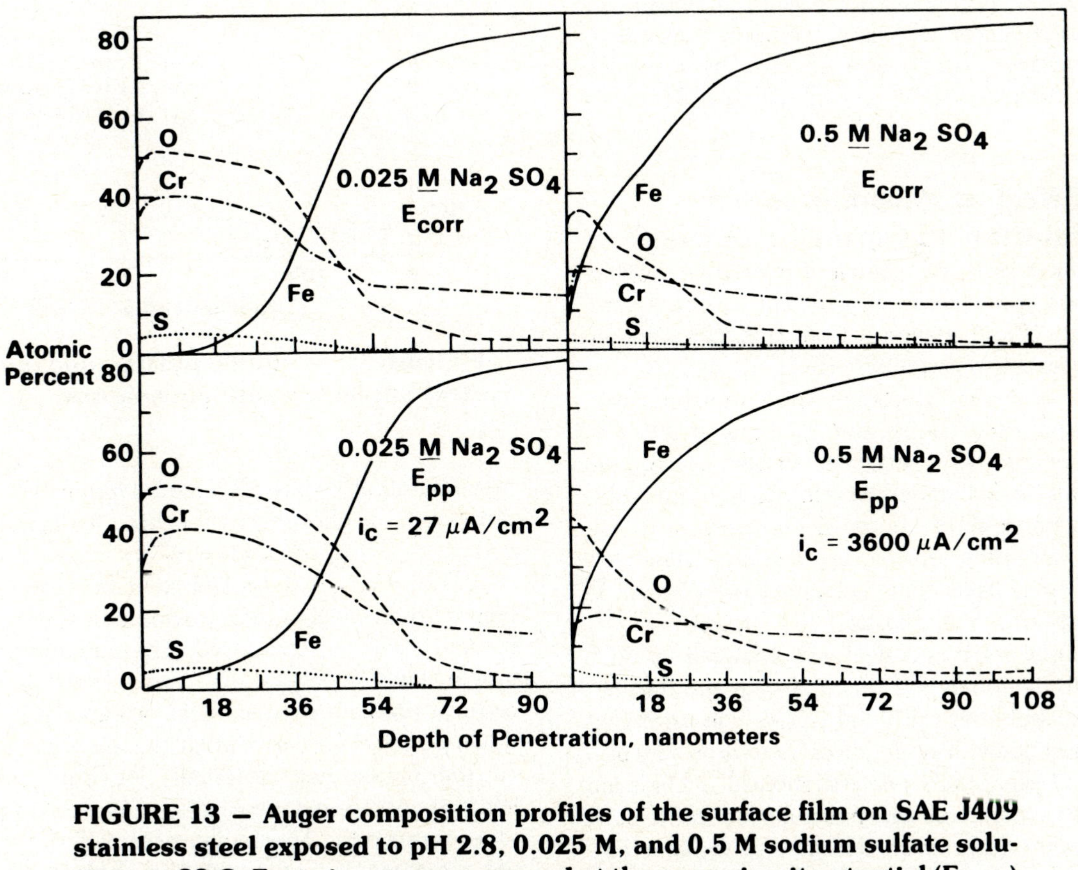

FIGURE 13 — Auger composition profiles of the surface film on SAE J409 stainless steel exposed to pH 2.8, 0.025 M, and 0.5 M sodium sulfate solutions at 82 C. Test pieces were exposed at the open circuit potential (E_{corr}) and polarized to the primary passive potential (E_{pp}).

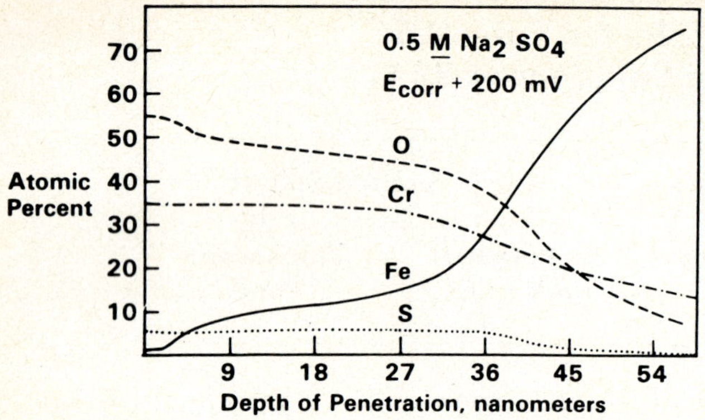

FIGURE 14 – Auger composition profiles of the surface film on SAE J409 stainless steel anodically polarized to 200 mV from the corrosion potential in pH 2.8, 0.5 M sodium sulfate solution at 82 C.

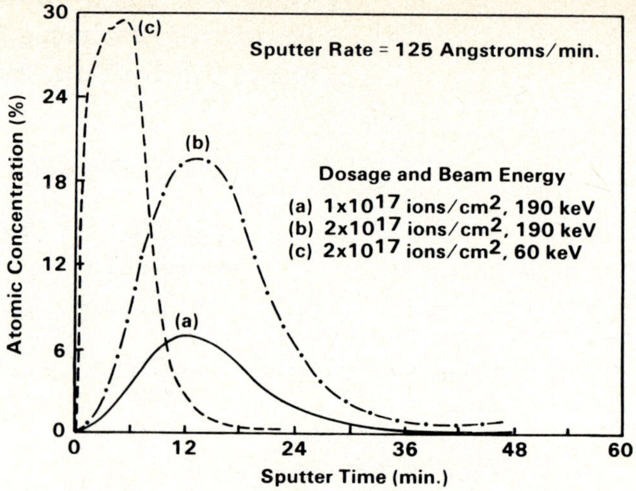

FIGURE 15 – Auger composition/depth profiles for steel implanted with phosphorus.

during which time the current decreased. However, after 30 min at 0.3 V in the 0.5 M solution both the film thickness and the current remained constant. Evidently, the less corrosive solution (0.025 M) permitted film growth during the 30 min period, whereas, the more corrosive solution (0.5 M) did not.

This series of experiments not only demonstrated the usefulness of the anodic polarization technique for determining the effects of solution parameters on corrosion characteristics, but it also showed how surface analyses can help to explain the observed polarization phenomena.

Effect of Phosphorus Ion Implantation on Corrosion of Steel[6]

This project explored the potential for improving the corrosion resistance of a metal by altering its surface composition with ion implantation.[(4)] Corrosion data were obtained using anodic polarization measurements.

Initially, mild steel specimens were implanted with ions of phosphorus, arsenic, and antimony at doses of 10^{15} ions/cm^2 and 10^{16} ions/cm^2. Anodic polarization measurements in a buffered borate solution (pH 8.5) showed that phosphorus improved the corrosion resistance more than either arsenic or antimony. Based on these results, the effects of implanting various concentrations of phosphorus into mild steel were studied.

Phosphorus-implanted steel was analyzed by AES and ESCA to determine the chemical state and distribution of the phosphorus. AES and ESCA high resolution spectra suggest that implanted phosphorus is present as a phosphide. Depth profile analysis showed a Gaussian distribution of phosphorus with maximum concentrations that varied with implantation dosage and voltage. This may be seen in Figure 15 where depth profiles show the distribution of phosphorus for three implantation conditions.

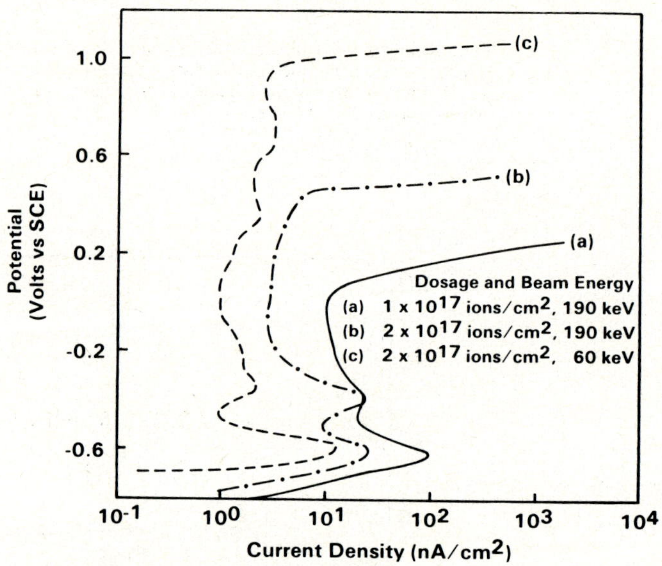

FIGURE 16 – Anodic polarization curves for steel implanted with phosphorus.

The effect of these implantations can be seen in Figure 16, in which anodic polarization curves for the implanted steels are compared. The buffered borate electrolyte, which contained 500 ppm NaCl, was used to cause a premature breakdown of passivity. The breakdown potential for each implanted specimen is indicated. At the breakdown potential, localized corrosion is initiated and a rapid increase in current takes place. The more positive the breakdown potential, the more resistant a material is to the initiation of localized corrosion. Comparing Curve (a) with Curve (b) shows that doubling the dosage of implanted phosphorus at an implantation voltage of 190 keV increased the breakdown potential by 0.40 V in the positive direction. Comparing Curve (b) with Curve (c) shows that reducing the implantation

[(4)]*Ion implantation* is a process for injecting an atomic species in the form of an ion beam into a solid material. The purpose of implanting ions is to alter the surface properties of the base material by changing its composition. The degree of chemical or physical change to the surface depends on the kind of ions, the number of ions, and the energy of the implanted ions.

voltage from 190 to 60 keV caused a shallower implantation and increased the breakdown potential by another 0.50 V in the positive direction. These results indicate that the pitting resistance of phosphorus implanted steel increases with phosphorus concentration and with its proximity to the surface.

The critical current, which is the first current density maximum on the polarization curve, determines how readily a metal will passivate; the smaller the current, the easier it is for the metal to passivate. In Figure 16, the critical current appears to decrease with phosphorus concentration at the surface, but subsequent studies with neon implantation indicated that the decrease in critical current was caused to a large extent by physical changes induced by the implantation process.

A set of phosphorus-implanted steel specimens were polarized to various pertinent potentials in buffered borate solution and then removed for Auger depth profile analysis. Comparisons of film thicknesses from these measurements were made with those from Auger depth profiles for unimplanted steel polarized to the same potentials. In this manner, the effect of phosphorus implantation on the surface oxide characteristics of steel during anodic dissolution and passivation could be studied. A typical anodic polarization curve for phosphorus-implanted steel is shown in Figure 17; the potentials where the individual polarization scans were terminated and specimens removed for analysis are indicated. The corresponding oxide film thickness are listed in Table 5. The potentials identified are: (1) the corrosion potential, E_{corr}; (2) the primary passive potential, E_{pp}; (3) the secondary passive potential, E_{sp}; (4) a potential in the midpassive region, E_{pass}; and (5) the same as (4) except that the specimen was held at that potential for one hour, E_{hold}.

According to the ESCA and Auger data, the corrosion films consisted largely of oxides of iron with little or no phosphorus present. The thickness of each film was calculated from the Auger depth profile and is expressed in angstroms. Table 5 shows that the oxide film at each potential on the implanted steel was considerably thicker than those on the unimplanted steel. In fact, the average thickness of the oxide film on implanted steel is about 2.4 times the average on unimplanted steel. In summary, the data show that phosphorus promotes the growth of thicker oxide films that are able to impart improved resistance to localized corrosion.

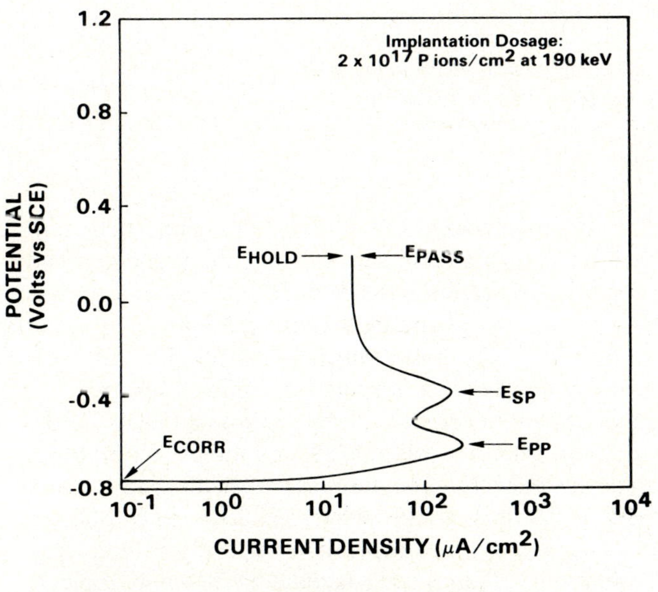

FIGURE 17 — Anodic polarization curve for phosphorus-implanted steel showing the potentials where specimens were removed for film thickness measurements.

TABLE 5 — Oxide Film Thickness Corresponding to Potentials Indicated in Figure 17

Potential	Film Thickness (Å) Implanted Steel	Unimplanted Steel
E_{corr}	190	100
E_{pp}	231	87
E_{sp}	174	76
E_{pass}	196	73
E_{hold}	207	94

Corrosiveness of Exhaust Gas Condensates[7]

The purpose of this investigation was to determine the corrosivity of exhaust gas condensates toward exhaust system materials, such as plain carbon steel, Type 1 aluminized steel, Galvalume,[(5)] and SAE 409 SS. The condensates selected for evaluation covered a range of compositions and pH values that resulted from a variety of engine operating conditions, fuels, and emission controls. Anodic polarization measurements were used as the primary means of evaluation. A cyclic immersion test was used in certain cases to evaluate longer term exposures under conditions that tended to stress oxide films.

Condensates were collected in 3.8-L (1-gal) batches from vehicles operated at constant speed on a chassis dynamometer at road loads. The test procedures and method of collection were discussed by Hunter.[8] Table 6 lists the condensates in order of increasing pH, and indicates the emission control system, the fuel, and the engine associated with each condensate. Table 7 lists the constituents in the various condensates.

The metals used for corrosion test specimens were (1) plain carbon steel, (2) Type 1 aluminized steel, which is carbon steel coated on both sides with an 8% silicon-aluminum alloy, (3) Galvalume, which is carbon steel coated on both sides with an alloy consisting of 55% aluminum, 1.5% silicon, and the balance zinc, and (4) SAE Type 409 stainless steel.

[(5)]Registered trade name.

TABLE 6 — Test Program Summary

Condensate pH	Emission Control System[1]	Fuel	Engine
2.3	no converter	Diesel	5.7 LD V-8 Olds (1980)
2.4	no converter	Diesel	5.7 LD V-8 Olds (1981)
2.6	DBM C3	Gasohol	2.5 L L-4 Pontiac (1981)
3.6	SBB (oxidizing)	Gasoline	3.8 L V-6 Buick (1980)
6.8	SBB C3	Indolene	3.8 L V-6 Buick (1981)
7.6	SBB C3	Gasoline	3.8 L V-6 Buick (1981)
7.8	SBB C3	Gasoline	3.8 L V-6 Buick (1981)
9.0	SBB C3	Gasohol	1.6 L L-4 Chevrolet (1981)

[1]Note: DBM = dual bed monolith converter, SBB = single bed bead converter, C3 = computer command control.

TABLE 7 — Analysis of Exhaust Gas Condensates (ppm)

Ident. pH	NH_3	$SO_4^=$	NO_3^-	Cl^-	$CO_3^=$	HCO_3^-
2.3	26	164	128	76	—	—
2.4	18	216	74	16	—	—
2.6	2	68	13	22	—	—
3.6	13	14	76	13	—	—
6.8	132	34	2	7	250	400
7.6	1039	549	<1	14	2035	1840
7.8	528	193	7	4	1250	1000
9.0	3803	199	<1	12	11,500	6200

Note: The above analysis should not be considered complete with regard to the near-neutral and basic condensates because some unidentified organic substances were detected in them. The analyses of the acid condensates are more complete.

Polarization measurements were made at a scan rate of 0.333 mV/s (1.2 V/h), and potentials were measured with respect to a saturated calomel electrode (SCE). Polarization measurements were started 60 minutes after immersion of the test specimen in the nitrogen-saturated condensate.

The cyclic immersion corrosion test procedure in this work was based on one developed by General Motors to evaluate bright metal decorative trim. By the use of automated equipment, the specimen (2.5 × 5 cm) is immersed for 2 seconds (s) in a corrosive solution, and then lifted out to dry under heat lamps for 100 s. For trim, 100 wet-dry cycles constituted a test, but for testing exhaust system materials in exhaust condensates, 1925 cycles were used in addition to overnight immersions to accelerate the corrosion.

A comparison of the corrosivity of exhaust condensates from engines with and without computer command control (C3) systems was made. To accomplish this comparison, anodic polarization measurements were made for four exhaust system metals in the pH 3.6, 6.8, and 7.8 condensates. As indicated in Table 6, the pH 3.6 condensate was generated with a single-bed bead (SBB), oxidizing converter without C3, and the pH 6.8 and 7.8 condensates were generated from systems with C3 and SBB converters using either indolene (6.8 pH) or gasoline (7.8 pH) as a fuel. The composition of the three condensates differed considerably, as shown in Table 7.

The anodic polarization curves of plain carbon steel in the three condensates are shown in Figure 18. It is evident from the curves that steel in the pH 3.6 condensate had the highest corrosion rates for the entire poten-

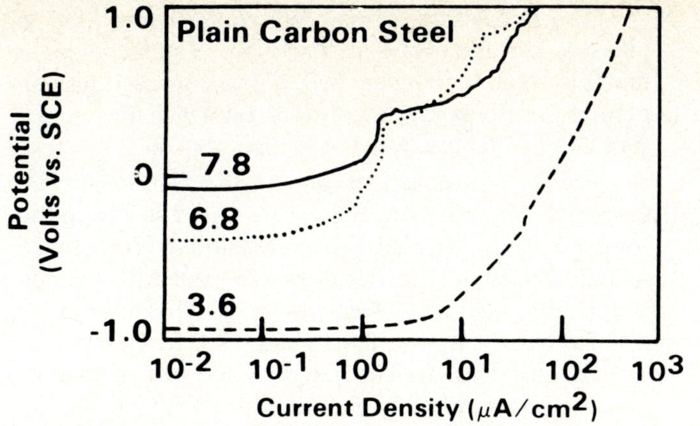

FIGURE 18 – Anodic polarization curves for plain carbon steel in exhaust gas condensates with pH values 3.6, 6.8, and 7.8.

tial range. In addition to differences in corrosion rates, there was a difference in the type of corrosion that occurred. In the pH 3.6 concentrate, there was a generalized attack of the entire surface but in the pH 6.8 and 7.8 solutions, there was pitting and crevice corrosion. The difference in the two types of corrosion may be seen in Figure 19 in which the generalized attack on the steel in the pH 3.6 condensate is shown at the top, and the pitting and crevice attack on the steel in the pH 6.8 condensate is shown at the bottom.

Cyclic immersion tests were performed to confirm the fact that pitting occurred in the pH 6.8 and 7.8 condensates, but it did not occur in the pH 3.6 condensate. Figure 20 shows photographs of carbon steel specimens (2.5 × 5 cm) exposed to the pH 6.8 and 7.8 condensates. Numerous pits are easily discernible in each case. The steel specimen exposed to the pH 3.6 condensate showed only general corrosion.

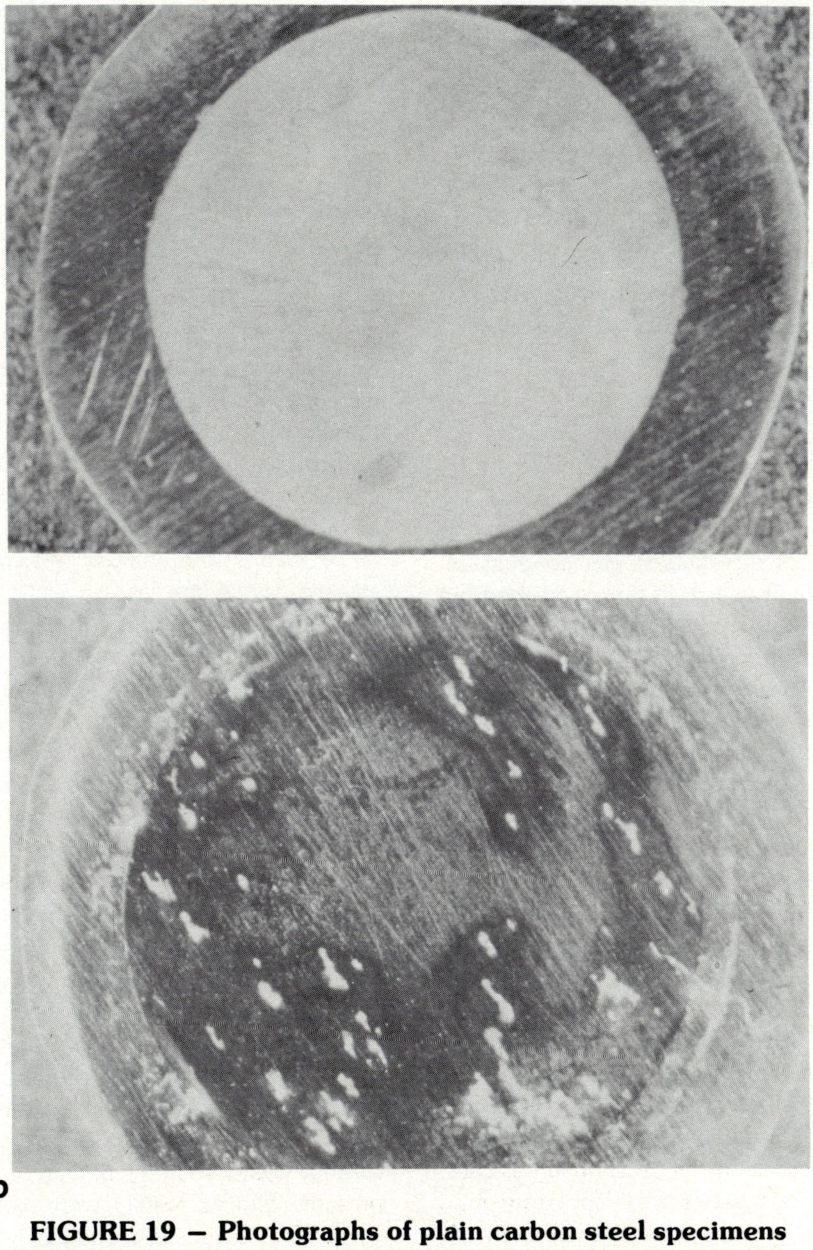

a

b

FIGURE 19 – Photographs of plain carbon steel specimens after anodic polarization in (a) pH 3.6 condensate (top) and (b) pH 6.8 condensate (bottom).

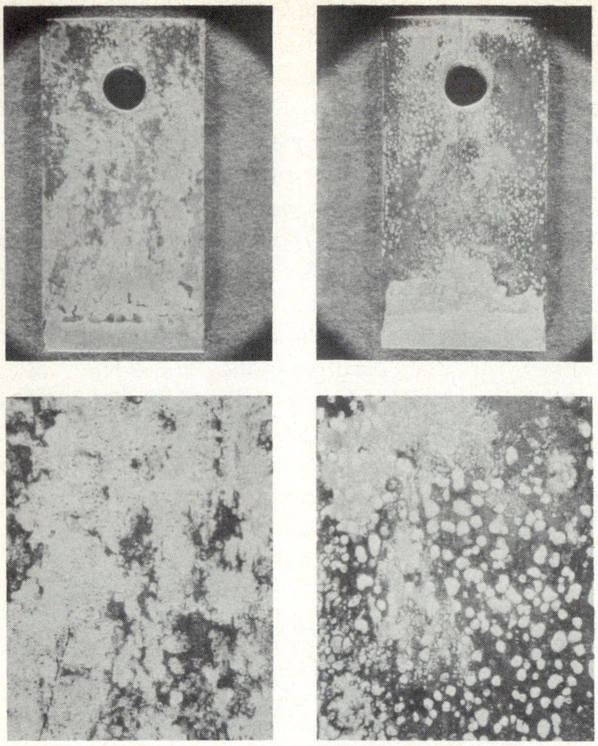

6.8 7.8

FIGURE 20 – Photographs showing (a) the entire steel specimen and (b) a closeup after cyclic immersion testing in the pH 6.8 and 7.8 condensates.

In addition to the steel, the corrosion behavior of stainless steel, aluminized steel, and Galvalume was investigated. It was concluded that:

1. The corrosivity of an exhaust gas condensate depends on its composition, which in turn is affected by variations in fuel, engine operating conditions, type of engine, and emission control system.
2. Acid concentrates are generally more corrosive than alkaline concentrates.
3. Pitting can occur in alkaline condensates.
4. Stainless steel was the most corrosion-resistant material tested.
5. Aluminized steel and Galvalume are about equally corrosion resistant.
6. Carbon steel was the least resistant material to acid condensates and was susceptible to pitting in alkaline condensates.

Corrosion of Metals in Ethanol[9]

The purpose of this study was to determine the corrosivity of the hydrated ethanol fuel used in Brazil on metals of the fuel system, and to determine the effectiveness of soluble corrosion inhibitors in preventing this corrosion. Electrochemical techniques and laboratory immersion tests were used to measure the corrosion rate of certain fuel system metals in ethanol contaminated with water and other impurities.

The polarization resistance technique was used to measure corrosion rates for this study. The technique is rapid, is accurate for either high or low corrosion system, permits the monitoring of corrosion rates with time, and is sensitive to changes in the corroding system.

Princeton Applied Research (PAR) Model 350 Corrosion Measurement System was used to obtain the corrosion data. Potentials were measured at 25 C using a saturated calomel reference electrode in a PAR electrode bridge tube, the tip of which was located 1 mm from the working electrode. The PAR Model 356 IR Compensation Module was used to measure the cell resistance, which was stored in the memory of the PAR Model 350 and used to correct all current-voltage curves for IR losses during data playback. Current was measured while scanning the potential from 20 mV cathodic to 20 mV anodic of the corrosion potential at a rate of 0.167 mV/s (0.6 V/h). The slope of the current-voltage curve at the corrosion potential was used to calculate the value of the corrosion current, I_{corr}, on the basis of the Stern-Geary equation[10]:

$$\frac{\triangle E}{\triangle I} = \frac{\beta_a \beta_c}{2.3\, I_{corr}\, (\beta_a + \beta_c)}$$

where $\triangle E/\triangle I$ = slope of curve (polarization resistance); β_a, β_c = anodic and cathodic Tafel constants. (A value of 0.1 V/dec was used for these constants in this study.); and I_{corr} = corrosion current.

In the initial phase of this project, six ethanol fuel samples were collected at six cities in Brazil. These fuels were analyzed, and the results are shown in Table 8. The presence of at least 5% water is due to the azeotrope that is formed during distillation. The other impurities are probably acquired during fermentation, distillation, storage, or handling of the fuel.

Based on this analysis, the corrosion rates of steel in ethanol containing water, acetic acid plus water, ethyl acetate plus water, chloride ions plus water, and a combination of all four contaminants were determined using the polarization resistance technique. Water was used at the 10% level because the additional water facilitated the electrochemical measurements and had very little effect on the corrosion rate. Corrosion rates were monitored periodically during the 30-h test period. Initially, corrosion rates decreased with time, but they became fairly constant for the last six hours of the test. The rates reported in Table 9 were measured at 30 hours.

The corrosion rates with water alone or in combination with acetic acid or ethyl acetate were small, and this was confirmed by the absence of visible corrosion at the end of the test. On the other hand, chloride ion caused a marked increase in corrosion rate, with visible accumulation of rust at pits formed on the steel surface. A very rapid increase in corrosion rate, with severe pitting and rusting, was observed when all of the contaminants were present. Similar results were obtained for Zamak 5[(6)] (SAE 925), a zinc die-casting alloy. The zinc did not pit as

[(6)]Registered trade name.

TABLE 8 – Analysis of Ethanol Fuel from the Following Six Locations in Brazil: (A) Bauru, (B) Natal, (C) Paulina, (D) Recife, (E) Ribeirao, and (F) San Paulo

		A	B	C	D	E	F
Water	%	6.5	7.0	5.8	5.7	6.4	6.2
Chloride Ion	ppm	10	12	6	0	6	4
Ethyl Acetate	%	0.01	—	0.06	0.01	0.02	0.01
Acetic Acid	%	0.01	0.01	0	0	0.01	0.03

TABLE 9 – Corrosion Rates of AISI 1020 Steel in Ethanol Containing Various Contaminants

Solution Composition H_2O	HAc	EtAc	Cl^-	Corrosion Rate nA/cm²
%	ppm	ppm	ppm	
10				7
10	100			27
10		100		26
10			10	551
10	100	100	10	2338

severely as the steel and the corrosion products were gray to white rather than rust color.

Based on the analysis in Table 8 and the corrosion rates in Table 9, the following solution was selected for testing inhibitors for ethanol fuels:

Ethanol	90%
Water	10%
Acetic Acid	100 ppm
Ethyl Acetate	100 ppm
Chloride Ion	10 ppm

Of the many commercial inhibitors tested, the majority provided adequate protection for the Zamak 5 but not for the steel. The only two inhibitors that projected both metals were Unicor[7] J, a dimer acid-type inhibitor commonly used in pipelines, and Promax Proal,[8] a blend of glycols and amines sold in Brazil for addition to the fuel tank.

Polarization resistance measurements were used to measure the corrosion rates of steel and Zamak 5 as a function of the concentration of inhibitors Unicor J and Proal.

[7]-[8]Registered trade name.

TABLE 10 – Corrosion Rates in the Standard Test Solution at Various Inhibitor Concentrations

Metal	ppm	Unicor J[1] Corrosion Rate nA/cm²	Promax Proal[2] Corrosion Rate nA/cm²
Zamak	0	2349	2349
	10	2212	
	25	76	
	50	54[3]	2032
	100	13[3]	
	500		59
	1250		19[3]
	2500		15[3]
Steel	0	2338	2338
	10	915	
	25	264[3]	
	50	103[3]	
	500		533
	1250		37[3]
	2500		12[3]

[1]Unicor J inhibitor supplied by UOP Process Division of UOP Inc.
[2]Promax Proal supplied by Promax Products Maximos, South America.
[3]No visual evidence of corrosion.

Table 10 lists the corrosion rates determined at the various inhibitor concentrations. Footnote No. 3 in the table indicates no visible corrosion of the metal surface and at that concentration the inhibitor is adequate. Based on these data, it would seem reasonable that inhibitors provide a viable method for preventing corrosion in hydrated ethanol fuel.

References

1. R. L. Chance, Corrosion, Vol. 33, No. 3, p. 108, 1977.
2. M. S. Walker, Materials Performance, Vol. 18, No. 4, p. 9, 1979.
3. Preparing, Cleaning, and Evaluating Corrosion Test Specimens, ASTM G1, ASTM Book of Standards, ASTM, Philadelphia, Pennsylvania.
4. R. L. Chance, Engine Coolant Testing: State of the Art, ASTM STP 705, W. H. Ailor, Ed., ASTM, Philadelphia, Pennsylvania, p. 270, 1980.
5. R. L. Chance, S. W. Gaarenstroom, Corrosion, Vol. 36, No. 2, p. 94, 1980.
6. M. S. Walker, R. L. Chance, Corrosion, Vol. 40, No. 6, p. 307, 1984.
7. R. L. Chance, R. G. Ceselli, Society of Automotive Engineers SP-538, The Study and Prevention of Corrosion, Paper No. 830585, Society of Automotive Engineers (SAE), Warrendale, Pennsylvania, p. 133, 1983.
8. J. E. Hunter, Jr., Society of Automotive Engineers SP-538, The Study and Prevention of Corrosion, Paper No. 830584, Society of Automotive Engineers (SAE), Warrendale, Pennsylvania, p. 125, 1983.
9. M. S. Walker, R. L. Chance, Proceedings of the 2nd Automotive Prevention Conference P-136, Paper No. 831828, p. 171, 1983.
10. M. Stern, A. L. Geary, J. Electrochem. Soc., Vol. 104, p. 56, 1957.

Predicting Galvanic Corrosion Using Electrochemical Techniques

R. BABOIAN*

Introduction

Electrochemical techniques for predicting galvanic corrosion are based on three types of measurements. These are: mixed potential, galvanic current, and polarization measurements.[1,2] Although each of these techniques has been used widely, careful experimentation is required to obtain useful information. In many cases, more than one technique is required to characterize a specific couple.

In general, it should be emphasized that in making the electrochemical measurement, results will be meaningless if the system is disturbed. Also, it is important that the measurements be conducted in the actual corrosive environment or in one that closely simulates that environment.

Potential Measurements

The use of potential measurements for predicting galvanic corrosion is based on the measurement of E_{corr}, the corrosion potential of metals. When the metals are arranged according to their potential measured in a particular environment (Table 1), the result is termed the *galvanic series*.[3]

In general, the corrosion rate of the more noble metal in a galvanic couple will be reduced and that for the more active metal will be increased. In some cases, the separation between the two metals in the galvanic series gives an indication of the magnitude of these effects.

When the polarization characteristics for metals are straightforward such as demonstrated in Figure 1 for monel, bronze, and Niresist, the galvanic series is useful in predicting not only the relative galvanic behavior of the metals, but also the magnitude of the effects of coupling the metals. The order and relative positions of these three metals in the galvanic series (Table 1) is correct for

*Texas Instruments, Inc., Attleboro, Massachusetts.

TABLE 1 — Arrangement of Metals and Alloys in a Galvanic Series Based on Potential Measurements in Seawater Velocity of Flow, 13 Ft/Sec: Seawater Temperature, 25 C (77 F) (from LaQue)

Zinc
Aluminum (Alclad 3S)
Aluminum 3S-H
Aluminum 61S-T
Aluminum 52S-H
Cast iron
Carbon steel
Stainless steel AISI 430 (17% chromium)
Ni-Resist cast iron (20% nickel)
Stainless steel AISI 304 (18% chromium, 8% nickel)
Stainless steel AISI 410 (13% chromium)
Ni-Resist cast iron (30% nickel)
Ni-Resist cast iron (20% nickel + copper)
Naval rolled brass
Yellow brass
Copper
Red brass
Composition G bronze
Admiralty brass
90-10 cupro nickel (0.8% iron)
70-30 cupro nickel (0.06% iron)
70-30 cupro nickel (0.47% iron)
Stainless steel AISI 430 (17% chromium)
Nickel
Stainless steel AISI 316 (18% chromium, 12% nickel, 3% molybdenum)
Inconel
Stainless steel AISI 410 (13% chromium)
Titanium (commercial)
Silver
Titanium (high purity from iodide)
Stainless steel AISI 304 (18% chromium, 8% nickel)
Hastelloy C
Monel
Stainless steel AISI 316 (18% chromium, 12% nickel, 3% molybdenum)

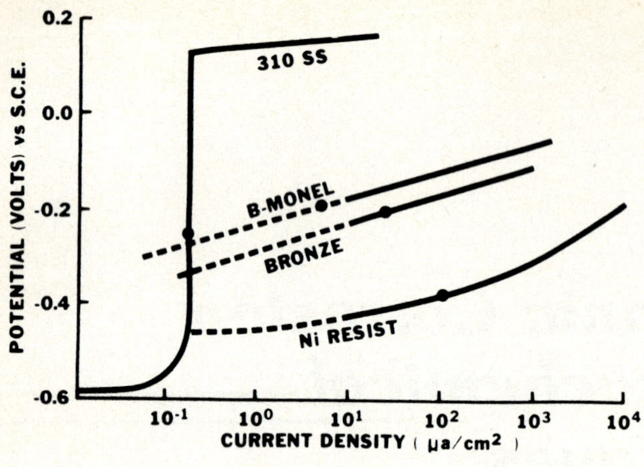

FIGURE 1 – Potentiodynamic polarization curves in 5% NaCl solution, 30 C.

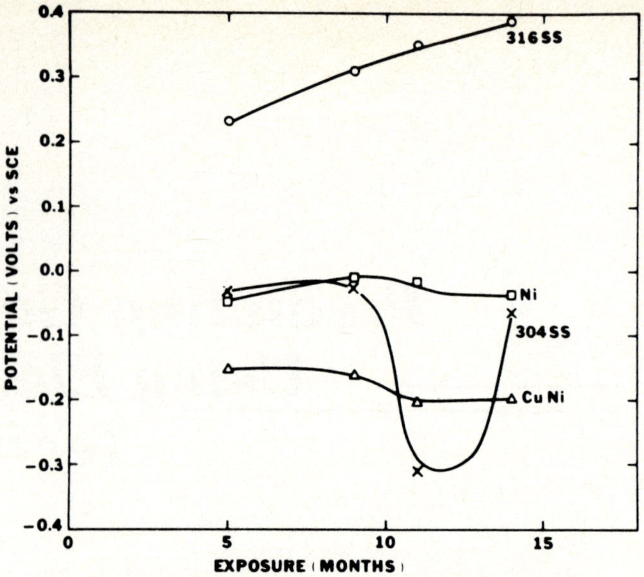

FIGURE 2 – Variation of corrosion potential with time for nickel, 90-10 copper-nickel, AISI 304 and 316 stainless steel in seawater.

the 5% sodium chloride environment. Measurement of the corrosion potential for these metals is sufficient.

There are several factors that lead to drawbacks in the use of potential measurements and the galvanic series. First, the formation of films on the metal surface will influence the corrosion rate so that over a wide range of potentials, the surface remains passive.[4] This is illustrated (Figure 1) by polarization curves for AISI 310 stainless steel (SS) in 5% sodium chloride solution. Its position in the galvanic series (according to E_{corr}) may indicate what effects it may have on other metals in a galvanic couple but the reverse is not necessarily true because of its wide passive potential range.[5]

The potential variation with time in a particular environment is another concern when constructing the galvanic series.[4] The relative position of metals in the series may change depending on the time of immersion. For example, in Figure 2, the measured corrosion potentials for AISI 304 and 316 SS vary considerably even after 10 or 15 months immersion in seawater. The relative position of nickel, AISI 304 SS, and copper-nickel alloy clearly depends on the time of potential measurement. Galvanic series have been constructed that take this behavior into account. Generally, a potential range is included so that overlap of metals in the series can be observed.

The polarization characteristics of metals vary according to the environment to which they are subjected. Metals that are easily polarized have a smaller effect than ones that are not readily polarized. Simple measurement of the corrosion potential does not provide information on the polarizability of the metal so that in many cases, their relative positions and their separation in the galvanic series are not a valid basis for predicting galvanic corrosion behavior.

Current Measurements

Measurement of galvanic currents between galvanically coupled metals is based on zero resistance ammeter techniques.[6] The basic zero resistance ammeter circuit, which has been used extensively, is shown in Figure 3. The galvanic current is measured by an ammeter, A,

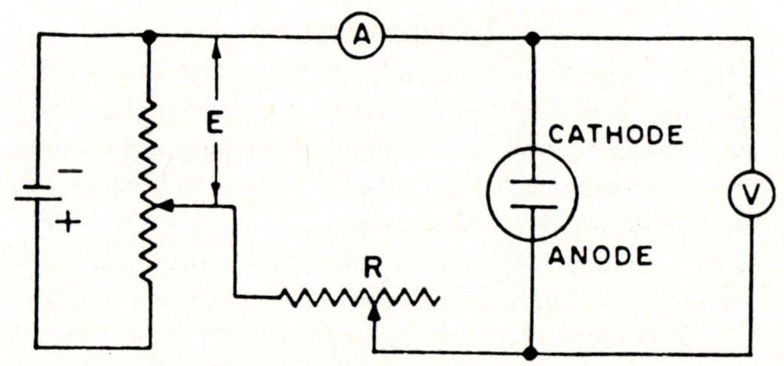

FIGURE 3 – Schematic circuit of elementary zero resistance ammeter.

by adjusting the voltage, E, or resistance, R so that the potential difference between the two elements is zero as indicated by the electrometer, V. Since short circuiting in a galvanic couple is indicated by zero potential drop, this current is the true short circuit current. For continuous recording of galvanic currents, the basic circuit is simplified to include a decade resistance box adjusted so that a recorder, set to 1 mV full scale, indicates the potential between the two elements. The galvanic current is calculated knowing the resistance and the potential.[7]

The data in Figure 4 for galvanic couples of copper with AISI 1006 carbon steel and AISI 409 and 430 SS were obtained using the basic zero resistance ammeter circuit. The magnitude of the galvanic current provides an indication of the severity of galvanic corrosion, which occurs in the specific 5% sodium chloride environment. Notice the slow drift to lower galvanic currents for AISI 1006 carbon steel with time. The sharp increase in galvanic current for AISI 409 SS is due to the initiation of localized corrosion. Initiation of localized corrosion is not indicated by the galvanic current for AISI 430 SS. This effect is observed, however, beyond the 60-hour test time

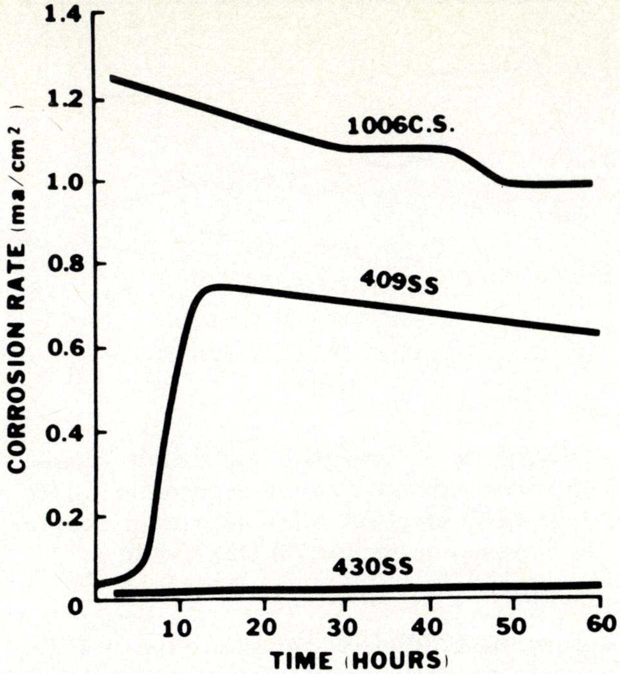

FIGURE 4 – Galvanic current versus time for AISI 1006 carbon steel and AISI 409 and 430 stainless steel coupled to copper in 5% sodium chloride solution (30 C).

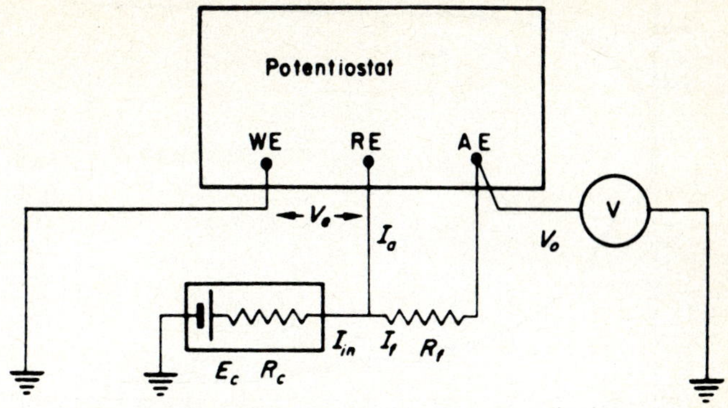

FIGURE 5 – Schematic circuit for use of a potentiostat as a zero resistance ammeter (Reference 11).

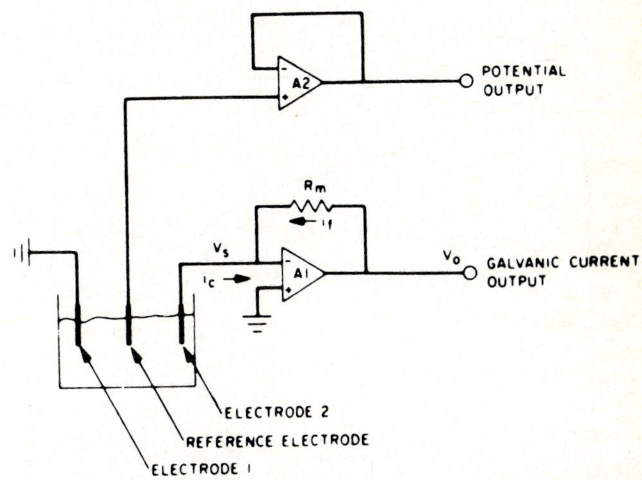

FIGURE 6 – Schematic circuit of a zero resistance ammeter using an operational amplifier (Reference 12).

shown in Figure 4.

The drawbacks of the basic zero resistance ammeter circuit shown in Figure 3 are: (1) During time periods in which the circuit is not balanced, polarization of the members in the galvanic couple occurs, leading to incorrect values and/or long recovery times, and (2) the continuous monitoring of the galvanic current requires constant attention. These disadvantages have been the focal point of new developments and improvements in the zero resistance ammeter techniques.

The development of self balancing, zero resistance ammeters allows continuous monitoring of galvanic currents. This was accomplished by electronic automatic adjustment of either R[8] or E[9] in the basic circuit. High input resistance voltage-sensing systems placed across the galvanic cell allow more accurate measurements in high resistance electrolytes.

More recent developments include the use of potentiostats incorporating the operational amplifier circuitry.[10] In the arrangement in Figure 5,[11] the control potential of the potentiostat is set to 0 V. One member of the galvanic couple is connected to the working electrode terminal while the other is connected to the reference electrode terminal. The auxiliary electrode terminal is connected directly to the reference electrode terminal, whereby the galvanic current is indicated directly by the potentiostat current meter, or it is connected through an external feedback resistor (R_f in Figure 5) and the galvanic current is measured by a voltmeter between the auxiliary and working electrode terminals. A null balance is thus maintained by means of the potentiostat solid-state operational amplifier circuit.

A commercial operational amplifier can be used as the voltage control and will give the same results as a potentiostat. The use of a simple operational amplifier as an alternate to the potentiostat was described by Lauer and Mansfeld[12] and by Henry and Wilde.[13] The advantages here include lower cost and smaller size. In Figure 6,[12] one member of the galvanic couple is connected, through ground, to the positive input of the amplifier. The other is connected to the negative of the high-gain operational amplifier. The members of the galvanic couple are maintained at zero potential difference and the galvanic current is derived from V_O, the voltage output of the amplifier and R_m, the feedback resistance (standard resistor). This arrangement allows measurements to be made in high resistivity electrolytes and continuous monitoring of the galvanic current using strip chart recorders. The reference electrode in Figure 6 is used only to measure the mixed potential of the galvanic couple.

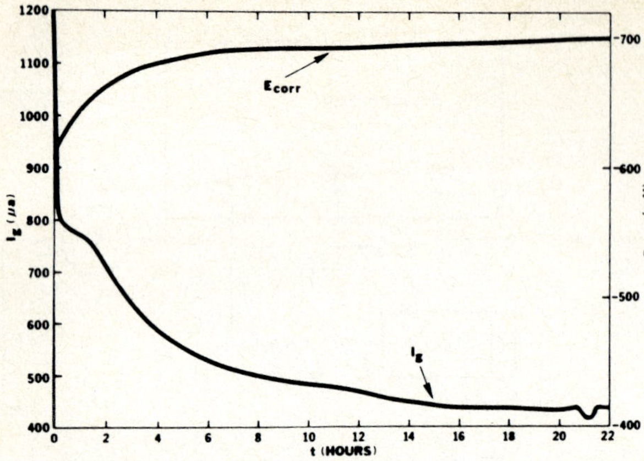

FIGURE 7 – Corrosion potentials (E_{corr}) and zero resistance ammeter currents (I_g) for OFHC copper-4130 steel couple in corrosive water (ASTM D2570-73).

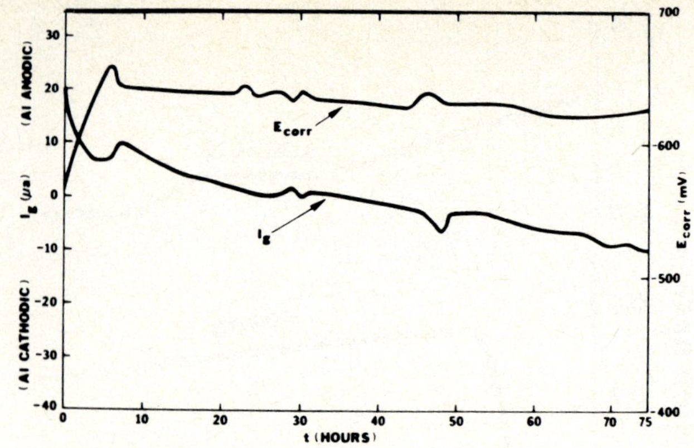

FIGURE 9 – Corrosion potentials (E_{corr}) and zero resistance ammeter currents (I_g) for AISI 4130 steel-AA 6061 aluminum couple in corrosive water (ASTM D2570-73).

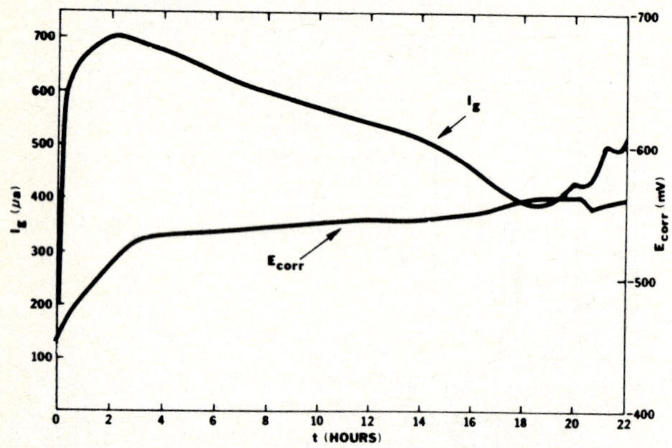

FIGURE 8 – Corrosion potentials (E_{corr}) and zero resistance ammeter currents (I_g) for OFHC copper-6061 aluminum couple in corrosive water (ASTM D2570-73).

Typical galvanic current data obtained with the operational amplifier circuitry are shown in Figures 7 through 9.

In Figure 7, the zero resistance current and the mixed potential are plotted vs time for steel and copper in ASTM[(1)] corrosive water (ASTM D2570-73). The data show that steel is anodic to copper in this environment. Both the mixed potential and galvanic current for this couple become stable after several hours.

In Figure 8, the data are plotted for aluminum and copper and aluminum is anodic. However, at the end of 18 hours, the galvanic current changes rapidly. This is due to the initiation of localized corrosion of aluminum in this environment.

The data for aluminum and steel are shown in Figure 9. Although the mixed potential remains quite stable with time, the galvanic current changes over the time period shown. Initially, the aluminum is anodic to steel. However, after 40 hours, the galvanic current changes sign, indicating that reversal takes place so that steel is anodic to aluminum in this environment.

Figures 8 and 9 show that time-dependent factors such as initiation of localized corrosion and anode-cathode reversal must not be overlooked and that long-term tests are frequently required. This is especially true when localized corrosion such as pitting is possible in the galvanic couple. Frequently, an induction period several weeks long is observed before galvanic pitting is initiated.[4]

The measured galvanic current is not always a measure of the true corrosion current because it is the algebraic sum of the currents resulting from anodic and cathodic reactions. When cathodic currents are appreciable at the mixed potential of the galvanic couple, the measured galvanic current will be significantly lower than the true corrosion current. Thus, large differences between the true corrosion rate calculated by weight loss and the rate obtained by galvanic current measurements have been observed. Calculation of the true corrosion current by correcting the measured galvanic current in some cases involves more work than actually measuring the corrosion rate by conventional weight loss techniques.

Polarization Measurements

One of the most important applications of the mixed potential theory of corroding systems is the treatment of galvanically coupled metals.[14] When two different corroding metals are coupled electrically in the same electrolyte, both metals are polarized so that each corrodes at a new rate. In Figure 10, the corrosion potentials and polarization parameters for uncoupled metals A and B are shown. Metal A is more noble than metal B in that the equilibrium potential is less negative. When the mixed potential theory is applied to the individual reactions (A/A^+, H_2/H^+, B/B^+, H^2/H^+), the uncoupled corrosion rates are $i_{corr,A}$ for metal A and $i_{corr,B}$ for metal B. When equal areas of metals A and B are coupled, the resultant mixed potential of the system $E_{corr,AB}$ is at the intersection where the total oxidation rate equals the total

[(1)]ASTM, Philadelphia, Pennsylvania.

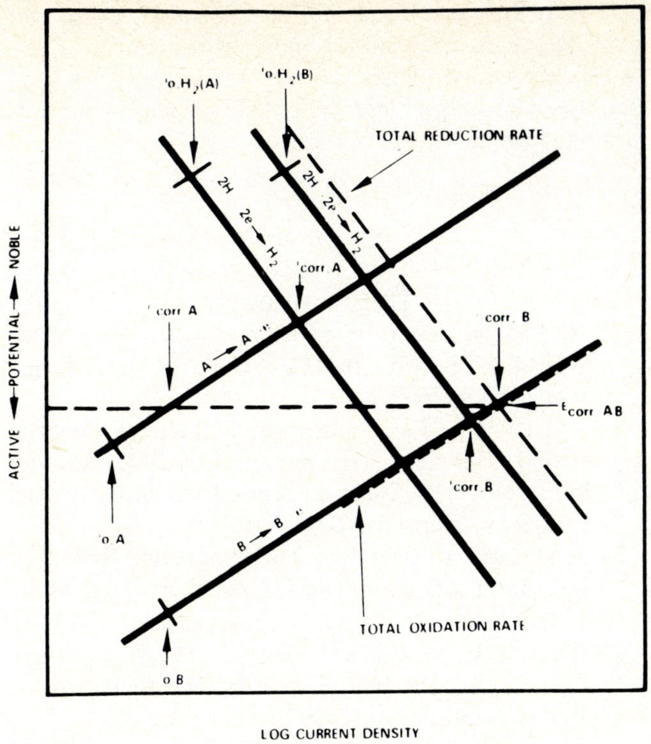

FIGURE 10 – Mixed potential behavior of galvanically coupled metals A and B.

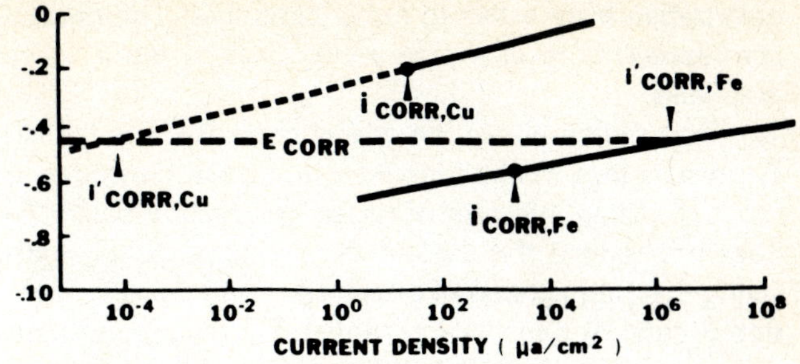

FIGURE 11 – Galvanic corrosion behavior of copper and iron in 5% sodium chloride solution (30 C) using polarization measurements.

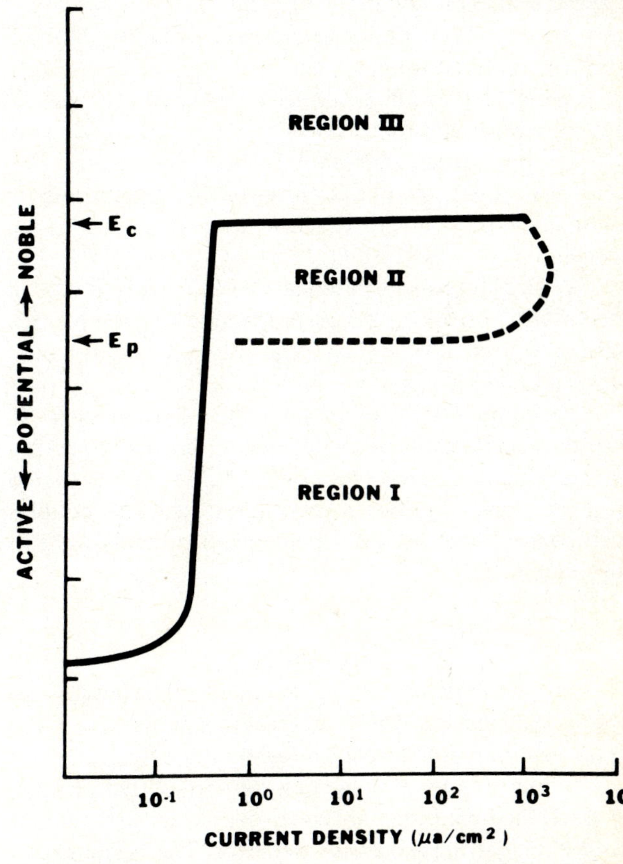

FIGURE 12 – Schematic potentiodynamic polarization curve for stainless steel in 5% sodium chloride solution showing three corrosion potential regions.

reduction rate. The oxidation rate of the individual coupled metals is such that metal A corrodes at a reduced rate, $i'_{corr,A}$, and metal B corrodes at an increased rate, $i'_{corr,B}$.

Use of polarization measurements in galvanic corrosion is based on this behavior. The specific information required to characterize a particular couple originates from anodic polarization curves for the individual metals in the galvanic couple and the mixed potential, E_{corr}, for the couple in the actual environment. Since the cathodic reactions also influence the mixed potential, measurements must be made duplicating the environment (such as geometry and area ratio) of the couple. It is important to note that this treatment can be applied to two[4,5,15] or more[16] metals in a galvanic couple.

In this treatment, the corrosion potential of the galvanic couple is measured, and the nature of corrosion in this potential region is determined for each metal from the individual polarization curves. For example, anodic polarization curves for DHP copper and AISI 1006 carbon steel in 5% NaCl solution are shown in Figure 11. The corrosion potentials of the isolated metals, obtained under actual corroding conditions in aerated 5% NaCl solution, are used to determine the rates of corrosion $i_{corr,Cu}$ and $i_{corr,Fe}$ from the polarization curves in Figure 11.

When copper and low carbon steel (LCS) are coupled galvanically under these conditions, the mixed potential is E_{corr}. From the polarization curves in Figure 11, the rate of corrosion of copper, $i'_{corr,Cu}$, is decreased and the rate of corrosion of iron, $i'_{corr,Fe}$, is increased.

Increasing the copper/LCS area ratio in the galvanic couple shifts the mixed potential (E'_{corr}) to less negative values so that galvanic corrosion of LCS is accelerated.

An important application of the use of polarization measurements in galvanic corrosion is in the prediction of localized corrosion. Polarization techniques and critical potentials have been used widely as a measure of pitting and crevice corrosion susceptibility of alloys in chloride solution.[17-20] For example, a schematic potentiodynamic polarization curve for a SS in sodium chloride solution is shown in Figure 12. The solid line represents anodic polarization from active to noble potentials. The dashed line represents reverse polarization back to more active potentials.

Polarization above the critical breakdown potential, E_c, results in a marked increase in current density as a result of pitting initiation. Reverse polarization below E_c does not reduce the current density appreciably, indicating that propagation of existing pits and crevice corrosion occurs. At the critical potential, E_p, the current density approaches zero. Below E_p, propagation of existing pits and crevice corrosion does not occur.

The corrosion behavior of SS can be thus separated into three potential regions: Region I, where protection from pitting and crevice corrosion is observed; Region II, where propagation of existing pits and crevice corrosion occurs; and Region III, where initiation of pitting and crevice corrosion occurs. Thus, galvanic couples of SS can lead to three performance categories in chloride-containing environments.

When the mixed potential for the galvanic couple between SS A and metal B lies in Region III, initiation of pitting and crevice corrosion on the SS occurs. This behavior is observed in the presence of stray currents or sometimes in galvanic couples with platinum group metals.

Where the mixed potential lies in Region II, this is the crevice corrosion and pit propagation region for SS. AISI 430 and 304 SSs undergo this type of corrosion when coupled to copper.

When the SS A and metal B mixed potential lies in Region I, protection from pitting and crevice corrosion occurs. Stainless steels, such as AISI 304, are protected from this type of localized corrosion in galvanic couples with carbon steel because their mixed potentials are in this region.

Summary

To successfully use electrochemical techniques to predict galvanic corrosion, measurements must be made in an environment closely simulating the actual one.

In many cases, simple measurement of the potential of each member in a galvanic couple is sufficient to predict the galvanic corrosion behavior. The resulting galvanic series of metals for a particular environment can be quite useful. However, more precise information is frequently required.

Measurement of galvanic currents can provide more useful information regarding galvanic corrosion. Recent developments with zero resistance ammeters allows continuous measurement of the galvanic current during true short circuiting conditions. This current, however, is not always equivalent to the corrosion current because it is the algebraic sum of the currents resulting from anodic and cathodic reactions. Thus, where cathodic currents are significant, the measured galvanic current may be appreciably smaller than the true corrosion current.

The use of polarization measurements in conjunction with the mixed potential of a galvanic couple can provide precise information regarding the behavior of galvanically coupled metals. This treatment has been used successfully for three member couples as well as for predicting galvanic localized corrosion.

References

1. R. Baboian, et al., Galvanic and Pitting Corrosion, ASTM STP 576, ASTM, Philadelphia, Pennsylvania, p. 5, 1976.
2. R. Baboian, "Electrochemical Techniques for Predicting Galvanic Corrosion," CORROSION/76, Paper No. 114, National Association of Corrosion Engineers, Houston, Texas, 1976.
3. F. L. LaQue, Corrosion Handbook, H. H. Uhlig, Ed., John Wiley and Sons, New York, New York, 1948.
4. R. Baboian, Localized Corrosion—Cause of Metal Failure, ASTM STP 516, ASTM, Philadelphia, Pennsylvania, p. 145, 1972.
5. R. Baboian, G. S. Haynes, Corrosion in Natural Environments, ASTM STP 558, ASTM, Philadelphia, Pennsylvania, p. 171, 1974.
6. R. H. Brown, R. B. Mears, Trans. Electrochem. Soc., Vol. 74, p. 510, 1938.
7. N. D. Greene, Experimental Electrode Kinetics, Rensselaer Polytechnic Institute, Troy, New York, 1965.
8. H. P. Godard, Corrosion, Vol. 7, No. 3, p. 93, 1951.
9. D. R. Maker, H. T. Francis, J. Electrochem. Soc., Vol. 102, p. 669, 1955.
10. J. Devay, B. Lenyal, L. Meszaros, Acta Chim. Hung., Vol. 62, p. 157, 1969.
11. J. A. von Fraunhofer, P. J. Staheli, Corros. Sci., Vol. 12, p. 767, 1972.
12. G. Laur, F. Mansfeld, Corrosion, Vol. 26, No. 11, p. 504, 1970.
13. W. D. Henry, B. E. Wilde, Corrosion, Vol. 27, No. 11, p. 479, 1971.
14. M. G. Fontana, N. Greene, Corrosion Engineering, McGraw-Hill Book Co., New York, New York.
15. R. Baboian, SAE Trans., Vol. 81, p. 1117, 1972.
16. R. Baboian, SAE Autom. Eng. Congr., Paper 710276, Detroit, Michigan, 1971.
17. J. Horvath, H. H. Uhlig, J. Electrochem. Soc., Vol. 115, p. 791, 1968.
18. M. Pourbaix, L. Klimzack-Mathieiu, C. Mertens, J. Meunier, C. Vanleugenhaghe, L. DeNuncky, J. Laverys, L. Neelemens, M. Warzee, Corros. Sci., Vol. 3, p. 239, 1963.
19. J. M. Defranoux, Corros. Sci., Vol. 3, p. 175, 1963.
20. M. Pourbaix, Corrosion, Vol. 26, No. 10, p. 431, 1970.

Section 6

Localized Corrosion

Use of Electrochemical Techniques for the Study of Crevice Corrosion in Natural Seawater

R. M. KAIN,* T. S. LEE,** and J. W. OLDFIELD***

Abstract

Several electrochemical techniques were used by the authors to study the crevice corrosion process of iron-base and nickel-base alloys in natural seawater and related chloride-containing environments. This paper reviews these techniques as they were applied to investigating the mechanism of crevice corrosion as well as to defining conditions for practical avoidance of crevice corrosion.

Introduction

Stainless steels (SSs) have traditionally had widespread use in seawater, and recent developments of more highly alloyed SSs for seawater service have furthered their use.[1,2] An important component in the development and application of these SSs has been experimental activity to characterize crevice corrosion behavior. This experimental work has involved simple immersion tests in the laboratory, field service tests in operating systems, and electrochemical evaluations using a variety of test methods. The tests have utilized the natural seawater environment as well as environments designed to simulate natural seawater and to accelerate crevice corrosion rates.

Several articles have been published in recent years that review the various experimental approaches used in the study of both crevice corrosion and pitting of stainless alloys.[3-9] Postlethwaite's[6] and Isseling's[7] recent reviews of electrochemical test methods present comprehensive discussions for a variety of materials and are highly recommended for further reading. Pitting test methods have some relevance to the subject of crevice corrosion when one considers that some environmental similarities exist in propagating active crevices and pits.[10] Additionally, some authors believe that the initial stage of initiation of corrosion within a crevice is in the form of pitting.[10,11]

Conventional electrochemical testing involving anodic polarization scanning techniques is useful in establishing relative rankings of different alloys according to some designated parameters, as shown schematically in Figure 1.[12] Similarly, the impact of other environmental variables, such as pH, and chloride concentration can be measured. Testing in this manner, while yielding valuable information, is sensitive in some cases to polarization scan rate, crevice geometry effects, and other experimental variables.[13,14] As such, correlation between the controlled electrochemical test and actual exposures are frequently limited to those materials susceptible to localized corrosion over a broad range of conditions and those most resistant to these conditions.[12]

In studying the behavior of a given SS or family of stainless alloys in seawater and related environments, several electrochemical techniques are used by the authors. Together with conventional immersion testing and mathematical modeling, these procedures enable a more complete characterization of the factors affecting crevice corrosion than any single test is likely to achieve. In this regard, the continued introduction of new and refined electrochemical test methods will further enhance

*LaQue Center for Corrosion Technology, Inc., Wrightsville Beach, North Carolina.
**National Association of Corrosion Engineers, Houston, Texas.
***Inco Alloy Products Laboratory Technology Centre, Birmingham, England.

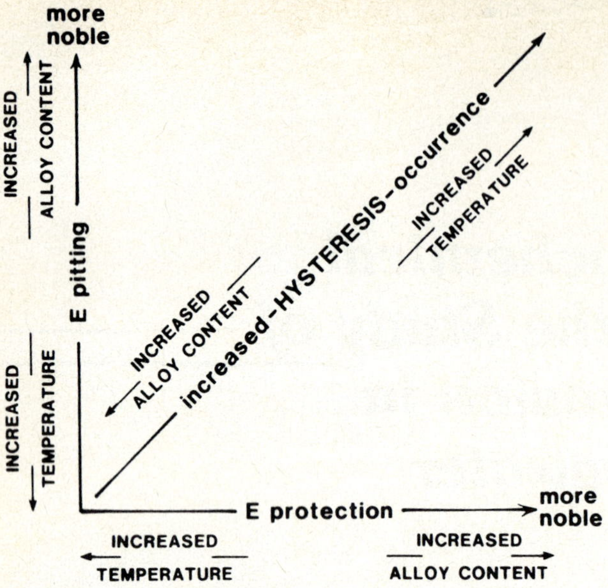

FIGURE 1 — Effect of increased alloy content and seawater temperature on various electrochemical parameters determined by cyclic polarization tests.

both the mechanistic and practical engineering approaches to the understanding and control of crevice corrosion.

The purpose of this paper is to provide examples of the utility of three electrochemical approaches: (1) anodic/cathodic polarization scanning; (2) remote crevice assembly testing; and (3) compartmentalized cell experiments. Results of natural seawater tests involving simple immersion methods will also be reviewed as supporting documentation of the electrochemical test data.

To facilitate establishing the relevance of any experimental method, an appreciation of the mechanism of a given corrosion phenomenon is important. To this end, a discussion of the crevice corrosion process and a brief review of a mathematical model of the process used by the authors forms a prelude to the experimental results.

Crevice Corrosion Mechanism

Several authors[15-17] have described the crevice corrosion mechanism process, which can be delineated in four progressive stages.

Initially [Figure 2(a)], a passive current flows over the entire metal surface. An occlusion creates an area in which oxygen availability is restricted and, as such, the passive current consumes all dissolved oxygen present in the electrolyte within this crevice. *Stage I* is hereby established.

Stage II of the process [Figure 2(b)] develops as the passive current at the metal surface continues in the confines of this deaerated electrolyte. The resulting introduction of metal ions (cations) through the passive film into the electrolyte has a twofold effect: (1) anions (primarily chloride ions) migrate into the crevice area from the bulk solution to effect a charge neutrality, and (2) hydrolysis reactions involving the metal ions result in an increase in acidity. These hydrolysis reactions take the form of:

$$M^{n+} + H_2O \rightarrow M(OH)^{(n-1)+} + H^+$$

Stage III [Figure 2(c)] shows the onset of accelerated corrosion, which occurs when the electrolyte within the crevice becomes sufficiently aggressive (high chlorides and acidity) to "break down" the passive film. This solution can be referred to as the *critical crevice solution*. *Stage IV*, or the continued propagation of crevice corrosion, then ensues.

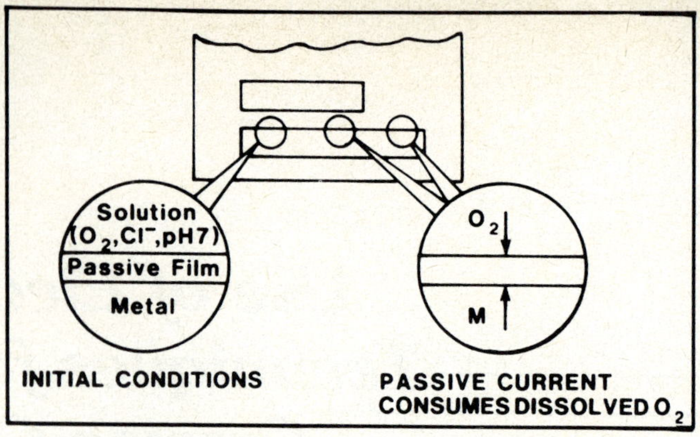

FIGURE 2(a) — Stage I of crevice corrosion.

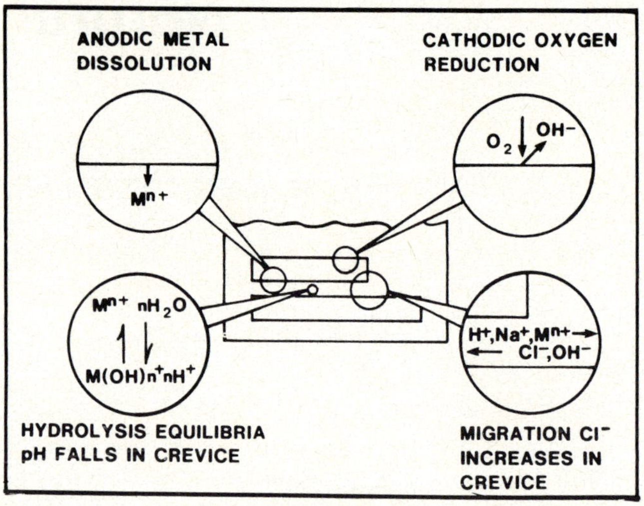

FIGURE 2(b) — Stage II of crevice corrosion.

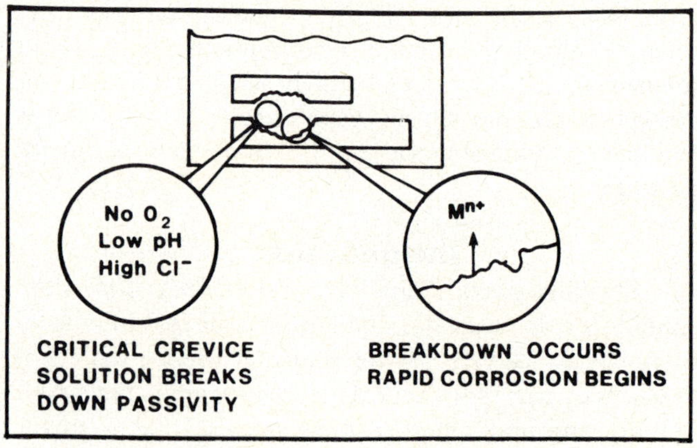

FIGURE 2(c) — Stage III of crevice corrosion.

Model of Crevice Corrosion Initiation. This four-stage mechanism has been used by Oldfield and Sutton[17,18] to provide a basis for a mathematical model of crevice corrosion. The initiation model uses several inputs to define the times required to progress through the first three stages of crevice corrosion and to define the nature of reactions occurring within the crevice environment. The inputs include:

(1) The oxygen content of the bulk solution, which is either known or measureable.
(2) The crevice gap (width or spacing between the metal surface and the crevice former) and crevice depth (distance from the mouth to the base of the crevice).
(3) The passive current density, which is experimentally determined by potentiostatic polarization of specimens at -100 mV (SCE) for one hour (typically 0.1 $\mu A/cm^2$ or less in seawater).
(4) The bulk solution chloride level and pH, which are known or measurable.
(5) The chemical equilibria, which describe the relevant hydrolysis reactions for the metal ions considered by the model (i.e., iron, chromium, and nickel).
(6) The alloy composition (i.e., iron, chromium, nickel, and molybdenum).
(7) The critical crevice solution in which the passive film breaks down.

The first output of the model yields the time required to deplete oxygen (t_d) within the crevice. Using this calculated time, the distance that oxygen can diffuse into the crevice is determined from the relationship $\delta_d = (\pi D t_d)^{1/2}$, where D is the diffusion coefficient for oxygen. This parameter defines the crevice depth for a given crevice gap, which must be exceeded in order for oxygen depletion to occur and for Stage II of the crevice corrosion process to begin.

The second stage of the model is the most complex. The key component in this stage is the calculation of the time required to decrease the pH from its bulk solution value to the critical value at which the passive film breaks down. This pH reduction is calculated from the rate of production of the various metal ions. It is dominated by the rate of chromium ion production, since it is the chromium hydrolysis reaction that controls the pH within the crevice.

The rate at which metal ions are produced within the crevice can be calculated by first determining the portion of the passive current (I_p) that corresponds to the production of ions of component metal "i." These variables are related by the equation:

$$\frac{I(i)}{I_p} = \frac{A(i) \cdot B(i)}{\sum_{j=1}^{n} A(j) \cdot B(j)}$$

where A represents the molar fraction of metal "i" in the alloy and B represents the charge of the metal ion.

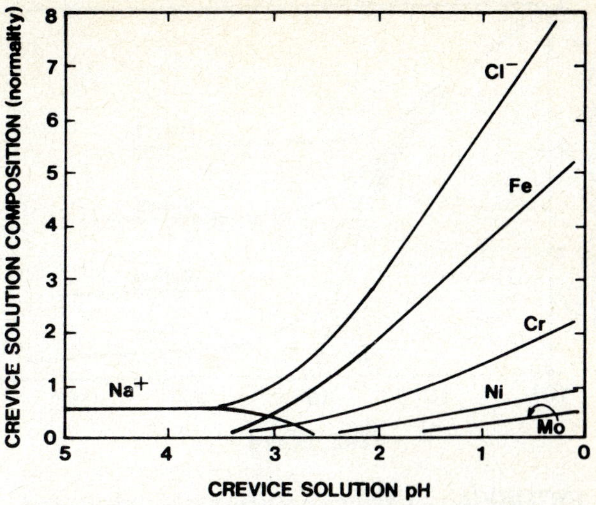

FIGURE 3 — Mathematical model prediction of crevice solution composition for AISI 316 SS in seawater.

Using this current, the rate of production of metal ions, R, in a crevice gap, x, can be calculated from the relationship:

$$R(i) = \frac{I(i) \cdot 1000}{F \cdot B(i) \cdot x}$$

Calculating the total amount of chromium required to decrease pH and knowing the rate of chromium production allow a calculation of the time required for the pH to fall by an increment, $\triangle$pH. This process is then repeated for each $\triangle$pH increment and the basic relationships between pH and chemical composition of the crevice and time are determined (Figure 3).

There are three adjustments to this pH-time relationship. One adjustment is made to correct the pH in the crevice to account for the chloride present.[19] Two additional adjustments to the time to reduce pH are made to take into account the mass transport in and out of the crevice. The migration adjustment takes into consideration the movement of cations out of the crevice and anions into the crevice as a result of current flow. The diffusion adjustment accommodates the movement of metal ions and chloride ions out of the crevice as a result of the concentration gradients that develop between the crevice solution and the bulk solution.

The net effect of these two factors is that the crevice solution will increase in acidity and chloride content until the chloride flux out of the crevice that results from diffusion equals the chloride flux into the crevice that results from migration. At this point, a steady state (i.e., limiting pH) is achieved. If the passive film breaks down in this solution, then crevice corrosion is possible for the particular crevice geometry considered.

Figure 4 shows the pH of a crevice solution as a function of time for crevices of varying depths. The shallower crevices generate less acid crevice solutions because of the greater influence of diffusion, while deeper crevices generate more aggressive solutions. The curve for a

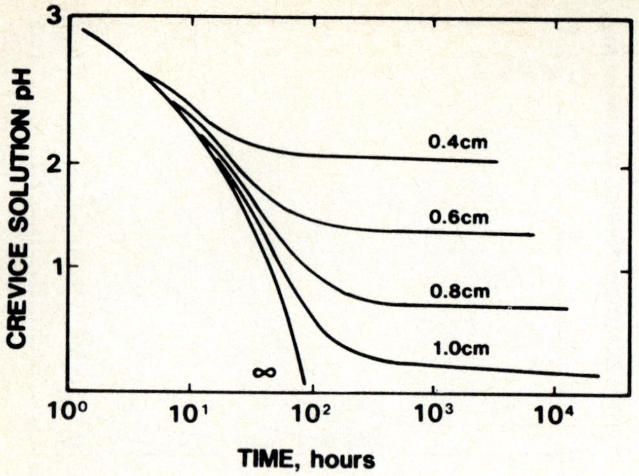

FIGURE 4 – Crevice solution pH calculated as a function of time for crevices varying in depth from 0.4 cm to infinity.

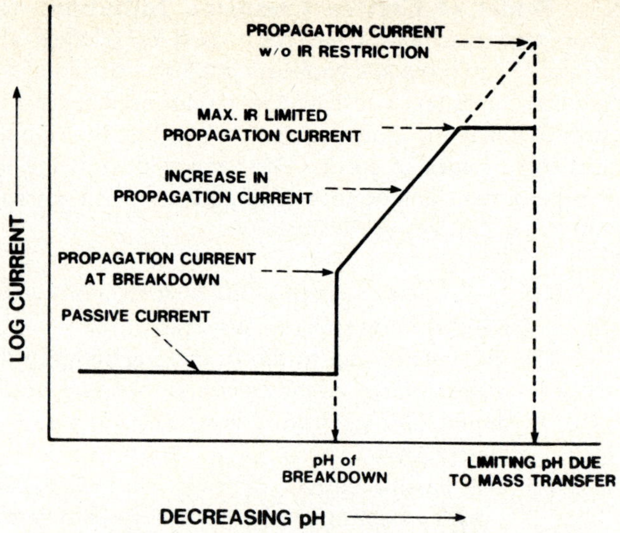

FIGURE 5 – Schematic illustration of anode current as a function of crevice pH.

crevice of infinite depth represents the conditions that can develop from migration in the absence of diffusion effects.

With this basic model, a number of calculations can be performed to allow examination of the relative effects of various factors on crevice corrosion initiation.[20-22]

Model of Crevice Corrosion Propagation. The final stage in the crevice corrosion process, and in practice the most practically relevant, involves propagation. This stage can be conveniently considered in terms of an electrochemical corrosion cell with an anode, the actively dissolving area within the crevice, and a cathode, usually sustaining oxygen reduction outside the crevice and/or hydrogen evolution within the crevicce. The rate of reaction is determined by the cell voltage, the electrolyte resistance, the cell geometry, and the relevant polarization relationships for the anodic and cathodic reactions.

The active area within the crevice, generated by the breakdown of the passive film, constitutes the anode. The anode parameters of interest are: (1) the nature of the initial active area within the crevice and its location, i.e., the nucleation kinetics; (2) the kinetics of growth of the corroding area, both lateral and in depth; and (3) the relationship between the anode potential and the anode current as a function of electrolyte composition, pH, and temperature.

Prior to breakdown of passivity and the onset of crevice corrosion, it is assumed that the current flows over the entire crevice area with an equal oxygen reduction current external to the crevice. Since this current is small, it is unlikely to be restricted by IR drop considerations down the crevice. Upon breakdown of passivity, both of these assumptions change. First, breakdown occurs over a discrete area within the crevice and secondly, the current increases by orders of magnitude. Thus, it becomes necessary to consider the IR drop down the crevice.

There is no simple way to predict the area where accelerated corrosion will first occur within a crevice. In practice, it will start in a small area and increase by some 2 or 3 dimensional growth mechanism. In the present model, it is assumed that the breakdown will occur in the last 1 mm of the crevice, furthest from the crevice mouth. With this definition of the active area and its relative position within the crevice, it is then possible to calculate the resistance from the mouth of the crevice to the active area.

Figure 5 shows a schematic illustration of the change in current within the crevice as a function of crevice solution pH. The current is assumed to increase instantaneously from the passive value of typically 0.1 $\mu A/cm^2$ to an active value of 10 $\mu A/cm^2$. It then increases according to a linear relationship between the log peak current and pH as determined from polarization data. The pH of the crevice solution eventually becomes limited by mass transfer considerations. This can occur either before or after the current is limited by the IR drop down the crevice. The maximum value of this IR-limited current is determined from the resistance down the crevice and the corrosion cell voltage simply by applying Ohm's law.

By definition in this model, the corrosion cell voltage is represented by the potential difference between the freely corroding passive potential of SS and the potential of the anodic current peak on the polarization curves in deaerated acid-chloride solutions. For austenitic SS in aerated seawater, this value is typically 275 mV, representing the difference between the active peak potential of −250 to −300 mV (SCE) and the passive potential of around 0 mV (SCE).

This approach to handling the propagation stage requires two additional input parameters to the model for each alloy considered. These are the slope of the log peak current vs pH curve and the corrosion cell voltage. This information, coupled with the value of resistance down the crevice and the previous inputs from the initiation model, allow calculations of anticipated initial propagation rates within defined crevice conditions. It is recognized that as propagation proceeds, several parameters such as cell voltage, crevice resistance, and crevice geometry will likely change in some interrelated fashion.

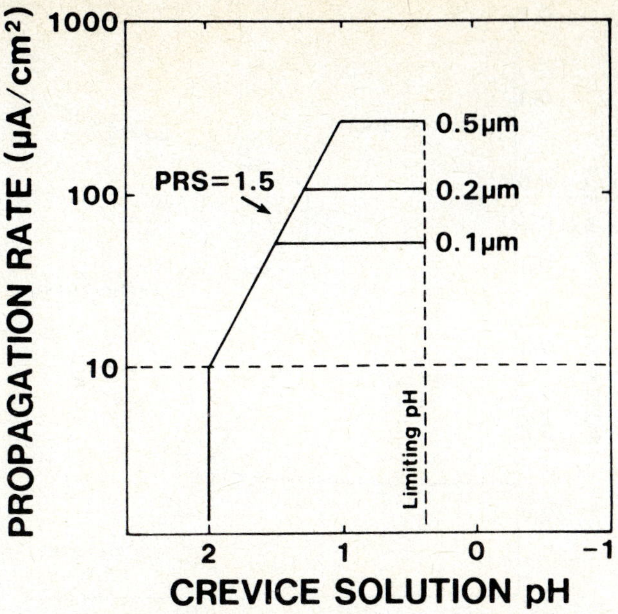

FIGURE 6 – Propagation rate as a function of crevice solution pH for a 20% Cr SS steel showing initiation of crevice corrosion at pH 2 and with a propagation rate current slope (PRS) of 1.5.

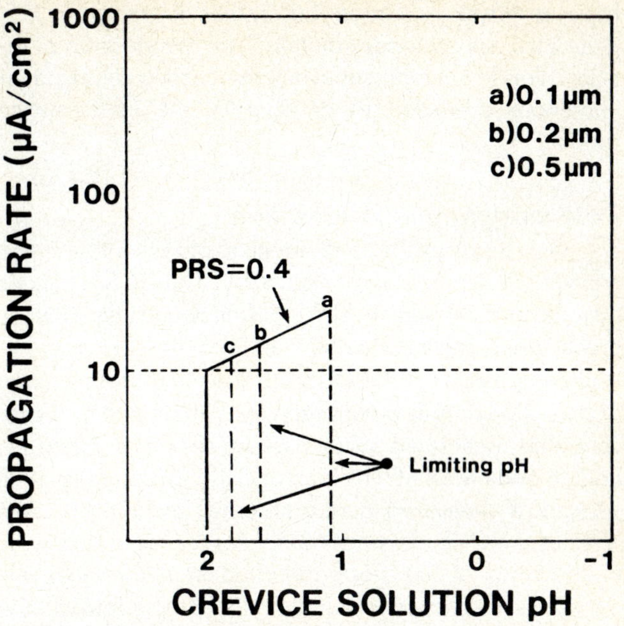

FIGURE 7 – Propagation rate as a function of crevice solution pH for a 20% Cr SS showing initiation of crevice corrosion at pH 2 and with a propagation rate current slope (PRS) of 0.4.

However, on a preliminary basis, this model can be used to calculate propagation rates of SSs as a function of various input parameters.[23] Figure 6 shows the calculated crevice propagation behavior for a SS (containing 20 wt% chromium), which will crevice corrode in a pH 2, 3 molar (M) chloride solution and has a peak current vs pH slope of 1.5. As in Figure 5, the sloping linear portion of the graph defines the current increase as a function of pH immediately following breakdown. The horizontal lines reflect the IR-limited propagation rates, as a function of three crevice gaps, which would be achieved for this particular alloy.

The data show that the propagation rate in this IR-limited condition increases with increasing crevice gap. This is opposite to the effect of crevice geometry on crevice initiation where a tighter crevice is more severe. In this particular case, however, the large gap results in a large volume of electrolyte in the crevice and, hence, a lower resistance. This lower resistance, with a fixed cell voltage, results in a higher allowable maximum current for propagation.

Figure 7 demonstrates the same type of data for a SS with the same initiation resistance as before but with a more shallow log current peak vs pH slope (by definition, a greater resistance to propagation than in the previous case). In this case, a value of propagation rate slope of 0.4 is shown. With this alloy, the IR-limited conditions are not reached for the gaps considered, and the limiting crevice solution pH is reached by mass transfer. In this situation, the tighter crevices result in the development of a lower pH crevice solution and a resultant higher propagation rate.

Experimental Procedures

Details of the various electrochemical tests and conventional test methods are described more thoroughly in the appropriate references. A brief synopsis of each procedure, however, is provided in the following sections.

Remote Crevice Assembly

In this test,[24,25] a physically separated but electrically connected pair of specimens is used. The smaller of the two specimens is sandwiched between two acrylic plates that create the crevice. Both the small creviced anode specimen and the larger uncreviced cathode specimen are exposed in the environment of interest. The measurement of current between these two specimens with a zero resistance ammeter provides an indication of time-to-initiation of crevice corrosion as reflected by a rapid increase in current. It also allows measurement of the propagation rate of corrosion in the crevice as indicated by the magnitude of current. The corrosion potential of the assembly can also be monitored as well as measurement of mass loss and depth of attack on the anode specimen at the conclusion of the test.

Compartmentalized Cell

This method[26] is similar to the remote crevice assembly with two specimens that are physically separated but electrically connected. With this technique, however, the two specimens are exposed in separate environmental compartments connected by an electrically/ionically conductive bridge. The large cathode specimen is exposed in one compartment containing the bulk environment of interest. The small anode specimen does not have a physical crevice but rather is exposed in a deaerated acid-chloride solution. The composition of this acid-chloride solution simulates the crevice environment conditions predicted to develop by the mathematical model.

Potentiodynamic Polarization

One input to the mathematical model is the critical crevice solution (CCS) for each alloy. The CCS is defined by a pH and chloride level that is sufficiently aggressive to break down the passive film and initiate corrosion.[18,27] The composition of the CCS is determined from a series of potentiodynamic polarization tests in a number of concentrated, deaerated solutions. The specific solution pH and chloride level at which an anodic current peak of 10 $\mu A/cm^2$ is measured during potential scanning is identified as the CCS. The test electrodes for these tests were in the form of flat stock mounted in a thermosetting epoxy resin. CCS criteria and details of the electrochemical procedures are described elsewhere.[18] Results of these tests are also used to establish characteristic pitting behavior and to develop propagation rate data.

Immersion Tests

Throughout the course of electrochemical experimentation, numerous immersion tests have been conducted in natural and synthetic media. For the most part, two types of crevice formers have been used to assess crevice corrosion behavior. The multiple crevice assembly[28] consists of a nonmetallic washer with a series of plateaus such that 20 separate crevice sites are created beneath each washer. Two washers are bolted to a test panel using a torque wrench to provide a consistent degree of initial tightness.[29] The second device is an acrylic washer, which creates a continuous, annular crevice area on each side of a test panel.

Results and Discussion

Potentiodynamic Polarization Tests

Effect of Alloy Mn and S Level. Decreased levels of manganese and sulfur in SS have been reported to contribute to improved pitting resistance.[30] Improved resistance to initiation of crevice corrosion by lowering the manganese content to 0.2% has also been reported.[31] Such reductions correspond to the same positive effect obtained by the addition of at least 1.5% molybdenum. Figure 8 compares the overall anodic behavior for two laboratory prepared heats of Type 316 SS with high and low levels of manganese and sulfur as determined in a deaerated, simulated crevice solution. As can be seen, a more noble pitting potential was clearly identified for the low manganese-low sulfur version. Whether or not a greater resistance to localized corrosion could be realized in practice with the low Mn-S alloy would depend on the precise potential exhibited by the material in the given environment.

To further assess the effects of Mn and S levels on expected seawater crevice corrosion behavior, additional potentiodynamic polarization and multiple crevice assembly tests were conducted.[32]

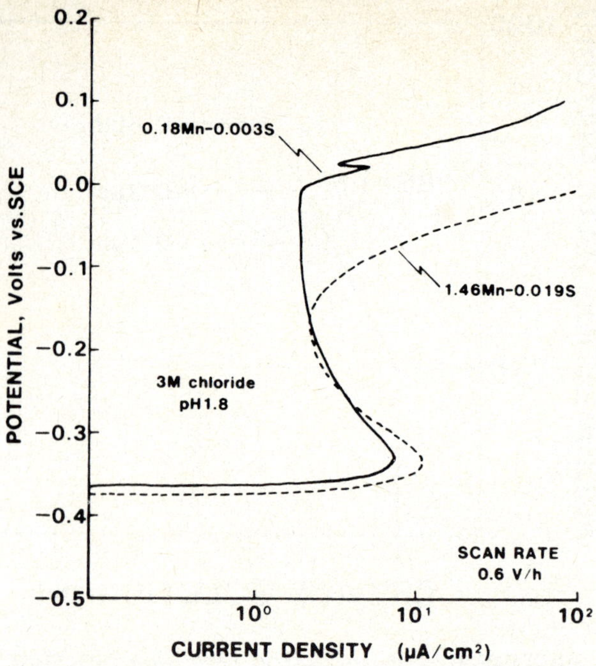

FIGURE 8 — Effect of Mn and S content on the anodic polarization behavior of Type 316 SS in a simulated crevice solution.

One test involved an electrochemical determination of the CCS composition required to initiate crevice corrosion. In this test, an alloy is defined as being more resistant if the CCS is at a lower pH/higher chloride value. Table 1 summarizes CCS values for several alloys of varied Mn and S levels.

In the case of Type 304 SS, relatively little difference in the CCS composition can be seen as a function of Mn and S levels. A slightly higher pH, however, is noted for the high Mn-high S version (i.e., less resistant). While generally similar findings were determined for Type 316 SS, a more substantial difference is shown between the CCS for the high Mn-high S heat and that for the other Mn-S combinations. Reducing Mn and S levels in the 20Cr-25Ni-4.5Mo-1.5Cu alloy had no significant effect on CCS composition.

The CCS values identified in Table 1 were used as inputs for mathematical modeling. Figure 9 gives modeling predictions for the times-to-crevice corrosion initiation as a function of crevice geometry. In this example, crevice depth (0.1 cm) is fixed and crevice gap or tightness for a nonmetal-to-metal assembly is varied. Alloy resistance is compared for high Mn-high S and low Mn-low S heats of Type 304 and 316 SS. Predictions for the two heats of the 20Cr-25Ni-4.5Mo-1.5Cu alloy having similar CCS values are typified by one curve.

These curves indicate that as crevice gap is increased, the time-to-initiation of crevice corrosion is increased. For each material, some crevice gap exists where crevice corrosion is no longer predicted. For both Type 304 and 316 SS, it can be seen that the high Mn-high S versions are predicted to be susceptible to corrosion over a wider range of tight crevices than the corre-

TABLE 1 — Critical Crevice Solutions for Stainless Steels

	Percent Content		Critical Crevice Solutions (CCS)	
Alloy	Mn	S	pH	Cl^- (M)
Type 304	0.16	0.005	2.00	1.5
	0.18	0.017	2.05	1.5
	1.20	0.006	2.00	1.5
	1.11	0.015	2.10	1.5
Type 316	0.18	0.003	1.70	3.5
	0.25	0.014	1.70	3.5
	1.48	0.004	1.65	3.5
	1.46	0.019	2.05	2.5
20-25-4.5-1.5	1.48	0.004	1.35	4.5
	0.18	0.013	1.30	4.5

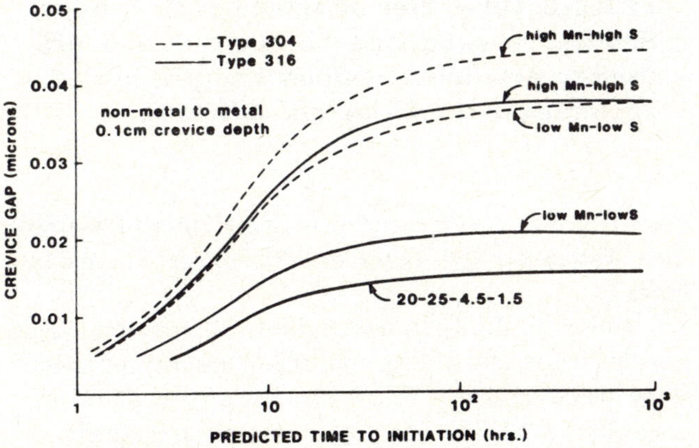

FIGURE 9 — Mathematical modeling prediction showing effect of alloy Mn and S content on resistance to crevice initiation as a function of crevice gap.

sponding low Mn-low S version. In this regard, the beneficial effects of decreased Mn and S content are more pronounced for Type 316 SS. As shown, the resistance of the low Mn-low S heat of Type 316 SS approaches the resistance of the higher alloyed SS.

The modeling predictions in Figure 9 also show that for extremely tight crevices (e.g., 0.01 μm gap), any effect of alloy content would be less noticeable. This was confirmed in actual tests conducted in 25 C, filtered seawater. Evidence of ongoing crevice corrosion beneath multiple crevice assemblies (MCA) fastened to 5 × 5 cm specimens was detected for all materials within three days of exposure. After ten days, all materials, regardless of Mn and S levels, exhibited substantial attack as indicated in Table 2.

Variations in depth of attack can be attributed to differences in the actual time-to-crevice corrosion initiation

TABLE 2 — Range of Crevice Corrosion Attack in Ten Day Multiple Crevice Assembly Tests in 25 C Filtered Seawater

	Depth of Attack (mm)		
Mn-S Variation	Type 316	Type 304	20Cr-25Ni-4.5Mo-1.5Cu
Low Mn-Low S	0.01 to 0.48	0.02 to 0.51	—
Low Mn-High S	0.02 to 0.48	—	0.05 to 0.56
High Mn-Low S	0.01 to 0.60	0.05 to 0.71	—
	0.01 to 0.38	0.10 to 0.58	
High Mn-High S	0.02 to 0.43	0.02 to 0.51	—
		0.09 to 0.38	

at individual sites beneath the multiple crevice assemblies, as well as site-to-site differences in the crevice solution concentration. Less severe crevice geometries than those created in the MCA test may allow for greater distinction of any positive effect of controlled Mn and S levels. However, since it is unlikely that crevice geometry can be controlled to the extent suggested by mathematical modeling, differences in behavior from one application to another should be expected.

Cast Alloy Comparisons. Multiple crevice assemblies also have been used[33] to evaluate the relative crevice corrosion resistance of 45 wrought and cast alloys. Six cast alloys from that study were selected for further evaluation.[34] A test matrix to assess the effects of crevice assembly torque level and test duration was applied in the evaluation of CF8M, CN7M, CN7MS, IN-862, cast alloy 625, and CW-12M-2. These alloys have similar chromium contents (18 to 21%), but range in molybdenum content from 2.5 to 18% and nickel content from 10 to 64%. To further qualify the seawater crevice corrosion test data and define its relevance to service experiences, additional electrochemical testing and mathematical modeling were employed.

Slopes of the pH vs anodic peak current plots have been used as an indicator of the resistance to crevice corrosion propagation.[11] The higher the value of the slope, the greater is the expected crevice corrosion attack. Table 3 gives the slope of plots from Figure 10 for the six alloys. CF8M and CN7MS exhibit the highest slope value and also show generally greater attack in the MCA tests than do CN7M and IN-862. Because no evidence of crevice initiation was found for cast alloy 625 and CW-12M-2, it is obviously difficult to assess the propagation resistance of these two alloys. The very low slope values indicated in Table 3, however, suggest that if initiation were to occur, any propagation for these two alloys would be considerably less than that described for the other materials.

Table 4 compares the average and maximum depths of attack for the various alloys in the immersion tests. Again, results of the initial 30-day test indicate somewhat greater resistance for the higher alloyed materials. Average values for CN7MS, CN7M, and IN-862 are substantially lower than that determined for CF8M.

TABLE 3 — Slopes of pH Versus Anodic Peak Current Plots[1]

Alloy	d(log I peak) / dpH
CF8M	1.9
CN7MS	1.8
CN7M	1.4
IN-862	1.4
Cast alloy 625	0.3
CW-12M-2	0.3

[1] After Figure 10.

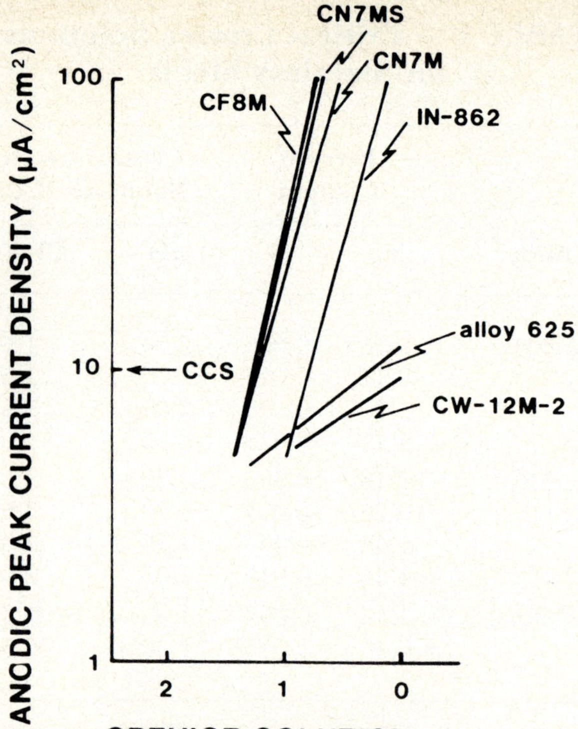

FIGURE 10 — Plot of anodic peak current density vs simulated crevice solution pH used to determine composition (pH) of CCS according to the 10 $\mu A/cm^2$ criterion.

The data describe severe attack for the four alloys that initiated under the described conditions. The progression of attack to the extent shown can be attributed not only to the rapid initiation but also to the limited area in which the crevice corrosion is confined. Also, it would appear that in the absence of accumulated biofouling, there was sufficient driving force from cathodic reduction reactions on the boldly exposed specimen surfaces to sustain the attack. Corrosion reported for CN7M and CF8M tested under similar conditions[35] are within the ranges of those determined in the present study.

The critical crevice solution data for the cast alloys indicate that a more acid solution is required for crevice corrosion initiation than for similar wrought materials.[11] Table 5 compares the present results for CF8M with those for the wrought counterpart, AISI 316 SS. The differences between the two product forms suggest that the cast version would be more resistant to initiation of crevice attack than the wrought material. This is supported by the accompanying multiple crevice assembly data describing the observed differences in initiation times.

Although the CCS tests indicate improved initiation resistance for the CF8M, the difference in propagation slope values suggests that it has less propagation resistance than AISI 316 SS. Again, the electrochemical indications are supported by the immersion test results. Distinct differences can be seen in the depth of attack measurements provided.

TABLE 4 — Crevice Corrosion Propagation Resistance of Cast Materials

Test Conditions	Depth of Crevice Attack (mm)											
	CF8M		CN7MS		CN7M		IN-862		Alloy 625		CW-12M-2	
	avg.	max.	avg.	max.	avg.	max.	avg.	max.	avg.	max.	avg.	max.
8.5 Nm												
30 days	0.97	3.77	0.59	3.82	0.50	2.33	0.42	1.22	0	0	0	0
60 days	+	+	+	+	+	+	0.63	1.75	0	0	0	0
90 days	+	+	+	+	+	+	+	+	0	0	0	0
2.8 Nm												
30 days	0.63	4.04	0.75	3.18	0.60	2.24	+	+	+	+	+	+
60 days	0.83	3.59	0.58	3.53	0.66	2.93	0.52	4.31	+	+	+	+

+ Not tested.

TABLE 8 — Crevice Corrosion of Stainless Steels in Natural and Synthetic Seawater

Alloys	Temperature (C)	Synthetic Seawater % Sites Attacked	Synthetic Seawater Max. Depth of Attack (mm)	Natural Seawater % Sites Attacked	Natural Seawater Max. Depth of Attack (mm)
AISI 304	25	31	0.11	42	2.91
AISI 316	25	38	0.08	4	1.12
AISI 304	50	31	0.15	33	0.18
AISI 316	50	28	0.13	38	0.08

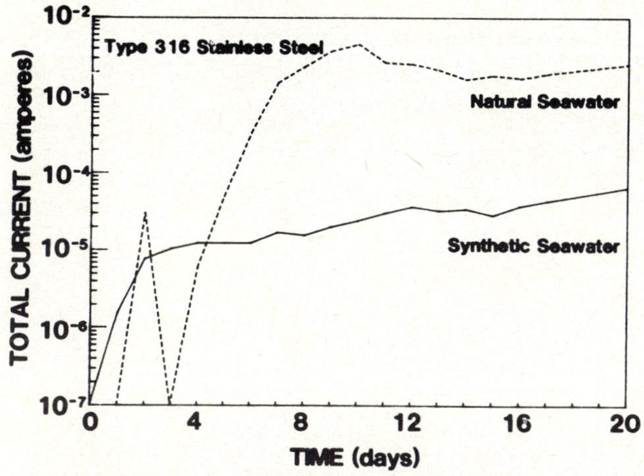

FIGURE 13 — Effect of natural versus synthetic seawater on crevice corrosion of AISI 316 SS.

Both the initiation and propagation behavior for AISI 304 and 316 SS in the present test are consistent with the reported multiple crevice assembly test data. This is especially true in describing the relatively minor difference in penetration shown previously in Table 6.

Effect of Environment Source. In addition to the chloride level, the nature of the chloride-containing solution is quite important in the crevice corrosion process. A comparison of crevice corrosion behavior in natural seawater with that in ASTM[1] synthetic seawater is shown in Table 8. The results are from multiple crevice assembly tests conducted for 30 days in solutions at 25 C.[38] The natural seawater was filtered to remove larvae of macrobiofoulers and both solutions were fully aerated and at a nominal velocity of 0.02 m/s. As can be seen, both alloys suffered over an order of magnitude greater attack in the natural seawater than in the synthetic seawater tests.

Figure 13 shows a similar phenomenon using the remote crevice assembly on AISI 316 SS in solutions at 30 C. The early rapid rise in current for the samples in synthetic seawater reveals that initiation of crevice corrosion occurs relatively consistently between the two environments. What is significantly different, however, are the rates of propagation in the two environments as evidenced by the current levels following initiation. Here it is obvious that the AISI 316 SS is corroding in natural seawater at a rate 10 to 20 times faster than that of the AISI 316 SS in the synthetic seawater.

Polarization tests have also been conducted to examine these differences and to investigate the possible role of oxygen reduction at the cathode surfaces.[39] Table 9 shows the average values of diffusion-limited current, Tafel slope, and exchange current density from tests in the stirred, aerated solutions. Nominally equivalent values of diffusion limited current were measured in both environments; however, the Tafel slope and exchange current density were somewhat lower in natural seawater.

Table 10 shows corrosion potentials measured after one hour of exposure in stirred, aerated solutions at 25 C. Of particular note is the more noble corrosion potential measured in the synthetic seawater for all three alloys. Using these data and those in Table 9, it is possible to calculate the oxygen reduction current at the corrosion potential as shown in Table 10. As can be seen, a significantly lower oxygen reduction current is evident in synthetic seawater. This is consistent with the observed much lower rates of crevice propagation in the synthetic environment. The implication is that anodic processes may control reaction rates in natural seawater with no limitations imposed by cathodic reactions. In synthetic seawater, however, oxygen reduction may be the rate controlling factor. With the lower oxygen reduction current evident in synthetic seawater, it is also conceivable that hydrogen evolution at cathodic areas within the crevice may be an important consideration.

Changes in Seawater Conditions. In most practical situations where stainless steels are used for seawater service, the environment is manipulated in one way or another. In the case of condenser technology, for example, seawater is (1) pumped at some velocity, (2) frequently chlorinated to minimize fouling, and (3) ultimately heated. Remote crevice assembly tests are particularly

TABLE 9 — Electrochemical Test Results in Natural and Synthetic Seawater

	Natural Seawater	Synthetic Seawater
Diffusion Limiting Current (A/cm²)	7.1×10^{-4}	7.1×10^{-4}
Tafel Slope (mV/decade)	86	111
Exchange Current Density (A/cm²)	3.8×10^{16}	5.0×10^{-14}

TABLE 10 — Corrosion Potentials and Corresponding Oxygen Reduction Currents in Natural and Synthetic Seawater

Seawater	Alloy	E_{corr} (V, SCE)	Oxygen Reduction Current at E_{corr} (10^{-7} A/cm²)
Natural	AISI 316	−0.200	3.77
	Alloy 904L	−0.195	2.52
	Alloy 6X	−0.180	1.69
Synthetic	AISI 316	−0.100	0.46
	Alloy 904L	−0.125	0.77
	Alloy 6X	−0.045	0.10

suited for investigating both the initiation and propagation responses to changes in environmental conditions. This has been accomplished in a temperature effect study where alloy 904L was exposed in refreshed, recirculated natural seawater and subjected to a series of incremental temperature changes.

As shown by the increase in current plotted (solid line) in Figure 14, during the initial period at 30 C, initiation of crevice corrosion is evident around day 7. While some pertubations in current are evident in the temperature steps to 35, 40, and 45 C, it is not until the temperature is increased to 50 C that a significant change occurs. At this temperature, the current is reduced to a value over two orders of magnitude lower than the peak current measured when the seawater was at 30 C. After several days at 50 C, the heaters were turned off and the seawater temperature was allowed to decrease to 30 C. Within two days, the current began to rapidly increase to the levels observed previously. The dashed line shows the behavior of a second specimen exposed at day 47 during the 50 C temperature period. While a small current peak was noted around day 60, no initiation of crevice corrosion occurred on this specimen until the seawater temperature was reduced to 30 C. At that time, the current increased to the same level as with the previous test.

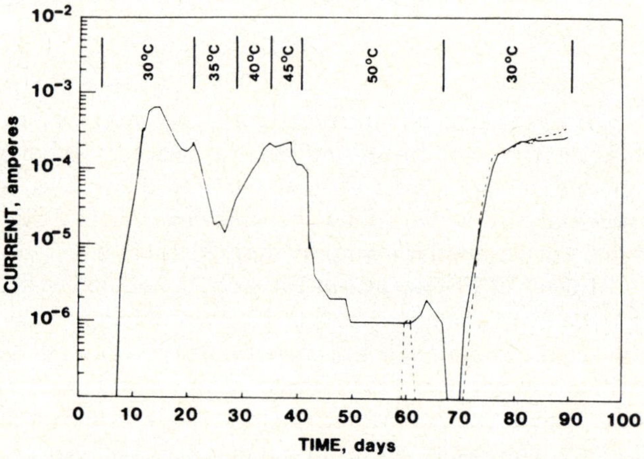

FIGURE 14 — Crevice corrosion propagation behavior as shown by remote crevice assembly current measurements in natural seawater at various temperatures.

The preceding results support earlier findings from remote crevice assembly tests for AISI 316 SS conducted at constant temperature within the normal ambient range and at 50 C.[24] In addition, multiple crevice assembly data presented in Figure 15 show that the range of attack in-

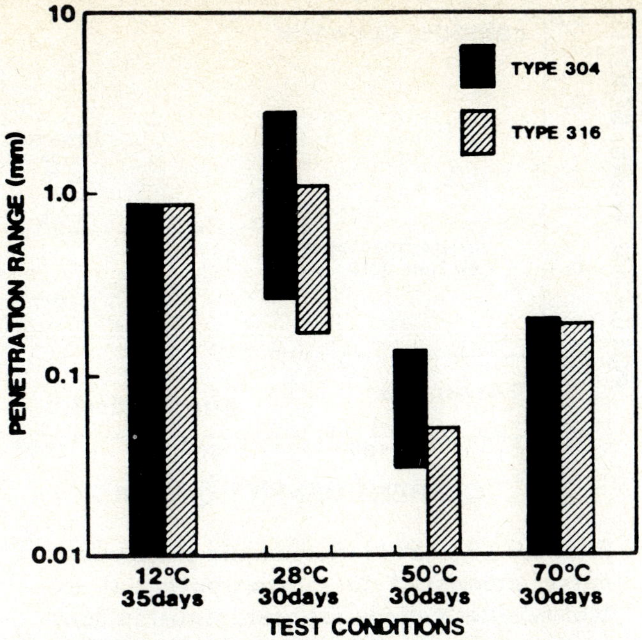

FIGURE 15 – Effect of natural seawater temperature on crevice corrosion behavior in multiple crevice assembly tests.

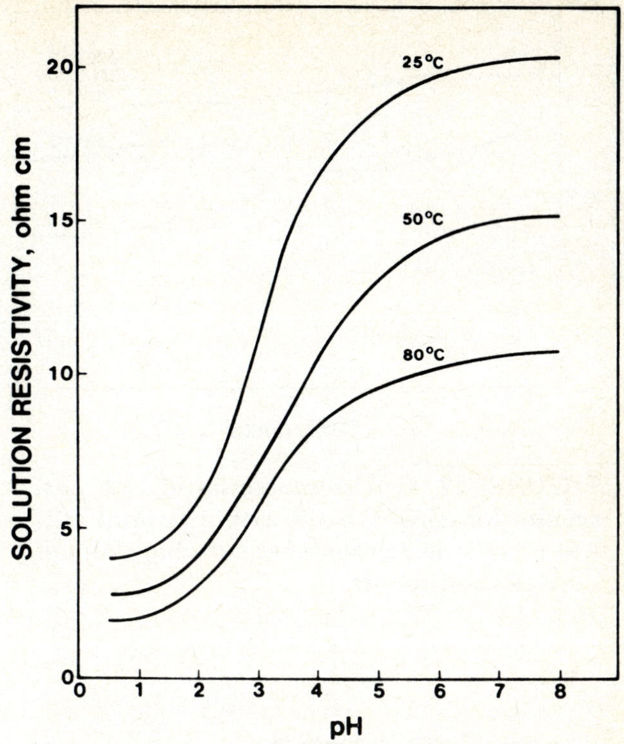

FIGURE 16 – Effect of temperature on resistivity of synthetic crevice solutions.

curred by AISI 304 and 316 SS is considerably lower at seawater temperatures of 50 and 70 C than at either 12 or 28 C.[20]

It is perhaps interesting to note that the difference in propagation behavior as a function of temperature appears to be unique for natural seawater. From Table 8 shown earlier, little difference in penetration was measured for multiple crevice assembly specimens exposed in 25 and 50 C synthetic seawater.

The exact mechanism for this observed phenomenon is not yet clear. Conventional electrochemical tests in natural seawater yield data that would traditionally imply that increased temperature should increase the extent of crevice corrosion. Previously reported results showed that as the seawater temperature is increased, there is an observed shift in both pitting and protection potentials to more active values and an increase in hysteresis on polarization curves[12] as summarized in Figure 1.

Measurements of resistivity of synthetic crevice solutions show a decrease in resistivity as the temperature is increased (Figure 16). This effect would be expected to enhance anodic reactions. Since this is not observed in practice with crevices in natural seawater, again it would appear that reactions at the cathodic surfaces may be rate controlling.

To examine this, the oxygen reduction currents were calculated from a series of polarization tests conducted to obtain corrosion potentials, diffusion-limited currents, Tafel slopes, and exchange current densities as a function of solution temperature. As can be seen in Table 11, the oxygen reduction current increased with temperature in both natural and synthetic seawater. These data, developed on specimens pre-exposed for one hour in solution, do not support the observed effect of temperature on crevice corrosion in the natural tests. It is recognized from other work, however, that the kinetics of oxygen reduction can change with exposure duration; this parameter is the subject of ongoing research.

TABLE 11 – Oxygen Reduction Current for Alloy 904L in Seawater at Various Elevated Temperatures

Seawater	Temperature (C)	Oxygen Reduction Current E_{corr} (10^{-7} A/cm^2)
Natural	25	2.52
	50	4.84
	80	13.8
Synthetic	25	0.77
	50	1.89
	80	73.5

Compartmentalized Cell Tests

Effect of Crevice Solution Concentration. As previously indicated, the ultimate crevice solution pH and chloride concentration can vary from one crevice geometry to another for any given material-bulk environment combination. It is, therefore, advantageous to examine alloy behavior over a range of possible crevice

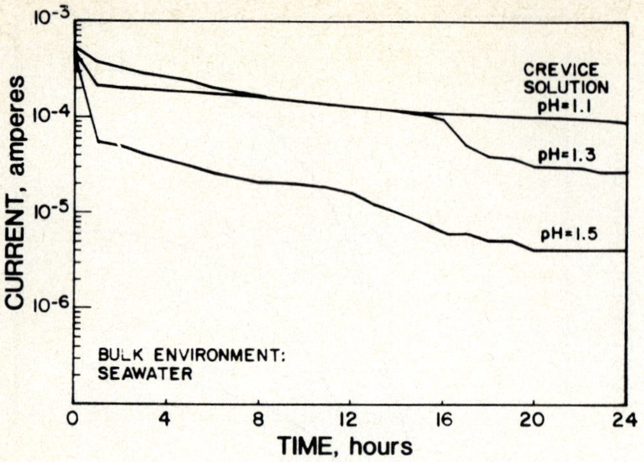

FIGURE 17 — Compartmentalized cell results for AISI 316 SS with a natural seawater cathode environment showing effect of crevice solution pH.

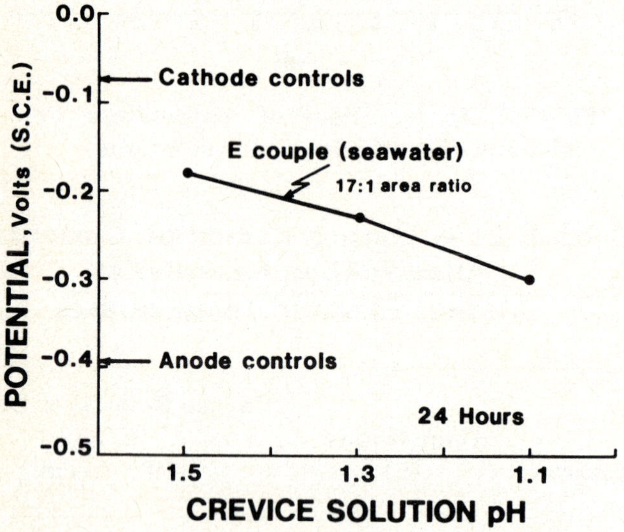

FIGURE 18 — Effect of crevice solution pH on the couple potential for AISI 316 SS in compartmentalized cell tests using natural seawater in cathode compartment.

solution concentrations.

In addition to the potentiodynamic polarization methods discussed earlier, the authors have utilized a compartmentalized cell approach to investigate the potential regimes where a crevice member will corrode (or passivate). In these experiments, polarization is induced by coupling the crevice member (exposed in simulated crevice solution) to a larger cathode member in the bulk environment.

Figure 17 describes the measured currents for AISI 316 SS in a series of tests utilizing crevice solutions in the pH range of 1.5 to 1.1 with corresponding chloride concentrations of 5 to 6 M. Recirculated natural seawater was selected as the cathode environment. While each of the curves describes initially high currents, substantial separation is evident after 24 hours.

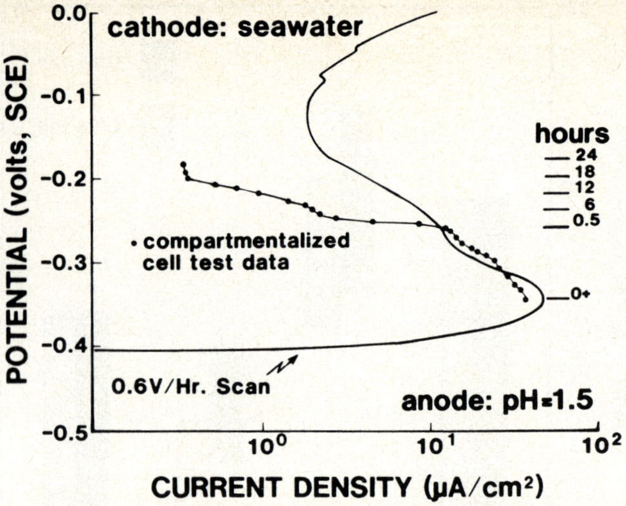

FIGURE 19 — Potentiodynamic polarization curve compared to current-potential exhibited after various exposure times in compartmentalized cell tests for AISI 316 SS in a pH 1.5 crevice solution.

Figure 18 shows the average potentials for freely corroding, crevice-free cathode members in natural seawater and those for freely corroding anode members in deaerated crevice solutions at the indicated pH value. Data points plotted on the graph identify the couple potentials measured in the bulk environment compartment. As can be seen, these potentials became more active with decreasing crevice solution pH. In general, these results are consistent with published potential data describing differences in potentials of active and passive SS in seawater.[40]

The result of coupling a cathode with the anode member in the more concentrated pH 1.1 solution resulted in a fivefold increase in anode mass loss as compared to the same experiment using the pH 1.5 solution. Similarly, the effects of polarization in the pH 1.1 solution produced a fivefold increase in mass loss as compared with a freely corroding member in the same deaerated solution.

Figure 19 compares the polarization behavior of the anode member in the pH 1.5 crevice solution (seawater cathode) with a potentiodynamic polarization curve developed at a scan rate of 0.6 V/h. Coupling of the anode and cathode members resulted in a shift to a potential corresponding to the anodic peak current maxima with subsequent rapid polarization to more noble potentials. The potential-current graphs obtained by the two methods are in excellent agreement up to a potential of −0.275 V (SCE). As would be expected, the longer exposure times in the compartmentalized cell resulted in lower passive currents at the more noble potentials.

In the case of tests in the pH 1.1 solution, the potential of the anode members remained at more active potentials [−0.3 V (SCE)] and, hence, the greater mass loss.

These results show that while aggressive crevice solutions were more concentrated than that required for

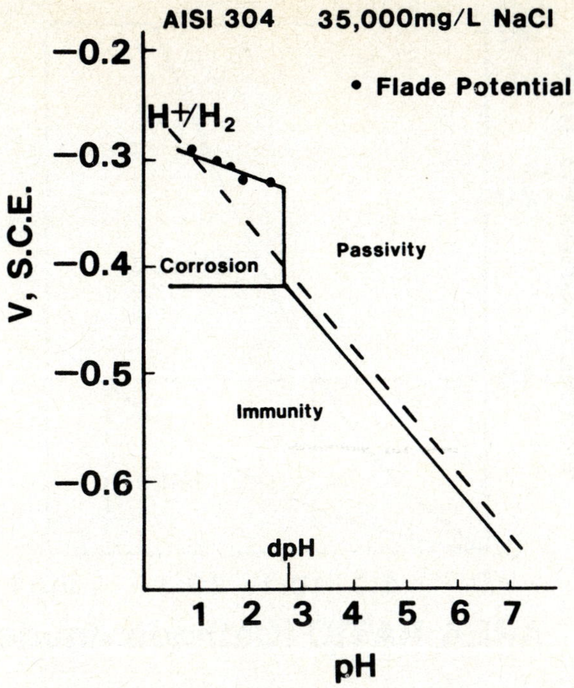

FIGURE 20 – Relative potential-pH domains for immunity, corrosion, and passivity of AISI 304 SS (after Azzerri, Mancia, and Tamba, Corros. Sci., Vol. 22, No. 7, 1982).

breakdown of passivity (CCS value), some anodic protection is possible as a result of galvanic effects of the cathode. Accordingly, the degree of propagation may be affected by the limiting pH within the crevice and the ability of the cathode to polarize the crevice into a low passive current regime.

These findings are consistent with recent work by others[19,41] who have also reported that anodic protection is possible in crevice solutions more concentrated than the depassivation pH or CCS. In Figure 20, for example, data for AISI 304 SS indicates that passivity can be established in crevice environments if the corrosion potential can be made more noble to the Flade potential.[41] For the present compartmentalized cell crevice environments, the Flade potential, allowing for some possible difference between AISI 304 and 316 SSs, corresponds to a potential of approximately −0.3 V (SCE).

Effect of Seawater Oxygen Concentration. The compartmentalized cell method has also been utilized to investigate the effect of bulk oxygen content on polarization of the anode in a deaerated crevice solution. In one series of experiments, anode members of alloy 904L were exposed to a simulated crevice solution containing 6 M chloride at pH of 0.8 at 25 C. Different mixtures of nitrogen and oxygen were bubbled in the natural seawater in the cathode compartment.

Table 12 describes the potential of the anodes after 24 hours as a function of the measured dissolved oxygen contents. At this pH level, deaeration of the seawater ($\leq$1 mg/L dissolved oxygen) resulted in the material remaining active with some low measure of corrosion currents. Increasing the oxygen concentration by purging with the various gas mixtures resulted in an increase in the anode potential to more noble values. Under these alloy-environment conditions, polarization to the passive state was described by decay in current densities to 1 $\mu A/cm^2$ or less.

Any agreement between these results and those from other tests, e.g., remote crevice assembly, would certainly depend on the similarity in actual crevice solution as well as any IR limitations associated with an applied crevice. The compartmentalized cell technique is shown, however, to be sensitive to detecting relatively small alterations of either the cathode or anode environments.

TABLE 12 – Effect of Seawater Oxygen Level on Corrosion Potential of Cathode and Anode Members in Compartmentalized Cell Tests[1]

Dissolved Oxygen Level (mg/L)	Potential (24 hours) E_{corr} Cathode (control) (V, SCE)	E_{corr} Cathode (coupled) (V, SCE)	E_{corr} Anode[2] (coupled) (V, SCE)
1	−0.225	−0.293	−0.295
1-2	−0.179	−0.254	−0.255
6-7	−0.111	−0.152	−0.158
12-13	−0.109	−0.111	−0.116

[1] Alloy 904L (crevice solution 6 M Cl^-, pH 0.8).
[2] E_{corr} of uncoupled "anode" controls = −0.315 to −0.320 V (SCE).

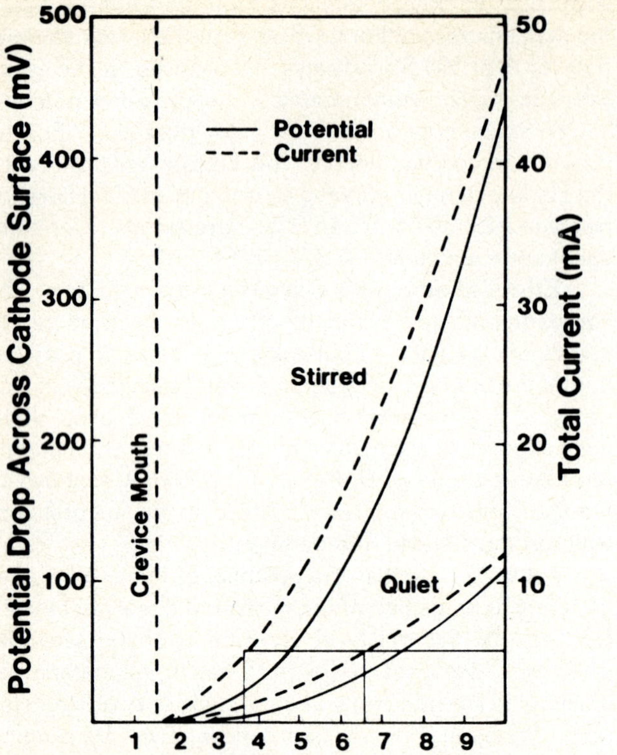

FIGURE 21 – Mathematical model prediction of potential drop and total oxygen reduction current as a function of distance across a cathode away from the crevice mouth.

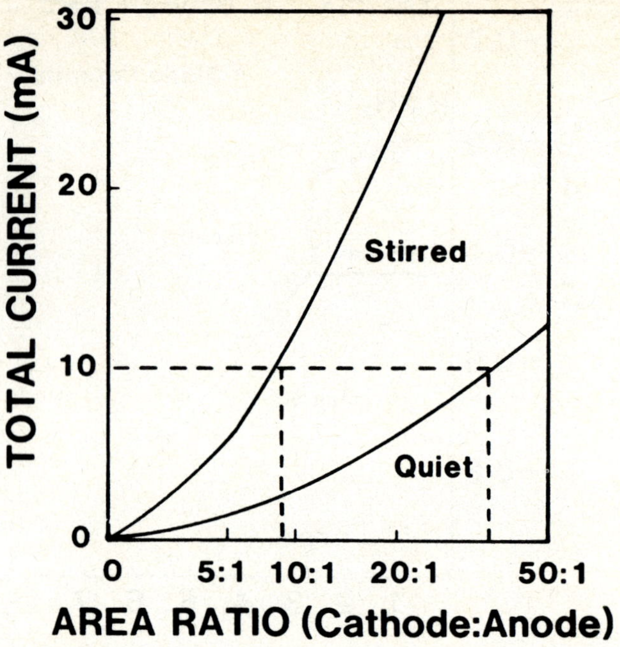

FIGURE 22 – Calculated oxygen reduction current as a function of cathode to anode area ratio.

Related Studies Requiring Electrochemical Testing

Effect of Velocity

Velocity is a particularly important parameter in seawater systems in that low design velocities can allow settlement of debris or fouling organisms that form crevice sites. Conversely, high velocities could be expected to remove some crevice formers from the system and flush other crevices to the extent that a detrimental acid-chloride solution may not form. The solution velocity could also be expected to influence reactions occurring at cathodic surfaces that are sustaining active crevice corrosion at local anodes. It is this aspect of cathodic oxygen reduction which has been recently modeled.

Figure 21 shows an estimation of the potential drop across a cathode and the total oxygen reduction current in both quiet and stirred seawater as a function of distance from the crevice mouth. This shows, for example, that in stirred seawater, a cathode surface extending 2 cm in radial distance beyond the crevice mouth would sustain a 5 mA total current. Therefore, if a total anodic propagation current was less than this magnitude, the crevice corrosion process would remain under anodic control with this size cathode. By contrast, with the same anode current, a cathode area extending 5 cm from the crevice mouth would be required in quiet seawater to retain the process under anodic control.

Using this concept, it is possible to estimate a cathode to anode area ratio at which crevice corrosion propagation might be restricted as a result of available oxygen reduction currents. This total available current is shown in Figure 22 as a function of the calculated area ratio for both stirred and quiet solution conditions.

From previous remote crevice assembly tests[24] with cathode areas over 400 times that of the crevice anode areas, a propagation current of approximately 1 mA/cm^2 was measured. It is reasonable to assume that no limiting cathodic current control occurred in those tests and that the measured current was reflective of anodic control. With a total crevice area of 10 cm^2, a total anode current of 10 mA would be possible. At that level, a cathode to anode area ratio of 9:1 in stirred solution and 35:1 in quiet seawater would be critical. Smaller cathode areas would be expected to limit the anodic propagation current as a result of limiting cathodic reduction currents.

Figure 23 shows experimental data from tests with the acrylic crevice assembly described earlier. The crevice area with this device is 10 cm^2. Thus, the above conclusions about critical area ratios are applicable. As can be seen, the most noticeable decrease in penetration or propagation is seen in stirred seawater when the cathode to anode area ratio is below 20:1. This compares to the predicted 9:1 critical ratio in stirred seawater. In quiet seawater, the critical ratio appears to be between 20:1 and 100:1 vs the predicted ratio of 35:1. The increase in seawater velocity increased the extent of attack by increasing the available oxygen reduction current at the intermediate area ratio. At the lowest area ratio of 4:1, where both tests would appear to be under cathodic control, there is a minimal effect of velocity. At the highest area ratio, where both tests appear to be under anodic control, there again is a minimal difference between stirred and quiet conditions.

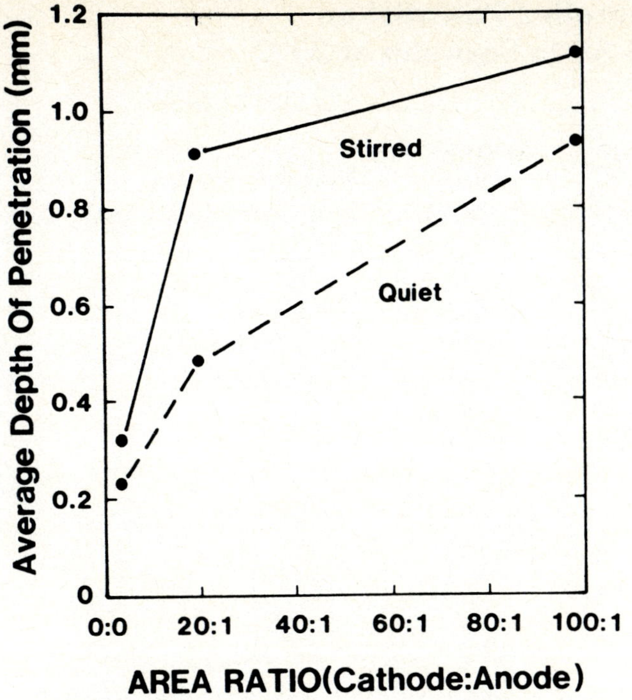

FIGURE 23 – Maximum depth of crevice attack in immersion tests in natural seawater with different cathode to anode area ratios.

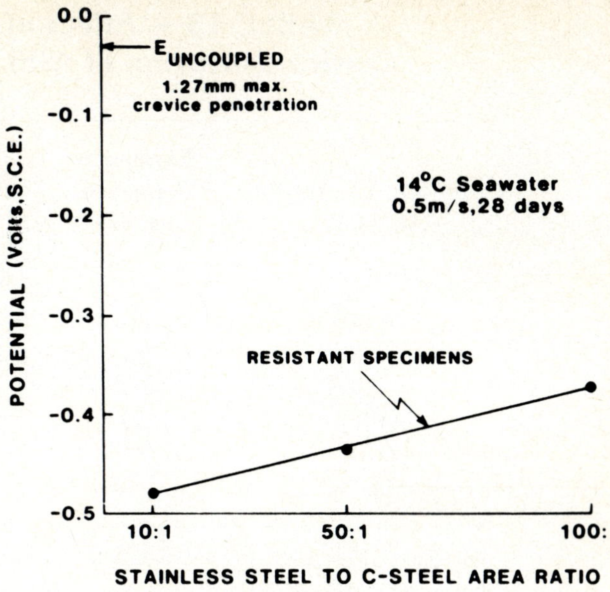

FIGURE 24 – Polarization of SS by coupling with carbon steel.

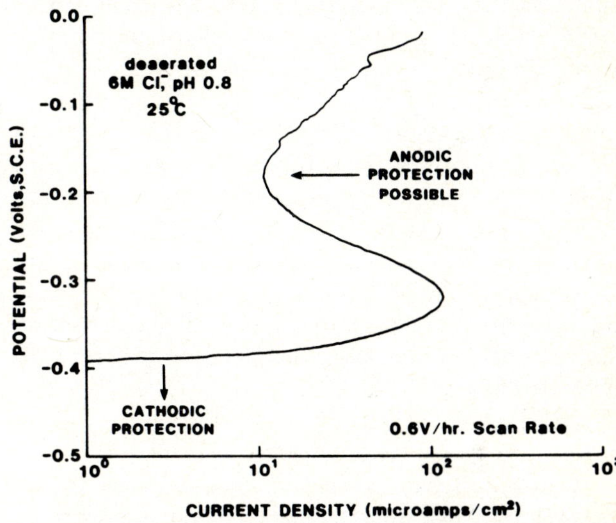

FIGURE 25 – Polarization behavior of AISI 316 SS in simulated crevice solution.

Protection of SS

Galvanic coupling with carbon steel has been previously recognized[42] as a viable means of affording protection to SS. A recent study[43] has quantified the effectiveness of this technique for control of crevice corrosion of AISI 304 and 316 SS in seawater. Carbon steel anodes that are 1/10 and 1/100 the area of the SS surface area have been shown to prevent crevice corrosion of AISI 316 SS in 28-day tests in low velocity seawater at 14 C. Unprotected samples corroded to a depth of up to 1.27 mm. Similar tests conducted in somewhat warmer seawater (i.e., 22 to 28 C) were also successful in controlling crevice corrosion for SS to steel area ratios of 50:1. While some initiation and slight penetration (0.02 mm maximum in 30 days) was measured at the 50:1 area ratio, the extent of penetration was more than two orders of magnitude less than that reported for uncoupled specimens.

Potential data reported[43] for the 14 C seawater test are plotted in Figure 24 as a function of the SS to carbon steel area ratio. Values for the unprotected specimens [−0.030 V (SCE)] approximate potentials reported elsewhere for AISI 316 SS undergoing crevice corrosion.[19,40] As can be seen, coupling with carbon steel resulted in an expected shift to more active potentials. Increasing the SS area reduced the degree of polarization in the active direction. These couple potentials are more active to, or closely approximate, the corrosion potential for AISI 316 SS exposed in deaerated, simulated crevice solutions (25 C) in the pH range 0.8 to 1.5.[19,23]

Figure 25 describes the anodic polarization behavior for AISI 316 SS as determined by potentiodynamic polarization tests in a deaerated, 6 M Cl^-, pH 0.8 solution.[22] As can be seen, any polarization resulting in potentials more negative than about −0.4 V (SCE) would cathodically protect AISI 316 SS in an aggressive crevice solution. Uncoupled material undergoing crevice corrosion could corrode in the active potential regime [i.e., −0.3 to −0.4 V (SCE), approximately] or at potentials more noble than the breakthrough potentials, i.e., −0.15 V (SCE) in this case.

In cases where polarization from coupling with carbon steel is not sufficient to adjust the potential of the SS to the cathodic region, reduction in crevice corrosion propagation is still possible. This could result if the crevice potential is maintained in the passive region or at low currents associated with potentials more active than the anodic peak. This was apparently the case where some slight penetration was observed in the warm seawater tests at the 50:1 SS to carbon steel area ratio.

TABLE 13 — Effect of Controlled Potential on Crevice Corrosion of AISI 304 SS in Seawater at 25 C

Controlled Potential (V, SCE)	Exposure Duration (days)	Average Current[(1)] (A)	Corrosion Rate[(2)] (μm/y)	Visual Evidence of Corrosion
−0.400	30	3.7×10^{-7}	0.5	no
−0.300	31	6.5×10^{-7}	0.9	no
−0.250	33	1.8×10^{-6}	2.5	no
−0.200	30	2.6×10^{-6}	3.7	no
−0.100	21	9.8×10^{-6}	13.9	yes[(3)]

[(1)]Average of 15 crevices during final 5 days of exposure.
[(2)]Based on average current.
[(3)]Based on staining and etched surface.

The preceding is further sponsored by earlier potentiostatic control tests reported for AISI 304 SS in natural seawater. Table 13 summarizes the results of those crevice tests covering the potential range of −0.1 to −0.4 V (SCE) and lasting upwards of 21 to 33 days. Under these conditions, no evidence of corrosion was found on any of the remote crevice assemblies polarized at −0.2 and −0.4 V (SCE). From the previous discussion, cathodic protection was provided at −0.4 V (SCE), while some degree of anodic protection was possible at those potentials corresponding to the passive potential regimes. Visual evidence of corrosion, together with an increase in average corrosion rate, were described for the test conducted at −0.1 V (SCE).

Summary

Crevice corrosion has been shown to be a complex phenomenon that can be treated as being composed of two discrete phases. The initiation and the propagation phases can be influenced by a wide range of factors, which can be classified as metallurgical, environmental, or geometrical in nature. The processes can be described mathematically, which allows convenient assessments of cause and effect relationships. This, in turn, enhances the understanding of factors influencing the processes and aids in the design of appropriate experiments to study crevice corrosion.

Application of relevant test methods allows a characterization of the crevice corrosion process and can yield data of practical engineering utility. The proper interpretation of results from tests that are designed to truly control the critical parameters affecting these results is equally important.

Relevant use and interpretation of results from electrochemical tests employing devices such as the remote crevice assembly and compartmentalized cell can expand understanding of many variables in the crevice corrosion behavior of engineering materials. It is possible to establish the relative effects of environmental factors such as bulk environment chloride level and dissolved oxygen content as well as effects of crevice solution pH on both crevice initiation and propagation. The unique nature of natural seawater in promoting a degree of crevice corrosion an order of magnitude more extensive than synthetic seawater can be clearly established with these techniques. The effect of increasing temperature of natural seawater on reducing the extent of crevice corrosion can also be documented in a unique manner.

A multifaceted approach to investigating crevice corrosion has been described which involves use of mathematical modeling and various compatible experimental methodologies. The continued evolution of this approach holds promise for achieving the ultimate goal of developing the capacity to accurately predict and define alloy crevice corrosion performance in engineering applications.

References

1. A. P. Bond, C. Bertoli, H. J. Dundas, "Corrosion Resistance of Stainless Steels in Seawater," Proc. of Symp. on Adv. Stainless Steels for Seawater Appli., Piacenza, Italy, February 1980.
2. J. R. Maurer, "Development and Application of New High Technology Stainless Alloys for Marine Exposures," Proc. of Symp. on Adv. Stainless Steels for Seawater Appli., Piacenza, Italy, February 1980.
3. B. F. Brown, Corrosion, Vol. 26, No. 8, p. 249, 1970.
4. G. N. Flint, Corrosion y Proteccion, Vol. 8, No. 23, 1977.
5. W. D. France, Localized Corrosion—Cause of Metal Failure, ASTM STP 516, p. 164, ASTM, Philadelphia, Pennsylvania, 1972.
6. J. Postlethwaite, Canadian Metallurgical Quarterly, Vol. 22, No. 1, p. 133, 1983.
7. F. P. Ijsseling, Brit. Corros. J., Vol. 15, No. 2, p. 51, 1980.
8. N. D. Greene, M. G. Fontana, Corrosion, Vol. 15, No. 1, p. 25t, 1959.
9. Z. Szklarska-Smialowska, Corrosion, Vol. 27, No.

6, p. 223, 1971.
10. H. H. Uhlig, Materials Performance, Vol. 22, No. 8, p. 35, 1983.
11. J. W. Oldfield, "Crevice Corrosion of Stainless Steels—The Importance of Crevice Geometry and Alloy Composition," presented at 19th Journees de Aciers Speciaux, Saint-Etienne, May 1980.
12. R. M. Kain, "Localized Corrosion Behavior of Natural Seawater, A Comparison of Electrochemical and Crevice Testing of Stainless Steel," CORROSION/80, Paper No. 74, National Association of Corrosion Engineers, Houston, Texas, 1980.
13. B. E. Wilde, Corrosion, Vol. 28, No. 8, p. 283, 1972.
14. P. E. Morris, Galvanic and Pitting Corrosion, ASTM STP 576, p. 261, ASTM, Philadelphia, Pennsylvania.
15. I. L. Rosenfeld, I. K. Marshakov, Corrosion, Vol. 20, No. 4, p. 115t, 1964.
16. M. G. Fontana, N. D. Greene, Corrosion Engineering, McGraw-Hill Book Co., New York, New York, p. 41, 1967.
17. J. W. Oldfield, W. H. Sutton, Brit. Corros. J., Vol. 13, p. 13, 1978.
18. J. W. Oldfield, W. H. Sutton, Brit. Corros. J., Vol. 13, p. 104, 1978.
19. J. Mankowski, Z. Szklarska-Smialowska, Corros. Sci., Vol. 15, p. 493, 1975.
20. R. M. Kain, T. S. Lee, "Crevice Corrosion of Stainless Steels in Ambient and Elevated Temperature Seawater," Proc. of the 5th Inter. Cong. on Marine Corros. and Fouling, Barcelona, Spain, May 1980.
21. T. S. Lee, R. M. Kain, J. W. Oldfield, Materials Performance, Vol. 23, No. 7, p. 9, 1984.
22. R. M. Kain, T. S. Lee, "Recent Developments in Test Methods for Investigating Crevice Corrosion," presented at the ASTM Inter. Symp. on Lab. Corros. and Standards, Bal Harbour, Florida, November 14-16, 1983.
23. J. W. Oldfield, T. S. Lee, R. M. Kain, Corrosion and Corrosion Protection, Vol. 81-8, Proceedings ESC, p. 213, 1981.
24. T. S. Lee, Electrochemical Corrosion Testing, ASTM STP 727, ASTM, Philadelphia, Pennsylvania, p. 43, 1981.
25. R. M. Kain, Materials Performance, Vol. 23, No. 2, p. 24, 1984.
26. R. M. Kain, T. S. Lee, "The Effect of Crevice Solution pH on Corrosion Behavior of Stainless Alloys," CORROSION/84, Paper No. 27, National Association of Corrosion Engineers, Houston, Texas, 1984.
27. J. R. Crolet, L. Seraphin, R. Tricot, Revere Metallurgie, p. 937, 1975.
28. D. B. Anderson, Galvanic and Pitting Corrosion, ASTM STP 576, ASTM, Philadelphia, Pennsylvania, p. 261.
29. T. S. Lee, R. M. Kain, "Factors Influencing the Crevice Corrosion Behavior of Stainless Steels in Seawater," CORROSION/83, Paper NO. 69, National Association of Corrosion Engineers, Houston, Texas, 1983.
30. J. Degerbeck, E. Wold, Werkst. und Korros., Vol. 25, p. 172, 1974.
31. J. Degerbeck, Werkst. und Korros., Vol. 29, p. 179, 1978.
32. R. M. Kain, T. S. Lee, J. R. Scully, "Crevice Corrosion Resistance of Type 316 Stainless Steels in Marine Environments," presented at 9th Inter. Cong. on Metallic Corros., Toronto, Canada, June 1984.
33. H. P. Hack, Materials Performance, Vol. 22, No. 6, p. 24, 1983.
34. R. M. Kain, "Crevice Corrosion Resistance of Several Iron Base and Nickel Base Cast Stainless Alloys in Seawater," CORROSION/82, Paper No. 66, National Association of Corrosion Engineers, Houston, Texas, 1982.
35. R. M. F. Jones, AFS Transactions, Vol. 81, p. 349, 1973.
36. R. M. Kain, Corrosion, Vol. 40, No. 6, p. 1984.
37. R. M. Kain, A. H. Tuthill, E. C. Hoxie, J. of Materials for Energy Systems, Vol. 5, No. 4, p. 205, 1984.
38. T. S. Lee, K. L. Money, Materials Performance, Vol. 23, No. 8, p. 28, 1984.
39. T. S. Lee, J. W. Oldfield, R. M. Kain, "The Role of Oxygen Reduction in Crevice Corrosion of Stainless Steels," presented at the 162nd ECS Meeting, Detroit, Michigan, October 1982.
40. F. L. LaQue, Marine Corrosion—Causes and Prevention, Chapter 6, Wiley Interscience, p. 179, 1975.
41. N. Azzerri, F. Mancia, A. Tomba, Corros. Sci., Vol. 22, No. 7, 1982.
42. G. D. Moller, "The Successful Use of Austenitic Stainless Steels in Seawater," 8th Offshore Technology Conference, Paper No. OTC 2699, 1976.
43. T. S. Lee, A. H. Tuthill, Materials Performance, Vol. 22, No. 1, p. 48, 1983.

Electrochemical Techniques for the Study of Pitting and Crevice Corrosion Resistance of Stainless Steels

P. LAU* and S. BERNHARDSSON**

Abstract

The resistance to pitting and crevice corrosion of stainless steels was investigated by potentiostatic and potentiodynamic techniques using a recently developed instrument that allows the study of the combined effect of potential and temperature.

The results of the pitting and crevice tests are presented in diagrams relating the critical temperature for corrosion to environmental parameters such as chloride content, pH value, and potential. A comparison with practical experience shows that these results can be used for materials selection.

Introduction

While in recent years advances have been made in using electrochemical testing techniques for the study of pitting and crevice corrosion in stainless steels, there is no general agreement on the use of these techniques. Although the applied testing techniques are easy to perform, they are often time-consuming, especially when temperature is one of the parameters to be tested. The variability in test results is often great, depending on a complex experimental situation where several variables are to be controlled simultaneously. In this paper, the applicability of some test techniques is demonstrated. A recently developed instrument is used.

Experimental

Instrument Concept

The instrumentation used in the experiments is shown in Figure 1. The Santron EMS[1] instrument is a potentiostat/galvanostat that can control the potential, current, temperature, and speed of a rotating electrode. Based on microprocessor techniques, experiments are conducted under software control. The application software built into the instrument contains several preprogrammed test methods and pertinent evaluations. The instrument concept is illustrated in Figure 2.

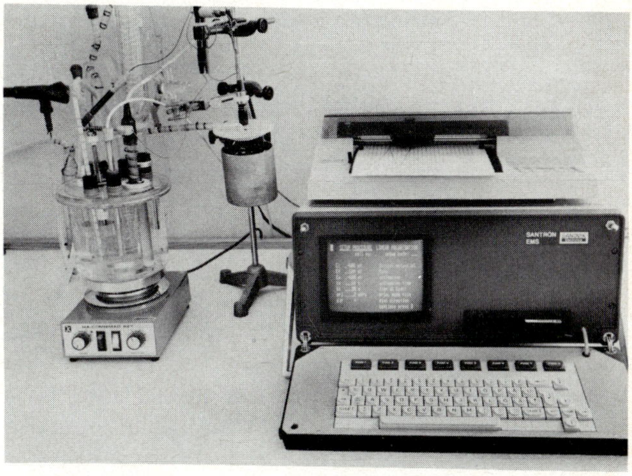

FIGURE 1 – Santron electrochemical measurement system (EMS) with measuring cell and plotter.

Measuring data are stored in random access memory (RAM) on board of each functional module for

*AB Sandvik Electronics, Sweden.
**AB Sandvik Steel, Sweden.

(1)Registered trade name.

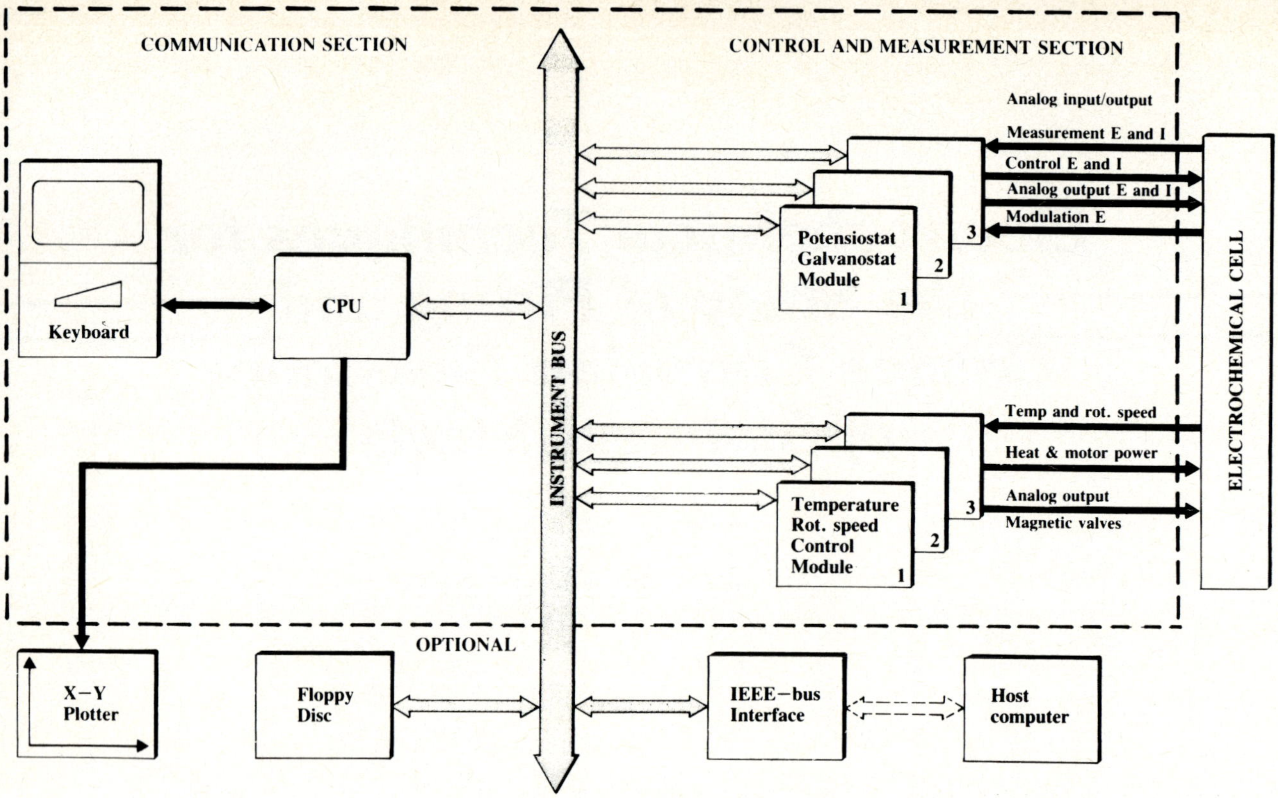

FIGURE 2 — The Santron EMS hardware block diagram. Using one central processing unit (CPU), an instrument bus and six processor based control modules connected to it, up to three cells are controllable independently.

access both during and after the experiment or can even be stored long-term on a built-in floppy disc. This renders the possibility of automatically repeating a test sequence with just one controlled parameter changed.

Other instrument features that result from microprocessor technology are: a real-time IR-compensation, an auto-ranging facility for current measurements, a signal-dependent auto-density storage, and an auto-calibration with correction with respect to all links in the measuring chain. All preprogrammed instrument actions are menu-driven and require no knowledge of programming.

Test Results

The Santron EMS and its predecessor, the Santron CTD 400,(2) have been and can be used for a number of different experiments. Some of them are exemplified in this paper.

Critical pitting temperature (CPT). The CPT concept was introduced by Brigham[1] and is increasingly used today for ranking materials as well as for materials selection. The test procedure is described in Figure 3. After immersion and temperature adjustment, a constant potential is applied to the test specimen. The cell current is monitored and used as a criterion for pitting occurrence.

If the current level that indicates pitting is not reached within the testing time, an increased temperature is applied and the polarization sequence is repeated. The experiment is terminated when, at a certain temperature, the criterion level is exceeded during a time greater than the delay time. The critical temperature has then been determined.

Figure 4 shows a temperature (time)-current chart obtained for a stainless steel (SS). The test specimen was cylindrical and had a diameter of 10 mm and a total area of 7.1 cm^2. It was placed vertically in the test solution. To avoid crevice attack at the mounting surface a polytetrafluoroethylene (PTFE) collar was used; during the experiment, the collar was filled with a gas (air, argon, or nitrogen). The specimen surface was ground with paper 600 grit at least 24 hours before the test. The test solution contained NaCl p. a. in distilled water. The pH value varied between 5.3 and 6.0.

In Figure 5, the resulting critical pitting temperatures for AISI 316 SS (17% Cr, 13% Ni, 2.7% Mo) are plotted vs the chloride content (logarithmic scale). Each point represents the average of six measurements. The critical pitting temperature decreases with increasing chloride content and with increasing test potential.

For correlation with practical experience, a large number of cases where heat exchanger tubes of AISI 316 SS failed were documented and compared with the CPT measurements at different potentials (Figure 6). Each point represents a reported failure resulting from pitting. The laboratory data at +300 mV SCE agreed best with the practical experience.

The pH dependence of CPT for AISI 316 SS was determined in an aqueous NaCl solution with 0.2% Cl^-. The pH was adjusted using HCl and NaOH. The results

(2)Registered trade name.

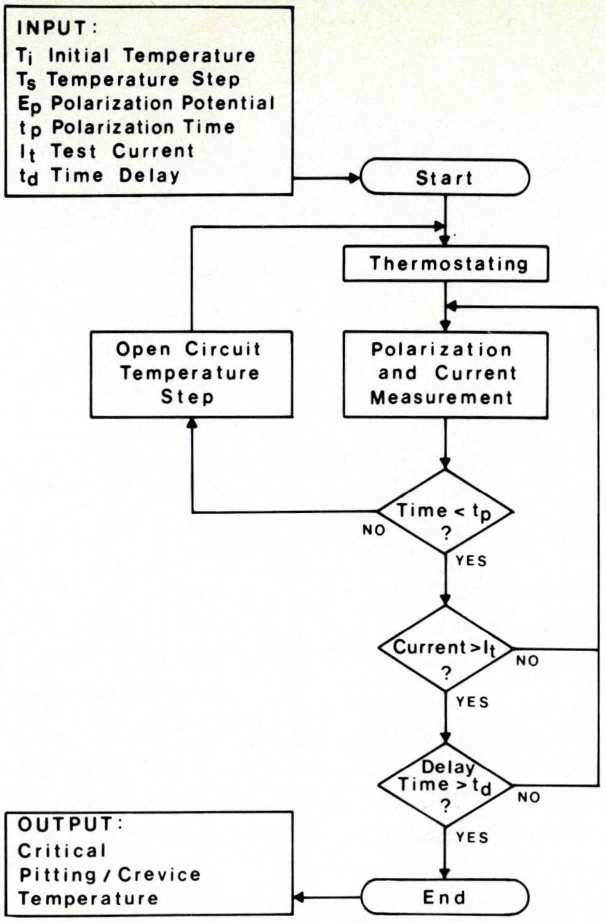

FIGURE 3 – Flow chart for CPT measurements. All control parameters can be chosen freely. Typical values are: Ti = 20 C; T_s = +5 C; E_p = 200, 300, 400 mV SCE; t_p = 15 min: I_c = 10 μA/cm²; t_d = 20 s.

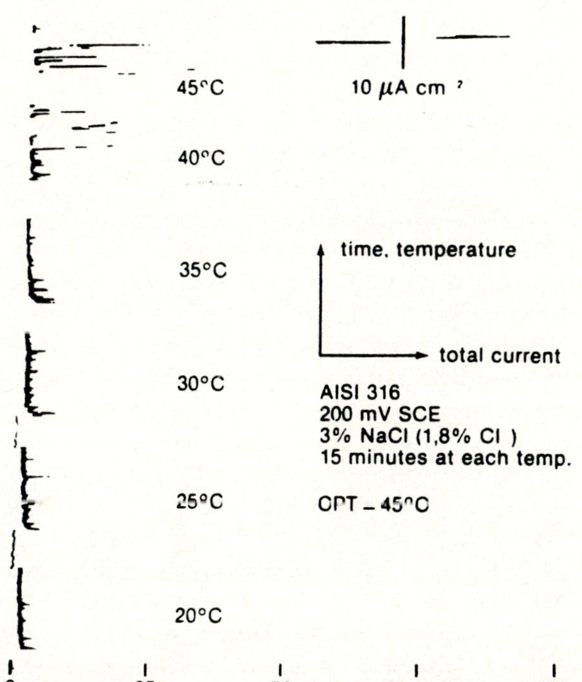

FIGURE 4 – A temperature-current graph from a CPT test.

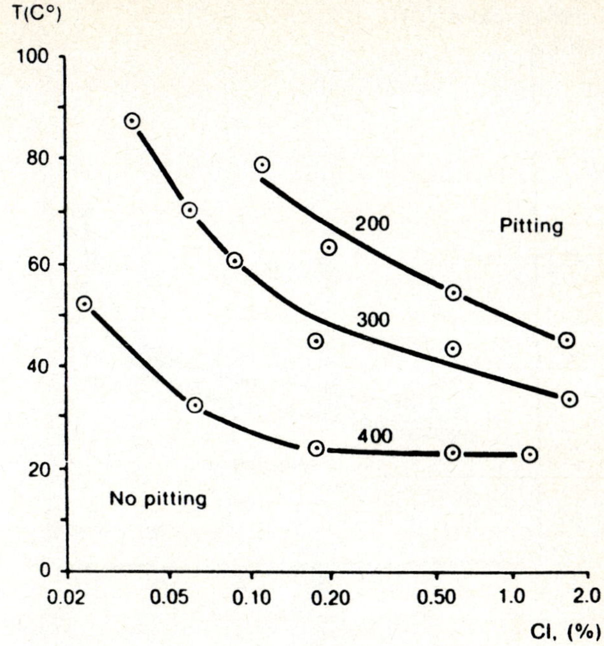

FIGURE 5 – CPT of AISI 316 SS at three different polarizing potentials. Each point represents the mean value of six measurements.

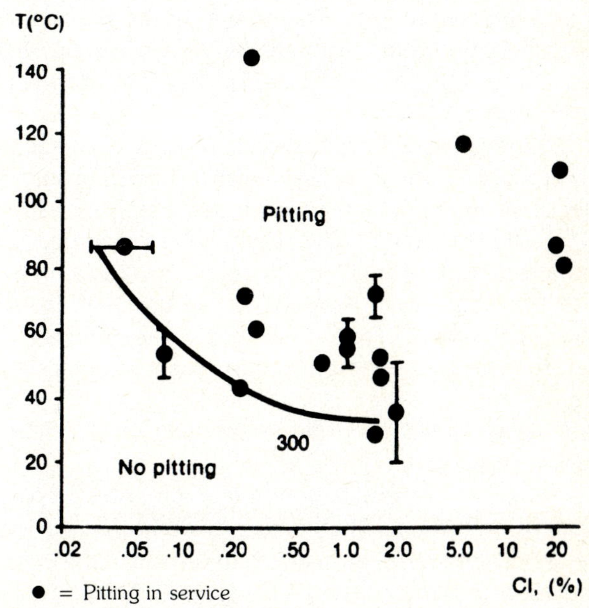

FIGURE 6 – Practical experience of AISI 316 SS compared with CPT measurements at 300 mV SCE.

at a test potential of +400 mV SCE are shown in Figure 7. The pH dependence is rather weak below pH = 10, but then increases markedly.

Critical crevice temperature (CCT). For the determination of crevice temperatures, the same accelerated laboratory method was used. The test was conducted in an aqueous solution with 3% NaCl and a pH of about 6. The crevices here were formed using square 10 × 10 mm polyacrylate spacers. The test comprised the following alloys: Alloy 825 (21.5% Cr, 42% Ni, 3% Mo,

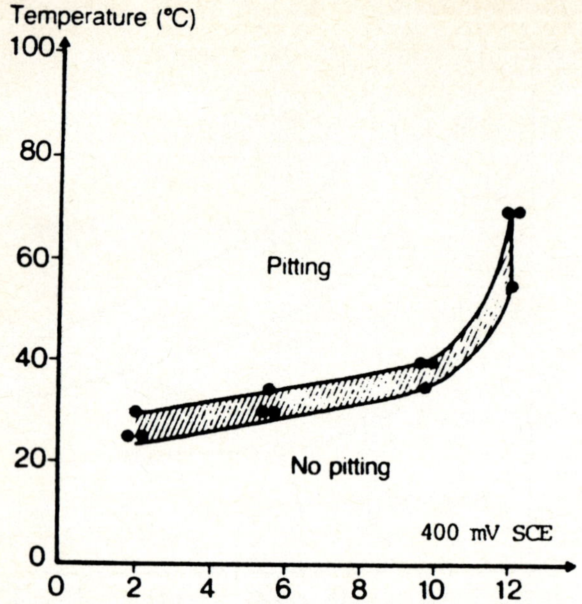

FIGURE 7 – pH dependence of CPT.

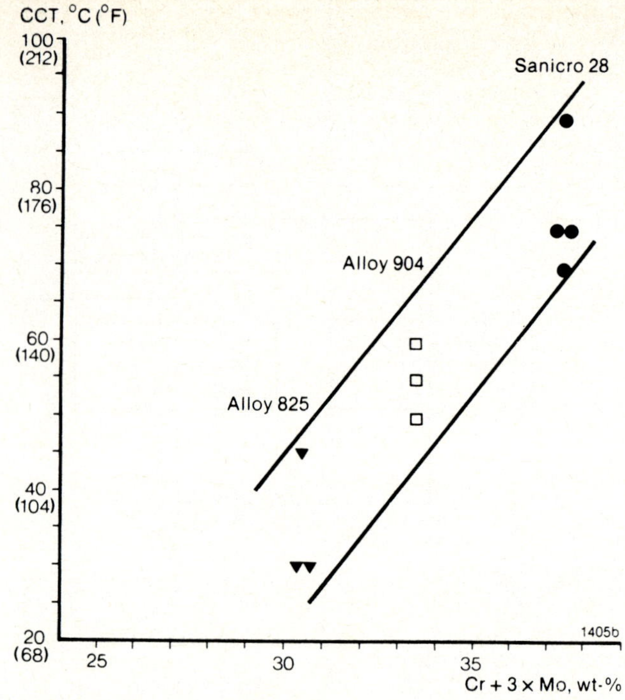

FIGURE 8 – Dependence of CCT on the Cr + Mo contents in the alloys: Polarization potential 200 mV SCE, testing time 15 min at each temperature, 3% NaCl, pH about 6.

Ti), Alloy 904L (20% Cr, 25% Ni, 4.5% Mo, 1.5% Cu), and Sanicro 28[(3)] (27% Cr, 31% Ni, 3.5% Mo, 1% Cu).

The result of the test is illustrated in Figure 8. The scatter indicated by the lines is probably due to less control of the crevice gap. The ranking of the alloys in this test agrees well with the parameter %Cr + 3%Mo, which is a measure of the pitting and crevice corrosion resistance of an alloy.[2]

At the University of Delaware, a large research program was performed testing crevice corrosion of stainless steels.[3] A number of alloys were tested, using the Santron CTD 400 instrument. The results were compared with immersion tests at 30 C for 30 days in seawater. In both tests, multiple crevice assemblies were used and the surface preparation of the test specimens was identical. The Santron tests were performed in synthetic seawater. The applied potential was set to 600 mV SCE, the critical current was 1000 μA, and the testing time was 20 min at each temperature.

Some of the results from the comparison are summarized in Table 1. The number of observed attacks in the 30-day test ranks the alloys with respect to the crevice corrosion resistance. The CCT values in Table 1 represent the average of at least 4 separate measurements. The maximum scatter was $\pm$2.5 C. The long and short term tests give a similar ranking.

Critical pH-values at various temperatures. The concept of a critical pH was introduced by Crolet[4]; it is used today to predict the behavior of an alloy in an acid crevice solution. The appearance of an anodic current peak in the dynamic polarization curve indicates the possibility of activating the alloy in that solution. The lower the pH necessary to induce an active peak, the more resistant the alloy is to crevice corrosion.

In these experiments, the test temperature as well as the pH of the test solution were varied. This alteration is done automatically to reduce the total testing time.

Normally, three polarization scans at three temperatures were performed before test piece and solution were renewed. The use of the same specimen can be justified since at low temperatures there is only a minor aging effect.

The following stainless steels were tested: AISI 316 SS, two heats of Sanicro 28, and one experimental alloy with a composition similar to Sanicro 28 (experimental alloy). All materials were tested in the quench-annealed condition, with the chosen annealing temperature being appropriate for every alloy. The specimens had a circular area of 1.1 cm^2 and were diamond polished.

The tests were conducted in aqueous 0.5 M chloride solution made from distilled and demineralized water and NaCl (p.a.). The initial pH was adjusted using HCl (p.a.) and varied between 0.3 and 3.8. Each scan started with 5-min activation at -600 mV SCE.

Figure 9 shows a typical polarization curve. The corrected anodic part is given by the dotted curve. Correction is simply performed by adding and subtracting, respectively, the Tafel line (dashed line) to the apparent anodic curve (continuous curve). The Tafel slope is calculated using the least square fitting method. Both calculations are performed after the experiment with data stored on the built-in floppy disc using one of the software options.

The maximum anodic current density I_{max} was evaluated as a function of the temperature and the pH value. The results are plotted in Figure 10. The scan sequence was repeated with three separate specimens. As shown for the experimental alloy, I_{max} increases with

[(3)]Registered trade name.

TABLE 1 — Crevice Corrosion Tests in Comparison

Alloy	Initiation Measuremements Navy tests(1)				Santron tests(2)		
	Sites	Rank	Sides	Rank	Failure Temperature C	F	Rank
A.L. 29-4C	0	1	0	1	90.0	194	1
Monit	0	2	0	2	67.5	153	2
Crucible SC-1	1	3	1	3	60.0	140	5
Ferralium 255	2	4	2	4	60.0	140	4
Haynes 20 Mod	6	5	2	5	47.5	117	7
A.L. 6X	11	6	4	6	57.5	135	6
254 SMO	18	7	5	7	62.5	144	3
904L	36	8	5	8	42.5	108	9
Jessop 700	47	9	5	9	45.0	113	8
Jessop 777	60	10	6	10	30.0	86	12
Carpenter 329	73	11	6	11	40.0	104	11
Nitronic 50	112	12	6	12	40.0	104	10

(1)Filtered seawater 30 C, 30-day tests, 3 panels, multiple crevice assembly: 20 serrations per side, 1.3716 cm^2 effective crevice area, 1:220 crevice-bold area ratio, 75 in.-lb torque.

(2)Synthetic seawater, 600 mV SCE applied potential, 1000 μA critical current, 20 min measuring time, multiple crevice assembly: 12 serrations per side, 0.6638 cm^2 effective crevice area, 1:40 crevice-bold area ratio, 75 in.-lb torque.

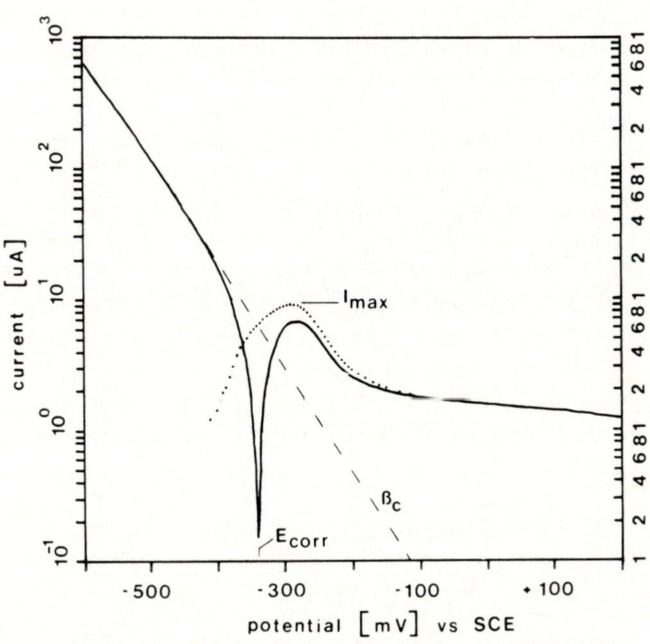

FIGURE 9 — Polarization curve for an experimental alloy. 0.5 M Cl^-, pH = 0.9, temperature = 15 C, scan rate = 20 mV/min, activation: 5 min at −600 mV SCE.

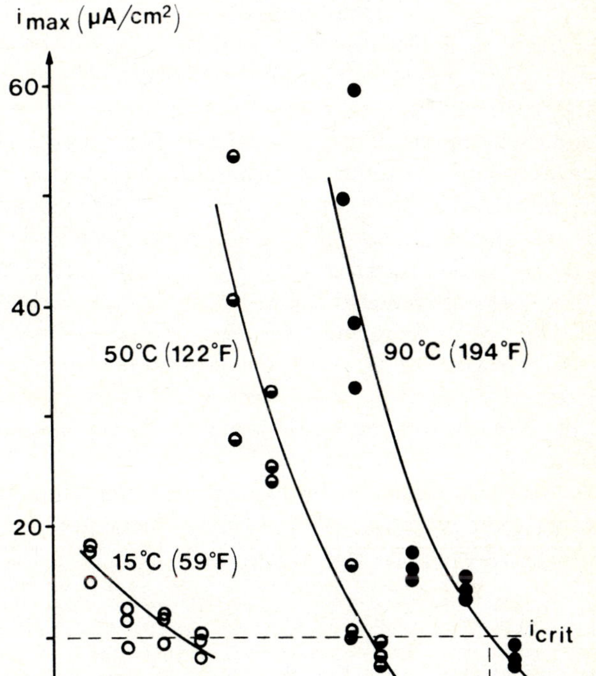

FIGURE 10 — pH dependency of I_{max} for an experimental alloy at three different temperatures.

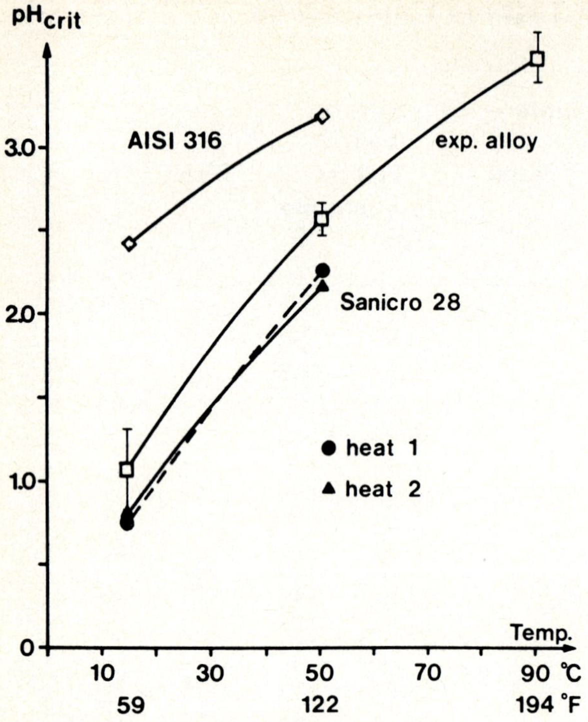

FIGURE 11 — Temperature dependence of the critical pH value for three steel grades.

increasing temperature and decreasing pH.

Figure 11 displays the temperature dependence of the critical pH values for three different alloys. For the occurrence of crevice corrosion, a critical current density level of 10 $\mu A/cm^2$ was chosen.

Discussion

Since CPT test results at +300 mV SCE show a good agreement with the practical experience of stainless steels, such data have been used to make materials recommendations for the industry. Test data exist for a number of alloys[5] and the effect of environmental parameters such as pH, as well as the contents of sulfate, phosphate, nitrate, and carbonate has been investigated. The materials recommendations have been very successful with no reported material failure.

The results from the CCT measurements show a good correlation to long-term seawater immersion tests. Replacing the immersion tests with CCT measurements of this kind means a drastic reduction in time and cost for the testing.

Figure 11 clearly shows that an increased temperature will increase the sensitivity to crevice corrosion.

The results presented in Figures 10 and 11 are difficult to correlate with practical performance. Combined with calculation of pH changes in crevices[6] at different temperatures, it would be possible to use these data to effectively rank these materials. Without such calculations, the ranking obtained is still of value.

It has been estimated that by using the automatized technique provided by the Santron EMS instrument, the time for measuring and evaluating polarization curves at different temperatures is reduced by a factor of 3 to 5. Additional benefits are reduced manual labor, automatic documentation, and ready-for-report output of the test results.

Conclusions

From the experimental results, it is obvious that temperature is an important environmental parameter that may strongly affect crevice and pitting corrosion. The comparison between laboratory results and practical experience and long-term tests, respectively, shows that the electrochemical measurements of CPT and CCT have several advantages. These parameters are easy to handle and provide valuable information in the process of material selection. These test methods can reduce both experimental time and the total amount of manual labor quite effectively. By using an instrument like the Santron EMS, not only can the accuracy in controlling and measuring be increased, but also the precision and efficiency in the evaluation can be improved.

References

1. R. I. Brigham, Corrosion, Vol. 28, No. 5, p. 177, 1972.
2. E.-H. Horn, D. Kuron, H. Grafen, Werkst. u. Korros., Vol. 30, p. 723, 1979.
3. N. S. Nagaswami, M. A. Streicher, "Comparison of Accelerated Laboratory Tests for Crevice Corrosion of Stainless Alloys," CORROSION/83, Paper No. 71, National Association of Corrosion Engineers, Houston, Texas, 1983.
4. J. L. Crolet, L. Seraphin, R. Tricot, Revue Metallurgie, p. 937, 1975.
5. S. Bernhardsson, R. Mellstrom, B. Brox, "Limiting Conditions with Regard to Pitting of Stainless Steels," 8th Int. Cong. on Metallic Corros., Mainz, Federal Republic of Germany, 1981.
6. S. Bernhardsson, L. Ericsson, J. Oppelstrup, I. Puigdomenech, T. Wallin, "Crevice Corrosion of Stainless Steels, Calculation of Concentration and pH Changes," 8th Int. Cong. on Metallic Corros., Mainz, Federal Republic of Germany, 1981.

Measurement of Pit Initiation and Propagation for Iron-Chromium-Nickel Alloys in Acid Chloride Environments by Electrochemical Techniques

P. E. MORRIS*

Abstract

The resistance of 24 Fe-Cr-Ni alloys to pitting corrosion was investigated in 26 nonoxidizing acid chloride environments at 60 C. Potentiodynamic polarization experiments were conducted at 60 V/h and corrosion potential vs time curves were determined in separate experiments after a brief cathodic pretreatment. Results from the electrochemical experiments were compared with results from 4- or 11-month immersion tests for 13 of the alloys in seven of the environments. Correlations were found between four electrochemical parameters and resistance to pit initiation and propagation. These correlations are illustrated in this paper for representative systems and they are discussed with regard to conditions that may exist on an alloy surface when pitting corrosion occurs.

Introduction

Electrochemical techniques have been widely used to determine pitting potentials (E_P) for Fe-Cr-Ni alloys in chloride-containing environments. Most of these studies have either ranked several materials according to E_P values[1-4] or have pointed out that variations in E_P values may occur with variations in the experimental procedure.[5-12] Few investigators have used electrochemical parameters other than E_P in an effort to correlate electrochemical behavior with resistance to pitting.[11,13] None of the studies that have been reported offer evidence that resistance to pitting corrosion can actually be predicted from electrochemical information.

The purpose of this study was to determine whether electrochemical techniques could be used to predict resistance to pit initiation and propagation for a wide range of Fe-Cr-Ni-Mo alloy compositions and for a wide range of nonoxidizing acid chloride environments. The results for representative systems show that certain relationships between four electrochemical parameters could be used to predict resistance to pit initiation. It was also found that other relationships between these four parameters could be used to predict resistance to pit propagation. A flow chart is presented to summarize the experimental procedures that were used in the study. The results are discussed in terms of a mechanism based on active and passive corrosion behavior, which may occur simultaneously on an alloy surface during pitting corrosion.

Experimental

An overview of the study is provided by the chart in Figure 1. A detailed analysis of the experimental results is presented in this paper for 5 of the 24 alloys and for 6 of the solutions as noted by the bottom rectangle in Figure 1. This sample population, which was limited to five alloys, provided examples of each category of behavior that was observed. The chemical analyses are given for the five alloys in Table 1 and a summary of the entire 26 test environments is given in Table 2.

Electrochemical test electrodes were prepared from 25.4 × 12.7-mm rectangular strips that were partially immersed to expose a known surface area, 50 to 80 mm^2. This partial immersion procedure was found to minimize interference from crevices, compared to that which occurred when a compression gasket-mounting procedure was used. All test electrodes were manually ground on dry 120-grit paper, washed and degreased

*International Paper Company, Mobile, Alabama.

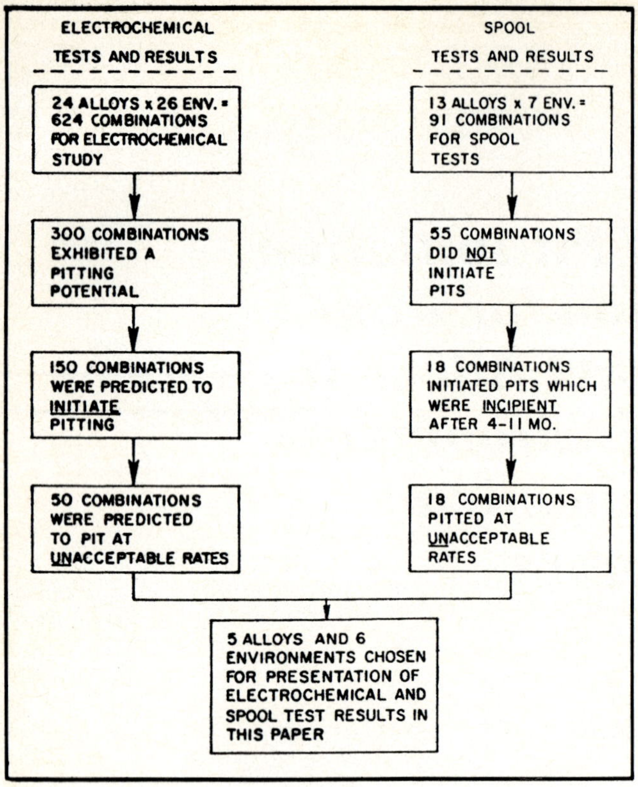

FIGURE 1 — Overview of the entire study.

within 5 minutes of each experiment. This surface preparation procedure was necessary to obtain reproducible electrochemical results.

Values for E_P, were determined from corrosion behavior diagrams (CBDs), which were obtained according to a previously published procedure.[14,15] Briefly, potentials were scanned at 60 V/h from approximately −1.40 V (all potentials were measured vs SCE) to a potential in the anodic current range that corresponded to an anodic current increase of one decade above the minimum passive current density. For systems that did not exhibit an E_P, this potential occurred in the oxygen evolution region (compare anodic curve segments A, B, and C in Figure 2). After reaching the maximum noble potential, the scan direction was immediately reversed, potentials were scanned back to the starting potential, and the potentiostat was then turned off. After 10 minutes, the corrosion potential (E_{CORR}) was recorded and a final scan was completed from E_{CORR} to the CBD starting potential at −1.40 V. The final cathodic polarization scan (which is not shown in Figure 2) was not used for the analyses of pitting resistance.

Corrosion potentials were also recorded in separate experiments that consisted of a 10-s cathodic charge at −1.4 V and subsequently monitoring E_{CORR} for either 1 or 24 hours

Immersion tests were performed on standard spool-type specimens. Circular specimens 56.7 mm in diameter with an 8.7-mm diameter hole in the center were mounted on a 7.9-mm diameter polytetrafluoroethylene (PTFE) insulated rod. Specimen thicknesses ranged from 1.3 to 1.6 mm. The 6-mm wide PTFE spacers that are normally used between specimens in this arrangement were omitted to minimize the changes for crevice corrosion, since the study focused on resistance to pitting. The 13 specimens were spaced approximately 6 mm apart on each rack and completely immersed in the test solutions. The spool specimens were tested with an as-received, i.e., mill-produced surface condition after being washed and degreased. Although the surface finish varied from alloy to alloy, they are considered to be realistic variations for usual field exposures. The immersion tests were conducted for either 4 or 11 months at 60 C.

After testing for either 4 or 11 months, the surfaces were examined first at 10X magnification using a binocular and then at magnifications up to 1000X using a scanning electron microscope (SEM). Pits that were just visible at 10X magnification were described as incipient (I) (after being confirmed to be a pit). Larger pits were characterized by penetration rates in mm/month, after measurement with a depth gage. If no evidence of pit initiation could be found with the SEM, then these specimens were characterized by (N). These results are given in Tables 3 and 4.

Results

Summary of Electrochemical Behavior

Three categories of anodic polarization behavior and six categories of E_{CORR} vs time behavior were observed as illustrated by the schematic diagrams in Figure 2. In the

TABLE 1 — Chemical Analyses for the Alloys Discussed in this Paper (wt%)

Alloy	Ni	Cr	Fe	Mo	C	Cu	Mn	S	Cb
AISI 304L	10	19	Bal	—	0.03	—	—	0.03	—
AISI 316L	12	17	Bal	2.5	0.03	—	2.0	0.03	—
Alloy 825	42	20	Bal	3.0	0.03	2.0	0.7	0.007	—
AISI 317L	13	19	Bal	3.5	0.03	—	2.0	0.03	—
Alloy 625	Bal	22	3.0	9.0	0.05	0.04	0.20	0.007	3.5

TABLE 2 — Summary of the Twenty-Six Test Environments Used for Electrochemical Testing

pH	Temperature (C)	Chloride Concentration (ppm)	Sulfate Concentrations (ppm)						
			<1 × 10⁰	1 × 10¹	1 × 10²	1 × 10³	1 × 10⁴	1 × 10⁵	Saturated
2	60	≤1 × 10⁰				X			
2	45	≤1 × 10⁰				X			
2	60	1 × 10¹				X			
2	60	1 × 10²				X			
2	60	1 × 10³	X (1)			X (1)			
2	60	1 × 10⁴	X (1)	X	X	X (1)	X (1)	X	X (1)
2	45	1 × 10⁴	X			X			
2	60	Saturated	X			X			
4	60	≤1 × 10⁰		X					
4	60	1 × 10¹		X					
4	60	1 × 10²		X					
4	60	1 × 10³	X	X					
4	60	1 × 10⁴	X (1)	X					
4	60	Saturated	X	X					

X Denotes that this combination was used for electrochemical testing only.
(1) Denotes that immersion tests were also conducted in this solution.

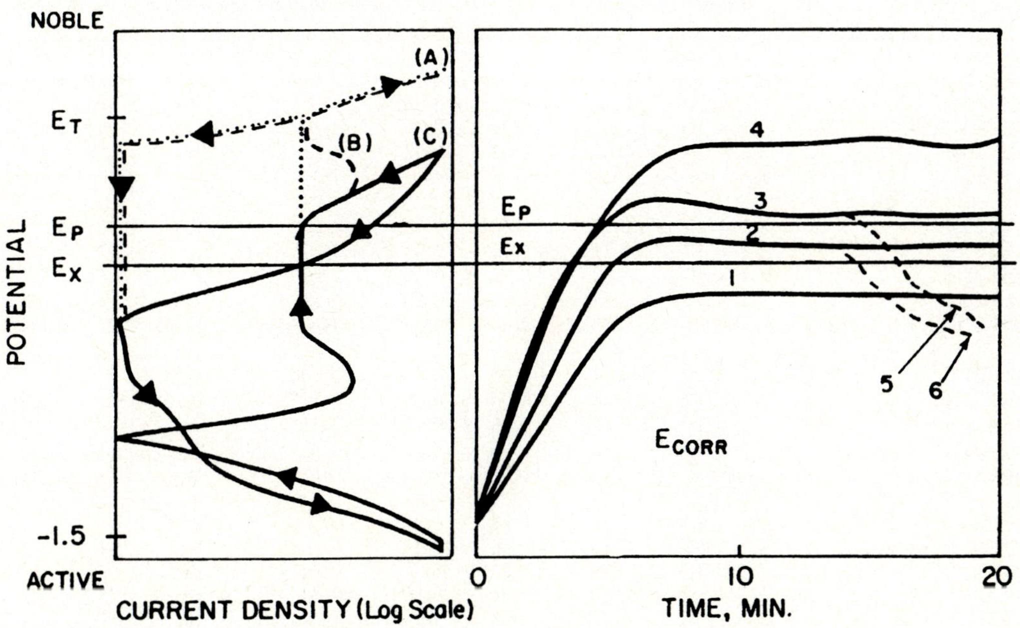

FIGURE 2 — Schematics of anodic polarization and E_{CORR} behavior variations that were observed.

left diagram, it can be seen that either: (1) no E_P occurred (see dotted line segment, A, which corresponds to oxygen evolution above the transpassive potential, E_T); (2) an E_P occurred with subsequent repassivation and normal transpassive behavior (see dashed curve segment, B); or (3) an E_P occurred and pits propagated at potentials noble to E_P (see solid curve segment, C). In case C, a crossover potential, E_X, is defined at the point where the forward and return curves cross. The occurrence of E_X means that a hysteresis "loop" was observed between the forward and return curves.

The values for E_{CORR} shifted rapidly into the noble direction immediately following the 10-s cathodic charge as illustrated by the right diagram in Figure 2. After ap-

TABLE 3 — Summary of Electrochemical and Immersion Test Results in 1 × 10³ ppm Chloride Solutions at 60 C

Alloy	pH	Sulfate Conc. (ppm)	E_{CORR} (V)	E_P (V)	E_X (V)	$\Delta\dot{E}/\Delta i$ (V/decade)	Pit Initiation Predicted?	Pit Propagation Rate (mm/mo)[1]
AISI 304L	2	$\leq 1 \times 10^0$	−0.09	−0.09	0.10	0.20	Yes	I
AISI 316L	2	$\leq 1 \times 10^0$	0.05	−0.04	0.34	0.20	Yes	I
INCOLOY alloy 825	2	$\leq 1 \times 10^0$	0.38	0.08	0.50	0.25	Yes	N
AISI 317L	2	$\leq 1 \times 10^0$	0.32	0.07	0.30	0.34	Yes	N
INCONEL alloy 625	2	$\leq 1 \times 10^0$	−0.17	None	NA	NA	No	N
AISI 304L	2	1×10^3	0.12	0.20	0.06	0.17	Yes	I
AISI 316L	2	1×10^3	0.20	0.24	0.18	0.15	Yes	0.06
INCOLOY alloy 825	2	1×10^3	−0.10	0.36	0.32	0.16	No	N
AISI 317L	2	1×10^3	−0.16	0.22	0.20	0.19	No	N
INCONEL alloy 625	2	1×10^3	−0.24	None	NA	NA	No	N

[1]Determined from 4- or 11-month immersion tests:

Key: N = no pit initiation, I = incipient pitting, and NA = not applicable.

TABLE 4 — Summary of Electrochemical and Immersion Test Results in 1 × 10⁴ ppm Chloride Solutions at 60 C

Alloy	pH	Sulfate Conc. (ppm)	E_{CORR} (V)	E_P (V)	E_X (V)	$\Delta E/\Delta i$ (V/decade)	Pit Initiation Predicted?	Pit Propagation Rate (mm/mo)[1]
AISI 304	2	$\leq 1 \times 10^0$	−0.11	−0.16	−0.05	0.15	Yes	0.14
AISI 304	4	$\leq 1 \times 10^0$	−0.14	−0.08	−0.08	0.09	No	N
AISI 304	2	1×10^3	−0.06	−0.06	0	0.16	Yes	>0.32 (P)
AISI 304	2	1×10^4	0.02	0.02	0.02	0.15	Yes	>0.11 (P)
AISI 304	2	Saturated	0.22	None	NA	NA	No	N
AISI 316L	2	$\leq 1 \times 10^0$	−0.14	−0.20	−0.06	0.15	Yes	0.02
AISI 316L	4	$\leq 1 \times 10^0$	−0.11	−0.06	−0.04	0.14	No	N
AISI 316L	2	1×10^3	0.03	−0.02	0.08	0.16	Yes	0.11
AISI 316L	2	1×10^4	0.18	0.18	0.12	0.12	Yes	>0.11 (P)
AISI 316L	2	Saturated	0.15	None	NA	NA	No	N
AISI 317L	2	$\leq 1 \times 10^0$	−0.01	−0.06	0.04	0.30	Yes	I
AISI 317L	4	$\leq 1 \times 10^0$	−0.13	0.01	0.08	0.25	No	N
AISI 317L	2	1×10^3	0.25	0.15	0.15	0.12	Yes	0.15
AISI 317L	2	1×10^4	0.32	0.26	0.16	0.13	Yes	0.12
AISI 317L	2	Saturated	0.10	None	NA	NA	No	N
INCOLOY alloy 825	2	$\leq 1 \times 10^0$	0.35	0.04	0.34	0.19	Yes	N
INCOLOY alloy 825	4	$\leq 1 \times 10^0$	−0.10	0.06	0.12	0.18	No	N
INCOLOY alloy 825	2	1×10^3	0.28	0.11	0.34	0.24	Yes	I
INCOLOY alloy 825	2	1×10^4	0.39	0.30	0.30	0.20	Yes	N
INCOLOY alloy 825	2	Saturated	0.23	None	NA	NA	No	N

[1]Determined from 4- or 11-month immersion tests:

Key: N = no pit initiation, I = incipient pitting, (P) = perforation, and NA = not applicable.

proximately 5 minutes, E_{CORR} achieved a value that was either: (1) active to E_X and E_P (see curve 1); (2) between E_X and E_P, where E_X is less than E_P (see curve 2); (3) noble to E_P, but near E_P (see curve 3); or (4) much more noble than E_P (see curve 4). In yet other cases, after approximately 10 minutes, the E_{CORR} values shifted again and out of the regions defined by (2) and (3) above, back to values active to E_P and E_X (see curve segments 5 and 6, respectively in Figure 2). In all of the above six cases, the most noble value that occurred for E_{CORR} was recorded and used for the analyses.

In cases where pits propagated at potentials noble to E_P during the CBD experiment (e.g., curve segment C in Figure 2), these anodic curves were examined to deter-

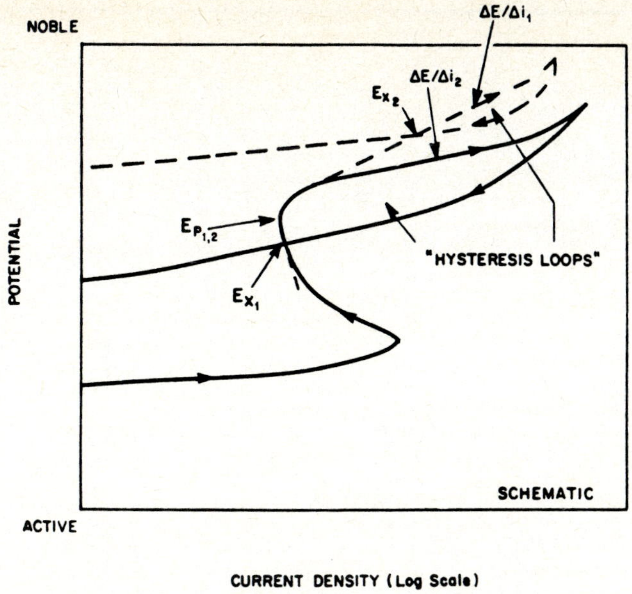

FIGURE 3 – Schematic of the anodic polarization parameters recorded for two systems that exhibit identical E_P values, but different values for E_X and $\triangle E/\triangle i$, respectively.

mine the three following parameters as illustrated in Figure 3: (1) E_P; (2) the slope $\triangle E/\triangle i$ (V/decade) of the linear segment; and (3) the potential at which the forward and return scan anodic curves cross, as defined by E_X. This information is listed in Tables 3 and 4. In many cases, systems had similar values for E_P but had large variations in the slope or E_X or both.

During the course of this study, the following variations were observed for the electrochemical parameters: (1) E_P varied from -0.20 to $+0.60$ V; (2) E_X varied from -0.20 to $+0.60$ V; and (3) E_X varied from (E_P − 0.20 V) to (E_P + 0.40 V).

Resistance in Pit Initiation

An examination of the anodic polarization and E_{CORR} vs time behavior patterns and their interdependences revealed correlations between the electrochemical test results and resistance to pit initiation. The complete procedure that was used for obtaining the measurements and analyzing them is outlined in Figure 4. The results for the five alloys are listed in Tables 3 and 4. First, no evidence of pit initiation was found for any system that did not exhibit an E_P. Secondly, systems that did exhibit an E_P, but where E_{CORR} never achieved a value as noble as that for E_P or E_X, exhibited no evidence of pit initiation.

Systems that exhibited an E_P as illustrated by curve C in Figure 2 and values of E_{CORR} that were equal to or noble to either E_P or E_X, were subject to pit initiation. However, the degree to which pits propagated for these systems was correlated with additional electrochemical parameters as discussed in the next section. Systems that exhibited E_P but subsequently repassivated (see curve B, Figure 2), and values of E_{CORR} that were equal to or noble to either E_P or E_X exhibited either incipient pitting or no evidence of pit initiation after 4 or 11 months of exposure.

Resistance to Pit Propagation

The primary indicator of resistance to pit propagation was found to be the value for $\triangle E/\triangle i$. This correlation is illustrated in Figure 5 by the 24 example systems that exhibited an E_P. Systems with $\triangle E/\triangle i$ values greater than 0.18 V/decade either exhibited no evidence of pitting or only incipient pitting after 4 or 11 months of exposure (note Zone III in Figure 5). Systems with the critical $\triangle E/\triangle i$ values of 0.17 to 0.18 V/decade may or may not pit at high rates of propagation (note Zone II in Figure 5). Values of $\triangle E/\triangle i$ less than 0.17 V/decade consistently exhibited high rates of pit propagation (note Zone I in Figure 5). It should be pointed out again that systems with $\triangle E/\triangle i$ values less than 0.18 V/decade did not pit if E_{CORR} was less than E_P or E_X, as denoted by the solid symbols in Zone I, i.e., when pit initiation was predicted not to occur in the first place.

Correlations were found between the resistance to pit propagation and certain relationships between E_{CORR} and E_P, as illustrated in Figure 6. All of the systems that exhibited poor resistance to pit propagation also exhibited E_{CORR} values that were between E_P and (E_P + 0.10 V), as indicated by the solid symbols contained within Zone II. The two data points in Zone I denoted by an asterisk represent systems for which E_X should be substituted for E_P in the comparison with E_{CORR} since E_X is active to E_P. All of the systems represented by data points in Zone III, i.e., for E_{CORR} greater than (E_P + 0.10 V) also exhibited $\triangle E/\triangle i$ values greater than 0.18 V/decade and good resistance to pit propagation. According to this analysis of E_{CORR}, systems that exhibited poor resistance exhibited a value for E_{CORR} between E_P (or E_X if E_X is less than E_P) and (E_P + 0.10 V). However, meeting this condition alone was not necessarily an indicator of poor resistance (note the open symbols in Zone II of Figure 6).

The relationship between E_X and resistance to pit propagation was investigated, using the pariety diagram in Figure 7. This analysis revealed that systems with poor resistance to pit propagation also exhibited E_X less than (E_P + 0.10 V) (note the solid symbols in Zones I and II). Systems with E_X greater than (E_P + 0.10 V) also exhibited $\triangle E/\triangle i$ greater than 0.18 V/decade and good resistance, as shown by the open symbols in Zone III of Figure 7.

The interrelationships between $\triangle E/\triangle i$, E_{CORR}, and E_X that were observed for systems exhibiting poor resistance to pit propagation are summarized in Figure 8. The three interlocking circles indicate that all three of the criteria were met for systems that exhibited poor resistance to pitting.

For example, if E_{CORR} is less than (E_P, and E_X), no pit initiation occurred. If E_{CORR} was greater than (E_P + 0.10 V), then good resistance to pit propagation was observed. However, in the case where E_P and E_X are less than E_{CORR}, which is less than (E_P + 0.10 V), i.e., Zone II in Figure 6, and $\triangle E/\triangle i$ was less than 0.18 V/decade but E_X was greater than (E_P + 0.10 V), i.e., Zone III in

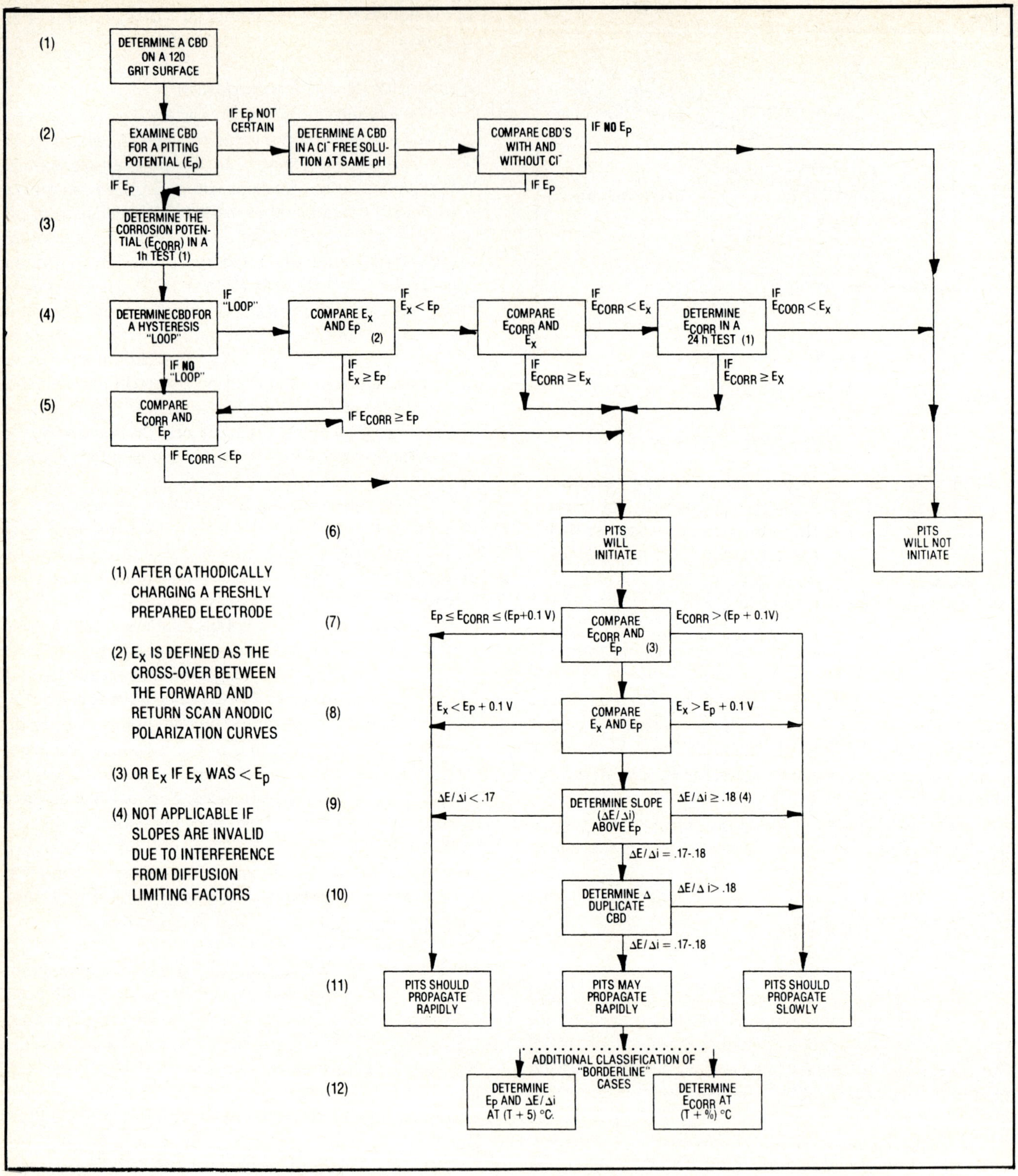

FIGURE 4 — Flow diagram for the prediction of resistance to pitting in nonoxidizing acid chloride environments using the procedures described in this paper.

Figure 7, then good resistance to pitting was also observed.

A similar scenario can be given for $\triangle E/\triangle i$ and for E_X. Consequently, it should be re-emphasized that failure to meet any one of the three criteria in Figure 8 was linked with good resistance to pitting.

Discussion of the Results

The conditions that are believed to exist on the surface of a pitting specimen are illustrated by the schematic diagram Figure 9(a). First, a passive film develops over most of the surface, with an associated E_{CORR} such as denoted by X. After pits have initiated at film defects, corrosion at the pit sites may be represented by a curve segment such as A, B, or C (i.e., for pits that cannot passivate at noble potentials). The curve segments A, B, and C, respectively, represent increasing tendencies

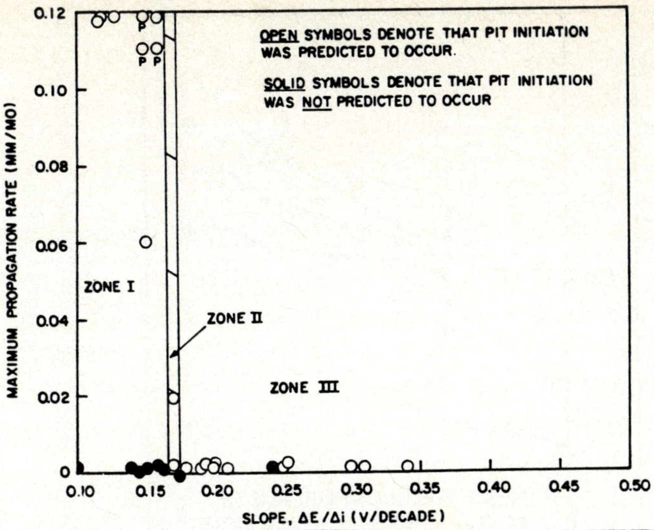

FIGURE 5 — Relationships between the slope of the anodic curve segment noble to Ep and pit propagation rates that were obtained from spool tests for systems that exhibit an Ep.

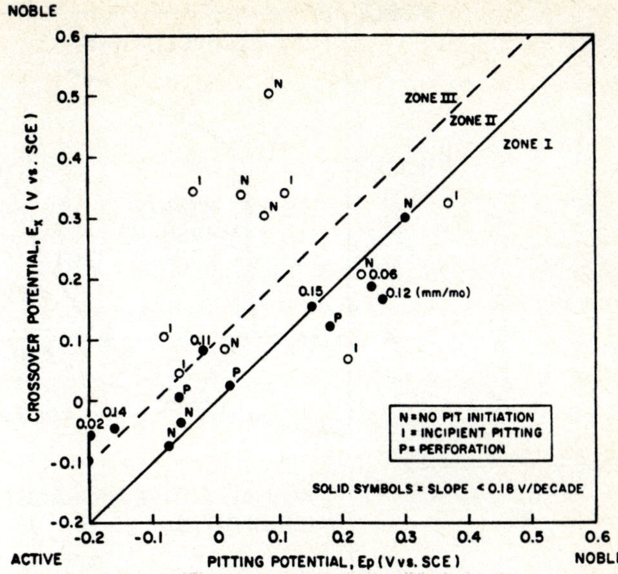

FIGURE 7 — Relationships between the values for the pitting potential, the crossover potential, and △E/△i and pit propagation rates.

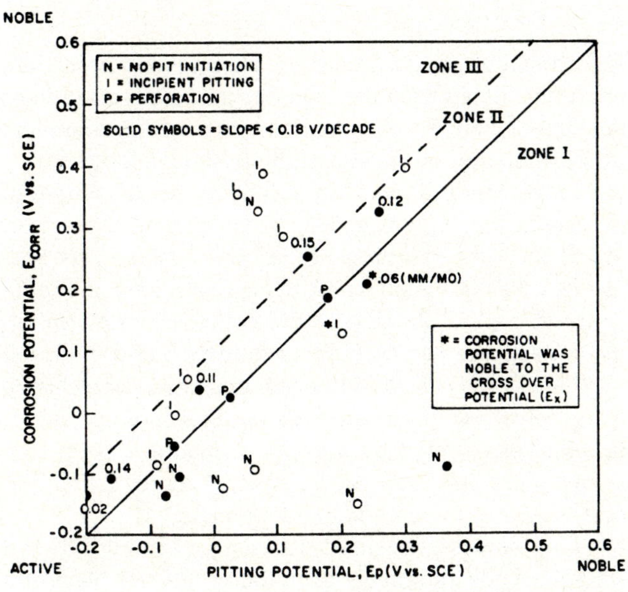

FIGURE 6 — Relationships between the values for the pitting potential, the corrosion potential, and △E/△i and pit propagation rates.

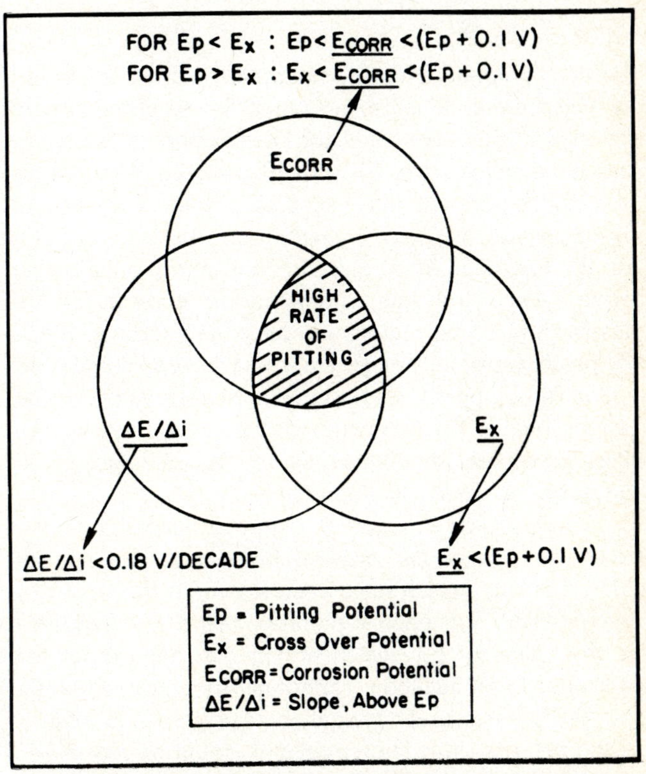

FIGURE 8 — Schematic of the three electrochemical requirements for high rates of pitting to occur on Fe-Cr-Ni alloys in nonoxidizing acid chloride environments using the procedures described in this paper.

toward active corrosion, i.e., higher rates that can be associated with the following variables: (1) increasing the solution temperature; (2) increasing the acid concentration; (3) increasing the chloride concentration; or (4) decreasing the alloy content.

During an electrochemical experiment, conditions (2) and (3) above are believed to occur within a corroding pit with increasing exposure time, without varying the environment outside of the pits. Also, an indication of a progression from theoretical curves A to B to C can be observed with increasing exposure time as a result of increasing surface area of corroding pits (i.e., as existing pits grow and/or new pits initiate). The short exposure time required for the 60 V/h polarization experiments was found to minimize interferences from these two particular time dependent variables.[9]

The measurement of conditions on a pitting surface

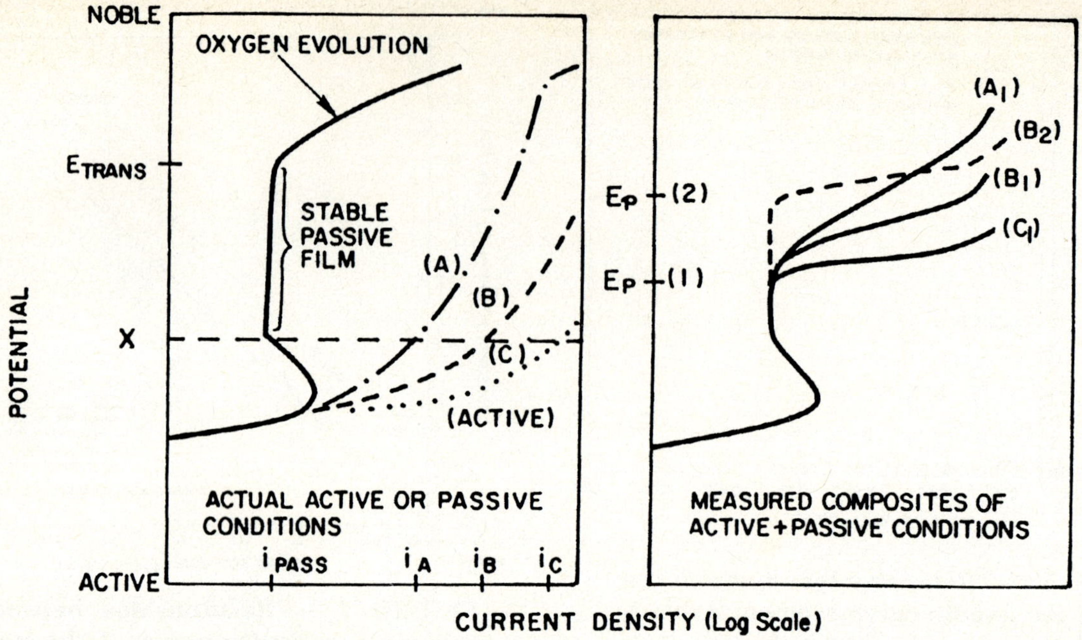

FIGURE 9 – (a) Schematic of anodic polarization characteristics at active and passive locations on a pitting surface and (b) schematic of the corresponding results that should be measured with the potentiostat.

during a potentiodynamic experiment can be illustrated by the schematic diagram in Figure 9(b). As potentials are scanned in the noble direction, dissolution rates are increased at defect sites. Chloride migration and hydrolysis of metal chlorides at defect sites leads to the formation of pits (active sites). Once "breakthrough" has occurred and an E_P has been detected, the measured polarization curve reflects the sum of the anodic currents at the passive and active regions on the pitting surface. Since the passive currents are generally low and constant in this region, the slope of the measured anodic curve should be proportional to the magnitude of the active current densities (compare theoretical curves A, B, and C to measured curves A_1, B_1, and C_1).

As illustrated by Figure 9, a group of alloys can have equivalent E_P values and yet have a wide range of $\triangle E/\triangle i$ values, which have been shown in this paper to correlate with resistance to pit propagation. Consequently, the value for E_P, which provides a range order for resistance to pit initiation, cannot be used independently to provide an accurate evaluation of resistance to pitting. In the present study, no correlation could be found between the value for E_P alone and the resistance to pit propagation.

The relative magnitude of the theoretical curves A, B, and C in Figure 9 should be reflected by the size of the hysteresis loop that is created by the return scan curve and consequently the value for E_X (see Figure 3). Systems with pits corroding according to curve C should be more difficult to repassivate and should exhibit a larger hysteresis loop with a more active E_X when scanning back into the active direction than those corresponding to curves A or B. Consequently, the use of E_X (in cases where E_X is less than E_P) for comparison with E_{CORR} is legitimate according to the schematic diagrams in Figure 9. It should be noted, however, the scanning procedure and potential at which the scan direction is reversed must be kept constant for each experiment in order to prevent artificial changes in the size of the hysteresis loops.[1]

In this study, it was found that the measurement of meaningful E_{CORR} values was far more difficult than for any of the other electrochemical parameters. This is illustrated by the variations of E_{CORR} with time shown in Figure 2. In the absence of the cathodic pretreatment, systems represented by curve segments 5 and 6 tended to exhibit E_{CORR} values that never became as noble as E_P or E_X, which would give a false prediction of no pitting. Systems represented by curve 4 in Figure 2 tended to behave more like those represented by curves 2 and 3 in the absence of the cathodic pretreatment.

The values measured for E_{CORR} should be an average of the potentials over the entire test electrode surface. Thus, higher noble E_{CORR} could easily be linked with low active current at pit sites or low densities. Also, since a passive film covers most of the surface, the influence of active potentials at pit sites on the overall average potential may be more difficult to detect. On the other hand, for systems with highly active pitting, the E_{CORR} will be a function of exposure time, and will be weighted toward more active potentials (compare regions 5 and 6, Figure 2). This condition could lead to a false prediction that pits will not initiate. In this regard, it was found that the measurement of E_{CORR} values, during the early stages of passive film development (i.e., after the cathodic pretreatment) was essential in order to predict resistance to pitting corrosion.

Summary

An electrochemical procedure has been developed that permits accurate predictions of resistance to pitting for Fe-Cr-Ni-Mo alloys in nonoxidizing acid chloride environments. Using this procedure, three requirements must be fulfilled as illustrated in Figure 8, before poor resistance to pitting is predicted. Failure to meet any one or more of these three requirements is an indication that high rates of pit propagation should not occur. These three requirements are:

(1) $\triangle E/\triangle i$ less than 0.18 V/decade; and
(2) E_P less than E_{CORR} less than (E_P + 0.10 V), or E_X less than E_{CORR} less than (E_P + 0.10 V) if E_X less than E_P; and
(3) E_X less than E_P + 0.1 V.

The value for $\triangle E/\triangle i$ was found to be the most important parameter with regard to resistance to pit propagation.

References

1. B. E. Wilde, J. S. Armijo, Corrosion, Vol. 23, No. 1, p. 208, 1967.
2. J. Horvath, H. H. Uhlig, Electrochem. Soc., Vol. 115, p. 791, 1968.
3. Z. Szklarska-Smialowska, Corrosion, Vol. 27, No. 6, p. 223, 1971.
4. Z. Szklarska-Smialowska, "The Pitting of Iron-Chromium-Nickel Alloys," Localized Corrosion, R. W. Staehle, B. J. Brown, J. Kruger, A. Agrawal, Eds., National Association of Corrosion Engineers, Houston, Texas, 1974.
5. W. Schwenk, Corros. Sci., Vol. 5, p. 245, 1965.
6. H. P. Leckie, J. Electrochem. Soc., Vol. 117, p. 1152, 1970.
7. A. Broli, H. Holtan, M. Midjo, Brit. Corros. J., Vol. 8, p. 175, 1973.
8. E. A. Lizlovs, A. P. Bond, Corrosion, Vol. 31, No. 6, p. 219, 1975.
9. P. E. Morris, Galvanic and Pitting Corrosion-Field and Laboratory Studies, ASTM, STP 576, ASTM, Philadelphia, Pennsylvania, p. 261-275, 1976.
10. P. E. Manning, Corrosion, Vol. 36, No. 9, p. 468, 1980.
11. R. C. Scarberry, E. L. Hibner, J. R. Crum, "Assessment of Pitting Potential Measurements and Severely Corrosive Environments," Paper No. 245, CORROSION/80, National Association of Corrosion Engineers, Houston, Texas, 1980.
12. H. C. Man, D. R. Gabe, Corros. Sci., Vol. 21, p. 323, 1981.
13. D. G. Tipton, R. M. Kain, "Effect of Temperature on the Resistance to Pitting of MONEL Alloy 400 in Seawater," Paper No. 36, CORROSION/80, National Association of Corrosion Engineers, Houston, Texas, 1980.
14. P. E. Morris, R. C. Scarberry, Corrosion, Vol. 28, No. 12, p. 444, 1972.
15. J. G. Stoecker, O. W. Siebert, P. E. Morris, "Practical Applications of Potentiodynamic Polarization Curves in Materials Selection," Materials Performance, Vol. 22, p. 13, 1983.

History and Mechanism of Polythionic Acid Stress Corrosion Cracking

HUGH H. HOROWITZ*

Abstract

The problem of polythionic acid stress corrosion cracking (SCC) of sensitized stainless steels has been recognized since the 1940s. Recent evidence indicates that the active corrodents causing SCC are the thiosulfate and tetrathionate anions. They function by adsorbing on the steel surface and promoting anodic dissolution of the chromium depleted steel. In electron spectroscopy for chemical analysis (ESCA), the adsorbate exhibits peaks corresponding to the thiosulfate itself. Cracking inhibitors function not by desorbing the thiosulfate but by promoting passivation.

History

The problem of polythionic acid stress corrosion cracking (SCC) has plagued the petroleum industry for over forty years, and related problems have recently surfaced in the nuclear industry as well. Considerable progress has been made in recent years in understanding the phenomena associated with the cracking process. These will be reviewed here and some further experimental facts recently discovered will be presented.

The SCC phenomenon consists of intergranular cracks forming near room temperature in sensitized austenitic stainless steels, under stress, in the presence of air, water, and a sulfide scale on the interior walls of the vessel or pipe. *Sensitization* refers to a heat treatment of the steel in the range of 400 to 700 C causing chromium-rich metal carbides to precipitate at the grain boundaries and depleting chromium content there to the point where corrosion resistance is lost (< 12% Cr). Sensitization was first noted in the heat affected zones (HAZ) around welds, but has also been caused by the process conditions themselves, as well as by fabrication heat treatments.

The problem was first reported in a fluidized bed catalytic cracker during World War II.[1] In 1956-1957, it was first ascribed to polythionic acids ($H_2S_{3-6}O_6$) by Dravnieks and Samans.[2] The problem has been observed in various types of hydrodesulfurization units, in sulfur plants, catalytic crackers, cokers, and residuum treatment units of various types.

More recently, related problems have appeared in the nuclear industry. In the Girdler process for the production of deuterium oxide, for example, multiple fractional distillations of water are carried out in the presence of H_2S; when the equipment is opened to the atmosphere, similar conditions to those in petroleum refineries exist.[3] In pressurized boiling water reactors (BWRs), solutions of sodium thiosulfate, included as quenching agents for radioactive iodine, were found in separate studies[4,5] to be extremely potent crack promoters, acting very similarly to the closely related tetrathionate, $S_4O_6^{=}$, which is a member of the polythionate family. A recent review summarizes both the science and the technology[5]; it hypothesizes that the key corrodent is elemental sulfur, and that the role of the sulfur compounds is simply to act as donors for the elemental sulfur.

Although the problem has never been completely solved, various steps have been taken in the petroleum industry to control or minimize it. For example, steel containing low levels of carbon or components such as titanium or niobium to reduce the activity of carbon, may be used to prevent sensitization. Secondly, air or moisture may be excluded during turnarounds by using special atmospheres. Thirdly, solutions of an inhibitor, sodium carbonate, may be poured into reactor vessels or piping to stop the corrodent action.

Dravnieks and Samans devised a method for simulating polythionic acid SCC in the laboratory using Wackenroder's liquid, an acidic mixture of sulfur compounds formed by the reaction of H_2S with SO_2-saturated

*Exxon Research & Engineering Co., Florham Park, New Jersey.

water. Recently, workers at the University of Roorkee[7] have found that of all the polythionic acids in this mixture, only the tetrathionate was active in promoting SCC. In addition, the nuclear industry work cited above showed that the strong acidity of Wackenroder's liquid (resulting from the strong polythionic acids as well as the free sulfuric acid) is not necessary. Cracking occurred at pH 5, where the strong acids exist exclusively as double-charged anions, $S_2O_3^=$ and $S_4O_6^=$. Similarly, it was shown in a previous publication[8] that the reaction that is said to give rise to polythionic acids in the petroleum industry, namely the reaction of iron sulfide scale with air and water, generates weakly acidic solutions. Most of the product is Fe_2O_3 and elemental sulfur, but small but appreciable amounts of thiosulfate and possibly tetrathionate form in solution, probably as their iron salts. The presence of residual FeS acts to prevent strong acid build-up in any case. Nevertheless, the solution formed by the FeS-O_2-H_2O reaction was able to cause cracking of sensitized stainless steel in slow strain rate tests.

In the same previous publication,[8] it was reported that the above active anions did not act as sulfide did to promote hydrogen entry into steel. In electrochemical hydrogen permeation experiments on a low alloy steel in acetic acid-sodium chloride solution, saturation with H_2S caused about 70% of the cathodically produced hydrogen atoms to pass through the steel membrane, whereas with an equivalent concentration of $Na_2S_2O_3$ or $Na_2S_4O_6$ only 2 to 4% permeation was observed. Thus, the surface poisoning reaction induced by sulfide is not duplicated by thiosulfate or tetrathionate. On the other hand, both of these anions did stimulate the anodic dissolution of low alloy steels, giving rise to an enhanced anodic peak in voltammetric scans in the active-passive region. These observations suggested that these two anions function by promotion of anodic dissolution processes within the crack rather than by promotion of hydrogen embrittlement (HE). As to the similarity of action of the two, it was noted that the theoretical potential for the reduction of $S_4O_6^=$ to $S_2O_3^=$ is above the corrosion potential of steel in the absence of air and that the reduction could actually be carried out on a steel electrode. Thus, the true corrodent at the crack tip could be $S_2O_3^=$ in all cases. Thiosulfate is known to be a complexing agent for metal ions including ferric, and it was verified that it formed a complex with ferrous ions as well, albeit a weak one. However, the hypothesis that soluble complex formation plays a role in SCC was later abandoned in favor of adsorption on the surface. The tendencies of anions to participate in the two phenomena are of course related: both adsorption and complexation involve overlap of metal and ligand orbitals involving either metal ions or edge atoms of the metal lattice.

Subsequent work on the cracking process itself using the slow strain rate method showed that cracking was accompanied by an anodic current whose charge corresponded to a substantial amount of metal reacted. (Figure 1[9] illustrates this point.) This experiment was conducted using potentiostatic control under a nitrogen atmosphere. Under "natural" cracking conditions with air

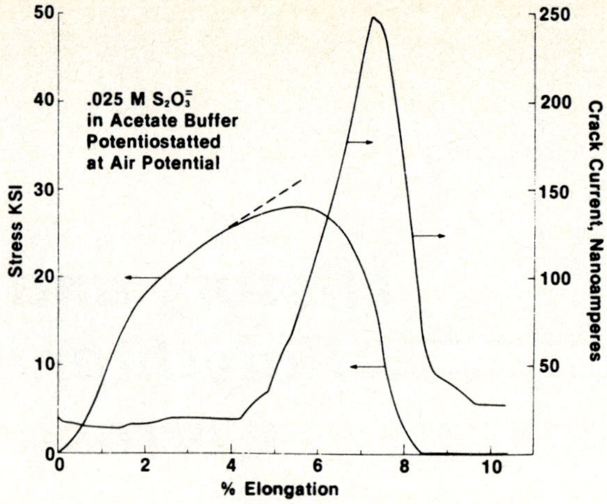

FIGURE 1 —Cracking in slow strain rate test accompanied by anodic current.

present, the anodic process in manifested by a sharp reduction in electrode potential, beginning at crack inception and ending after rupture.

The cracking process was related to an increase in anodic current prior to passivation, seen in voltammetric scans on low alloy steels in the presence of thiosulfate or tetrathionate (Figure 2). The low alloy steels were intended to simulate roughly the chromium-depleted alloy in the grain boundary regions of sensitized stainless steels. The relationship between the stimulation of the anodic process in the scans and the cracking process was solidified when it was found that four added anions, which inhibited SCC of sensitized stainless in the presence of thiosulfate, also inhibited the current increase in the scans, while four other anions permitted both the cracking and the anodic stimulation. This eight-for-eight correlation reinforced the concept of a relation between the two phenomena. The inhibiting anions were carbonate, borate, phosphate, and nitrite. The noninhibiting anions were acetate, sulfate, nitrate, and chloride.

It was then determined that the effect of thiosulfate and tetrathionate on anodic dissolution was catalytic, and could not involve complexation of the dissolved iron. First, more iron could be dissolved from an AISI 1018 carbon steel than there were $S_2O_3^=$ ions in the entire anodizing solution. Second, and more impressive, a sample of AISI 1018 carbon steel or 7 Cr-0.5 Mo steel could be exposed to 0.001 molar (M) $Na_2S_2O_3$, or $Na_2S_4O_6$ solution, then washed and rinsed in distilled, deionized water and still exhibit the promotional effect of $S_2O_3^=$ in a solution containing no $S_2O_3^=$ (Figure 3). This evidence proves that the thiosulfate ion is chemisorbed strongly on the steel surface, where it interferes with the passivation process. It may interact with the freshly precipitated passivating oxide so as to change its morphology from a homogeneous protective coating, to a granular, less protective one. In support of this idea, separate model experiments showed that ferrous hydroxide, precipitated in the presence of sodium thiosulfate, yielded a black granular substance rather than the usual green gelatinous material.

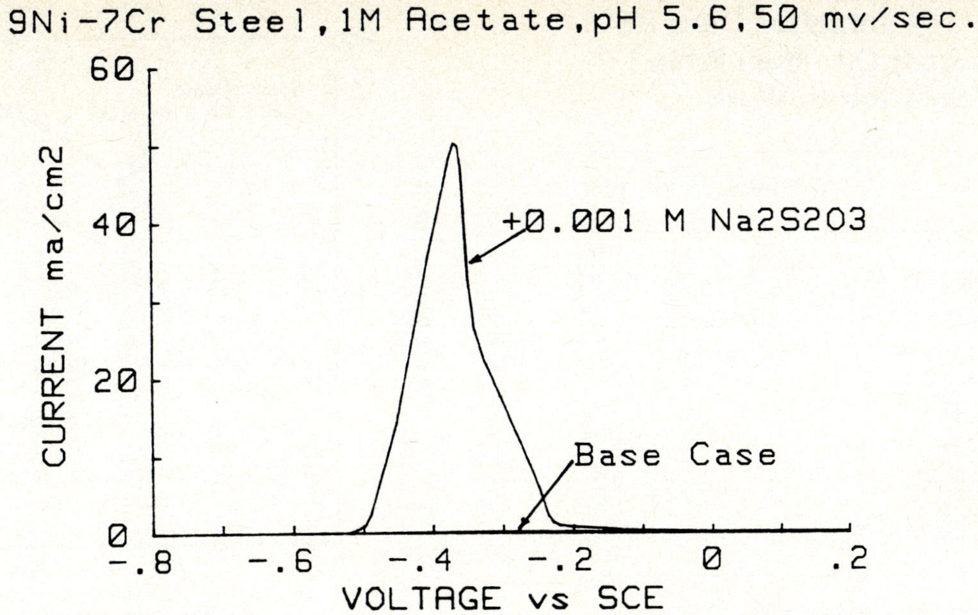

FIGURE 2 — Anodic stimulation by thiosulfate.

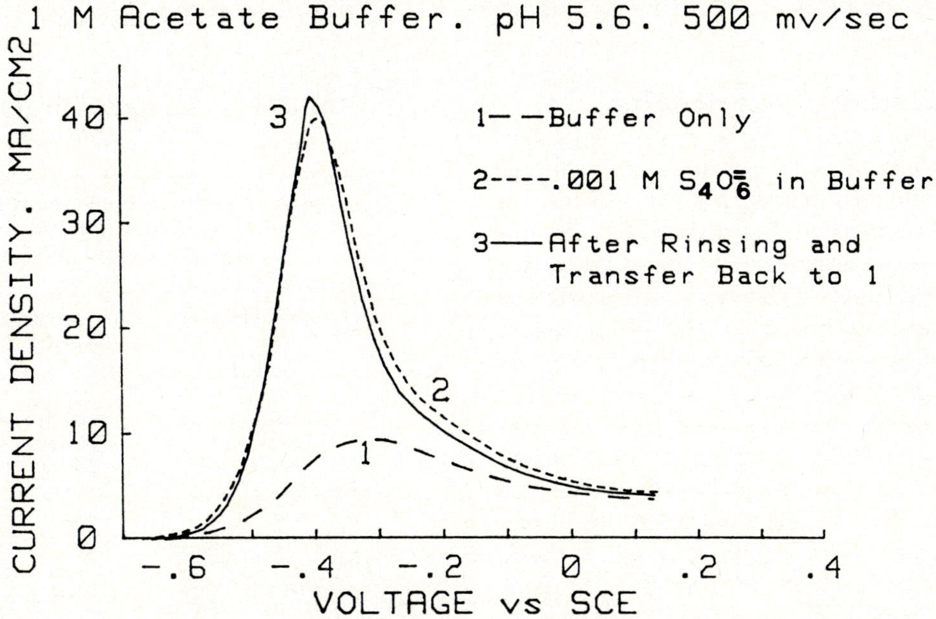

FIGURE 3 — Adsorption of tetrathionate on 7-Cr steel.

It was next shown that the anions that inhibit cracking do not do so by causing the desorption of the adsorbed thiosulfate. After thiosulfate adsorption, if the steel was rinsed in an inhibitor solution rather than in distilled water, the anodic stimulation effect still remained, when the metal surface was scanned voltammetrically in an acetate buffer. Only when the scan was conducted in an inhibitor solution was the anodic stimulation caused by the adsorbed thiosulfate eliminated (Table 1). The inference is that the inhibiting ions must be helping to bolster or repair the imperfect film created by the adsorbed thiosulfate, probably by the formation of insoluble iron salts or mixed oxides or hydroxides. These ions are known to be able to perform this function.[10]

The picture emerging from this is that cracking is caused by an anion that can adsorb on the chromium-depleted metal in the grain boundaries and accelerate its anodic dissolution.

One apparent complication of the simple connection between anodic stimulation and crack propagtion was provided by Newman, Sieradzki, and Isaacs[11] who showed that the rate of crack propagation was sometimes far too high to be explained by the maximum anodic current density observed in scratching tests. Their data indicated that a large fraction of the crack growth was due to mechanical rupture. The electrochemical reaction followed and helped to continue the cracking process by advancing the zone of high stress in the metal and causing embrittling phase transformations.

In this study, data are presented on the effect of

TABLE 1 — Inhibitor Does Not Desorb $S_2O_3^=$; Must Be in Solution to Work

	Anodic Peak, mA
1. No Additive	5,7
2. +.001 M $S_2O_3^=$	45,53
3. 60 Seconds in .05 M $Na_2B_4O_7$	
4. No Additive	51
3′ .001 M $S_2O_3^=$ +.05 M $Na_2B_4O_7$	<.01

Inhibitor plugs holes in oxide film.

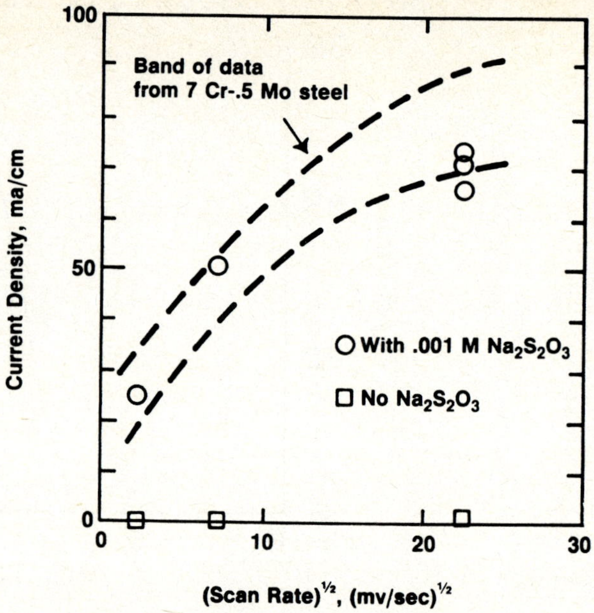

FIGURE 4 — Stimulation of anodic dissolution by thiosulfate; 9 Ni-7Cr steel in 0.9 M Na acetate + 0.1 M acetic acid.

nickel in the alloy on the thiosulfate anodic stimulation effect and ESCA analyses of the chemical nature of the adsorbed sulfur compound.

Experimental and Results

Whereas previous studies of the anodic stimulation effect and the adsorption of thiosulfate were conducted on a carbon steel and on a 7 Cr-0.5 Mo steel, the present work used a steel containing 9% nickel and 7% chromium to simulate more accurately the grain boundary region of a sensitized AISI 304 stainless steel. Prepared as a billet, this material was rolled to 50% of its original thickness and annealed under argon at 800 C 1 hour (h). It was machined into a 5/8 in. (1.6 cm) diameter disk and mounted in a PAR[1] holder so that 1 cm^2 was exposed. Scans were run in 0.9 M sodium acetate — 0.1 M acetic acid buffer (pH 5.6), with or without the addition of 0.001 M $Na_2S_2O_3$. The anodic stimulation effect was clearly in evidence as shown in Figure 3. Comparison with previous curves obtained on 7 Cr-0.5 Mo steel shows that the effect of the nickel content is primarily to reduce the base case current densities. With thiosulfate present, the results were quite comparable (Figure 4). Thus, the presence of nickel in the alloy does not eliminate the anodic stimulation by thiosulfate, but rather enhances it.

Adsorption of thiosulfate also occurred fairly tenaciously on the 9 Ni-7 Cr steel as illustrated in Table 2. In the presence of 0.001 M $Na_2S_2O_3$, anodic peak heights of about 60 mA/cm^2 were observed three times at a scan rate of 500 mV/s. This high scan rate was used in adsorption studies to avoid the loss of adsorbate, which occurred slowly at high potentials. After the last scan, the electrolyte was removed and the electrode was rinsed with distilled-deionized water and air dried. It was then immersed in fresh electrolyte containing no thiosulfate where it produced a peak of 33 mA/cm^2 in a voltammetric scan. While there was some loss of activity, it remained

[1]Registered trade name.

TABLE 2 — Adsorption of Thiosulfate on 9 Ni-7 Cr Steel

Anodic Voltammetric Scans at 500 mV/s in 0.9 M Na Acetate + 0.1 M Acetic Acid. Peak at −0.30 ± .04 V/SCE.

Electrode Treatment	Concentration $Na_2S_2O_3$	Peak Height (mA/cm^2)
Immersed	.001 M	58.5
Repeat	.001 M	61.5
Repeat	.001 M	63.5
Rinse and Air-Dry	None	33.3
Rinse and N_2-Dry	None	33.6
Rinse and Air-Dry	None	28.0
Hold in Air 2 Hours	None	22.5
Repolish	None	3.0[1]

[1]No distinct peak.

far more active than the original alloy in the base electrolyte: peak height < 1 mA/cm^2. This activity persisted for several more treatments without further exposure to dissolved thiosulfate, as shown. These results are typical of adsorption runs conducted on the other steels. The stability of the adsorbate makes it possible to carry out ESCA analyses in a vacuum environment without fear of changing the adsorbate drastically during transfer from the electrochemical cell to the vacuum system.

ESCA analyses were conducted on the surfaces of the 9 Ni-7Cr and the 7 Cr-0.5 Mo steels after adsorption of thiosulfate and rinsing in distilled deionized water. The surfaces showed some sulfur in the presence of oxygen, iron, chromium, and molybdenum or nickel, plus contamination by sodium and silicon. However, the sulfur was present entirely as sulfide and SO_X. There was no

TABLE 3 — ESCA Analysis of Thiosulfate Adsorbate

Adsorption from 0.9 M Na Acetate, 0.1 M Acetic Acid + .001 M $Na_2S_2O_3$ at Room Temp.			
Steel Type	**9 Ni-7 Cr**		**7 Cr-0.5 Mo**
S on Surface, % of Total Peak Heights	2.3[1]		3.1[2]
Sulfate Peak			
Binding Energy CV	168.5		168.5
% of Total S	87		55
$Na_2S_2O_3$ Binding Energy[3]		168.5	
Sulfide Peak			
Binding Energy ev	160.9		161.5
% of Total S	13		45
$Na_2S_2O_3$ Binding Energy[3]		161.4	

(1)Other elements: O, Si, Na, Fe, Ni, Cr, C
(2)Other elements: O, Si, Na, Fe, Cr, Mo, C
(3)From "Handbook of X-ray Photoelectron Spectroscopy," C. D. Wagner, W. M. Riggs, L. E. Davis, J. F. Moulder, G. E. Muilenberg, Eds., Perkin Elmer Corp., Physical Electronics Div., Eden Prairie, Minnesota, 1979.

peak corresponding to elemental sulfur. Sodium thiosulfate itself shows two peaks in ESCA, one in the sulfide range, the other in the SO_X range (Table 3). Thus, it is not possible to state at the present time whether the thiosulfate has chemisorbed with cleavage to sulfide and residual SO_X, or has chemisorbed intact as a thiosulfate moiety. The sulfur often seen in stress corrosion cracks may come from subsequent acidification of the thiosulfate, which produces sulfur and SO_2. In this case, however, where there is an excess of buffer at pH 5.6, no acidification could occur and the adsorbate can remain in its original form. Stress corrosion cracking of sensitized stainless steel does occur in acetate buffer, indicating that this is the active form of the adsorbate.

Discussion

The catalytic effect of adsorbed thiosulfate and tetrathionate helps explain how concentrations in the sub-ppm range can produce SCC, as in Reference 4. It is interesting that the adsorbate can remain on the surface while the metal under it is being dissolved. Such behavior has been observed with sulfide monolayers adsorbed from the gas phase,[12] which were able to catalyze the dissolution of about 10^4 metal atoms per adsorbed sulfur into dilute H_2SO_4, while retaining at least 80% coverage by the end of the voltammetric scan. The implication is that the adsorbed atoms are able to transfer from one metal atom to another as they catalyze their dissolution.

The behavioral similarities of adsorbed sulfide and thiosulfate are striking, and certainly suggest a relationship between the two. The hydrogen permeation data and the ESCA analyses presented here indicate that the adsorbates obtained from thiosulfate and tetrationate may be similar to but not identical to that from adsorbed sulfide. However, further work is required to resolve this point.

As to the overall mechanism of polythionic stress corrosion cracking, it seems prudent to consider it as a case of strain- (or stress-) assisted active path dissolution. Supporting this are innumerable micrographs of fracture surfaces showing intergranular corrosion perpendicular to the crack face. There is, as well, the enhanced anodic dissolution rates or delayed repassivation seen on chromium-depleted alloy steels here and in prior references, which have been related here to the cracking process by their mutual sensitivity to inhibiting anions. Following Reference 11, anodic dissolution can be visualized inside the crack removing metal until defects reach the surface or until embrittling phase transformations occur in front of the crack tip. These induce mechanical cracking, which re-invigorates the anodic dissolution process. The crack moves faster than Faraday's law allows, because most of the dissolution occurs along the walls of the crack, which are kept unpassivated by the catalytic action of the adsorbed corrodent. The larger surface area of the crack sides relative to the cross section of the tip keeps the actual current density in the crack within the bounds of the current densities observed in the laboratory, and well below that required by the rate of movement of the crack tip.

Summary and Conclusions

1. The stress cracking of sensitized stainless steels can be attributed to the action of thiosulfate and tetrathionate anions. Strong acidity is not necessary.

2. Cracking is associated with an anodic dissolution process, which reacts substantial amounts of metal from the intergranular region.

3. The corrodent anions also stimulate strong anodic dissolution processes in voltammetric scans on chromium-depleted alloys. The same anions that inhibit the cracking process also inhibit the anodic dissolution process, thereby strengthening the evidence for a relation between the two phenomena.

4. Thiosulfate and tetrathionate function by adsorbing on the metal surface and catalyzing the anodic dissolution reactions. The adsorbate shows in ESCA (or x-ray photoelectron spectroscopy) sulfur peaks that correspond to sulfide and sulfate, as in thiosulfate itself, but no elemental sulfur.

5. Inhibitors that retard cracking function not by desorbing the sulfur anions, but by reinforcing passive film formation.

6. The overall cracking process may be described as strain- (or stress-) assisted active path dissolution, in which microcracks occurring mechanically stimulate metal dissolution of the bare crack walls, whose repassivation is delayed by the action of the corrodents. The dissolution process uncovers defects, or causes stress zone movement embrittling the metal, thereby inducing further microcracking and continuing the process.

Acknowledgments

The author expresses his appreciation to the Electrochemical Society, Inc. for allowing the reproduction of Figures 1 and 2 and Table 1 of this paper from the Extended Abstracts of the 166th National Meeting of the Electrochemical Society, Volume 84-2, Paper No. 278, New Orleans, Louisiana, October 1984.

References

1. E. L. Hildebrand, Corrosion, Vol. 12, p. 73, 1956.
2. A. Dravnieks, C. A. Samans, Proc. Amer. Petr. Inst., Vol. 37, p. 100, 1957.
3. R. L. Tapping, P. A. Lavoie, R. D. Davidson, Materials Performance, Vol. 22, No. 9, p. 43, 1983.
4. S. Dhawale, G. Cragnolino, D. D. Macdonald, EPRI Project RP 1166-1 (FCC 7805), 26-32, Electric Power Research Institute (EPRI), Palo Alto, California, 1980.
5. H. S. Isaacs, B. Vyas, M. W. Kendig, Corrosion, Vol. 38, No. 3, p. 130, 1982.
6. G. Cragnolino, D. D. Macdonald, Corrosion, Vol. 38, No. 8, p. 406, 1982.
7. S. Ahmad, M. L. Mehta, S. K. Saraf, I. P. Saraswat, Corrosion, Vol. 38, No. 6, p. 347, 1982.
8. H. H. Horowitz, Corros. Sci., Vol. 23, p. 353, 1983.
9. H. H. Horowitz, "Chemisorption and Catalysis in Polythionic/Thiosulfate Stress Corrosion Cracking," J. Electrochem. Soc., Vol. 132, p. 2064, 1985.
10. L. L. Shreir, Corrosion Control, Vol. 2, Newnes-Butterworths, London, p. 18:41, 1976.
11. R. C. Newman, K. Sieradzki, H. S. Isaacs, Met. Trans., p. 2015, 1982.
12. P. Marcus, A. Tessier, V. Oudar, Corros. Sci., Vol. 24, p. 258, 1984.

Section 7

Corrosion Monitoring

On-Line Corrosion Monitoring—A Current View

J. C. BOVANKOVICH*

Abstract

On-line corrosion monitoring can be used to (1) provide an early warning of serious corrosion conditions whereby corrective action can be taken before damaging effects result, (2) diagnose the cause of the corrosion problems, (3) determine the effectiveness of a corrosion problem solution, and (4) conduct process monitoring. The operating principles of the most widely used techniques are described. Several examples of where on-line corrosion monitoring has been used are described.

Introduction

The cost of corrosion in the United States is estimated to be 4.2% of the gross national product[1]; about 15% of this is preventable. This translates to $153 billion in cost with $23 billion preventable. Using on-line corrosion monitoring technology with other complementing measurements in chemical and petrochemical processes can help reduce the corrosion bill. With these potential savings in mind, corrosion engineers should consider the potential application of on-line corrosion monitoring for reducing corrosion-caused shutdowns and extending plant campaigns. It will also help in better plant operations.

Definitions

In order to understand the purpose of on line corrosion monitoring, several definitions are necessary for clarity. First, *corrosion* is the destruction or deterioration of a material from reaction with its environment. *Corrosion monitoring* is the systematic measurement of corrosion of an item or piece of equipment. *On-line corrosion monitoring* is the real-time measurement of the corrosion performance of metallic materials, using a probe or sensor and an instrument.

*Engineering Dept., E. I. DuPont de Nemours and Co., Inc., Wilmington, Delaware.

Aim

The aim of corrosion monitoring is to assist in understanding the corrosion process and to obtain information for use in controlling corrosion or corrosion consequences.

Methods of Corrosion Monitoring

The principle direct methods of corrosion monitoring include electrical resistance corrosion measurement, instantaneous corrosion rate measurement, corrosion coupons, radiography, ultrasonics, galvanic current measurement, and "telltale" holes.

Direct Methods of On-Line Corrosion Monitoring

Of the list of corrosion monitoring methods, currently two direct methods have been widely accepted for practical on-line monitoring. They are the *electrical resistance* and *instantaneous corrosion rate* methods. The latter method is also called *linear polarization resistance* or *polarization resistance*. Both methods are described in some detail in a NACE technical committee report[2]; the latter technique is treated in detail by Mansfeld.[3]

On-Line Corrosion Monitoring System Components and Functions

An on-line corrosion monitoring system consists of a probe or sensor, extension cable, instrument, and readout. *Sensors* serve as the interface with the process and the metal for measuring corrosion rate. Sensors can be designed to be installed permanently: (1) where a shutdown is required for removal (Figure 1); (2) in a bypass

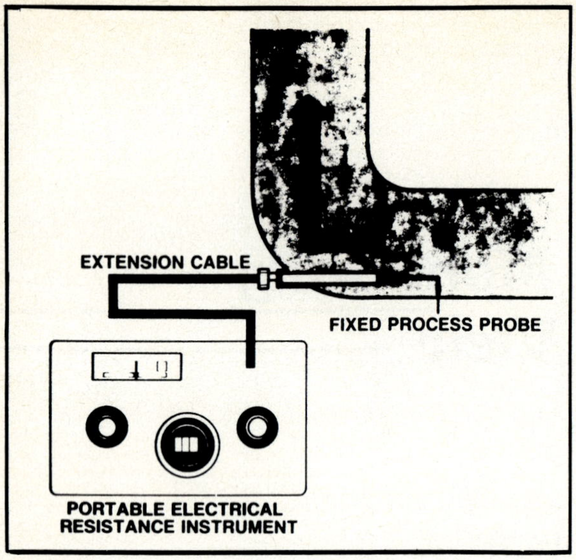

FIGURE 1 – Fixed electrical resistance process probe.

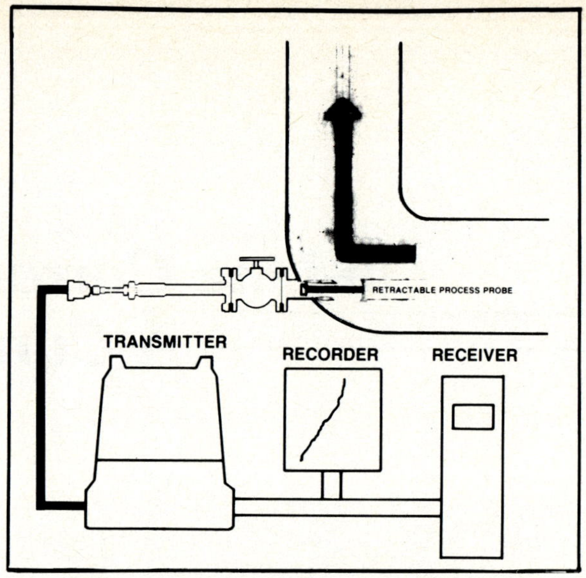

FIGURE 3 – Retractable process probe.

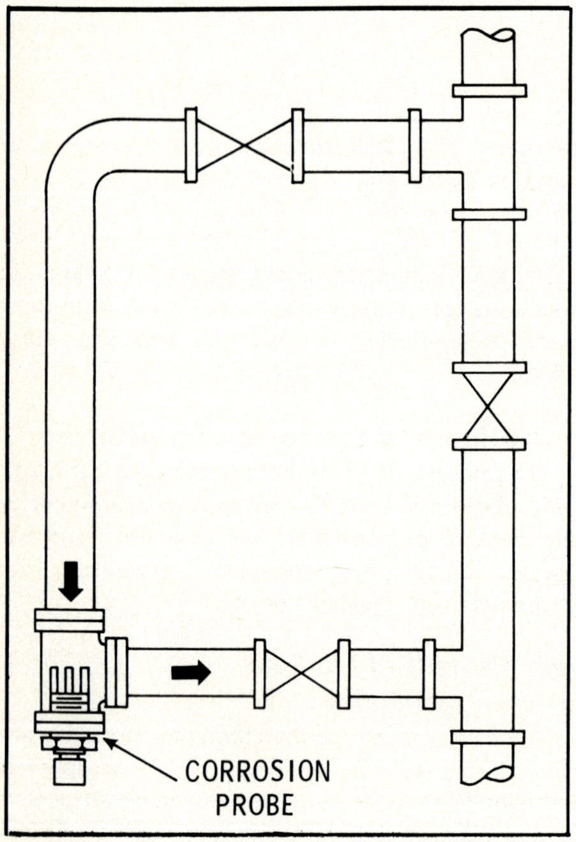

FIGURE 2 – Fixed process probe in bypass piping.

FIGURE 4 – Portable electrical resistance instrument.

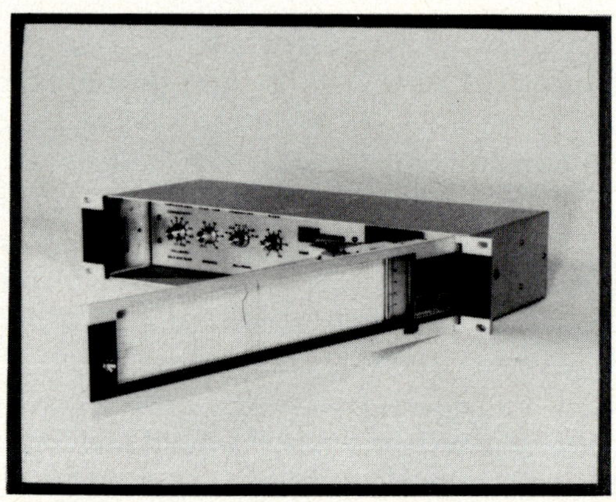

FIGURE 5 – Continuous instantaneous corrosion rate instrument.

arrangement, where the section holding the sensor can be blocked from the process stream by redirecting it away from the sensor by valves (Figure 2); and (3), while a plant is running, by insertion and retraction through a small pipe chamber, with valves and a packing gland (Figure 3).

Next in the system is an *extension cable* attached to a sensor. The cable can be part of the measuring instrument if it is portable, or it can be a long extension cable leading back to an instrument in the control room.

The *instrument* can be portable, where periodic measurements are made by usually going to the sensor (Figure 4). The instrument can be continuously operating, usually located in the control room (Figure 5).

Last in the system is the *readout,* which can be either the instrument meter display, a strip chart recorder, or a computer.

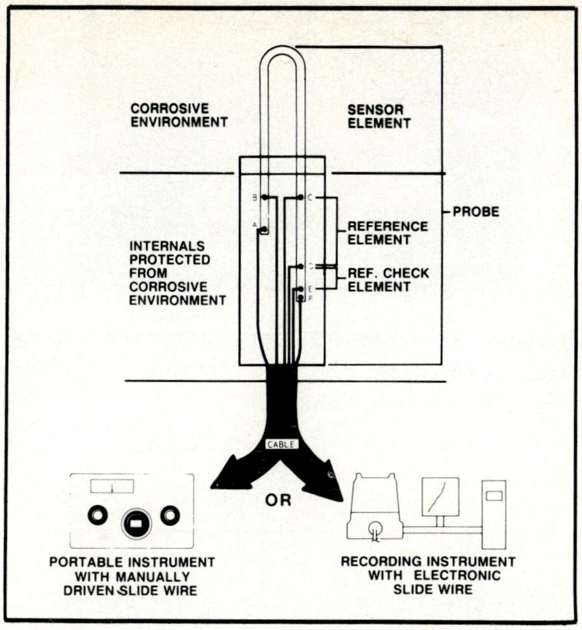

FIGURE 6 — Schematic diagram of electrical resistance corrosion measuring system.

Principle On-Line Corrosion Monitoring Methods

The principle on-line corrosion monitoring methods are the electrical resistance (ER) technique and the instantaneous corrosion rate (ICR) technique.

The electrical resistance or electrical corrosion coupon technique developed by Dravnieks and Cataldi[4] uses a low-resistance Kelvin electrical bridge circuit (Figure 6). The bridge resistance elements, except the slide wire or variable resistance, are made of the metal of interest. One resistance element of the probe is exposed to the process environment, either gas, liquid, or vapor. The remaining bridge resistance elements are encapsulated within the probe body and serve as reference or comparison elements. One encapsulated resistance element is used for comparing with the exposed resistance element. The other encapsulated resistance element is used as a check to ensure that the reference element maintains its integrity throughout the life of the sensor. As the exposed resistance element thins down because of corrosion, its electrical resistance increases. It is this resistance change that is measured and related to corrosion. The principle is similar to a corrosion coupon, but instead of making weight measurements to measure corrosion, electrical resistance measurements are made to measure the cross section. The electrical resistance probe is connected to a portable or continuously read instrument. Electrical resistance probes can be used in pipelines, piping systems, bridge decks, process vessels, and other process equipment. They can be used for measuring corrosion in the atmosphere, and in conductive and nonconductive solutions, like organic chemicals. They can be used for measuring corrosion in any environment up to 5000 psi (34 MPa) (gage) at 750 F (399 C).

The instantaneous corrosion rate, or ICR method, is based on electrochemical principles elucidated by Stern and Geary.[5] Metal electrodes are exposed to the process stream. These electrodes can also serve as classical, weight loss corrosion coupons. The ICR technique requires that the corrosive media have some electrical or solution conductivity. The technique may be used in a nonconductive fluid which becomes conductive before corrosion occurs, such as an anhydrous chemical which becomes corrosive when contacted by water. Sensors are available for environments up to 4000 psi (28 MPa) (gage) at 500 F (260 C).

By comparing the two principle techniques of on-line corrosion monitoring to corrosion coupon monitoring, the corrosion engineer can economically justify application of this technology and pave the way toward acceptance. Criteria for acceptance should at least include general corrosion rate measurement, localized corrosion measurement, usefulness, data retrieval rate, and corrosion rate calculations.

Corrosion coupons and the electrical resistance technique are widely accepted; corrosion coupons are accepted even more. The instantaneous corrosion rate technique, however, is accepted with some reservation.

For general corrosion rate measurement, corrosion coupons are the standard. Usually, the electrical resistance technique has good correlation with this form of corrosion. With the instantaneous corrosion rate technique, correlation depends on the application and environment.

When measuring localized corrosion, pit depth measurements on weight loss coupons are the standard. The electrical resistance technique can give poor correlation with pitting. However, there is good correlation for crevice corrosion. The instantaneous corrosion rate technique may provide some qualitative correlation with localized corrosion, depending on the extent of attack.

Weight loss coupons are useful in most environments. Electrical resistance sensors can also be used in most environments. The instantaneous corrosion rate method is limited to fluids with some conductivity when corrosive.

The data retrieval rate with coupons is slow. With the electrical resistance technique, data is obtained at a moderate rate. With the instantaneous corrosion rate technique data is retrieved rapidly, almost in real-time.

For calculating corrosion rates for coupons, dimension and weight measurements have to be made before exposure. Cleaning and weight measurement are required after exposure. All of this must be done and calculated for one data point.

With the electrical resistance technique, the data is logged and must be calculated by differentiation to obtain corrosion rate. This is done by dividing the resistance change by time. When the instantaneous corrosion rate technique is calibrated, the corrosion rate is obtained directly without calculation.

Why Use On-Line Corrosion Monitoring?

There has been a trend toward more integration of chemical and petrochemical plants where an unexpected failure in one facility may have repercussions in another facility at the same plant site. These new plants are large single-line process units which must achieve long continuity of operation to be economical. In addition to the financial incentives, there is increasing concern for personnel safety and environmental impact effects. On-line corrosion monitoring can serve to give additional confidence and ensurance against an unexpected occurrence by providing sentinels that complement other process measurements. With monitors or sentries in place, an alarm can be sounded so that corrective action can be taken before corrosion effects become serious.

Where and When to Use It?

On-line corrosion monitoring is applied when the performance of materials of construction may be jeopardized by serious corrosion. It may also be applied when one wants to diagnose a corrosion problem by real-time corrosion measurement, when the problem occurs, and relating this measurement to process conditions that may include process stream sampling and analyses. It may also be effective when one wants to determine the effectiveness of a corrosion problem solution. Lastly, where one wants to provide a process control measurement directly relatable to corrosion measurements.

Application

An example of the application of on-line corrosion monitoring will be given for each of the uses cited above—performance of materials of construction, diagnosis of a corrosion problem, determination of the effectiveness of a corrosion problem solution, and process control.

The first example where on-line corrosion monitoring was used for following the performance of materials of construction involved a Gulf Coast area plant. During early process planning, the potential for serious corrosion was recognized so that on-line corrosion monitoring was considered in all phases of planning, from the pilot plant operation through the operating plant. To monitor the performance of the chosen materials of construction, the plant routinely measured and logged corrosion measurements from 20 electrical resistance probes installed in carefully selected locations. These probes were an integral part of the overall corrosion monitoring program, which included corrosion coupons, ultrasonic thickness measurements, and visual examinations.

In at least one instance, plant personnel were able to recognize a serious corrosion problem in one of the distillation columns and to alert operations and maintenance so that plans could be made to correct the problem during an orderly shutdown. Scheduled repairs were made so that a corrosion surprise was avoided.

An example of using on-line corrosion monitoring to diagnose a corrosion problem involved one midwestern plant. There was severe off-gas compressor rotor corrosion. The compressor was used to compress a wet organic chemical vapor. By using on-line corrosion monitoring, a corrosion event was recognized and arrangements made for process stream chemical analyses whereby the contaminants were identified. A course of corrective action was initiated to minimize the contaminants and thereby keep the compressor rotor corrosion to a minimum.

FIGURE 7 – On-line corrosion monitoring process in vessel and complement of instruments in analyst's shack.

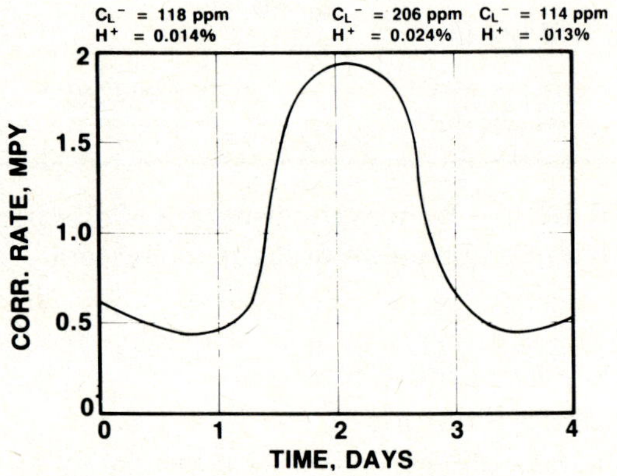

FIGURE 8 – Corrosion rate vs time and chemical analyses for corrosion event.

Figure 7 is a double photograph showing probes installed in a small vessel near the compressor. Also shown is the cmplement of instantaneous corrosion rate and electrical resistance technique instruments and recorders in an analyst's shack. The probes were connected to the instruments with extension cables.

Arrangements were made so that when a corrosion excursion or event was detected, process stream samples were taken and analyzed. Figure 8 shows the corrosion event being detected with the instantaneous corrosion rate measuring instrument and the analytical results of chemical analyses for acid and chloride. The higher corrosion rate of 2 mpy (0.051 mm/y) vs 1 mpy (0.025 mm/y) was associated with higher than usual amounts of both acid and chloride contaminants. Using these data to diagnose the corrosion problem, the plant instituted a program to solve it.

The next example applies to using on-line corrosion monitoring to determine the effectiveness of a corrosion problem solution. At another Gulf Coast area plant, consideration was given to increasing working pressure and temperature in some distillation columns. Before making the process change, the potential for adverse corrosion effects on materials of construction and operation of other pieces of equipment was considered. To satisfy these concerns for materials, a series of tests was con-

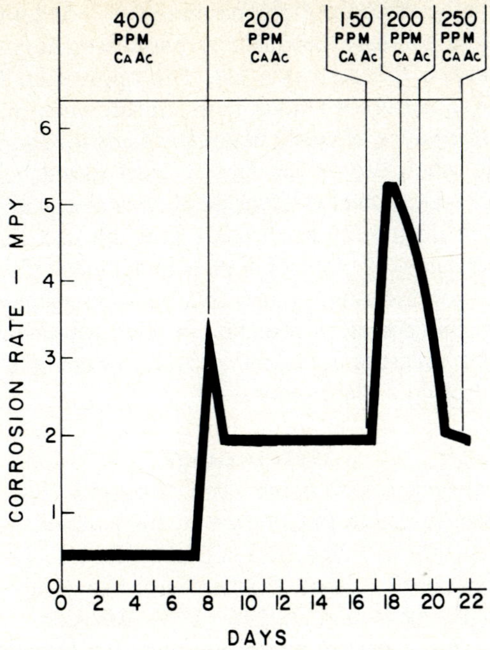

FIGURE 9 — Still corrosion rate vs days.

ducted in the plant. The tests included the installation of on-line corrosion monitoring systems to follow in real-time process changes and their effects on corrosion.

At existing conditions, the corrosion rate on stainless steel was 2 mpy (0.051 mm/y). At the new conditions of elevated pressure and temperature, the corrosion rate was 18 mpy (0.46 mm/y), a rate much too high for the existing equipment. Since an organic acid was considered a principle cause of corrosion, it was believed that the addition of calcium acetate would react with the acid and reduce corrosion rate.

Instantaneous corrosion rate measurements provided a unique method for following the reaction and optimizing the amount of calcium acetate required to remove the acid. Figure 9 is a plot of corrosion rate vs days for the plant test campaign. It should be remembered that without the calcium acetate addition, the corrosion rate is 18 mpy (0.46 mm/y), well beyond the values shown on the ordinate. The campaign started at 400 ppm calcium acetate addition where the corrosion rate was less than 1 mpy (0.025 mm/y). At 200 ppm, the corrosion rate became stable at 1 mpy (0.025 mm/y). When 150 ppm was tried, the corrosion rate began to increase excessively. To return to a tolerable corrosion rate of 2 mpy (0.051 mm/y), calcium acetate was increased in steps from 150 to 200 ppm, then to 250 ppm. There was no need to demonstrate that after recovery the acceptable corrosion rate could be maintained with a 200 ppm addition of calcium acetate.

The last application of on-line corrosion monitoring was for process control at a midwest plant making an acid salt. The reaction goal was to produce the product with a balance of reactants. If either reactant was provided in excess, the plant ran the risk of having an emissions problem in addition to taking an economic penalty for wasted reactants. A pH measuring system with the sensor in the reactor effluent was supposed to be used to control the

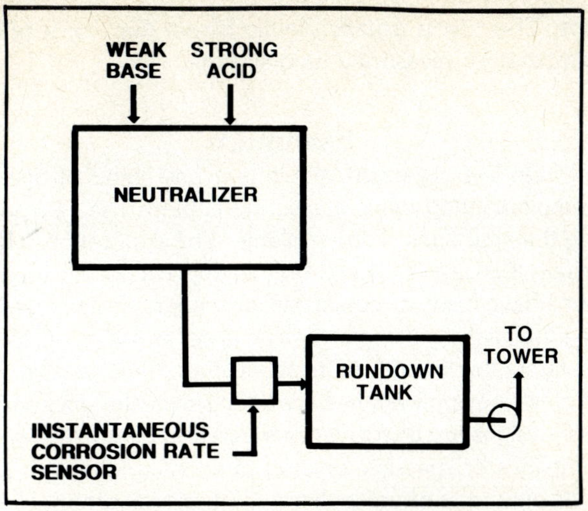

FIGURE 10 — Acid salt plant.

reaction. The reactor effluent at 85 to 140 C with excursions to 185 C was causing difficulties with the pH measuring system. An instantaneous corrosion rate measuring system was substituted for the pH measuring system (Figure 10). The empirical relationship of instantaneous corrosion rate to pH was used by the operators to control the reaction. With the reaction under proper control the plant averted emissions problems and realized substantial savings by not wasting reactants.

Costs

Off-the-shelf electrical resistance technique corrosion measuring sensors for use in plants range from $315 to $625 each, depending on the alloy and probe type, i.e., whether fixed or retractable. Portable instrument costs range from $1650 to $1840. Transmitter-type instruments costs range from $995 to $1650, not including current loop power supply or receivers; these can be provided by the transmitter vendor as a unit or separately by the user. These options permit the user to have his computer receive, handle, and display the data in a manner compatible with his system. When considering multichannel installations, instruments incorporating multiplexers and microcomputers should be considered.

Costs for off-the-shelf instantaneous corrosion rate probes for plant use range from $185 to $700 (not including the cost of electrodes), depending on the alloy and probe type, i.e., whether fixed or retractable. Electrodes range in cost from $10 to $225 per probe set, depending on the electrode alloy.

A continuously reading instantaneous corrosion rate monitoring instrument ranges in cost from $1635 to $2100. These instruments are usually line-powered and have readouts compatible with a variety of recorders and computers. Multichannel instruments incorporating multiplexers and common chassis are available and should be considered when several channels would be monitored.

When submitting costs for project approval one must consider the cost of a complete system loop including probe, instrument, extension cable, installation, and add-ons. The cost is usually slightly higher than for a pH or temperature measuring loop system.

Summary

On-line corrosion monitoring has come of age for complementing other measuring systems used for sensing the vital signs of these plants. The applications given above illustrate the breadth of its application. Its applications have been successful since standard designs are being the vital signs of plants. The applications given above illustrate the breadth of its application. On-line corrosion monitoring applications have been successful since standard designs are being incorporated into plant planning, it simplifies or speeds corrosion problem solutions, it permits unique or novel problem solutions not otherwise obtainable, and it has proven cost effective in most applications.

Successful applications are reached through the cooperation of corrosion engineers and plant individuals willing to experiment with and evaluate new technology. Most of the problems associated with new technology introduction have been solved so that reliable, safe systems can now be specified, installed, and made to operate with a minimum of difficulty.

Since this technology is relatively new, a successful on-line corrosion monitoring system requires a commitment of money, time, and personnel to realize its full cost effectiveness. The commitment usually involves someone at the plant to see that things get done locally. Backup and follow-through assistance are available from experts and vendors acquainted with the technology.

The application of on-line corrosion monitoring will contribute to plant productivity improvement and will help reduce the corrosion bill. The application of this technology deserves serious consideration when monitoring performance of costly materials of construction, particularly when service life can be jeopardized by plant upset or leak problems. Consideration is also warranted: (1) when diagnosing a corrosion problem and studying events related to plant operating conditions, (2) when determining the effectiveness of a problem solution to a serious plant corrosion condition; and (3) when real-time corrosion measurements can provide meaningful input for controlling a plant process.

References

1. "Economic Effects of Metallic Corrosion in the United States. A Report to Congress by the National Bureau of Standards," Special Publication 511-1, National Bureau of Standards, Gaithersburg, Maryland, 1978.
2. NACE Publication 3D170 (1984 Revision), "Electrical and Electrochemical Methods for Determining Corrosion Rates," National Association of Corrosion Engineers, Houston, Texas, 1984.
3. F. Mansfeld, "The Polarization Resistance Techniques for Measuring Corrosion Currents," Advances in Corrosion Science and Technology, M. G. Fontana, R. W. Staehle, Eds., Plenum Press, Vol. 6, p. 163, 1976.
4. A. Dravnieks, H. A. Cataldi, Corrosion, Vol. 10, No. 7, p. 224, 1954.
5. M. Stern, A. L. Geary, J. Electrochem. Soc., Vol. 104, p. 56, 1957.

Electrochemical Characteristics of the NILU/SCI Atmospheric Corrosion Monitor

S. HAAGENRUD,* J. F. HENRIKSEN,* and R. WYZISK**

Abstract

The NILU/SCI[(1)] monitor estimates the instantaneous atmospheric corrosion rate by recording and integrating the current generated in electrolytic cells by atmospheric exposure. The instrumentation and data handling is completely automated, recording also the time of wetness (TOW) as the time when the cell current exceeds a fixed "wet" threshold value. The electrochemical equations governing the measuring system are solved by the finite difference method, and used to calculate the potential/current distribution in the cell. The low cell current output is caused by high ohmic drop, small working parts of electrodes, and high nonrecorded corrosion of electrodes. The method is well suited for TOW recordings, and could still be used for qualitative estimation of instantaneous corrosion rates.

Introduction

The rate of atmospheric corrosion of metals is a complicated function of a great number of factors such as the surface wetness, temperature, amount of air pollutants, nature of the material, and the amount, structure, and composition of corrosion products on the metal surface. To obtain a better understanding as to how these parameters affect the atmospheric corrosion, much effort has been devoted to modifying electrochemical techniques for determination of the instantaneous atmospheric corrosion rate and/or the time of wetness (TOW).[1] Most of these methods make use of galvanic cells that generate potential/current across the electrodes when a humidity film is formed. Kucera and Mattson at the Swedish Corrosion Institute (SCI)[2] were the first to develop this technique for practical measurements, which included development of electrolytic cells and an electronic integrator for the continuous recording of TOW and cell current (Figure 1). This work was also taken up by the Norwegian Institute for Air Research (NILU), and through a Nordic research program, SCI and NILU used the method in several studies.[3]

On exposure to the outdoor atmosphere, the cell current varied over a wide range depending on precipitation and humidity conditions, as shown in Figure 1(b). Typical current range for steel cells was 10^{-8} to 10^{-4} A. From the cell current response, the TOW was defined as the time when the cell current exceeded a fixed threshold value. In most studies, this value was set at 10^{-6} A.

Although the principle of the measuring technique is simple, the interpretation of the measured current in electrochemical terms, and its connection with the corrosion rate, is a complex matter. In principle, the cell current could be interpreted by either using electrochemical kinetic terms, or by trying to find an empirical relation between the cell current and the atmospheric corrosion rate. In all the studies conducted thus far, the empirical approach has been chosen, while interpretation in terms of electrochemical kinetics is the main subject of this paper.

Conventionally, the empirical relation between the cell current and the real corrosion rate was expressed by the cell factor (CF). This factor was very low in most cases. For carbon steel cells, the CF varied from 0.05 to 0.15, while for zinc cells, it was higher and varied more.[3] The CF was fairly constant for measurements at one site, but depended to some extent on the degree of pollution

*Norwegian Institute for Air Research, Lillestrom, Norway.
**Institute for Inorganic Chemistry, Wroclaw, Poland.
(1)Norwegian Institute for Air Research (NILU); Swedish Corrosion Institute (SCI).

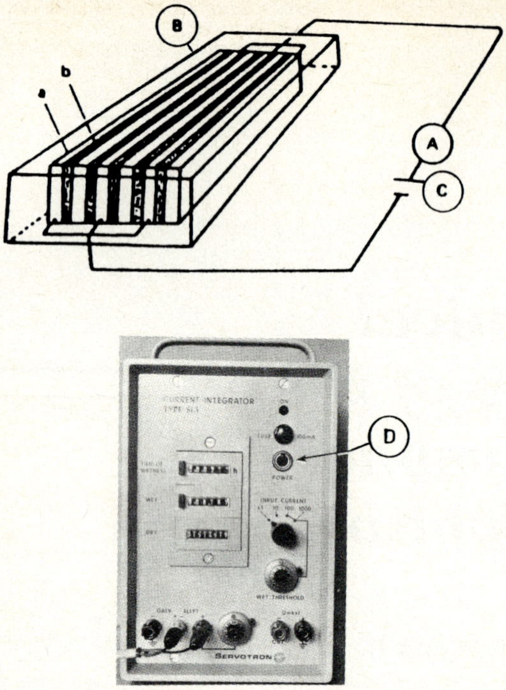

FIGURE 1(a) — General arrangement of electrochemical device for measurement of atmospheric corrosion: (A) zero resistance ammeter, (B) electrochemical cell of electrolytic type, (a) electrodes, (b) insulators, (C) external emf, and (D) electronic current integrator and TOW recorder.

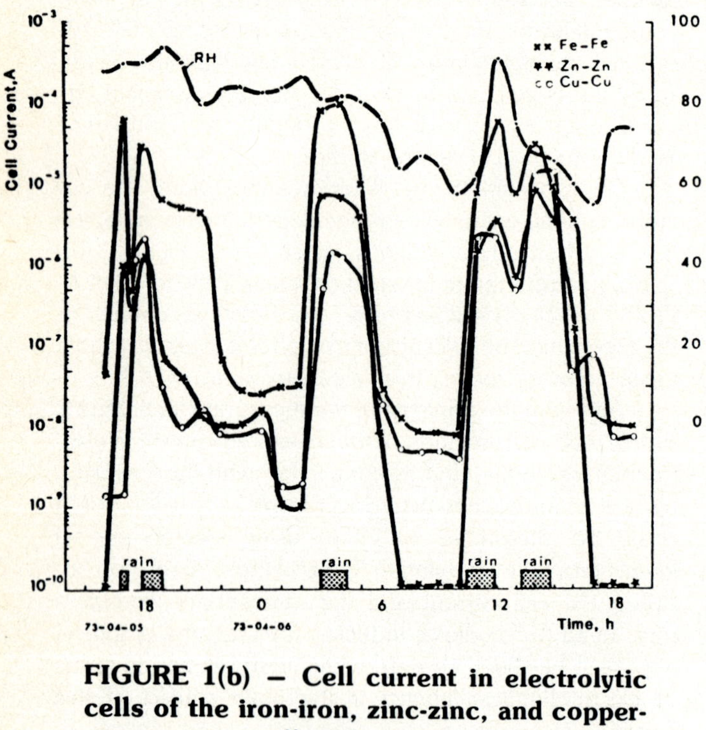

FIGURE 1(b) — Cell current in electrolytic cells of the iron-iron, zinc-zinc, and copper-copper types, all with an imposed potential of 100 mV, and the relative humidity in Stockholm over the period of April 5-7, 1973.[1]

of the atmosphere. This was a serious limitation of the approach.

From the earlier experience, further development was aimed at:

1. Automatic measuring equipment and data analysis.
2. Smaller, cheaper, and more easily produced cells.
3. Interpretation of the measured cell current in terms of electrochemical kinetics.

As the results showed that the method could be used for continuous TOW monitoring, the work with items 1 and 2 was given high priority. The task was fulfilled in 1982/83 by the construction of the instrument "NILU WETCORR"[(2)] (time of WETness and COrrosion Rate Recorder), named after its main functions. This work has been published recently,[4] and it will be briefly described below.

The purpose of the present study was:

1. To develop and describe an electrochemical model for the cell that can be used for calculation of: (1) the electrochemical terms involved and (2) the optimum cell construction in terms of geometrical dimensions.

2. To compare the calculated corrosion from the model with field and laboratory corrosion data.

3. To use the model calculations to investigate the following hypothesis for the low current output: (1) most of the impressed potential is consumed by high ohmic potential drop in the film, and (2) high resistance in the humidity film makes the electrode current distribution so uneven that only a minor part of the electrode is "active."

NILU WETCORR

The development of automatic instrumentation began in 1980, initiated by the need for an automatic multichannel current integrator for study of the corrosivity of different parts of the car body.[5] The instrument is shown in Figure 2(a). The instrument has six independent channels with voltage sources and current integrators for which the impressed potential and wet threshold level can be chosen separately. The impressed potential can be either DC or AC at requested frequency.

The integrated current is recorded on either a "dry" counter, or on a "wet" counter when the wet current threshold is exceeded. The TOW is then simultaneously recorded on another counter. The integrating time interval can be chosen in either minutes for "wet" current recording, or in hours for "dry" current recording. The time interval changes automatically from hours to minutes when the current in one cell exceeds a wet threshold value. The integrator records either the accumulated current, or the current for each separate sampling interval.

In the car study, the TOW was mapped at twelve different points of the car body as a function of driving and weather conditions. The electrolytic cell used was the old

---

[(2)]Registered trade name.

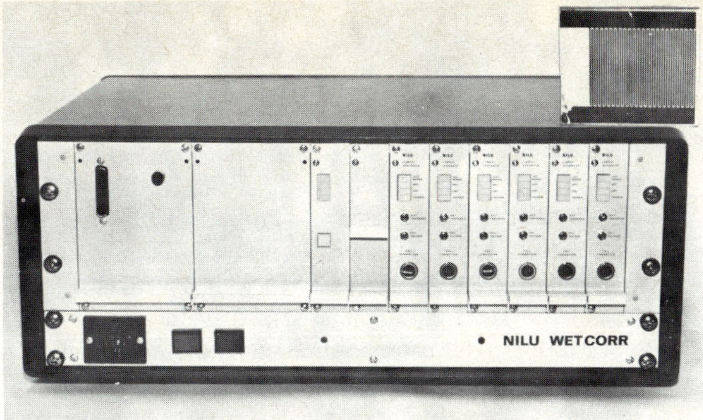

FIGURE 2 – NILU WETCORR and Cu/Cu TOW sensor.

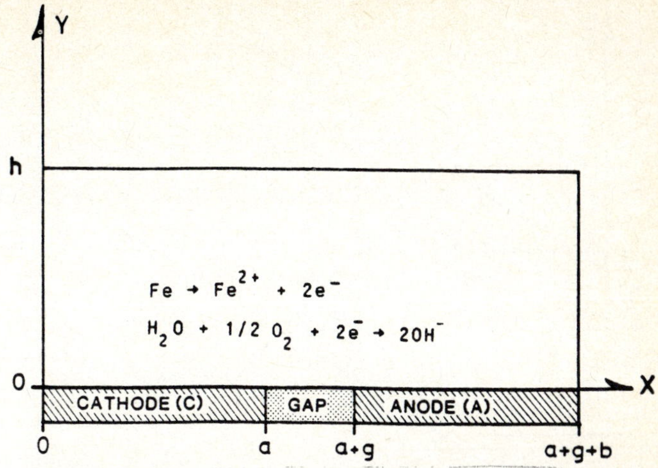

FIGURE 3 – Schematic diagram of the two dimensional electrochemical cell with insulation gap (g) and electrolyte thickness (h).

Fe-type used earlier,[3] which is also the model cell used in the present paper. However, the NILU WETCORR method uses cheap cells produced by the well-developed film technique from the electronic industry [Figure 2(b)]. This is the same type of cell as used by Sereda.[6] In principle, the sensor consists of the metal in question. Producers were found to supply cells of Cu/Cu, Fe/Fe, Zn/Zn, and Al/Al, each at the price of about $10 USA. Some of the measurements conducted in the process of developing the TOW sensors were recently published.[4]

Description of the Model

Analytical Model

The analysis performed follows the methodology as used by Doig and Flewitt.[7] They applied a numerical finite difference method for the calculation of electrode potential distribution within the electrolyte above a coplanar, linear, semi-infinite corrosion couple. Similar to their model, the NILU/SCI cell [Figure 1(a)] was given the geometrical form shown in Figure 3. Since the cell consists of several long, thin parallel anodes and cathodes, it can be assumed that the electrodes are semi-infinite, and that only half of the electrode width affects the potential distribution of the model cell. The next half is assumed to be a part of the next parallel cell. Then the electrode surface lies in the xz plane and is infinite in the $\pm z$ direction. With the abscissa along the cell surface, and the origin at the middle of one of the electrodes, the couple consists of a cathodic element of width "a," an insulation gap of width "g," and an anode of width "b," covered by an electrolyte of depth "h" in the $+y$ direction. The electrolyte is bounded by perfect insulators at $y = h$ and $x = 0$ and $x = a+g+b$. The symmetry of the electrode configuration constrains the current flow to the xy plane such that the analysis is reduced to a two-dimensional problem.

Assuming the cell reactions to be under activation control with logarithmic Tafel polarization behavior, the net current on each electrode follows the equation[8]:

$$I = I_{corr}\left(\exp\frac{E - E_{corr}}{\alpha_{an}} - \exp\frac{E_{corr} - E}{\beta_{cat}}\right) \quad (1)$$

Equation (1) is valid under the assumptions that only one anodic and one cathodic reaction take place on each electrode, and that the anodic and cathodic reactions take place randomly on the surface without any IR drop involved in the reactions.

Since there is no accumulation or loss of ions in the bulk of the electrolyte, the distribution of electrode potential within the electrolyte must satisfy Laplace's equation

$$\nabla^2 E = 0 \quad (2)$$

with the following boundary conditions:

At anode and cathode surface

$$\frac{\delta E}{\delta y}\bigg|_{\substack{o \le x \le a \\ y = o}} = \frac{I_C(x)}{C}$$

$$\frac{\delta E}{\delta y}\bigg|_{\substack{a+g \le x \le a+g+b \\ y = o}} = \frac{I_A(x)}{C} \quad (3)$$

where C is the conductivity of the electrolyte.

Since there is no current flow perpendicular to the insulating boundary, nor through the insulation:

$$\frac{\delta E}{\delta x}\bigg|_{\substack{x = o \\ o \le y \le h}} = 0$$

$$\frac{\delta E}{\delta x}\bigg|_{\substack{x = a+g+b \\ o \le y \le h}} = 0$$

$$\frac{\delta E}{\delta y}\bigg|_{\substack{y = h \\ o \le x \le a+g+h}} = 0 \quad (4)$$

$$\frac{\delta E}{\delta x}\bigg|_{\substack{x = a \\ y = o}} = 0$$

$$\frac{\delta E}{\delta x}\bigg|_{\substack{x = a+g \\ y = o}} = 0$$

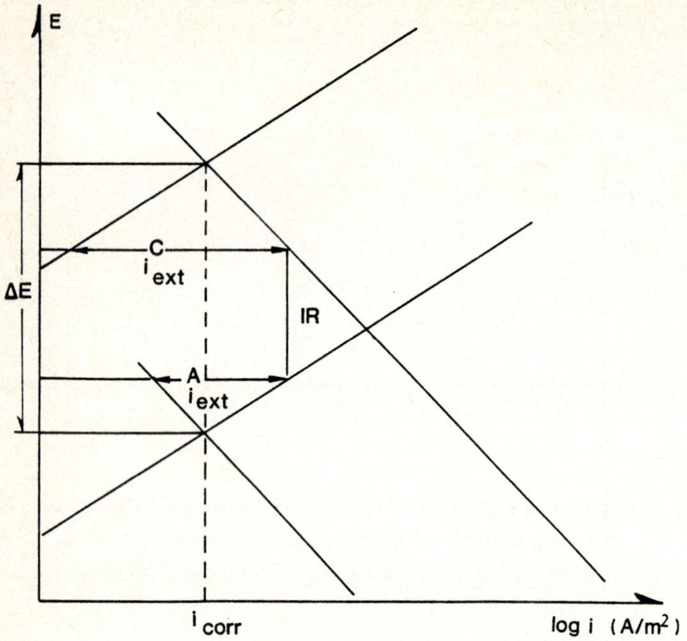

FIGURE 4 – Stern diagram of the polarized electrochemical cell with an impressed potential $\triangle$E assuming that the Tafel slopes and the corrosion current have the same values on both the anode and the cathode.

$$\frac{\delta E}{\delta y}\Big|_{\substack{y=o \\ a \leq x \leq a+g}} = 0 \qquad (5)$$

Having found the right expressions for the net cathodic current, $I_{C(x)}$, and net anodic current, $I_{A(x)}$, from Equation (1), these boundary conditions can be used to solve Laplace's equation.

When anode and cathode in the model cell are of the same material, it can be assumed that the Tafel slopes (α and β) and the corrosion current (i_{corr}) had the same values on both the anode and the cathode. Taking into account the impressed potential $\triangle$E, and the IR drop, it is possible, in the same way as for galvanic couple, to express the steady-state potential of the two electrodes in the electrolytic cell as (Figure 4):

$$E_{corr} \text{ and } E'_{corr} = E_{corr} + \triangle E$$

Taking into account all locations along the electrodes, the final equations governing the polarization characteristics of the electrodes are:

$$|I_C(x)|_{o \leq x \leq a} = I_{corr}\left(\exp\left(\frac{E'_{corr} - E_x}{\beta}\right) - \exp\left(\frac{E_x - E'_{corr}}{\alpha}\right)\right)\Big\{ |I_A(x)|_{a+g \leq x \leq a+g+b} =$$

$$I_{corr}\left(\exp\left(\frac{E_x - E_{corr}}{\alpha}\right) \qquad (6)\right.$$

$$\left. - \exp\left(\frac{E_{corr} - E_x}{\beta}\right)\right) \qquad (7)$$

These two equations are valid as long as the cell reactions are under activation control. Under atmospheric conditions with a very thin humidity film varying in thickness from practically zero to about 1000 μm, either the anodic or the cathodic reaction will be diffusion controlled at low and high thicknesses, respectively. To take into account the diffusion controlled state of the reaction, the equations are modified as follows:

$$|I_C(x)| = I_{corr}\exp\left(\left(\frac{E'_{corr} - E_x}{\beta}\right)\left(1 - \frac{|I_C(x)|}{I^{lim}_{cat}}\right)\right) - I_{corr}\exp\left(\left(\frac{E_x - E'_{corr}}{\alpha}\right)\left(1 - \frac{I_C(x)}{I^{lim}_{an}}\right)\right) \qquad (8)$$

$$I_A(x) = I_{corr}\exp\left(\left(\frac{E_x - E_{corr}}{\alpha}\right)\left(1 - \frac{I_A(x)}{I^{lim}_{an}}\right)\right) - I_{corr}\exp\left(\left(\frac{E_{corr} - E_x}{\beta}\right)\left(1 - \frac{I_A(x)}{I^{lim}_{cat}}\right)\right) \qquad (9)$$

The Laplace equation has been solved by the finite difference method of successive over-relaxation using an iterative numerical procedure.[9,10] This required a rectangular cartesian grid net to be established over the region defined by the electrolyte. The grid contains 20 vertical and 39 horizontal lines. The potential values at the nodal points, defined by the intersection of the grid lines, were then determined by iterative solution of the system of finite difference equations associated with a numerical solution of Laplace's equation.

Model Input

The present computer program allows the variables listed below to be specified by the user. Values are given for the variables that have been kept constant for the calculations performed so far. The values have been obtained from the literature.[11-13] All numbers are given in SI units.

- Cathodic finger width: 0.25×10^{-3} m
- Cathodic Tafel slope: 0.220 V
- Cathodic exchange current: 0.1×10^{-4} A/m²
- Anodic finger width: 0.25×10^{-3} m
- Anodic Tafel slope: 0.123 V
- Anodic exchange current: 0.1×10^{-2} A/m²
- Anodic exchange potential: −0.559 V
- Insulating gap width: 0.2×10^{-3} m

- Cell length: 0.4 m
- Externally applied voltage: 0.1 V

For the following variables, values have been changed in different runs:

- Cathodic limiting current: 0.1, 0.2, 0.5, 1, 2, 5, 10 A/m^2
- Cathodic exchange potential: (1.228 − 0.059 pH) V. (Varies according to the chosen conductivity (pH) of electrolyte.)
- Anodic limiting current: 0.1, 0.2, 0.5, 1.0, 2.0, 5.0, 10 A/m^2.
- Electrolyte film thickness: Chosen to account for the thicknesses that occur during various humidity conditions in the atmosphere. These are: 90% RH (0.1 μm). 100% RH (1 μm), dew (5 to 10 μm), rain (50 to 100 to 1000 μm).
- Electrolyte conductivity: 0.228 S (pH 2.5), 0.0845 S (pH 3.2), 0.0043 S (pH 4.2), 0.008 S (pH 5.6).

These conductivities (and pH values) are chosen to account for the various pollutant concentrations occurring. By combination of various electrolyte film thicknesses and electrolyte conductivities, the various environmental scenarios can be accounted for.

Results

Program Output

The following variables are calculated and tabulated for each selection of input variables:

- Working part[3] of cathodic strip (%)
- Working part of anodic strip (%)
- Integrated cell current (A)
- Free corrosion current (A/m^2)
- Free corrosion potential (V)
- Integrated IR drop (V)

[3] The notation *working part* is introduced to express quantitatively the nonuniformity of the current distribution on the electrode. The working part of the electrode is defined as the percentage of the area through which 95% of the integrated cell current flows.

In addition, the program generates the: (1) potential and cell current distributions along the width of the electrodes and (2) corrosion current distribution along the width of the electrodes.

In this context, the term *corrosion current* is equivalent to the term *partial anodic current*. The cell current and corrosion current distribution are also shown graphically in Figure 5.

Corrosion Potential and Current at Various Limiting Diffusion Currents

Figure 6 shows the Stern diagram with the corrosion potentials and corrosion currents calculated for the various anodic and cathodic limiting diffusion currents chosen.

In atmospheric corrosion, the oxygen diffusion limiting current decreases and the anodic limiting diffusion current increases with increasing film thickness. The exact relationship is unknown, but the points in Figure 6 are calculated from the hypothetical relationship between film thickness and diffusion limiting currents shown in Table 1. This relationship is used for all calculations.

Working Part of the Electrodes

The working part of the electrodes as a function of film thickness is shown in Figure 7. The nonuniformity of current distribution is clearly seen. The working parts of the electrodes are different. The anodic working part goes through a minimum of about 60% at 10 μm and the cathodic working part decreases to about 10% at 1 μm [100% relative humidity (RH)] thickness.

IR Drop in the Electrolyte

In order to find the effective polarizing part of the impressed potential, the IR drop in the electrolyte must be calculated. Because of the nonuniformity of the cell current distribution, the IR drop had to be calculated by a numerical integration procedure along the length of the cell fingers. Figure 8 shows the integral IR drop for varying limiting diffusion currents with varying film thicknesses.

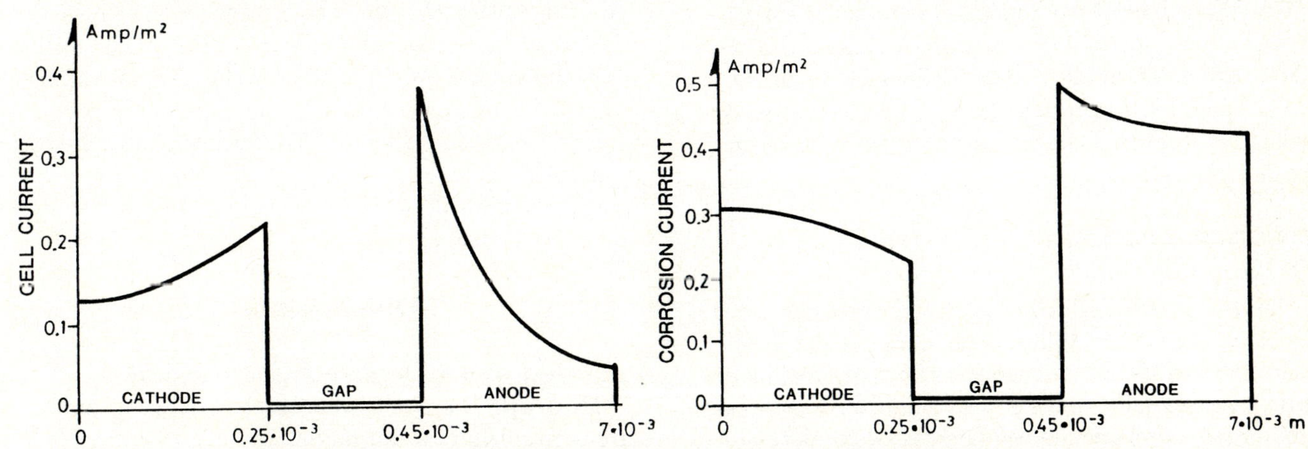

FIGURE 5 — The cell current and corrosion current distribution along the electrode surfaces at ΔE = 100 mV, pH = 4.2, and electrolyte thickness 5 × 10^{-5} μm and $i_{lim}^{O_2}$ = 0.6 A/m^2, i_{lim}^{Fe} = 3.0 A/m^2.

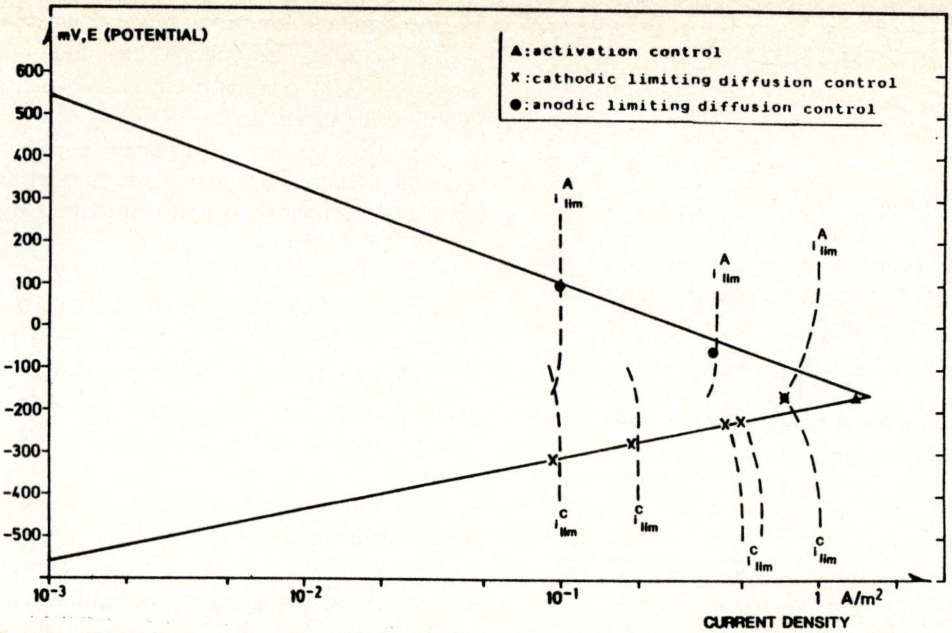

FIGURE 6 – Stern diagram at pH 4.2 with the calculated corrosion current and corrosion potential for various electrolyte thicknesses and corresponding anodic and cathodic limiting diffusion currents according to Table 1.

TABLE 1 – Hypothetical Relation Between Increasing Film Thickness and Anodic/Cathodic Limiting Diffusion Currents

Thickness μm	1	5	10	50	100	500	1000
i^{A}_{lim} (A/m²)	0.1	0.4	1.0	3.0	5.0	10.0	10.0
i^{C}_{lim} (A/m²)	2.0	1.3	1.0	0.6	0.5	0.2	0.1

The ohmic part of the impressed potential is always high. For a pH value of 4.2, it amounts to almost 100% below thickness 10 μm (dew). At a thickness of about 500 μm, it still amounts to about 55%, i.e., for all measuring cases the ohmic drop accounts for more than 50% of the impressed potential.

Cell Current and Corrosion Current at Various pH Levels

Figure 9 shows the cell current and corrosion current as a function of electrolyte conductivity at film thicknesses of 10 and 100 μm. As expected from the very high IR drop involved in the process, the cell current increases several orders of magnitude with increasing conductivity (decreasing pH). In contrast, the corrosion current that constitutes the main part of the current on the electrodes, and which is independent of the IR drop, increases much less with decreasing pH.

Cell Current and Film Thickness

The cell current generated as a function of film thickness is shown in Figure 10. Most of the calculations are for pH 4.2. As seen, the cell current increases several orders of magnitude when the film thickness increases. It goes through a maximum at about 100 μm and decreases thereafter as a result of limiting oxygen diffusion and less decreasing IR drop.

Discussion

Measured and Calculated Cell Current

IR drop. There is fair consistency between the cell current calculated by the model and the experimentally obtained cell current.[3] The experimentally obtained data have three striking characteristics. First, when the cell current exceeds the wet threshold, there is an abrupt increase in current over 2 to 3 orders of magnitude. Sec-

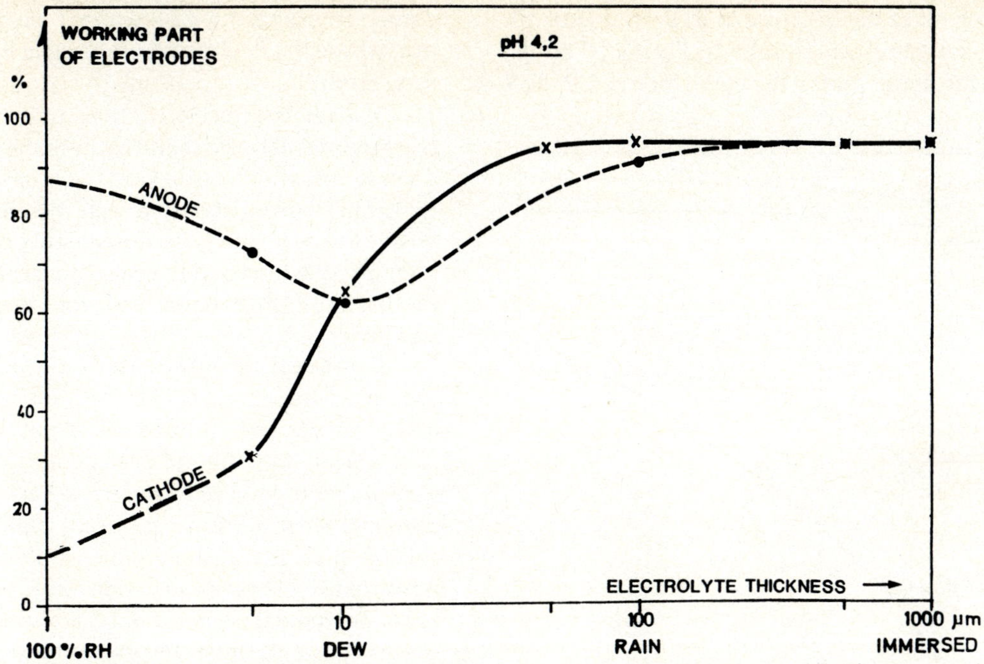

FIGURE 7 – Calculated working parts of the electrodes for various electrolyte thicknesses at pH 4.2, and corresponding anodic and cathodic limiting diffusion currents according to Table 1.

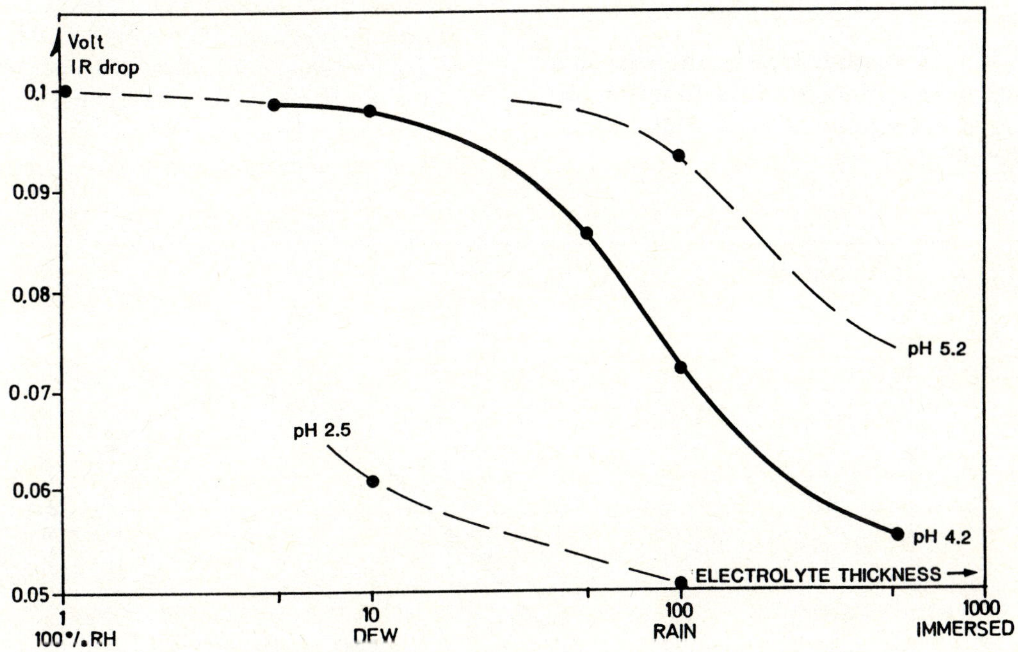

FIGURE 8 – Calculated IR drop in the cell as a function of electrolyte thickness at pH 2.5, 4.2, and 5.2.

ond, a more or less visible humidity film seems to be necessary to bring the current into the wet region. High relative humidities alone (>90%) do not cause the cell current to exceed the wet threshold. Third, in the wet region, the current seems to be much less dependent on the degree of wetness. The wet current rarely exceeds 20 μA, and cell current greater than 100 μA have never been experienced.

These observations can be explained by the model. Because of the very high IR drop involved for all humidity film thicknesses below 10 μm (Figure 8), only a small part of the electrode is active and very little current flows. When the film increases beyond 1 to 10 μm, the IR drop decreases rapidly and the current increases in the same way. At higher thicknesses, the cell current diminishes for two reasons: (1) decreasing oxygen diffusion (Table 1) and (2) leveling off of the decrease in IR drop (Figure 8). The effect of the latter can be seen in Figure 10 by the dotted line in the upper part of the curve for pH 4.2. This

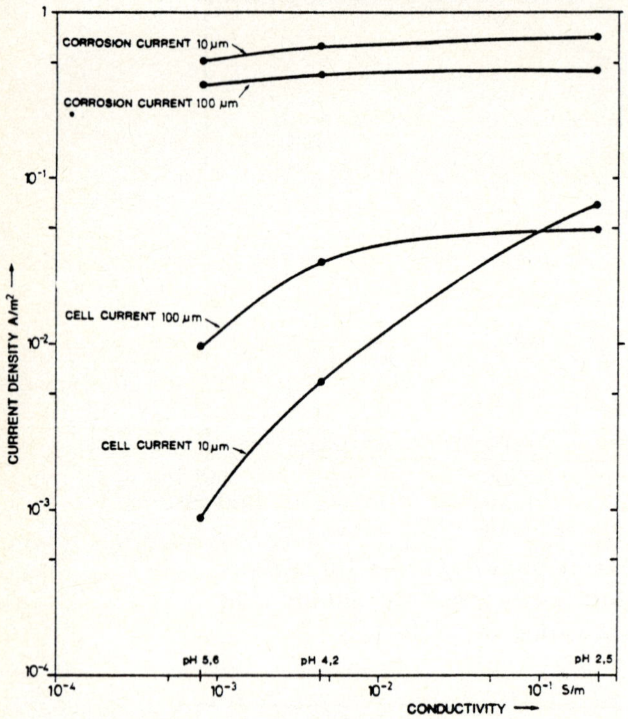

FIGURE 9 – Calculated cell currents and corrosion current densities as function of electrolyte conductivity at electrolyte thicknesses 10 and 100 μm.

curve represents the case without oxygen diffusion control, where the deviation from straight line is caused by IR drop control.

The relative position of the curves in Figure 10 will of course depend on chosen values of the variables, such as Tafel slopes and the true level of O_2 diffusion limiting currents and their relationship with the changing film thickness. However, the good agreement between the experimentally obtained and the calculated cell current confirms the validity of the model and the values chosen for the variables.

Conductivity (pH). Because of the very high IR drop involved at low thicknesses, the conductivity of the film is of decisive importance. Apparently, that would also indicate that the pH of the film is very important. A very wide range of pH values has been reported for thin humidity films.[15] From Figure 10, it should then be concluded that the cell current would exceed the wet threshold level over a very wide range of film thicknesses, as for example 0.3 μm for pH 2.5 and 20 μm for pH 5.6. This is contradictory to the experimentally obtained data, where "wet" current is never obtained below 95% RH. The contradiction between theory and experiment is due to the fact that the acidity on the corroding surface neutralizes very rapidly by the oxygen reduction. Calculations show that the acidity in thin films is neutralized in less than 1 s.[12] When it rains, the electrolyte is of course constantly renewed. On the other hand, pH in rainwater is seldom below pH 4; thus, it can be concluded that the range of pH occurring on corroding surfaces is probably very narrow. This is consistent with measurements showing small differences between the conductivity of

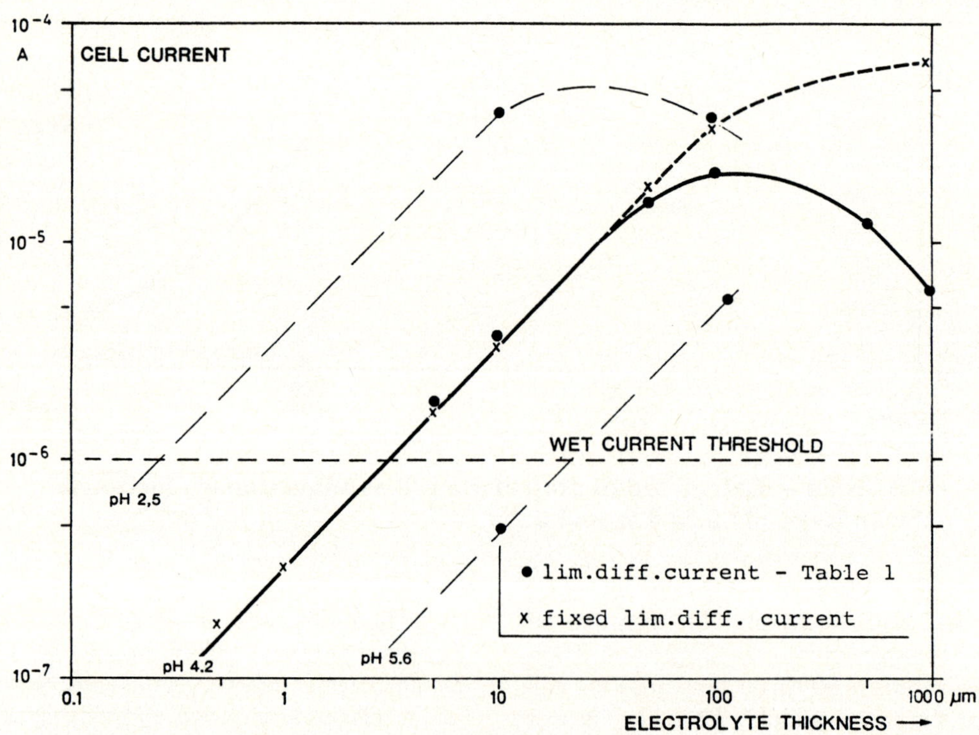

FIGURE 10 – Calculated cell currents as function of electrolyte thickness at pH 2.5, 4.2, and 5.2. The conventionally used wet current threshold of 10^{-6} A is marked.

precipitation in rural and urban atmospheres. In relation to Figure 10, it must then be concluded that for a constant film thickness, the range of possible cell currents is much less than what can be postulated. As a result, the range of humidities for exceeding the wet current threshold, is also much less. The method should thus be useful as a TOW meter.

Relation Between Cell Current and Corrosion Current

The low and relatively constant cell factor (CF) can now be explained (Figure 9). The cell current is controlled and minimized by the high IR drop, while the corrosion current, according to the mixed potential theory of the model, is not influenced. This results in a very low CF at low film thicknesses, and a considerably higher and more constant CF in the "wet" region. As can be seen from Figure 9, at pH 4.2 the:

$$CF = \frac{5 \times 10^{-3}}{5 \times 10^{-1}} \approx 0.01 \text{ at } 10\ \mu m$$

and

$$\frac{4 \times 10^{-2}}{4 \times 10^{-1}} \approx 0.1 \text{ at } 100\ \mu m$$

respectively.

The great variation in CF for "dry" and "wet" periods is of no significance for the experimentally determined cell factors for longer exposure periods. This is because the accumulated atmospheric corrosion is dominated by the corrosion occurring in the wet periods, according to:

$$C = \sum_{i=1}^{i=n} \tau_i v_i \qquad (10)$$

where: C = accumulated corrosion, τ_i = time of wetness for each individual wet period, and v_i = corrosion rate corresponding to each wet period.

If the accumulated atmospheric corrosion rate is now considered to be determined by the real wet periods with visible film thickness, then the CF calculated from Figure 9 is in very good agreement with the experimentally obtained CF of 0.05 to 0.15 for monthly and yearly exposures.[3] In addition, the observed slight dependency of CF on the pollution level or conductivity[3] can also be explained from the curves in Figure 9. At low thicknesses, the conductivity dependency is high; at high thicknesses, it is low.

These findings imply that the NILU/SCI measuring principle cannot be used for exact quantification of the corrosion rate in each individual wetness period. However, having now explained the level and variation of CF, the method can then confidently be used to evaluate the instantaneous corrosion rate with a fair degree of accuracy. Although the measurements and data handling are automatic, any quantitative evaluation of individual periods must be done separately. Simultaneous recordings of rain and relative humidity will be valuable in such evaluations, as would studies of each variable under controlled conditions.

Comparison of Corrosion Current with Weight Loss Data

Previously reported monthly corrosion data for rural and industrial atmospheres range from 15 to 110 g/m^2.[3,15] With TOW = 35%, the mean corrosion rates for the wetness periods will be about 43 and about 315 g/m^2, respectively. Calculated as corrosion current, this is equivalent to i_{corr} between 0.15 and 1.0 A/m^2. This is in very good agreement with the corrosion currents obtained from Figure 9.

Future Applications

As reported earlier, the technique is suitable for TOW measurements. It should also be valuable for comparative studies of different alloys, and for studies of protective properties of paint coatings. With the presently developed theory for the cell function, the technique should also be very useful as a continuous atmospheric corrosion rate monitor. However, the exact quantitative evaluation of each wet period requires additional environmental data, such as values for conductivity and film thickness.

With regard to further development of the method, the presently used principle involves AC impressed potential to counteract the electromotive force being generated by DC polarization. In addition, the electrolytic cell is further developed, as shown in Figure 2. Further development of the cell will depend on the main purpose of study. The output of the cell can be increased by decreasing the gap and the width of the fingers and increasing the total length of the fingers. This seems to be purposeful for TOW sensors. For atmospheric corrosion rate sensors, the thickness of the fingers and the gap between them should be kept as great as practical to ensure proper functioning. With the presently obtained production procedure, and the analytical model developed, cheap cells of optimum dimensions can be produced as required.

Conclusions

1. The results obtained by the model are in agreement with experimentally obtained data, and serve to explain the observations.

2. The low cell current output is caused by high ohmic drop and small working parts of the electrodes.

3. The cell factor varies very much with the conductivity and thickness of the electrolyte, being $<<1\%$ in dry periods and about 1 to 20% in wet periods. As a result of the high ohmic drop, the nonrecorded corrosion of the electrodes is always considerable.

4. The method can be used for recording time of wetness (TOW) and for a fairly accurate evaluation of the continuous atmospheric corrosion rate.

References

1. F. Mansfeld, "Electrochemical Methods for Atmos-

pheric Corrosion Studies," Symp. on Atmos. Corros., Hollywood, Florida, p. 139, 1980.
2. V. Kucera, E. Mattsson, ASTM STP 558, ASTM, Philadelphia, Pennsylvania, p. 239, 1974.
3. S. E. Haagenrud, V. Kucera, J. Gullman, "Atmospheric Corrosion Testing with Electrolytic Cells in Norway and Sweden," Symp. on Atmos. Corros., Hollywood, Florida, p. 669, 1980.
4. S. E. Haagenrud, J. F. Henriksen, T. Danielsen, A. Rode, "An Electrochemical Technique for Measurement of Time of Wetness," Third Inter. Conf. on Dura. of Build. Mats., Helsinki, Finland, 1984.
5. R. Ericsson, S. E. Haagenrud, J. F. Henriksen, "Simultaneous Measurements of Corrosivity and Environment on Different Parts of the Automotive Body," CORROSION/82, National Association of Corrosion Engineers, Houston, Texas, 1982.
6. ASTM G 01.04 Draft proposal "Standard Recommended Practice for the Measurement of Time of Wetness on Surfaces Exposed to Wetting Conditions as in Atmospheric Corrosion Testing," ASTM, Philadelphia, Pennsylvania.
7. T. Doig, T. E. J. Flewitt, J. Electrochem. Soc., Vol. 126, No. 12, p. 2057.
8. F. Mansfeld, "The Polarization Resistance Technique for Measuring Corrosion Currents," Advances in Corrosion Science and Technology, Vol. 6, 976, p. 163.
9. B. A. Carre, Computer J., Vol. 4, p. 73, 1961.
10. R. Wycisk, S. E. Haagenrud, J. F. Henriksen, Calculation of Potential and Current Distribution in Electrochemical Cells," NILU-report in preparation.
11. F. Mansfeld, Corrosion Kinetics Under Thin Electrolyte Films, Abstract No. 121, Electrochemical Society Fall Meeting, Pittsburgh, Pennsylvania, 1978.
12. S. E. Haagenrud, PhD Thesis. Dept. of Metall. and Metals Working, Technical University of Norway, February 1972. (Norwegian.)
13. J. A. Gonzales, E. Otero, C. Cabanas, J. M. Bastidas, Brit. Corros. J., Vol. 19, No. 2, p. 89, 1984.
14. R. Grauer, P. J. Moreland, G. Pini, A Literature Review of Polarization Resistance Constant (β) Values for the Measurement of Corrosion Rate, National Association of Corrosion Engineers, Houston, Texas, 1982.
15. V. Kucera, E. Mattsson, "Atmospheric Corrosion," in: F. Mansfeld, Corrosion Mechanisms. M. Decker, Inc. (In print.)

Corrosion Monitoring by Galvanostaircase Polarization

S. T. HIROZAWA*

Abstract

The galvanostaircase polarization (GSCP) method has a distinct advantage over potential-controlled polarization methods, particularly when applied to metals that show a strong tendency to passivate, e.g., aluminum. GSCP eliminates the induction effect; hence, it also eliminates current hysteresis. Therefore, the critical potentials and the passivation current obtained from a GSCP curve have practical utility in the monitoring of various types of corrosion. For example, the breakdown potential appears to be related to the general corrosion rate, while the protection potential appears to be related to localized corrosion. In most antifreeze coolants tested, the passivation current ($i_{pass'n}$) appears to be related to corrosion accompanying physical depassivation, i.e., when the passive film is destroyed by sand abrasion, cavitation, or nucleate boiling at heat-rejecting Al surfaces. The use of $i_{pass'n}$ provides an added dimension in the assessment of corrosion inhibition.

Introduction

Potential-controlled polarization methods have two very important drawbacks, namely, the induction effect and current hysteresis.[1] The induction effect is responsible for the positive bias and nonreproducibility in the determination of the breakdown or pitting potential (E_b or E_{pit}), as well as the scan rate sensitivity of E_b.[1-6] E_b obtained by potentiodynamic polarization (PDP) did not correlate well with service conditions[6] and other laboratory tests.[5]

Current hysteresis, commonly observed in cyclic PDP, is a direct result of the induction effect.[7,8] The extent of current hysteresis is affected by the point of reversal of the voltage; the higher the current at the point of reversal, the larger the hysteresis loop.[9] The area of the loop is equivalent to charge density if time is substituted for potential.[8] Rudd and Scully[9] observed that the protection potential (E_{prot}) is shifted in the negative direction by increasing the size of the hysteresis loop while Wilde[10] found that the value of E_{prot} varies linearly with the log of the coulombs passed. Several attempts were made to obtain a more reliable E_{prot}. Pesall and Liu[11] developed the potentiostatic scratch method, which appeared to be the best approach at that time. Syrett[12] pointed out that the scratch method suffers the disadvantage that the passive film may not be broken in the favored pitting sites. Sinigaglia, et al.,[13] found that the scratch method was not precise when the alloy contained particularly favorite sites for pit nucleation. Barbosa and Scully[14] found that E_{pit} cannot be accurately determined using the decrease of the repassivation rate near E_{pit}, since they found pitting to occur at potentials for which the repassivation rate was high. Therefore, the scratch method is ideal for alloys that are relatively free of inclusions. To overcome the inclusions problem, Syrett[12] developed the pit propagation rate (PPR) method for the determination of E_{prot} but Sinigaglia, et al.,[13] found the method to be too tedious and difficult.

A galvanostaircase polarization (GSCP) method was developed,[1,7,8,15] which combines the advantages of the quasistationary galvanostatic method (Broli, et al.[4]), the scratch potentiostatic method (Pesall and Liu[11]), and the PPR curve method (Syrett[12]). The advantages of GSCP are: (1) rapid, accurate, and precise (agrees with scratch test on pure Al), (2) eliminates induction time, (3) eliminates current hysteresis, (4) charge density is programmed precisely and kept at a minimum, (5) noise information in voltage transients signals commencement of pitting, (6) corrosion is localized at the most susceptible sites, and (7) two critical potentials are obtained: E_{prot} is reproducible and meets the usual definition of E_{pit}, while E_b is nonreproducible because it is related to the degree

*BASF Corp., Wyandotte, Michigan.

of passivation. The higher the E_b (relative to E_{prot}), the greater the inhibition against general corrosion but the greater the tendency toward crevice corrosion. Because of the elimination of the induction effect, E_b obtained by GSCP is more negative than E_b determined by potentiodynamic or quasistationary potentiostatic polarization.

The previous works on GSCP[1,5,7,8,15] were concerned with the localized corrosion on Al. The purpose of this paper is to demonstrate the usefulness of the passivation current obtained from a GSCP diagram in assessing the probability of accelerated corrosion accompanying abrupt depassivation, e.g., pump cavitation, sand abrasion, and nucleate boiling at heat-rejecting Al surfaces.

Experimental

A programmable timer was used to control a Princeton Applied Research Company (PARC)[(1)] Model 175 universal programmer to generate a voltage staircase signal. This voltage staircase signal was fed to the PARC Model 173 potentiostat/galvanostat and, under the "CONTROL I" mode, the resulting current staircase signal was applied to the aluminum working electrode. The potential transients resulting from the applied current staircase were recorded. The steady-state potentials at the end of the up-steps were extrapolated to zero time (or current) to obtain the breakdown potential (E_b). The steady-state potentials at the end of the down-steps were extrapolated to zero current to obtain the protection potential (E_{prot}). The method for the determination of the passivation current ($i_{pass'n}$) and the depassivation current (i_{depass}) will be described later.

The preparation of the aluminum test specimens and the test media were described earlier.[5]

Results and Interpretation

Galvanostep Method[5]—Concentration Effect

The development of the GSCP method in this laboratory was preceded by the use of the galvanostep (stationary galvanostatic) method. Figure 1 shows the galvanostep curve of an experimental coolant C at three concentrations. The steady-state voltages at each current step are plotted against current. E_b for the 45% antifreeze solution is 1.4 V, but E_b drops precipitously for the 25% solution. No further change is seen at 15%. However, note that the current required to initiate passivation ($i_{pass'n}$) increases when the antifreeze concentration is decreased from 25 to 15%. It appeared that this $i_{pass'n}$ phenomenon might expand the usefulness of GSCP in screening of inhibitors and optimization of coolants.

The galvanostep method has several disadvantages. It is time-consuming and requires human judgment to determine when steady-state is achieved. To save time, only 7 points were determined between 0 and 1000 $\mu A/cm^2$. In order to improve the resolution in the determination of $i_{pass'n}$, it was essential to obtain more points particularly between 500 and 1000 $\mu A/cm^2$. Around 50 steps in each direction would be ideal. The galvanostep method would have been much too time-consuming.

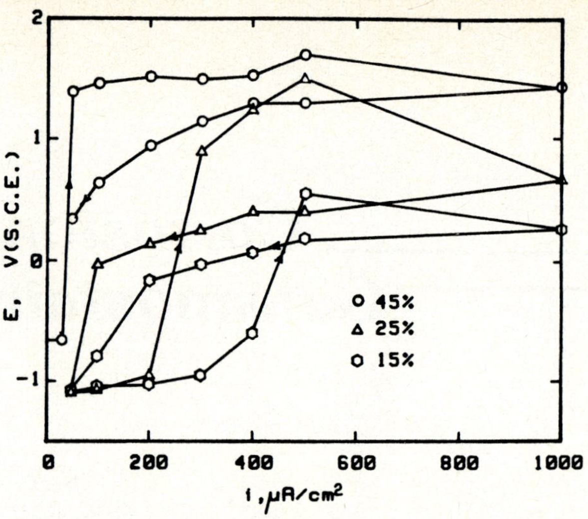

FIGURE 1 — Effect of concentration on the galvanostep polarization curves of AA 3003 Al in experimental antifreeze C at room temperature.

GSCP was developed to overcome these disadvantages of the galvanostep method.

Temperature Effect

Figure 2(a) shows the dilution effect of a commercial coolant (A) on the GSCP curves at room temperature. Dilution lowers E_b in a passivating medium, but dilution has no effect on E_{prot}. The latter is observed in nitrate-free coolants (reasons given in Reference 16). Figure 2(b) shows the GSCP curves of the 25 and 15% solutions at 82.2 C. The 15% solution is unable to passivate the Al up to 1000 $\mu A/cm^2$, while the 25% solution requires 350 $\mu A/cm^2$ to initiate passivation.

Figure 3(a) shows the GSCP curves for a nitrate-containing experimental coolant B. Note that both E_b and E_{prot} are affected by dilution—this is attributed to the nitrate effect.[5,15,16] Figure 3(b) shows the GSCP curves for the same solutions run at 82.2 C. For the sake of clarity, only the up-step curves are given. At this elevated temperature, the 45% solution required 347 $\mu A/cm^2$ and the 25% solution required 910 $\mu A/cm^2$ to initiate passivation, while the 15% solution showed no sign of passivation up to 1000 $\mu A/cm^2$. By keeping the dilution constant, it is possible to determine the critical concentration of a key inhibitor or the critical pH of a given environment using $i_{pass'n}$.

pH Effect

Figure 4 shows the effect of pH on the GSCP curves of AA[(2)] 3003 Al in 25% ethylene glycol solution containing 1% $NaNO_3$ and 100 ppm NaCl. The pH 11 curve shows how $i_{pass'n}$ and i_{depass} are determined by extrapolation. The $i_{pass'n}$ and i_{depass} values obtained from Figure 4 are plotted in Figure 5. The i_{depass} curve generally follows the solubility curve of Al as a function of pH. However, since nitrate is present, Al should be stable at low

[(1)]Princeton Applied Research Co. (PARC), Princeton, New Jersey.

[(2)]Aluminum Association (AA), Washington, DC.

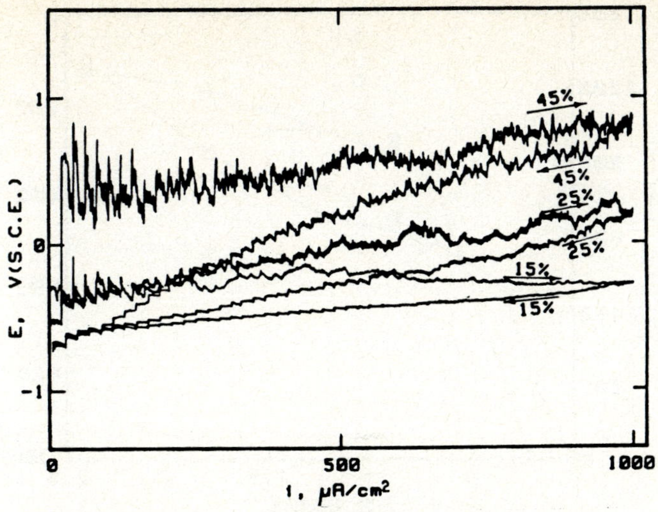

FIGURE 2(a) – Effect of concentration on the GSCP curves of AA 3003 Al in commercial antifreeze A containing no nitrate; room temperature; 20 μA/cm^2; 1 min steps.

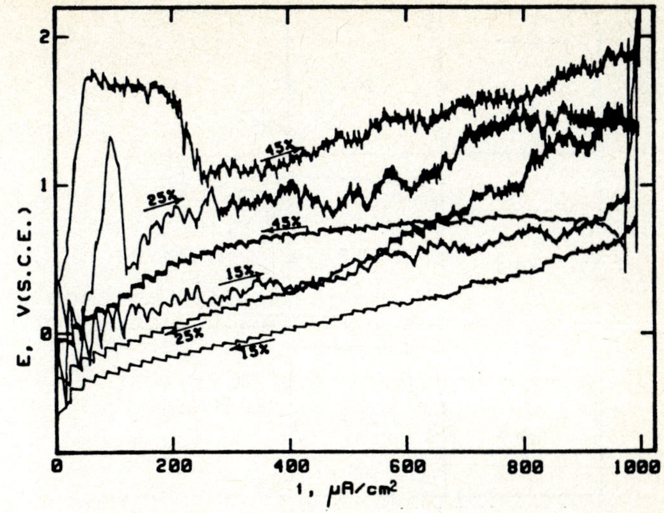

FIGURE 3(a) – Effect of concentration on the GSCP curve of AA 3003 Al in a nitrate-containing commercial antifreeze B; room temperature; 20 μA/cm^2; 1 min steps.

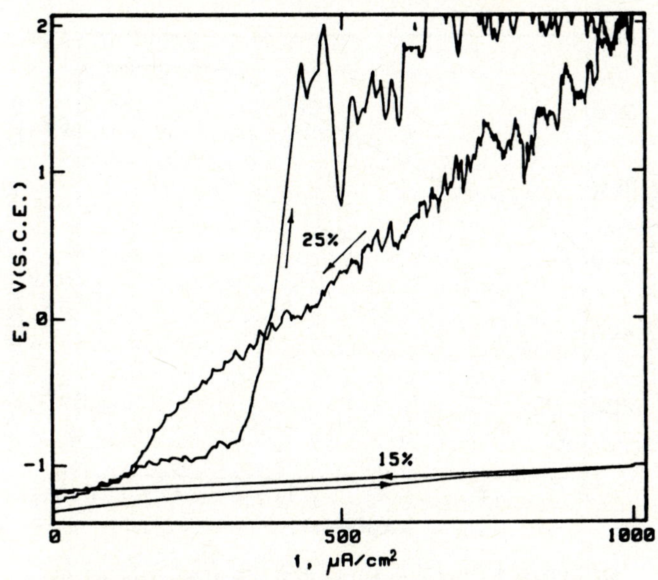

FIGURE 2(b) – Effect of temperature and concentration on the GSCP curves of AA 3003 Al in commercial antifreeze A containing no nitrate; 82.2 C; 20 μA/cm^2; 1 min steps.

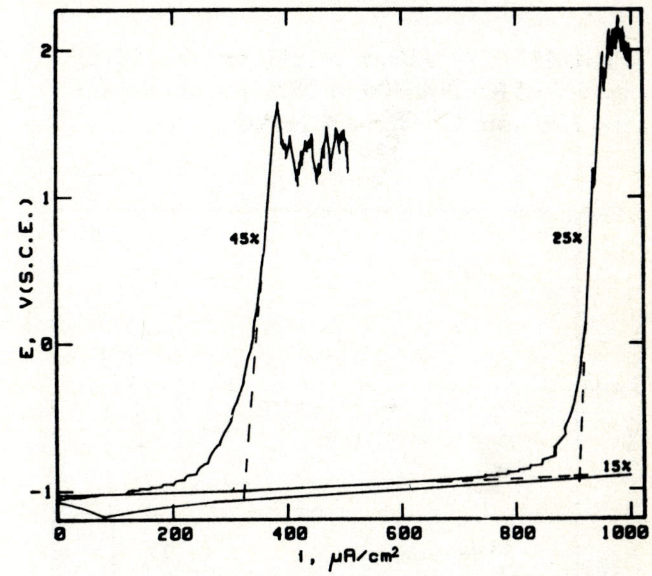

FIGURE 3(b) – Effect of temperature and concentration on the GSCP curves of AA 3003 Al in a nitrate-containing commercial antifreeze B; 82.2 C; 20 μA/cm^2; 1 min steps.

pH as indicated by the $i_{pass'n}$ curve. As pH is increased, $i_{pass'n}$ started to increase at pH > 10. Therefore, it appeared that the $i_{pass'n}$ is more sensitive than i_{depass} in detecting weakness in the corrosion inhibition of coolants in the neutral to basic region. $i_{pass'n}$ was selected for the correlation studies in the evaluation of antifreeze coolants.

Correlation of $i_{pass'n}$ with Al Pump Cavitation Corrosion

Curve A in Figure 6 shows the $i_{pass'n}$ as a function of borax concentration in a 15% experimental coolant D (benzoate/nitrite type). Curve B shows the perforation time for an Al water pump (modified ASTM[3] D2809—the 100 h test duration was expanded to 1000 h) as a function of borax concentration. A high $i_{pass'n}$ probably means that a high rate of generation of the components of the corrosion reaction product exceeding its dissolution rate is necessary to passivate the Al. Therefore, under abrupt depassivation by the cavitation process, repassivation is very difficult and the probability for pump perforation is increased. The perforation time increased as $i_{pass'n}$ (and borax level) decreased, until at 0.2% borax, the pump did not perforate but failed the test with a rating of 2. At 0% borax, the test was passed at a rating of 8.

[3]ASTM, Philadelphia, Pennsylvania.

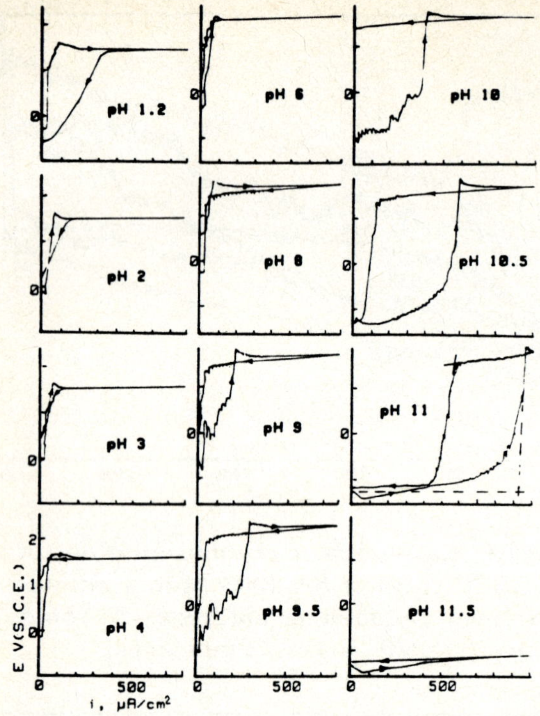

FIGURE 4 – Effect of pH on the GSCP curves of AA 3003 Al in 25% glycol containing 100 ppm Cl^- and 1% $NaNO_3$.

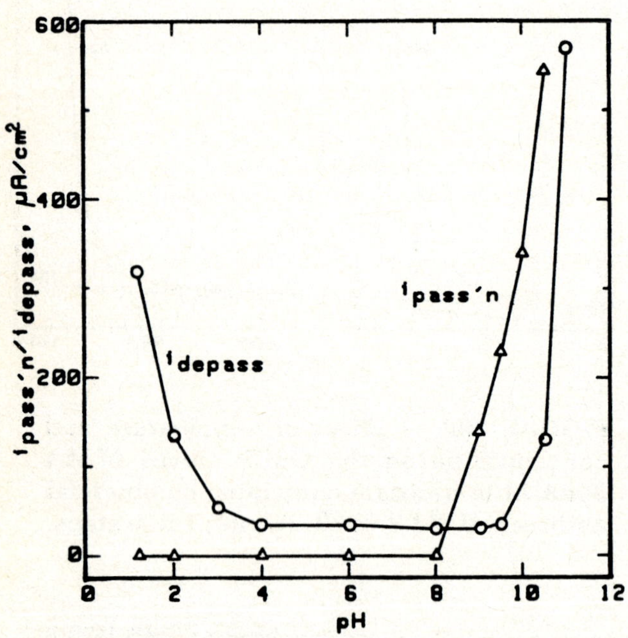

FIGURE 5 – Effect of pH on i_{depass} and $i_{pass'n}$ for AA 3003 Al in 25% glycol containing 100 ppm Cl^- and 1% $NaNO_3$.

The two points that deviate from their respective curves belong to the same experimental coolant where the level of the inhibitors other than borax was increased by 50%. The increase in the inhibitor levels had a very harmful effect: it caused a large positive deviation in $i_{pass'n}$ with a corresponding large negative deviation in the perforation time.

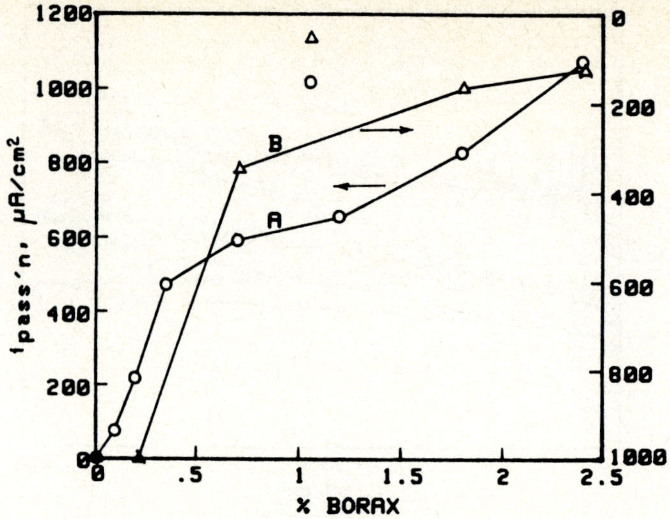

FIGURE 6 – $i_{pass'n}$ and Al pump cavitation corrosion as a function of borax concentration in experimental Coolant D.

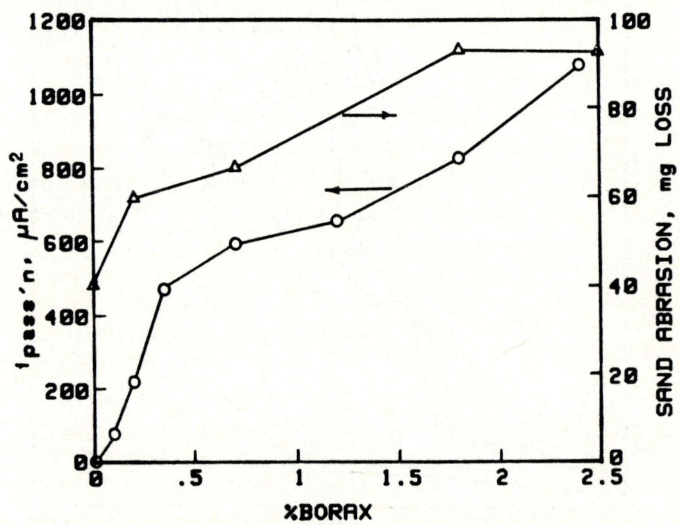

FIGURE 7 – $i_{pass'n}$ and sand abrasion of Al as a function of borax concentration in experimental Coolant D.

Correlation of $i_{pass'n}$ with Sand Abrasion of Al

Figure 7 shows the borax effect on $i_{pass'n}$ and sand abrasion of Al radiator tubes (in-house test) in experimental coolant D. Again it seems that there is a good correlation between $i_{pass'n}$ and sand abrasion of Al.

Correlation of $i_{pass'n}$ with Hot Surface Corrosion of Al

Walker has shown that corrosion is accelerated at heat-rejecting Al surfaces[17] while Kawamoto and Suzuki have shown that borax is detrimental to heat-rejecting Al surfaces.[18] Figure 8 shows the effect of borax (0%, 0.2%, and 0.7%) on the corrosion at heat-rejecting surface of Al in experimental coolant D. The corrosion rate

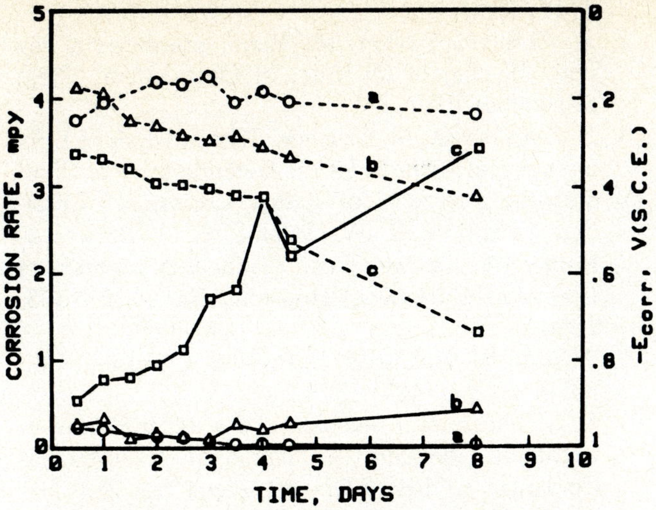

FIGURE 8 – $i_{pass'n}$ and corrosion of heat-rejecting Al surface as a function of borax concentration in experimental Coolant D. Borax: ○ = 0%, △ = 0.2%, □ = 0.7%. Solid Line = corrosion rate; broken line = E_{corr}.

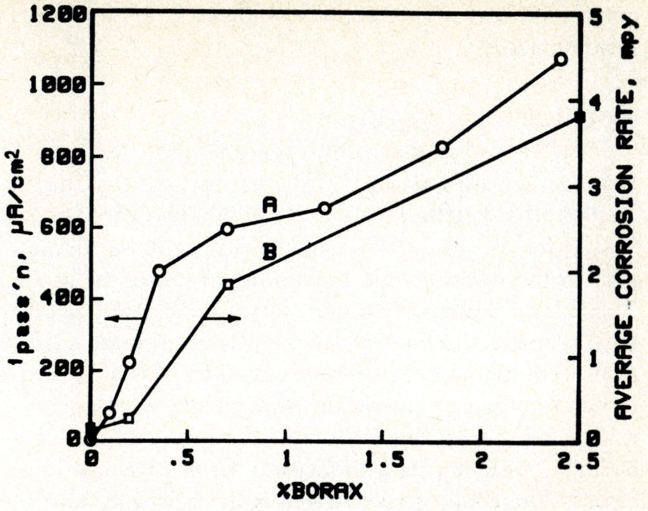

FIGURE 9 – $i_{pass'n}$ and average hot surface corrosion rate during first week as a function of borax concentration.

was determined by the potentiostaircase polarization resistance method[8] at 135 C in a pressurized cell. The temperature was returned to ambient overnight and during weekends. At 0% borax (Figure 8, curves a, $i_{pass'n}$ = 0 μA/cm²), the corrosion rate remains very low even after a long period at ambient temperature. (Frequently after a weekend at room temperature, the surface depassivates upon heating as indicated by a sharp rise in corrosion rate accompanied by a sharp drop in E_{corr}.) At 0.2% borax (Figure 8, curves b, $i_{pass'n}$ = 80 μA/cm²), E_{corr} drops and the corrosion rate rises although it is still under the arbitrary specification of 1 mpy. At 0.7% borax (Figure 8, curves c, $i_{pass'n}$ = 600 μA/cm²), the corrosion rate is much higher and after a long period at ambient temperature there is a significant drop in E_{corr} and rise in the corrosion rate.

The corrosion rates in Figure 8 are averaged over the 8-day test period and compared to $i_{pass'n}$ as a function of borax concentration in Figure 9. The good correlation between $i_{pass'n}$ and hot surface corrosion is very similar to the examples on pump cavitation and sand abrasion shown in Figures 6 and 7, respectively.

Discussion

E- vs i-controlled Polarization

E-controlled polarization techniques are passive in nature, i.e., one applies a voltage across a passive film and waits for film breakdown or charge breakthrough. At times, this wait can be very long. In PDP, one is never certain whether the points on the curve are at steady state regardless of how slow the scan rate is. Under highly passivating conditions, e.g., Al in modern antifreeze coolants, E_b is much too noble and highly scan rate-dependent because of the induction effect.[5,16] The scratch method[11] is a very special E-controlled method that eliminates the induction effect by removing a portion of the protective film. This method can determine E_{prot}, but the determination of E_b is precluded.

Polarization techniques that are i-controlled are active in nature, i.e., one does not wait for the breakdown of the protective film, but rather programs the breakdown and the corrosion rate and its change at any desired schedule. GSCP is an example of i-controlled polarization.[1] Because GSCP circumvents the induction effect, E_b is much less noble than that determined by quasi-stationary potentiostatic polarization (QSP) and PDP.[16] The E_{prot} (and not E_b) determined by GSCP satisfies the usual definition of E_{pit}. This was confirmed by potentiostatic (not point by point) corrosion tests.[16] For pure Al, E_{prot} determined by GSCP agrees with the scratch method.[1,9] For metals with cathodic inclusion, GSCP would be preferred over the scratch method because, based on intuition, GSCP should attack the most vulnerable sites.

$i_{pass'n}$

E_b and E_{prot} of Al reside in the pitting/crevice domain bounded by −600 mV at the low end and 1.55 V vs SCE at the upper end in the presence of 100 ppm Cl^- at room temperature.[15,16] When the temperature is raised to 82.2 C, E_b and E_{prot} may drop around 50 mV for well-inhibited coolants, but for inadequately inhibited coolants, the apparent E_b drops precipitously out of the pitting crevice domain to around −1 to −1.3 V. At this low potential range, localized corrosion is of no concern because of the high rate of corrosion over the entire specimen. As the current is raised, the potential may remain low or rise in some cases. The rise of potential is interpreted as the initiation of passivation at a certain rate of formation of corrosion products which exceeds its dissolution rate. The current necessary to initiate the passivation is called the *passivation current*. Using the same logic, i_{depass} is the current below which passivity can no longer be maintained. i_{depass} is usually lower than $i_{pass'n}$ (see Figure 4); this is probably due to hysteresis involving formation and dissoultion of the corrosion product.

Application of $i_{pass'n}$

In the earlier part of the present work, the movement of E_b from 1.6 to −1.3 V was used to assess the inhibition of the general corrosion of Al. The variation of $i_{pass'n}$ from 0 to >1000 $\mu A/cm^2$ adds another dimension to the electrometric monitoring of corrosion of Al. This added dimension is particularly significant because $i_{pass'n}$ appears to be correlated to the accelerated corrosion accompanying abrupt depassivation of the Al components in the cooling system, e.g., pump cavitation corrosion, sand abrasion of the radiator tubes, and nucleate boiling at heat-rejecting surfaces of the engine head and block. When the protective film is destroyed by one of these processes, the coolant must be able to repassivate the bare metal rapidly; a rapid rise in potential upon application of the first current step would indicate this ability. Such a system would have zero passivation current by definition. Increasing $i_{pass'n}$ would indicate increasing inability to repassivate the bare metal thereby increasing the corrosion rate by the three mechanisms. In the case of pump cavitation, the results could be catastrophic because corrosion is highly localized in the high pressure region where the vapor bubbles collapse.

The examples given in this paper show how $i_{pass'n}$ can provide a rapid inexpensive means of evaluating and optimizing coolants. The duration of the GSCP test is 10 to 50 minutes, depending on the magnitude of $i_{pass'n}$. The long, expensive test methods are used on coolants that have been selected by the GSCP screening test.

Conclusion

GSCP circumvents the induction effect and current hysteresis observed in the E-controlled polarization diagrams. E_b and E_{prot} determined by GSCP are very meaningful and practical. The movement of E_{prot} along the potential axis correlates with localized corrosion while the movement of E_b correlates with the general corrosion. As the aggressiveness of the environment (temperature, dilution of inhibitors, and aggressive compounds) increases, the apparent E_b "bottoms out," i.e., as the aggressiveness is further increased, there is no further change in E_b. However, as the aggressiveness of the environment is further increased, $i_{pass'n}$ was found to correlate quite well with three types of corrosion of Al in a benzoate/nitrite type experimental coolant in automotive cooling system, which involve abrupt depassivation.

Acknowledgment

The author thanks Dan Coker, Cheryl Cabbil, and Lester Wilson for the laboratory work; Dan Coker for the modification of the potentiostat into a galvanostaircase apparatus; and John Conville for the critical review of this manuscript.

References

1. S. T. Hirozawa, J. Electrochem. Soc., Vol. 130, p. 1718, 1983.
2. A. Broli, H. Holtan, M. Midjo, presented at the 25th Meeting of Int. Soc. of Electrochem. (ISE), September 1974.
3. S. Smialowska, M. Czachor, "The Analysis of Electrochemical Methods for the Determination of Characteristic Potentials of Pitting Corrosion," Localized Corrosion, R. W. Staehle, B. F. Brown, J. Kruger, A. Agrawal, Eds., National Association of Corrosion Engineers, Houston, Texas, p. 353, 1974.
4. A. Broli, H. Holtan, K. L. Prestud, Corrosion, Vol. 30, No. 12, p. 427, 1974.
5. S. T. Hirozawa, "Mechanism for the Inhibition of Localized Corrosion of Aluminum in Antifreeze Coolants," CORROSION/82, Paper No. 263, National Association of Corrosion Engineers, Houston, Texas, 1982.
6. P. E. Manning, Corrosion, Vol. 36, No. 9, p. 468, 1980.
7. S. T. Hirozawa, "Galvanostaircase Polarization: Elimination of Hysteresis and the Incubation Period," paper presented at the Research Conference at CORROSION/83, April 1983.
8. S. T. Hirozawa, "Current Versus Voltage Hysteresis: Effect on Electrometric Monitoring of Corrosion," Laboratory Corrosion Tests and Standards, ASTM STP 866, G. S. Haynes, R. Baboian, Eds., ASTM, Philadelphia, Pennsylvania, p. 108, 1985.
9. W. J. Rudd, J. C. Scully, Corros. Sci., Vol. 20, p. 611, 1980.
10. B. E. Wilde, "On Pitting and Protection Potentials: Their Use and Possible Misuses for Predicting Localized Corrosion Resistance of Stainless Alloys in Halide Media," Localized Corrosion, R. W. Staehle, B. F. Brown, J. Kruger, A. Agrawal, Eds., National Association of Corrosion Engineers, Houston, Texas, p. 342, 1974.
11. N. Pesall, C. Liu, Electrochim. Acta, Vol. 16, p. 1987, 1971.
12. B. C. Syrett, Corrosion, Vol. 33, No. 6, p. 221, 1977.
13. D. Sinigaglia, B. Vinventini, G. Taccani, G. Salvago, G. Dallaspezia, J. Electrochem. Soc., Vol. 130, p. 991, 1983.
14. M. Barbosa, J. C. Scully, Corros. Sci., Vol. 22, p. 1025, 1982.
15. S. T. Hirozawa, "Study of the Mechanism for the Inhibition of Localized Corrosion of Aluminum by Galvanostaircase Polarization," paper presented at the Int. Conf. on Corros. Inhib., May 1983.
16. S. T. Hirozawa, Electrochem. Soc. Meeting, Paper No. 48, 1982; in publication.
17. M. S. Walker, GMR-1272, General Motors Research Publication, March 1973.
18. S. Kawamoto, H. Suzuki Mitsubishi, Heavy Industries Technical Review, Volume 11, No. 2, p. 102, 1974.

Proper Electrical Resistance Corrosion Probe Span Selection

G. L. COOPER*

Abstract

Selection of the correct electrical resistance probe span is critical for optimum results when using this system for corrosion monitoring. The probe span is determined by the sensing element geometry and thickness. An element that is insensitive to the nominal system corrosion rate will require long time periods before a corrosion rate can be calculated. An element of high sensitivity conversely may have an unacceptably short operation life.

After selection of a correct sensing element span, a compromise to a higher or lower sensitivity may be necessary because of mechanical considerations.

Another selection criterion is the type of corrosion monitoring desired. In some cases the determination of an actual corrosion rate is less important than the ability to quickly observe changes in corrosivity.

History of Electrical Resistance Corrosion Monitoring

The operating details and principles of electrical resistance (ER) probes have been described in previous papers and articles.[1-3] Correlation with corrosion rates as determined by other methods has been verified.[4]

Acceptance of the ER method grew quickly and soon there were over 100 corrosion monitors in active field use monitoring corrosion in the petroleum industry.[5] Their use in other industries expanded as well, until today applications can be found in production,[6] refining,[7] chemical processing,[8] cathodic protection,[9] and atmospheres.[10]

Most electrical resistance probes are read with a portable instrument. Portable units were the first to be developed and remain relatively popular. As an industry average, a probe reading frequency of about two to three times per month is normal.

*Rohrback Cosasco Systems, Santa Fe Springs, California.

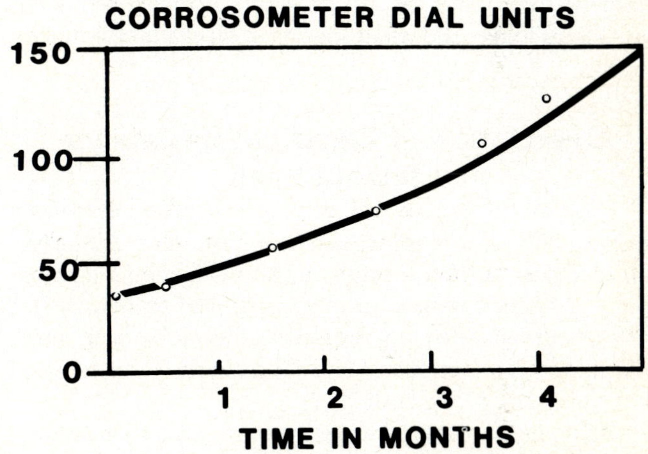

FIGURE 1 – Corrosion rate calculated in months using ER probe.

Continuous recording or AC-powered electrical resistance instruments were originally developed as electromotive or electronically driven versions of the portable instrument.[4,11] Later, more modern types of instruments made full use of microprocessors and digital electronics.[12]

Most important to the development of ER probes and instruments has been the user's desire to obtain corrosion data in the shortest possible time. This desire has been the primary motivation behind the latest generation of instrumentation.

In the earliest days of probe usage, a speedy response to changes in corrosivity might be measured in months or weeks (Figures 1 and 2); however, this was not adequate. It became necessary for corrosion monitoring instruments to read in days (Figure 3) with the ultimate goal of shortening this time period to hours or minutes.

With the available instrumentation and a suitable probe, response times measured in minutes are now

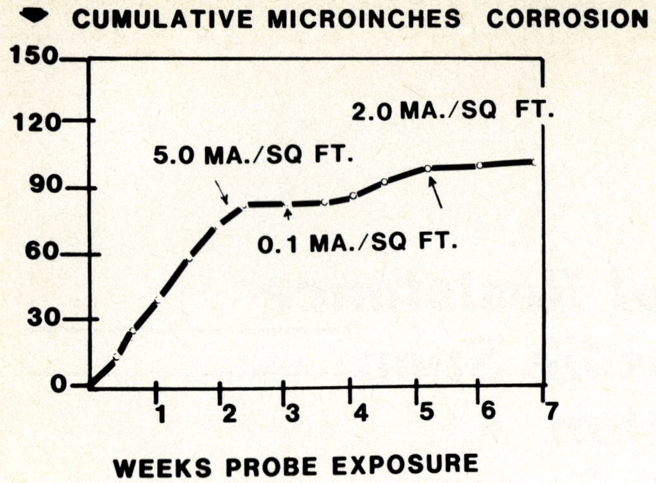

FIGURE 2 – Corrosion rate calculated in weeks using ER probe.

possible. The relationships for selection of a suitable probe, or more correctly *probe span,* are not widely understood, and this has led to field problems. It is the intent of this paper to unscramble those relationships, describe them, and then to provide a practical series of guidelines to be used when specifying electrical resistance probes.

Description of Electrical Resistance Element Span

While ER probes are commonly called *resistance probes,* this term is something of a misnomer. Actually, an electrical resistance probe has three elements (Figure 4): (1) a measurement element, exposed to corrosion; (2) a reference element, located in the probe body and not exposed to corrosion; and (3) a check element, also located in the probe body.

Rather than a direct measure of resistance on an exposed element, a ratio measurement is obtained:

$$\frac{R_M}{R_R} \qquad (1)$$

where R_M = measure element resistance, and R_R = reference element resistance.

The check element is similarly compared to the reference element and acts as a verification that the probe end seals have not failed. If they were to do so, this would destroy the measurement ratio's validity.

An electrical resistance instrument is used to quantify the changed ratios via a display or meter. These instruments are based on the wheatstone bridge principle.

$$\frac{R_1}{R_2} = \frac{R_3}{R_4} \qquad (2)$$

where (typically): R_1 = measurement element resistance, R_2 = reference element resistance, R_3 = an instrument-based reference resistance, and R_4 = an instrument-based variable resistance.

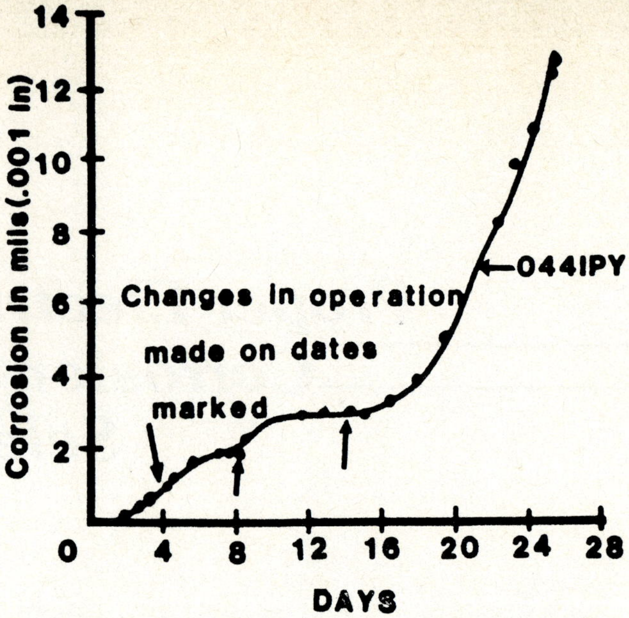

FIGURE 3 – Corrosion rate calculated in days using ER probe.

Metal Loss: Mr/Rr(time2)–Mr/Rr(time1)

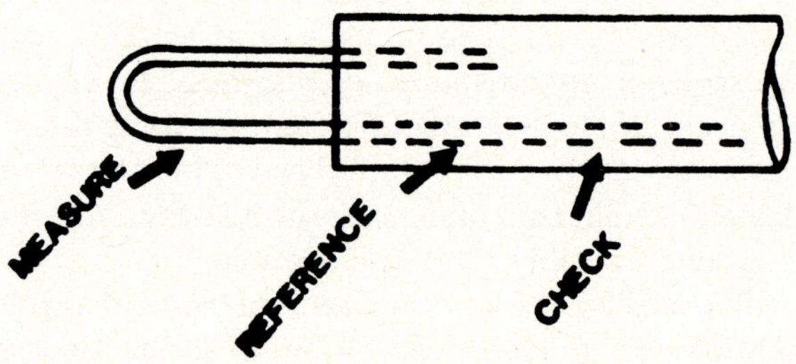

FIGURE 4 – Electrical resistance probe cross section illustrating the location of the measure, reference, and check elements.

Thus, if one side of the equation (R_1/R_2) is changed via corrosion to R_1 (measure element), the other side of the equation must be changed proportionally to bring both sides back into balance.

In actual practice, this proportional change to R_3/R_4 is accomplished over an arbitrary range of 0 to 1000 dial divisions. Zero dial divisions would theoretically be a new uncorroded probe surface, while 1000 dial divisions would correspond to the end of the measure element life or span.

It has been found that it is not practical to continue to linearize ratios of resistance change beyond one half of the measurement element thickness. This fact determines the available span of measurement for an electrical resistance probe. It is one half of the measure element's original thickness. The measurement spans of most commercially available measuring element types are listed in Table 1.

These sensing element spans are further illuminated in Figures 5 through 7.

TABLE 1 – Measurement Element Span

Element Type	Original Thickness (in.)	Span (in.)
W40 Wire loop	0.040	0.010
W60 Wire loop	0.060	0.015
W80 Wire loop	0.080	0.020
T4 Tube loop	0.004	0.002
T8 Tube loop	0.008	0.004
T10 Cylinder	0.010	0.005
T20 Cylinder	0.020	0.010
T50 Cylinder	0.050	0.025
S2 Strip loop	0.002	0.0005
S4 Strip loop	0.004	0.001
S8 Strip loop	0.008	0.002
S1 Strip flush	0.001	0.0005
S2 Strip flush	0.002	0.001
S4 Strip flush	0.004	0.002
S8 Strip flush	0.008	0.004
S10 Strip flush	0.010	0.005

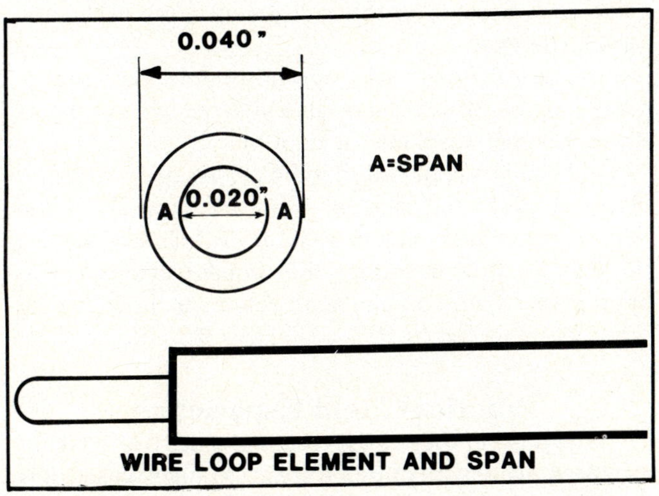

FIGURE 5 – Wire ER probe span with 0.040 in. (1 mm) original thickness.

The measurement element span (also called *life*) is the available metal for measurement. It is, as stated, one half of the original material thickness. It is directly analogous to the measurement span of a 1 meter stick—1000 millimeters. Changes in penetration along this span are measured via changing (increasing) dial divisions, just as changes in length of the 1 meter stick are noted using graduations of 1 millimeter each.

Selection of a Span Compared to a Known Corrosion Rate

There are three parameters that determine how rapidly a corrosion rate can be calculated with an electrical resistance probe. These are:

1. System corrosion rate [mpy (25 μm)]
2. Element span [mils (25 μm)]
3. Instrument resolution (% accuracy of full scale)

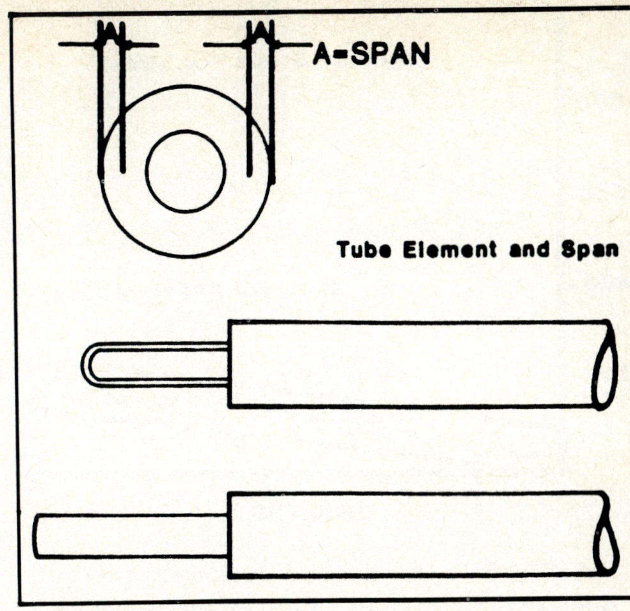

FIGURE 6 – Tube or cylindrical element ER probe span.

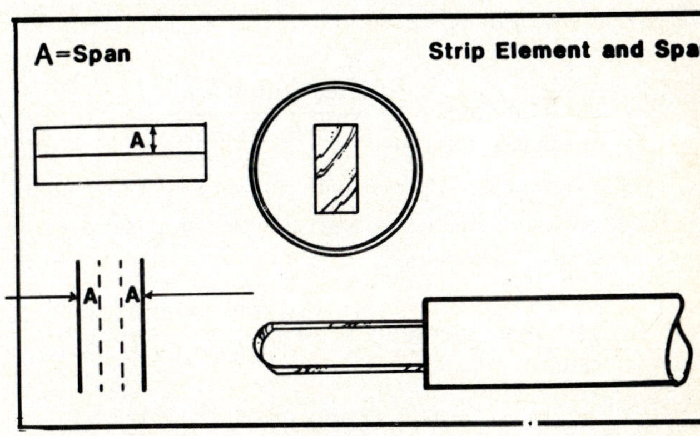

FIGURE 7 – Strip element (loop and flush types) ER probe span.

(A fourth parameter, frequency of probe reading, may be included with automatic instruments, but it is omitted here for clarity.)

Knowing these three parameters, a correct measurement element span can be selected by following these steps:

1. The first step should be to determine element life for a given span at the observed corrosion rate. This can be done by dividing the probe span, expressed in mils, by the corrosion rate (estimated if unknown).

$$\text{Life in years} = \frac{\text{Span (mils)}}{\text{Corrosion rate (mpy)}} \quad (3)$$

2. The second step is to determine the daily change in divisions (remembering a total of 1000 are available):

$$\text{Daily change in divisions} = \frac{1000\ \text{(Divisions)}/365}{\text{Life in years}} \quad (4)$$

3. The third and final step is to compare the answer from step 2 with the instrument resolution. This final step

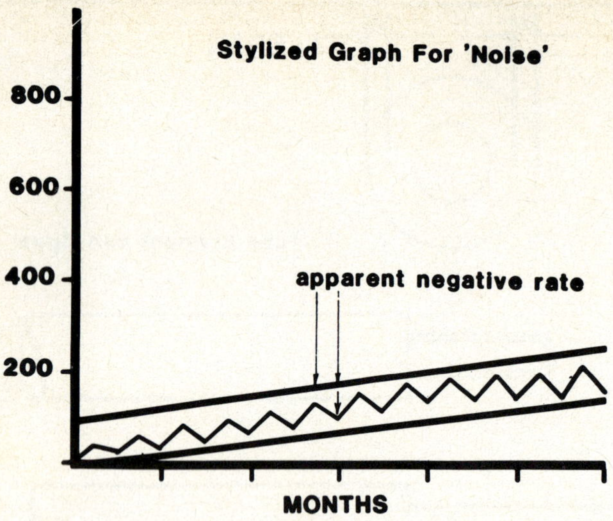

FIGURE 8 – Upward and downward movement in dial divisions within an instrument's resolution limits.

Selecting ER Probe Span

Span = Measurable Element Thickness

1. $\dfrac{\text{Span } (0.001")}{\text{Corrosion Rate (Mp)}} \times 12$ = Months Service

2. $\dfrac{1000}{\text{Months Service}}$ = Monthly Dial Change

3. $\dfrac{\text{Monthly Dial Change}}{30}$ = Daily Change

4. A/ If Step 3 ≤ 3 Divisions select more sensitive span.
B/ If Step 3 ≈ 3 Proceed.
C/ If Step 3 ≥ 6 Divisions select less sensitive span.

NOTES:

1. Based upon 1% Ins. Resolution.
2. Based upon 4-5 day interval.
3. Based on 6-12 months probe life.

FIGURE 9 – Suggested steps to selecting an ER probe span.

will determine how rapidly corrosion rates could be calculated using an ER system:

$$\text{Time (in days) to reliable data} = \frac{\text{Resolution limit}}{\text{Daily change}} \qquad (5)$$

As an example, a probe with a span of 10 mils (250 μm) will be used and compared to corrosion rates of 1 mpy (A) and 20 mpy (B).

$$\text{A. } 10 \text{ years element life} = \frac{10 \text{ mils (span)}}{1 \text{ (mpy)}} \qquad (6)$$

$$0.27 \text{ divisions per day change} = \frac{1000 \text{ (Divisions)}/365}{10 \text{ (years element life)}} \qquad (7)$$

$$\text{B. } 0.5 \text{ years element life} \quad \frac{10 \text{ mils (span)}}{20 \text{ (mpy)}} \qquad (6)$$

$$5.48 \text{ divisions per day change} = \frac{1000 \text{ (Divisions)}/365}{.5 \text{ (years element life)}} \qquad (7)$$

To complete the example, a comparison of these daily changes with the resolution limits of an instrument is needed. For clarity of illustration, a portable instrument will be used. Typical resolution limits for these machines are (±)5 divisions or 10 divisions total.

In case A:

$$\frac{10}{0.27} = 37 \text{ days to reliable change} \qquad (8)$$

In case B:

$$\frac{10}{5.48} = 1.8 \text{ days to reliable change} \qquad (8)$$

Practically, this means that in case A, one would have to wait more than a month to calculate corrosion rates when using a portable meter, while in case B, a rate could be calculated in less than two days. Attempts to perform rate calculations in less time than this could involve error because of the inclusion of noise in the observed changes.

Noise is defined as the upward or downward movement in divisions that can occur within the limits of an instrument's own resolution (Figure 8).

It becomes apparent that the proper selection of an electrical resistance element span will take all of the parameters mentioned into account. By following steps 1 through 3, a potential user of ER probes can arrive at an appropriate element span and determine how rapidly corrosion rates can be derived (Figure 9).

Practical Field Experience

The push to shorten ER response times has led to an increased awareness of probe span.[6] In many cases, it is the critical aspect of a successful application of electrical resistance monitoring.

If an element is insensitive to the general corrosion rate (i.e., too thick) it will take longer time periods to determine rate. Conversely, an element that is too sensitive (i.e., too thin) will have an unacceptably short life.

For many years, downward shifts in dial divisions, obtained while reading electrical resistance probes, have been blamed on conductive coatings. These coatings have been theorized as having temporarily attached themselves to the probe and then dropped off.

The source for the coatings was thought to come from two main areas: (1) corrosion products such as FeS occurring from areas upstream of the probe, and (2) an oxide layer of corrosion products from the probe element itself.

In field applications, it has been found that when conductive coatings deposit on the element, they tend to stay on the probe. The result is the readings tend to drop in a negative direction until they reach zero (Figure 10).

Even when coatings have been analyzed for chemical composition and found to have a high metallic con-

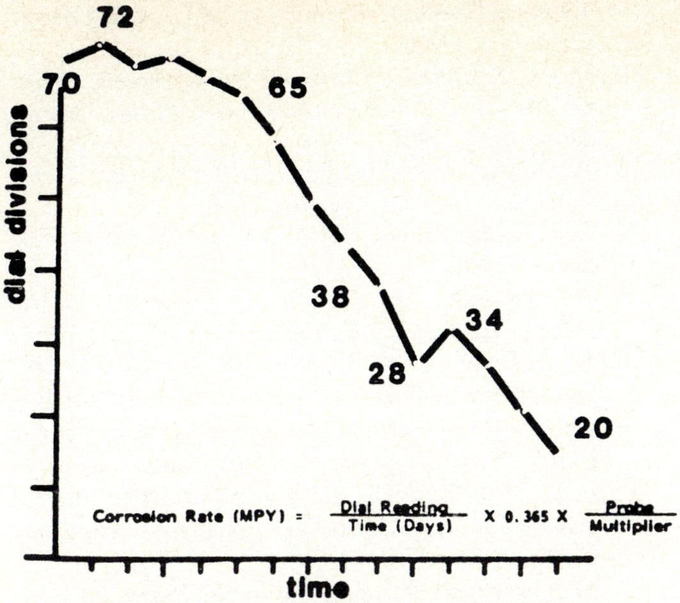

FIGURE 10 – Declining readings from an ER probe with a sensing element plated by conductive coating.

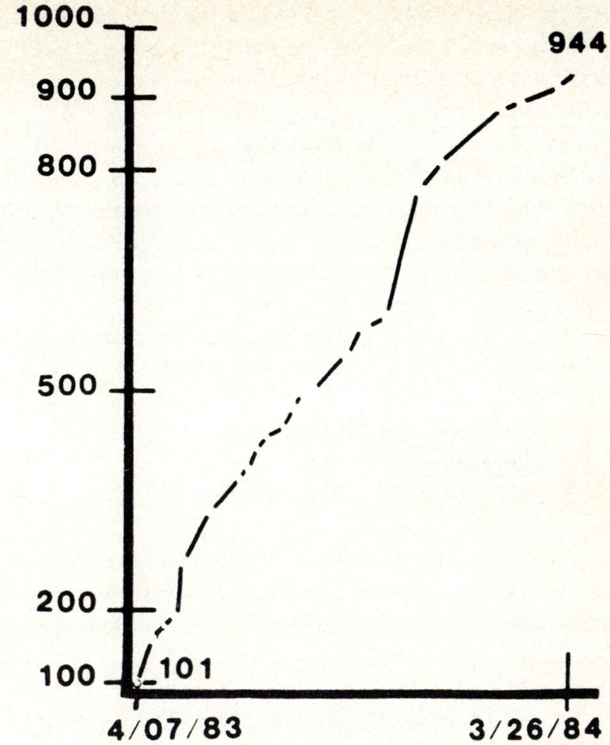

FIGURE 11 – Increasing readings in divisions from an ER probe coated with 96.4% Fe (as FeS) and 0.7% Cr (as CrS).

tent, they do not necessarily affect readings, as is illustrated by the readings taken from a probe (Figures 11 and 12) that had a coating composed of 96.4% Fe as FeS and 0.7% Cr as CrS.

In contrast, a AISI 316 SS element probe dropped to a zero reading when coated with the following:

Element	Percentage
Ni	76
Co	0.33
Mo	0.31
Cu	0.17
Fe	0.43
Sb	0.55
Si	0.014
Mg	0.0062
Mn	0.0051
Pb	0.027
Cr	0.062
Ag	0.00088
Ti	0.015
Al	0.0029
Ca	0.0053

The nickel, present as an almost homogeneous coating, became a preferential current path causing the probe to appear as if it had been plated.

Instrument Selection and Impact on Probe Span Selection

In the example listed for probe span selection, a portable electrical resistance instrument was used. Different instruments have differing resolutions. Microprocessor powered units have resolutions as fine as 0.1% of full scale. This is a hundredfold improvement over the

FIGURE 12 – Coated area of ER probe element from Figure 11.

portable unit. Varying the resolution limit in step 3 accordingly would take this variation into account when selecting a probe span.

Summary

Probe measurement element span may be the critical item in a successful electrical resistance corrosion monitoring application. The span of an electrical resistance corrosion probe is one half of the measurement element's original thickness.

When selecting a probe element, three parameters should be considered in consecutive steps:

1. Corrosion rate (in mpy)
2. Element span (in mils)
3. Instrument resolution (% of full scale)

An improper probe element may have an unacceptably short life when it is too sensitive. Conversely, when it is insensitive to the ongoing corrosion rate, it may require a lengthy period before a corrosion rate can be calculated.

The result of an insensitive probe element may be noise, which can cause erroneous rate calculations. Coatings of conductive materials have been considered the primary cause of noise or downward readings defections; however, it is now thought that improper sensing element span is the chief cause.

Instrument selection plays a role in rapid rate determinations from electrical resistance probes. Microprocessor or full time AC powered variants may provide a hundredfold decrease in response times when compared with portable units.

References

1. G. H. Rohrback, "Determination of Corrosion Rates in a Variety of Different Processes by Means of the Corrosometer," presented at the Sixth Western Regional Division Conference of NACE, Long Beach, California, 1956.
2. D. H. Stormont, "Corrosion Rates Directly Measured by New Resistance Method," Oil and Gas J., January 21, 1957.
3. A. Dravneiks, A. Freedman, E. Trosckinski, Corrosion, Vol. 14, No. 4, p. 29, 1958.
4. E. C. Winegartner, Corrosion, Vol. 16, No. 6, p. 265+, 1960.
5. A. Dravnieks, A. Freedman, "Electrical Probes Monitor Corrosion," Petroleum Refiner, July 1958.
6. S. Galbraith, L. A. Disbrow, K. A. Van Buskirk, "Installation and Use of Automated Electrical Resistance Probe Systems to Monitor Corrosion in the Eastern Operating Area of the Prudhoe Bay Oil Field," CORROSION/84, Paper No. 239, National Association of Corrosion Engineers, Houston, Texas, 1984.
7. H. Lee Craig, "Catalytic Reformer Corrosion Control," CORROSION/84, Paper No. 129, National Association of Corrosion Engineers, Houston, Texas, 1984.
8. D. R. Bergstrom, Materials Performance, Vol. 20, No. 9, p. 17, 1981.
9. H. J. Fromm, Materials Performance, Vol. 16, No. 11, p. 21, 1977.
10. G. Brown, M. Lunn, C. Miller, C. Shelor, "Corrosion Protection in Pulp and Paper Mill Control Room Environments," 1982 TAPPI Engineering Conference, TAPPI, Atlanta, Georgia.
11. J. H. Lanaman, L. Azar, "A New Instrument for Measurement of Corrosion by Electrical-Resistance Techniques," CORROSION/73, Paper No. 5, National Association of Corrosion Engineers, Houston, Texas, 1973.
12. G. L. Cooper, "Modern Microprocessor Electrical-Resistance Instruments," CORROSION/82, Paper No. 173, National Association of Corrosion Engineers, Houston, Texas, 1982.

Subject and Author Indexes

Subject Index

Author Index